Herbal Nutraceuticals

Herbal Nutraceuticals

Products and Processes

Edited by

Santosh Kumar Upadhyay
Department of Botany
Panjab University
Chandigarh, India

Sudhir Pratap Singh
Department of Industrial Biotechnology
Gujarat Biotechnology University
Gandhinagar, India

WILEY

This edition first published 2025

Registered Office(s)
John Wiley & Sons, Inc., 111 River Street, Hoboken, NJ 07030, USA
John Wiley & Sons Ltd, The Atrium, Southern Gate, Chichester, West Sussex, PO19 8SQ, UK

For details of our global editorial offices, customer services, and more information about Wiley products visit us at www.wiley.com.

Wiley also publishes its books in a variety of electronic formats and by print-on-demand. Some content that appears in standard print versions of this book may not be available in other formats.

Library of Congress Cataloging-in-Publication Data Applied for:
Hardback ISBN: 9781394241545

Cover Design: Wiley
Cover Image: © Lallapie/Shutterstock

Set in 9.5/12.5pt STIXTwoText by Straive, Pondicherry, India

Printed and bound by CPI Group (UK) Ltd, Croydon, CR0 4YY

C9781394241545_011024

Contents

About the Editors

Dr. Santosh Kumar Upadhyay is currently working as an assistant professor of Botany at the Panjab University, Chandigarh, India. He has been working in the field of Plant Biotechnology for more than 16 years. His present research focuses in the area of functional genomics for stress tolerance. He is involved in the bio-prospecting and characterization of various important proteins from plant biodiversity, and defense and stress signaling genes in bread wheat. His research group has developed a pool of vital genes and long noncoding RNAs associated with abiotic and biotic stress tolerance and signaling in plants. He has also established the method for genome editing in bread wheat using CRISPR-Cas system and developed a tool SSinder for CRISPR target site prediction. His research contribution led to the publication of more than 70 research papers in leading journals of international repute. Further, there are more than five national and international patents, 50 book chapters, and 12 books in his credit. In recognition of his substantial research record, he has been awarded NAAS Young scientist award (2017–2018) and NAAS-Associate (2018) from the National Academy of Agricultural Sciences, India, INSA-Associate Fellow (2023), and INSA Medal for Young Scientist (2013) from the Indian National Science Academy, India, NASI-Young Scientist Platinum Jubilee Award (2012) from the National Academy of Sciences, India, and Altech Young Scientist Award (2011). He has also been the recipient of the prestigious DST-INSPIRE Faculty Fellowship (2012) and SERB-Early Career Research Award (2016) from the Ministry of Science and Technology, Government of India. Dr. Upadhyay also serves as a member of the editorial board and reviewer of several peer-reviewed international journals.

Dr. Sudhir Pratap Singh is currently Professor at the Department of Industrial Biotechnology, Gujarat Biotechnology University, Gandhinagar, Gujarat, India. He has been working in the area of molecular biology and biotechnology for more than a decade. Currently, his primary focus of research is gene mining, biocatalyst engineering, and characterization of novel enzyme variants for the production of high-value functional biomolecules from low-cost feedstock. He has generated metagenomic data resources from diverse ecological niches and ethnic food samples and has characterized novel genes encoding enzyme variants with desirable catalytic properties for the biosynthesis of functional sugar molecules such as D-allulose, turanose, trehalose, and trehalulose. Further, he has achieved enzymatic production of prebiotic molecules, such as fructooligosaccharides, glucooligosaccharides, 4-galactosyl-Kojibiose, xylooligosaccharides, levan, dextran, type III resistant starch, etc. Prof. Singh has published more than 75 research articles (SCI) and 8 books (edited). Further, he has 12 patents (granted) to his credit as an inventor. He has been conferred the TATA Innovation Fellowship (DBT, Govt. of India), Professor SB Chincholkar Award (BRSI), Young Scientist Award (IBA), SBS-MKU Genomics Award (BRSI), and Professor Hira Lal Chakravarty Award (ISCA, DST, Govt. of India). His team was awarded Gandhian Young Technological Innovation Award (SRISTI) in 2019. He is a life member of the National Academy of Sciences, India. Further, Prof. Singh is the Fellow of the National Academy of Agricultural Sciences (FNAAS) and the Fellow of the International Society of Energy, Environment, and Sustainability (FISEES).

List of Contributors

Hanan A. Alatawi
Department of Biological Sciences, University College of Haqel, Tabuk University, Tabuk
Saudi Arabia

Mahra Mohammad Moosa Ali Alblooshi
Health Sciences Division, Abu Dhabi Women's College, Higher Colleges of Technology, Abu Dhabi, UAE

Omar S. Al-Odat
Department of Biomedical Sciences, Cooper Medical School of Rowan University
Camden, NJ, USA

Department of Chemistry and Biochemistry
Rowan University, Glassboro, NJ, USA

Amenah S. Alotaibi
Depatment of Biology, Faculty of Science
Genomic & Biotechnlogy, University of Tabuk, Tabuk, Saudi Arabia

Osama Aloudat
Department of Biomedical Sciences, Cooper Medical School of Rowan University
Camden, NJ, USA

Shamshul Ansari
Health Sciences Division, Abu Dhabi Women's College, Higher Colleges of Technology, Abu Dhabi, UAE

Anuradha
Plant Biotechnology Laboratory, Department of Biotechnology, Delhi Technological University, New Delhi, India

Mohd Aznan Md Aris
Faculty of Medicine, International Islamic University Malaysia, Kuantan Campus
Kuantan, Pahang, Malaysia

Rajeshwar Kamal Kant Arya
Department of Pharmaceutical Sciences
Sir J. C. Bose Technical Campus
Bhimtal, Kumaun University, Nainital
Uttarakhand, India

Shavkatjon Azizov
Faculty of Life Sciences, Pharmaceutical Technical University, Tashkent, Uzbekistan

Navneeta Bharadvaja
Plant Biotechnology Laboratory, Department of Biotechnology, Delhi Technological University, New Delhi, India

Vikas Bhatt
Devsthali Vidyapeeth College of Pharmacy
Rudrapur, Uttarakhand, India

Dheeraj Bisht
Devsthali Vidyapeeth College of Pharmacy
Rudrapur, Uttarakhand, India

Andrii Boiko
Department of Management and Economy of Pharmacy and Drug Technology of Postgraduate Faculty, Danylo Halytsky Lviv National Medical University
Lviv, Ukraine

Andrian Boiko
Department of Therapeutic Dentistry Periodontology and Dentistry FPDE, Danylo Halytsky Lviv National Medical University
Lviv, Ukraine

Anoma Chandrasekara
Department of Applied Nutrition, Wayamba University of Sri Lanka, Makandura
Gonawila, Sri Lanka

Shikha Chauhan
Department of Food Technology and Nutrition, School of Agriculture, Lovely Professional University, Phagwara
Punjab, India

Priyanku Pradip Das
Department of Pharmaceutical Chemistry School of Pharmaceutical Sciences, Shoolini University, Solan, Himachal Pradesh, India

Jeevitha Devaraju
Division of Horticulture, School of Agricultural Sciences, Karunya Institute of Technology and Sciences
Coimbatore, India

Abishek Dinesh
Department of Food Technology and Nutrition, School of Agriculture, Lovely Professional University, Phagwara
Punjab, India

Sashya Diyapaththugama
Department of Applied Nutrition, Wayamba University of Sri Lanka, Makandura
Gonawila, Sri Lanka

Daniel A. Guirguis
Department of Biomedical Sciences, Cooper Medical School of Rowan University
Camden, NJ, USA

Arun Kumar Gupta
Department of Food Science & Technology Graphic Era (Deemed to be University)
Dehradun, Uttarakhand, India

Graphic Era Hill University
Dehradun, Uttarakhand, India

Noraihan Mat Harun
Faculty of Medicine, International Islamic University Malaysia, Kuantan Campus
Kuantan, Pahang, Malaysia

Ridar Hendri
Faculty of Fisheries, Riau University
Pekanbaru, Indonesia

Soraya Ismail
Faculty of Medicine, International Islamic University Malaysia, Kuantan Campus
Kuantan, Pahang, Malaysia

Irwandi Jaswir
Department of Chemistry, Padang State University, Padang, Indonesia

International Institute for Halal Research and Training, International Islamic University Malaysia, Kuala Lumpur, Malaysia

Avinash Kumar Jha
Department of Food Technology and Nutrition, School of Agriculture, Lovely Professional University, Phagwara
Punjab, India

Subash C. Jonnalagadda
Department of Chemistry and Biochemistry Rowan University, Glassboro, NJ, USA

Aroma Joshi
Department of Food Science & Technology Graphic Era (Deemed to be University)
Dehradun, Uttarakhand, India

Shruti Kabra
Department of Food and Nutrition, Telangana Mahila Viswavidyalayam, Koti, Hyderabad Telangana, India

Rajan Katoch
Biochemistry Laboratory, Department of Chemistry & Biochemistry, CSKHPKV Palampur, Himachal Pradesh, India

Kiran
Department of Genetics & Plant Breeding CSKHPKV, Palampur, Himachal Pradesh, India

Sapna Avinash Kondalkar
Regional Ayurveda Research Institute Gwalior, Madhya Pradesh, India

Deepak Kumar
Department of Pharmaceutical Chemistry School of Pharmaceutical Sciences Shoolini University, Solan, Himachal Pradesh, India

Vijay Kumar
Himalayan School of Biosciences, Swami Rama Himalayan University, Dehradun Uttarakhand, India

Kok-Song Lai
Health Sciences Division, Abu Dhabi Women's College, Higher Colleges of Technology, Abu Dhabi, UAE

H. Lalhlenmawia
Department of Pharmacy, Regional Institute of Paramedical and Nursing Sciences Zemabawk, Aizawl, Mizoram, India

Widya Lestari
Faculty of Dentistry, International Islamic University Malaysia, Kuantan Campus Kuantan, Pahang, Malaysia

Swee-Hua Erin Lim
Health Sciences Division, Abu Dhabi Women's College, Higher Colleges of Technology, Abu Dhabi, UAE

Iryna Lozynska
Department of Biochemistry, Danylo Halytsky Lviv National Medical University Lviv, Ukraine

Roman Lysiuk
Department of Pharmacognosy and Botany Danylo Halytsky Lviv National Medical University, Lviv, Ukraine

Madhu
Department of Botany, Panjab University Chandigarh, India

Ajay Kumar Meena
Regional Ayurveda Research Institute Gwalior, Madhya Pradesh, India

Farid Menaa
Department of Biomedical and Environmental Engineering, California Innovations Corporation, San Diego, CA, USA

Niranjan Mishra
Applied Phycology and Biotechnology Division, CSIR-Central Salt and Marine Chemicals Research Institute, Bhavnagar Gujarat, India

Sandhya Mishra
Applied Phycology and Biotechnology Division, CSIR-Central Salt and Marine Chemicals Research Institute, Bhavnagar Gujarat, India

Aniket More
Department of Food Technology and Nutrition, School of Agriculture, Lovely Professional University, Phagwara Punjab, India

Sonia Morya
Department of Food Technology and Nutrition
School of Agriculture, Lovely Professional
University, Phagwara, Punjab, India

Rupasree Mukhopadhyay
Department of Genetics and Biotechnology
Telangana Mahila Viswavidyalayam, Koti
Hyderabad, Telangana, India

Bindu Naik
Department of Food Science & Technology
Graphic Era (Deemed to be University)
Dehradun, Uttarakhand, India

Arno Neumann
BET Bioscience Extraction Technologies Inc.
Abbotsford, British Columbia, Canada

Patrick A. Newport-Ratiu
Department of Biomedical Sciences, Cooper
Medical School of Rowan University
Camden, NJ, USA

Department of Chemistry and Biochemistry
Rowan University, Glassboro, NJ, USA

Dedi Noviendri
Bioprocess and Molecular Engineering
Research Unit, Kulliyyah of Engineering
International Islamic University Malaysia
Kuala Lumpur, Malaysia

Serhii Oliinyk
Nobel – Ukraine Pharmaceutical Company
Kyiv, Ukraine

Petro Oliinyk
Department of Disaster Medicine and Military
Medicine, Danylo Halytsky Lviv National
Medical University, Lviv, Ukraine

Manoj K. Pandey
Department of Biomedical Sciences, Cooper
Medical School of Rowan University
Camden, NJ, USA

Kanti Bhooshan Pandey
BDIM Division, CSIR-Central Salt and Marine
Chemicals Research Institute, Bhavnagar
Gujarat, India

Gopika S. Pillai
Department of Food Technology and
Nutrition, School of Agriculture, Lovely
Professional University, Phagwara
Punjab, India

Ashwini Prabhu
Division of Cancer Research and Therapeutics
(CaRT), Yenepoya Research Centre, Yenepoya
(Deemed to be University), Mangalore
Karnataka, India

Panneer Selvam Ranchana
Division of Horticulture, School of
Agricultural Sciences, Karunya Institute of
Technology and Sciences Coimbatore, India

Sarvesh Rustagi
Department of Food Technology, SALS
Uttaranchal University, Dehradun
Uttarakhand, India

Nicole K. Schmalbach
Department of Biomedical Sciences, Cooper
Medical School of Rowan University
Camden, NJ, USA

Karishma Sebastian
Division of Horticulture, School of
Agricultural Sciences, Karunya Institute of
Technology and Sciences, Coimbatore, India

Freny Shah
Applied Phycology and Biotechnology
Division, CSIR-Central Salt and Marine
Chemicals Research Institute, Bhavnagar
Gujarat, India

Fereidoon Shahidi
Department of Biochemistry, Memorial
University of Newfoundland, St. John's
Newfoundland, Canada

Alok Sharma
Regional Ayurveda Research Institute
Gwalior, Madhya Pradesh, India

Manjula Belagihalli Siddalingappa
Division of Horticulture, School of Agricultural Sciences, Karunya Institute of Technology and Sciences
Coimbatore, India

Sudhir Pratap Singh
Department of Industrial Biotechnology
Gujarat Biotechnology University
Gandhinagar, India

Mehak Thakur
School of Biological and Environmental Sciences, Shoolini University, Solan
Himachal Pradesh, India

Neelam Thakur
Biochemistry Laboratory
Department of Chemistry & Biochemistry
CSKHPKV, Palampur
Himachal Pradesh, India

Ankur Tripathi
College of Dairy & Food Technology
Agriculture University, Jodhpur
Rajasthan, India

Santosh Kumar Upadhyay
Department of Botany, Panjab University
Chandigarh, India

Sharvary Arun Vichare
Department of Food Technology and Nutrition
School of Agriculture, Lovely Professional University, Phagwara, Punjab, India

Michael Wink
Institute of Pharmacy and Molecular Biotechnology, Heidelberg University
Heidelberg, Germany

Shikha Yadav
Department of Pharmacy, Galgotias University, Noida, Uttar Pradesh, India

Gabriella Yao
Department of Biomedical Sciences, Cooper Medical School of Rowan University
Camden, NJ, USA

Preface

Nutraceuticals are nutritive pharmaceuticals of health significance. It could be fortified food product or dietary supplement. Nevertheless, the health benefits of nutraceuticals are over and above the basic nutritional function. The recommended level of consumption of nutraceuticals not only maintains good health but also functions against life-style-related acute and chronic metabolic disorders. Thus, nutraceutical bioactive molecules help in disease prevention and promote optimal health, quality, and longevity of life. The medical uses of nutraceuticals in disease like depression, stress, cardiovascular and gastrointestinal disorders, cancer, cholesterol, obesity, and diabetes have been intensively studied. These bioactive molecules of high value are obtained from natural biosphere resources. Nutraceuticals can be derived from animals, plants, microorganisms, and fungal sources. Herbal bioactives comprise a variety of phytochemicals like polyphenols, alkaloids, terpenoids, and derivatives thereof. Many traditional literatures support the utilization of herbs as natural remedies to handle the metabolic and physiological disorders. Herbs may be categorized into food, medicinal, and poisonous types, according to the active and functional constituents they harbour.

The current book entitled “Herbal Nutraceuticals: Products and Processes” enlisted the numerous important herbal nutraceuticals from different plants sources including horticultural and agricultural important crops, as well. The book started with the overview of herbal nutraceuticals and covered diverse aspects of their application for the human health. Various chapters included in the book described the medicinal importance of various herbal constituents as alternative medicines, their importance in health benefits in the modern day life style, their antidiabetic and anticarcinogenic potentials, and antioxidant properties. In addition, the nutraceutical potential of various cereals, pseudo cereals, legumes, fruits, and oils have been specifically discussed in various chapters. It is a well-known truth that nothing is absolutely beneficial in the world. The positive and negative implications are associated with each substance. And therefore, at the end chapters describing the safety concern and toxicity evaluation have been included. Overall, *Herbal Nutraceuticals: Products and Processes* is a comprehensive book for the diverse knowledge and application for human health benefits.

Santosh Kumar Upadhyay
Department of Botany
Panjab University, Chandigarh, India

Sudhir Pratap Singh
Department of Industrial Biotechnology
Gujarat Biotechnology University, Gandhinagar, India

1

Plants Based Nutraceuticals: An Overview

Dheeraj Bisht[1], Vikas Bhatt[1], Rajeshwar Kamal Kant Arya[2], Shikha Yadav[3], Deepak Kumar[4], Sudhir Pratap Singh[5], and Santosh Kumar Upadhyay[6]

[1] *Devsthali Vidyapeeth College of Pharmacy, Rudrapur, Uttarakhand, India*
[2] *Department of Pharmaceutical Sciences, Sir J. C. Bose Technical Campus, Bhimtal, Kumaun University, Nainital, Uttarakhand, India*
[3] *Department of Pharmacy, Galgotias University, Noida, Uttar Pradesh, India*
[4] *Department of Pharmaceutical Chemistry, School of Pharmaceutical Sciences, Shoolini University, Solan, Himachal Pradesh, India*
[5] *Department of Industrial Biotechnology, Gujarat Biotechnology University, Gandhinagar, Gujarat, India*
[6] *Department of Botany, Panjab University, Chandigarh, India*

1.1 Introduction

The term nutraceuticals evolved from a combination of nutrition and pharmaceuticals and was coined by Stephen De Felice in 1989. According to him, nutraceuticals are food commodities or fractions that grant health benefits and clinical advantages (Durazzo et al. 2020). The word nutraceuticals, is not so popular worldwide and is many of time it is confused by the term dietary supplements, which is an important need for the health of humankind. However, in-depth studies of both terms manifest some root differences, such as the fact that nutraceuticals are prominent in treating several illnesses besides being important dietary supplements (Bommakanti et al. 2023). US FDA is usually not in favor of the term nutraceuticals, which is generally denoted as food-emerged derivatives with some other values specified by the elementary nutritional fractions that exist in that specific food item. Another misleading term for nutraceuticals is functional food, which can be described as any food being synthesized or manufactured employing scientific intelligence, with or without competent information on why or how it is being used (Tsiaka et al. 2022).

Moreover, when functional food helps to manage and treat ailments other than anemia, it is categorized as a nutraceutical (Jalgaonkar et al. 2019). These incorporate a large variety of products, including dietary supplements, genetically altered products, isolated nutrients, plant products, and processed cereals. Modifications in lifestyle patterns bring about influential changes in the everyday dietary regime that may lead to the emergence of newly developed lifestyle abnormalities such as type-2 diabetes and obesity. Other than these, cardiac disorders, namely stroke and ischemic heart dysfunction, are the most common reasons for mortality in the entire world. Nutraceuticals are reported as appropriate medications for these factors to cure such lifestyle

Herbal Nutraceuticals: Products and Processes, First Edition. Edited by Santosh Kumar Upadhyay and Sudhir Pratap Singh.

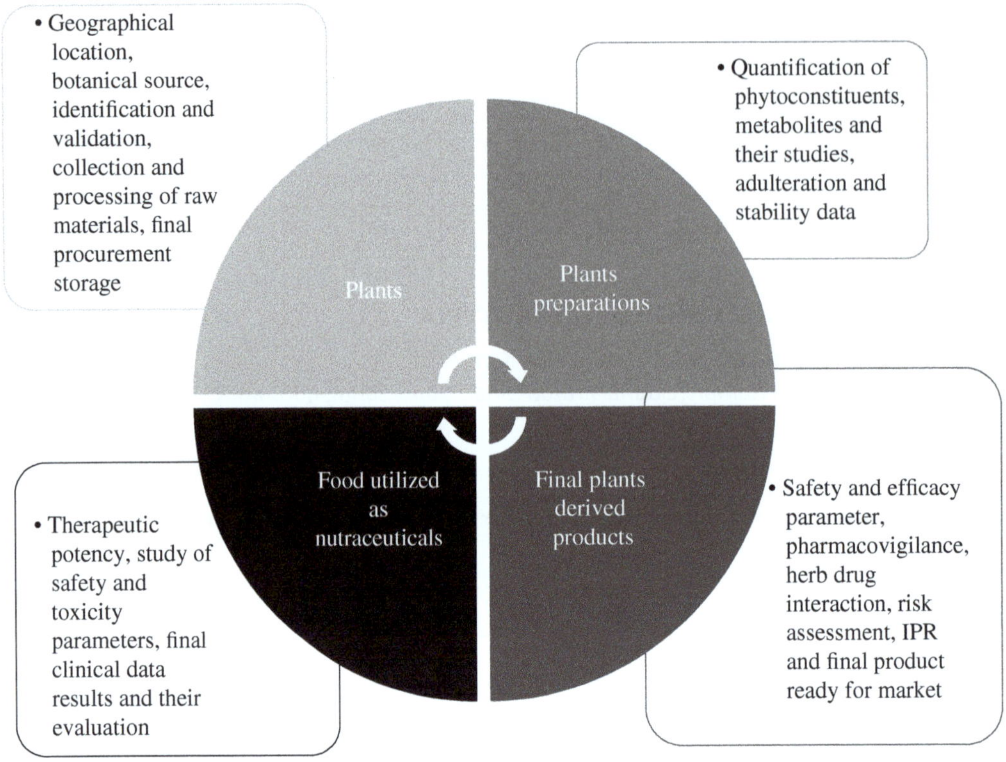

Figure 1.1 Identification and validation of plant-based nutraceuticals. *Source:* Mukherjee et al. (2016); Donno (2020).

abnormalities, hence attaining worldwide acceptability expeditiously. Besides these lifestyle abnormalities, nutraceuticals are utilized for the cure of various types of clinical manifestations, namely hypercholesteremia, allergic problems, sleep disorders, inflammations, immunodeficiencies, depressive situations, malignancies conditions, hypertension, cardiovascular diseases (CVDs), and anemic disorders (Jalgaonkar et al. 2019; Tsiaka et al. 2022).

It can be suggested that nutraceuticals have very efficacious and potent outcomes and marvelous health benefits that make them crucially attractive to consumers. In the last few years, the significance and importance of plant-based nutraceuticals have evolved, and the development of nutraceutical has revealed that it is crucial for human beings to have sound health considerations for better wellness and fitness. This work focuses mainly on the therapeutically active medicinal phytoconstituents derived from various parts of plants worldwide to cure and prevent various ailments. These phytochemicals obtained from plants could be used as nutraceuticals, further exploring their numerous remedial benefits and the theme behind their remarkable and valuable impacts on human health (Figure 1.1) (Mohammad and Imran 2019; Shahidi 2012).

1.2 Historical Background Behind Nutraceutical Investigations

The father of current remedies, Hippocrates (460–377 BC), suggested the root stone for modern-time nutraceuticals via the fabled quote, "Let food be thy medicine and medicine by the food." He was recognized as the founder of the idea that specific foods could also be a remedy for curing and controlling a disorder besides the drug moieties (Nwosu and Ubaoji 2020). Roman physician Galen used his expertise and knowledge based on his profession to develop and prepare dietary regimens that could be used to manage the health aspect of the whole population (Malve and Bhalerao 2023).

At the beginning of the nineteenth century, nutritional investigations were started by Francois Magendie. His investigations were based on practical facts and raised the question of whether foods lacking nitrogen could give nutrition. This provided an essential and substantial idea to consider besides proteins, fats, carbohydrates, and minerals to attain healthy and adequate nutrition (Nwosu and Ubaoji 2020; Puri et al. 2021). Further, similar investigations and results obtained by many investigators led to the development of crucial nutrients like vitamins. Medicinal plants such as Vinca and Taxus have significant cytotoxic effects, and the phytoconstituents obtained from these plants are used in cancer treatment. Similarly, the phytochemicals obtained from ginseng are used as an immunostimulant and chemotherapeutically active drug. As an herbal remedy, it has been well-recognized in China for over 2000 years (Arai et al. 2016).

On the other hand, the Ayurveda, an excellent compilation for Indian health science, reveals strong authentication of food used as nutraceuticals to cure and manage ailments. In the same way, Egyptians have established the clinical significance of numerous spices like garlic, onion, turmeric, fennel, coriander, cumin, and even reputed equivalent or very costly metals such as gold. A ship crew was described as having significant mortality due to scurvy, and these obstacles were resolved by the intake of a vitamin C-rich diet. Some investigation outcomes were seen in patients suffering from goiter, where iodine-containing rich salt played a crucial role. Hence, it can be stated that with time, scientific information has evolved, and food habits were logically and rationally modulated for the cure or prevention of disorders, and this may be a good sign that has led to the current development of nutraceutical (Andrew and Izzo 2017; Durazzo et al. 2020).

1.3 Pervasive Requisition for Nutraceuticals

Present-time working patterns and deranged lifestyles have led to new ailments, usually termed lifestyle deformities. The reason behind these types of dysfunction is the alteration in improper and haphazard routine of life, innutritious and inadequate diet, consumption of more alcohol, fast foods containing unhygienic high chain fatty acids, no physical exercise, disturbed biological clock of life, improper body posture, excessive use of computer and networking, sitting several hours in one position, insufficient sleep, rest and extreme stress and tension mainly (Jain et al. 2022; Malve and Bhalerao 2023). Hence, the intake of nutraceuticals in present medical and pharmaceutical science provides the best feasible solution for managing sound lifestyles for humans. Despite managing lifestyle-associated problems, these are considered and are well-recognized for predisposition to diverse clinical situations. Many nutraceuticals available in the market assist in stopping the transformation of minor lifestyle abnormalities into serious ailments (Chopra et al. 2022). The wide acceptability of nutraceuticals went ahead in 1980 when researchers were involved in the efficiency assessment of such kinds of ingredients accompanied by their role in mass media. Other reasons like a sudden increase in extreme clinical costs, enhanced life demands, and augmented health consciousness by individuals have also led to consumer acceptability of such ingredients (Anand and Bharadvaja 2022; Shahgholian 2022).

1.4 Classification of Nutraceuticals

Based on their origin, nutraceuticals are divided into two categories: (i) traditional and (ii) nontraditional nutraceuticals.

1.4.1 Traditional Nutraceuticals

These are food-based nutraceuticals that we use in our routine life; nature provides us with various types and sources of nutraceuticals that we consume daily, such as fruits, milk, vegetables, grains, meat, fish, and eggs (Bhaskarachary 2016; Bhat and Bhat 2011; Dixit et al. 2021; Upadhyay and Singh 2021, 2023). These are taken in raw or cooked form. Traditional food improves health and provides nutrients, and it has several medicinal properties, such as anti-inflammatory and antioxidant properties (Alkhatib et al. 2017). Traditional food includes tomatoes, fermented milk, lentils, soybeans, kidney beans, nuts, rice, wheat, and oranges. Rice is the major food component worldwide; it contains high starch and very low fat, salt, and sugar content and improves gut microbiota health (Umadevi et al. 2012). Traditional nutraceuticals include prebiotics, dietary fibers, antioxidant spices, polyunsaturated fatty acids (PFAs), and polyphenols.

1.4.1.1 Prebiotics

In 1995, Gibson and Roberfroid defined prebiotics as "a nondigestible food ingredient that beneficially affects the host by selectively stimulating the growth and activity of one or a limited number of bacteria in the colon, thus improving host health" (Hutkins et al. 2016). WHO/FA identified prebiotics as a nonviable food component that confers health benefit(s) to the host associated with microbiota modulation (FAO/WHO 2001). In 2008, prebiotics got a new definition: the "selectivity 'or the' potency" of a prebiotic indicated by the stimulation of specific gut microbiota (De Vrese and Schrezenmeir 2008). In general, prebiotics are used by the gut microbes as a source of nutrients for their metabolic activities and proliferation. Therefore, prebiotics became important in functional food supplement preparations (Swennen et al. 2006). Established prebiotics include lactulose, galactooligosaccharides, polydextrose, fructooligosaccharides, mannan-oligosaccharides, and inulin. Other new prebiotics include polyphenols, bacterial pectic oligosaccharides, isomalto-oligosaccharides, exopolysaccharides, sugar alcohols, resistant starch, and xylooligosaccharides (Drakoularakou et al. 2004). A few prebiotics are unrefined barley, soya beans, raw oats, yacon, breast milk, nondigestible oligosaccharides, and nondigestible carbohydrates (Davani-Davari et al. 2019). But they can also be found in food sources like almonds, chia seeds, flaxseeds, garlic, oats, dandelion greens, onion, barley, artichoke, and many other herbal sources (Kaur et al. 2021). However, most prebiotics are synthesized from complex polysaccharides through enzymatic digestion (Chapman et al. 2011). Thus, we can say herbal nutraceuticals have a symbiotic relationship with beneficial gut microbiota, thus playing a part in the probiotic function. For example, gut microbiota ferments dietary flaxseed fibers, which exhibit a prebiotic effect, and these fibers result in the modification of gut microflora and prove health beneficial.

1.4.1.2 Dietary Fibers

Dietary fiber, also known as roughage, is a portion of plant-derived content that is not entirely degraded by the digestive enzymes of human beings. These differ in chemical contents and are categorized by their viscosity, fermentability, solubility, and ionization, and fibers are processed in the human body in a specific way. Dietary fiber has mainly two portions: soluble fiber and insoluble fiber, which are constituents of plant-based food, namely cereals, grains, legumes, fruits, vegetable seeds, and nuts, utilized as nutraceuticals (Shahidi 2012). A balanced and proper diet, rich in dietary fiber intake, is mainly incorporated with very sound health aspects and reduces the cause of various harmful and detrimental ailments. Dietary fiber also comprises nonstarch polysaccharides and other kinds of phytoconstituents of plants such as lignins, inulin, dextrin, starch, cellulose, oligosaccharides, beta-glucans, pectins, and chitins (Bojarczuka et al. 2022; He et al. 2022). Some important sources of dietary fibers utilized as nutritional fibers are summarized in Table 1.1.

Table 1.1 List of dietary fibers used as nutraceuticals.

Serial number	Phytoconstituents	Water-insoluble nutritional fiber is used as nutraceuticals
1	Starch	Rich amylose heat, raw bananas, potatoes, barley, corn
2	Cellulose	Maximum plant vegetables, fruits, and cereals
3	Hemicelluloses	Legumes, cereals, timber, bran
4	Pentose	Oat, rye
5	Hexose	Barley, wheat
6	Xanthan gum	Formation of Xanthomonas bacteria from sugar components
7	Lignin	Cereals, vegetables, filaments of garden beans
8	Chitin	Exoskeleton of crustaceans and insects, also found in fungi
Water-soluble dietary fibers		
1	Arabinoxylan	Psyllium
2	Pectin	In the skin of the fruit, commonly quines and apples, vegetables
3	Alginic acid	Found in algae
4	Raffinose	Legumes
5	Carrageenan	Found in red algae
6	Polydextrose	Found in synthetic polymer
7	Fructans	Substitute in a few plants' taxa starch as depo of carbohydrate

Source: Puri et al. (2021); Bojarczuka et al. (2022).

1.4.1.3 Antioxidant

Antioxidants are usually employed for scavenging free radicals from body cells, and their significance in prohibiting free radicals and preventing oxidative stress. These tiny components exist in natural plants utilized as nutraceuticals; the phytochemicals present in these plants are rich in antioxidant containing property and are evidenced to be very useful and appreciable for everyone for the cure and management of neurological, cardiovascular, dermatological, and ophthalmic problems. The results of these nutraceuticals present in plants have very good therapeutic benefits for human health (Barbieri et al. 2020). Glutathione is found in several phytoconstituents of naturally occurring plants, and it is a popular nutraceutical that contains the amino acids glutamic acid, cysteine, and glycine. Alpha-lipoic acid (ALA) is also used as an antioxidant. ALA is widely accepted as a "universal antioxidant" because it can elicit substantial and efficacious antioxidant effects. Carotenoids are abundantly occurring natural pigments in naturally existing medical compounds, and so far, more than 600 carotenoid constituents have been validated and well-identified as being consumed as nutraceuticals. Among them, β-Carotene has outstanding and noteworthy clinical advantages. Carotenoids are beneficial pigments that contribute to the defense and safety of plants toward photooxidative mechanisms. They possess prominent antioxidants scavenging singlet molecular oxygen as well as peroxyl radicals. In the human organism, carotenoids are part of the antioxidant defense system. They interrelate and combine with other antioxidants to give good health benefits; the combination of carotenoids is better reported than singularly used constituents (Stahl and Sies 2003).

1.4.1.4 Spices

Spices are most widely employed as auspicious and prominent aromatic, coloring, and fragrance-containing constituents of beverages and are most widely consumed for their nutraceutical properties. Antimicrobial, antibacterial, antiviral, antifungal, carminative, antipyretic, analgesic, wound healing properties, antioxidant, digestive, laxative, purgative, cardiotonic, antiemetic properties of spices possess enormous kinds of therapeutic applications, and the phytoconstituents present in these spices are used for their remarkable nutraceutical effect (De and De 2018; Shenoy et al. 2022; Upadhyay and Singh 2021).

Antineoplastic, antihypertensive, anti-inflammatory, antihypercholesterolemic, antidiabetic, and antimalarial are all activities of nutraceuticals that have foremost significance for human health. Some of the most popularly used nutraceuticals present in the form of phytoconstituents in medicinal plants like curcumin exist in turmeric, allicin, the active constituent present in garlic, capsaicin present in red pepper, fiber, saponin exists in fenugreek, and gingerol present in the ginger plant, piper have eminent health and medicinal benefits as well, as reported evidently by research studies (Heghes et al. 2022; Kumar et al. 2016; Kumar Paswan et al. 2021).

1.4.1.5 Polyunsaturated Fatty Acids

PFAs are a subcategory of fatty acids distinguished and specified by a backbone of two or above two C=C double by dietary sources, mainly including drying oils. A few PFAa are called essential because our body does not synthesize them, they are consumed through nutritional sources including drying oils. Polyunsaturated fats include flax seed, walnuts, sunflower, avocados, omega-3, and omega-6 fatty acids; soyabean oil is the safest polyunsaturated acid and most beneficial for aspects of human health and is considered an essential fatty acid that is required for body cell growth and for performing brain function of the human body. The constituents in these PUFA are availed as nutraceuticals and are vital for performing human body functions (Saini et al. 2021; Ummalyma et al. 2019).

1.4.1.6 Polyphenols

Polyphenols are also used as nutraceuticals and are small chemical constituents obtained from plant-derived sources. They are used through several dietary sources. However, they are very easy to attain through a balanced and proper diet, such as vegetables, fruits, leaves, flowers, and spices. Currently, over 8000 kinds of polyphenols are reported principally, including quercetin-like flavonoids and catechin present in fruits. Berries, nuts, apples, blackberries, and cocoa contain phenolic acids, which have excellent therapeutic benefits for human health, and they help to cure various human ailments (De Araújo et al. 2021; Silva and Pogačnik 2019).

1.5 Popularly Found Herbs Utilized as Nutraceuticals

Some of the popularly used herbs and their phytochemicals are used as nutraceuticals, and their remedial utility is mentioned below (Table 1.2) (Bommakanti et al. 2023; Castro-Castaneda et al. 2022; Dixit et al. 2021; Gonçalves et al. 2018; Jha et al. 2021; Komala et al. 2022; Lestari 2021; Puttasiddaiah et al. 2022; Sharma et al. 2022; Upadhyay and Singh 2021, 2023; Williamson et al. 2020).

1.5.1 Nontraditional Nutraceuticals

These nutraceuticals are synthesized, or semi-synthesized food products made by using artificial means such as genetic engineering, or natural food products fortified with nutrients to improve the health benefits of nutraceuticals (Sharma et al. 2022). They are further classified as (i) recombinant, and (ii) fortified nutraceuticals (Singh and Sinha 2012). Mineral or vitamin-fortified cereals and β-Carotene fortified rice are examples of nontraditional nutraceuticals (AlAli et al. 2021).

Table 1.2 List of herbs consumed as nutraceuticals.

Serial number	Popular name	Botanical name	Phytochemicals	Remedial utility
1	Bael	Unripe fruits of *Aegle marmelos* Corr. (Rutaceae)	Marmelosin	Prevention of dysentery and diarrhea, appetizer, and digestive
2	Senna	Dried leaves of *Cassia Angustifolia* (Leguminosae)	Sennosides A and B	Laxatives and purgatives
3	Turmeric	Rhizomes of *Curcuma longa* (Zingiberaceae)	Curcumin	Antibacterial, anti-inflammatory, antiseptic, anticancer
4	Asafetida	Oleo gum resins of *Ferulaasafoetida L.* (Umbelliferae)	Umbellic acid and ferulic acid	Expectorant, digestive, appetizer, and stimulant
5	Onion	Dried bulbs of *Allium cepa* (Liliaceae)	Alliin and allicin	Antibacterial, antiatherosclerosis, hypoglycemic and antihypercholesteremic
6	Garlic	Dried bulbs of *Allium sativum* (Liliaceae)	Allicin and alliin	Antibacterial, antigout, antihypercholesteremic anti-inflammatory, nervine stimulant
7	Ginger	Dried rhizomes of *Zingiber officinale* (Zingiberaceae)	Zingerols and zingiberene	Antipyretic, anti-inflammatory, expectorant, antibacterial, antiplatelet, antihyperglycemic, used in chronic bronchitis
8	Liquorice	Dried roots of *Glycyrrhiza glabra* (Leguminosae)	Liquritin and glycyrrhizin	Expectorant, antiallergic, anti-inflammatory
9	Aloes	Dried juice of leaves of *Aloe barbadensis* Mill. (Liliaceae)	Aloesin and aloins	Antiviral, wound healing effects, antiplatelet, anti-inflammatory, some preparation aloe reduces blood cholesterol level, emollient
10	Brahmi	Herbs of *Centella asiatica* (umbilliferae)	Madecassoside and asiaticoside	Memory enhancer, nervine tonic, antistress, antianxiety
11	Ashwghandha (Indian ginseng) or Indian Winter Cherry	Roots of *Wthaniasomnifera* (Solanaceae)	Withanolides, Withaferin, anaferine, anahygrine	Adaptogen, treats stress induced ailments, dementia, and Parkinson's disorder, lowers blood sugar levels and fats
12	Arjuna	Dried stem bark of *Terminalia arjuna* Rob, Combretaceae	Tryptophan, tyrosine, histidine, cysteine, triterpenoids, and sterols	Cardiotonic, lowers blood cholesterol levels, used to treat atherosclerosis, cardiomyopathy

(*Continued*)

Table 1.2 (Continued)

Serial number	Popular name	Botanical name	Phytochemicals	Remedial utility
13	Rauwolfia	Dried roots and rhizomes of *Rauwolfia serpentine*, Apocyanaceae	Reserpine, ajmalicine, ajmaline, yohimbine	Antihypertensive, used to treat diarrhea, malaria, and male infertility
14	Cinchona	Dried barks of *Cinchona officinalis*	Quinine, quinidine, cinchonine, cinchonine, quinic acid	Antimalarial, appetizer
15	Strophanthus	Dried ripe seeds of *Strophanthus kombe* olivApocyanaceae	Strophanthin, Quabain	Used to treat cardiac failure, antioxidant, analgesic, wound healing effects
16	Dioscorea	Dried rhizome of many species of *Dioscorea compositae, D. villosa, D. deltoidea*	Phenanthrene, purine derivatives, spirostane, sesquiterpene	Used to treat rheumatoid arthritis and asthma, preparation of progesterone and other steroids
17	St. John Wort	Flowering shrub of *Hypericum perforatum*	Hypericin, naphthodianthrones	Antidepressants, used to treat mild to moderate depression
18	Echinacea	Flowering parts of *Echinacea purpurea* (L.), Asteraceae	Chicoric acid, cynarine	Used to treat urinary tract infections, allergic rhinitis
19	Giloy	Climbing shrub (*Tinospora cordifolia*) that grows on other trees, from the botanical family Menispermaceae	Berberine or berberine hydrochloride Giloin, tinosporanacetate, tinosporalacetate, tinosporidine,	It may help to treat diabetes, high cholesterol, and high blood pressure, to treat conditions like dengue fever, and immunologically active

1.5.2 Recombinant Nutraceuticals

Genetically engineered nutraceuticals are called recombinant nutraceuticals in biotechnology, and genetic recombination techniques are utilized for manufacturing the nutraceuticals. Energy-boosting nutraceuticals such as cheese, fermented starch, vinegar, bread, multivitamin corn, golden rice, mustard, and golden kiwi are synthesized using this technique (Sharma et al. 2022; Upadhyay 2021, 2024).

1.5.3 Fortified Nutraceuticals

Fortified nutraceuticals are fruit juice or food products in which the nutrients are mixed or added to improve the health benefits of nutraceuticals. Vitamin D3-fortified milk and calcium-fortified orange juice are examples of fortified nutraceuticals. They improve the working efficiency of patients. Fortified folic acid has been used since 1998, and in 2016, FDA-approved folic acid-fortified corn flour, a fortified food product with folic acid, which showed improved efficiency in cancer treatment (Gupta et al. 2010).

1.6 Utility of Nutraceuticals in Cure of Different Diseases

Nutraceuticals improve health and well-being and regulate immunity to prevent and treat many diseases and health problems. Nutraceuticals can be used to treat many disorders, as outlined below (Figure 1.2).

1.6.1 Nutraceuticals for Cardiovascular Diseases

CVDs have risk factors that can be effectively addressed with nutraceutical intervention. Nutraceuticals have promising potential to treat cardiovascular disorders, as has been supported by sustainable data. Cardiovascular disease primarily affects blood arteries and heart function. CVDs have a significant contributor to global mortality, and are responsible for around 30% of annual deaths worldwide. Dietary supplements are effective for managing and preventing cardiovascular illnesses. They can be categorized as those used for treating arrhythmias, congestive heart failure, angina, hypertension, and hyperlipidaemias. Below are some nutraceuticals and dietary supplements used to treat and prevent CVDs (Jalgaonkar et al. 2019; Tsiaka et al. 2022).

1.6.1.1 Allicin and Allin

There is a correlation between elevated levels of plasma triglycerides and blood cholesterol in individuals with ischemic heart disease and atherosclerosis. *Allium sativum* exhibits anti-hyperlipidemic effects by promoting the elimination of cholesterol and its metabolites through fecal matter while simultaneously decreasing the formation of cholesterol within the body. This phenomenon helps to attain a more advantageous high density lipoprotein (HDL) to low density lipoprotein (LDL) ratio. If allicin and alliin are not broken down by gastric acids, they can affect cholesterol levels. The efficacy of garlic supplementation on blood cholesterol levels was evaluated

Figure 1.2 Displays the use of nutraceuticals and dietary supplements in different disorders.

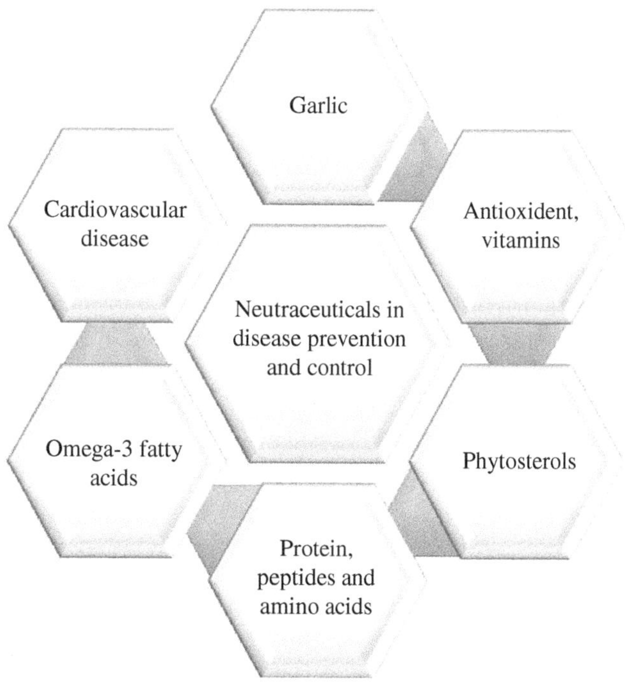

using thirteen placebo-controlled trials, including 781 participants. In addition to its anti-hyperlipidemic actions, garlic exhibits inherent antihypertensive characteristics (Helal et al. 2019; Kausar et al. 2023).

1.6.1.2 Omega-3 Fatty Acids

Marine sources yield PFA. The utilization of docosahexaenoic acid (DHA) and marine omega-3 eicosapentaenoic acid (EPA) is paramount in managing and preventing CVDs. The Diet and Reinfarction Trial (DART), involving 2033 men who had suffered a heart attack, revealed that the utilization of fish oil supplements resulted in a 29% reduction in mortality rates over two years. The ingestion of fish oil led to a notable reduction of 45% in untimely fatalities, a 30% decline in cardiovascular ailments, and a 20% decrease in overall mortality. In recent clinical trials, omega-3 fatty acids have decreased the likelihood of cardiac arrhythmias and improved the well-being of persons with atherosclerosis-induced plaque formation. The enhancement of electrical stability in cardiac cells is facilitated by omega-3 fatty acids, resulting in an extended relative refractory period, and perhaps assisting in managing arrhythmias (Barry and Dixon 2021; El-Beltagi et al. 2022).

1.6.1.3 Soya Isoflavones

Soy proteins and soy isoflavones are considered vital nutrients with potential medicinal advantages, such as antihyperlipidemic, antihypertensive, antihyperglycemic, antioxidant, anticancer, anti-inflammatory, antiobesity, and neuroprotective properties. These properties have been scientifically supported, establishing a foundation for the observed associations. According to clinical study data, the consumption of soy protein has been found to reduce blood cholesterol levels in human subjects (Ohta et al. 2022). According to the Unites State Food and Drug Administration (USFDA), consuming 25 g of soy proteins or isoflavone daily can effectively lower blood pressure in postmenopausal women. Soy proteins benefit serum lipid levels, especially in those with elevated cholesterol levels. According to a study, those who adhered to a diet characterized by a low saturated fat intake exhibited a decreased prevalence of coronary heart disease. The lipid profiles were not affected by soy isoflavone (Moller et al. 2008). A study conducted by Abaco and Abalon observed a significant reduction in the LDL to HDL ratio after consuming novel soy products abundant in isoflavones, cotyledon soy fiber, and soy phospholipids (Liao et al. 2021).

1.6.1.4 Proteins, Peptides, and Amino Acids

Hypertension is linked to cardiovascular illnesses. Angiotensin converting enzyme (ACE) inhibitors are a primary treatment for the illness, although they can cause adverse effects including hypotension, high potassium levels, reduced kidney function, coughing, and skin rashes. ACE inhibitors are naturally present in casein and whey protein obtained from milk. Animal studies have shown that milk-derived proteins have antihypertensive effects. Similar findings have been documented in clinical trials, showing a statistically significant decrease in blood pressure (Majeed et al. 2022; Mannucci et al. 2021).

1.6.1.5 Antioxidants and Vitamins

Antioxidants are also used as supplements in chronic conditions like cardiovascular disorders and cancer. They inhibit LDL cholesterol oxidation by neutralizing the harmful impact of free radicals. Vegetables, fruits, fish, and fixed oils are rich sources of antioxidant and vitamins that help in preventing the free radicals generation. Some epidemiologic research has shown that coronary heart disease (CHD) patients who consume high amounts of antioxidants experience lower rates of morbidity and mortality. Antioxidant supplements containing vitamins C and E aid in CHD prevention. Supplementation with ß-carotene can lead to adverse effects and is therefore not advised. The National Health and Nutrition Examination Survey-I cohort study found that increased vitamin C intake decreases the risk of CHDs. The study spanned over 10 years and involved American men and women aged 25–74. Participants were randomly assigned various combinations of 10 nutritional supplements for five years (Poli et al. 2021).

1.6.1.6 Phytosterols

Phytosterols structurally resemble to the cholesterol. They impede the absorption of cholesterol in the intestine by competing with cholesterol.

They naturally occur in vegetable oils, seeds, nuts, cereals, and wood pulp (de Melo et al. 2018). Consuming phytosterols resulted in higher liver absorption of LDL, decreased blood LDL levels, and decreased cholesterol absorption. Research has shown that consuming plant sterols can decrease LDL cholesterol by up to 15%. Plant sterols are extracted from natural grains, including soy, sunflower, and maize. Research has shown that consuming 2–3 g/day of plant sterols/stanols can lower LDL cholesterol levels by up to 20%. However, the effectiveness may vary across individuals (Jones et al. 2021).

1.6.1.7 Alkaloids

Rauwolfia serpentine is a commonly used ayurvedic and homeopathic medicine for reducing blood pressure. It contains alkaloids (ajmaciline, ajmaline, and reserpine) that lower blood pressure (Kausar et al. 2023). Studies show that Ajmaciline reduces blood pressure by increasing leucine and choline levels and decreasing tyrosine levels (Feng et al. 2014). The aspidocarpine alkaloid obtained from *Aspidosperma desmanthum* also lowers blood pressure (Monteiro et al. 2022).

1.6.2 Nutraceuticals in Radiotherapy and Chemotherapy-based Cancer Treatment

Cancer is a major threat to public health, causing around 10 million deaths annually worldwide. Research indicates a significant correlation between fruit eating and a lower risk of several cancers (Thakur et al. 2024). Certain types of cancer exhibit considerable resistance to chemotherapy and radiotherapy, hence diminishing the efficacy of systemic cytotoxic chemotherapy and radiotherapy in enhancing patient survival (Emran et al. 2022). Various combination medications offer a viable approach to treating cancer (Bayat Mokhtari et al. 2017). Several botanical species and organic supplements have been identified as effective in mitigating the negative consequences of radiotherapy and chemotherapy (Maiuolo et al. 2021). Probiotics and vitamins improve the efficacy of traditional cancer treatment by neutralizing the cancer cells (Di Napoli et al. 2023). Hence, it is recommended to employ these interventions in combination with radiotherapy or chemotherapy to mitigate adverse effects and improve the effectiveness of treatment. (Ahmad et al. 2021). The nutraceutical industry has transitioned from promoting health to preventing illnesses. Numerous herbs and phytochemicals currently employed have demonstrated safety and efficacy as nutraceuticals, which can impede cancer proliferation, alleviate the adverse effects of radiotherapy and chemotherapy, and augment the effectiveness of these therapeutic interventions (Puri et al. 2021). The impact of phenolic compounds derived from Aglianico Grape pomace on colorectal cancer cell lines at various stages of development was investigated by Caponio et al. through an *in vitro* digesting model. The extract derived from Aglianico GP showed notable effects on cellular proliferation, apoptosis, and several cellular processes. Both HT29 and SW480 cells exhibited a notable elevation in the concentrations of Bax, the Bax/Bcl-2 ratio, and caspase-3. The significant total phenolic content (TPC) and antioxidant activity seen in the digested extract of Aglianico GP were attributed to anthocyanins, phenolic acids, and flavonoids, as determined using Ultra High-Performance Liquid-Chromatography-Diode-Array (UHPLC-DAD) analysis. Zhang et al. (2021) conducted a study investigating the combined effects of chrysin and apigenin on colorectal cancer (CRC) by inhibiting the p38-Mitogen-Activated Protein Kinase (P38-MAPK/AKT) pathway. Applying apigenin and chrysin at a concentration of 25 μM in both cell lines resulted in a notable decrease in clones, inhibition of migration and invasion, and facilitation of apoptosis. In addition, chrysin and apigenin significantly reduced the activation of p-P38 and p-AKT. The tumor-inhibiting effects of apigenin and chrysin were effectively mitigated by anisomycin, a P38 agonist. The combination of apigenin and chrysin, at a concentration of 25 μM each, effectively suppressed the growth and spread of CRC cells by reducing the activity of the P38-MAPK/AKT pathway (Calvani et al. 2020).

1.6.2.1 Curcumin

Curcumin, also known as diferuloyl methane, is derived from turmeric (*Curcuma longa*). Curcumin is considered a potent nutraceutical for treating cancer. Preclinical research on curcumin indicates that it can inhibit the development of cancer in various types, including pancreatic, colorectal, prostate, gastric, and hepatic cancer. It has shown effectiveness in suppressing cancer progression at all stages, including angiogenesis, metastasis, and proliferation. It is more successful when used in conjunction with chemotherapy and radiotherapy for treating cancer (Malve and Bhalerao 2023; Nwosu and Ubaoji 2020).

1.6.2.2 Ginger and Genistein

Ginger is a nutraceutical with antimutagenic, antioxidant, and anti-inflammatory properties that can reduce the negative effects of radiotherapy and chemotherapy. These are the characteristics.

Ginger contains compounds that offer beneficial radio-protective effects. Ginsenoside Rf and ginseng, along with their polysaccharides, help reduce the side effects of cancer treatment regimens and have been reported to decrease the chance of cancer recurrence by 50% (Nwosu and Ubaoji 2020).

Genistein is a powerful isoflavone with potential anticarcinogenic effects. In vitro research has demonstrated that many components exhibit their antitumor properties only at elevated unattainable concentrations through regular dietary intake. Therefore, achieving the intended effect at the tumor site is challenging, highlighting the need to consider the distribution method *in-vivo* investigations and clinical trials. Ensuring that the natural components are nontoxic is crucial when formulating a remedy. Some substances are more effective at a young age, and genistein is one of them, as research has shown (Figure 1.3) (Calvani et al. 2020).

1.6.3 Nutraceuticals for Prostate Cancer Treatment

Prostate cancer (PCa) is the most prevalent malignancy and ranks as the second largest cause of cancer-related deaths among American men. Men with ancestors from the United States and Africa exhibit the highest mortality rates from PCa in comparison to those of European ancestry. Despite the effectiveness of current treatment techniques, there is always a risk of developing resistance to the disease and progressing to metastasis or other complications over time. Thus, there is a need for more efficient and nonharmful therapy methods to address these significant challenges and ensure appropriate care and treatment for this illness. Various safe nutraceuticals are available as effective anti-PCa therapies. Using nutraceuticals in medical design can enhance precision, reduce chemotherapy toxicities, lower disease resistance, and treat diseases in various stages. The text below discusses specific nutraceuticals that could potentially be employed as treatment (Hsiao et al. 2019; Li et al. 2014).

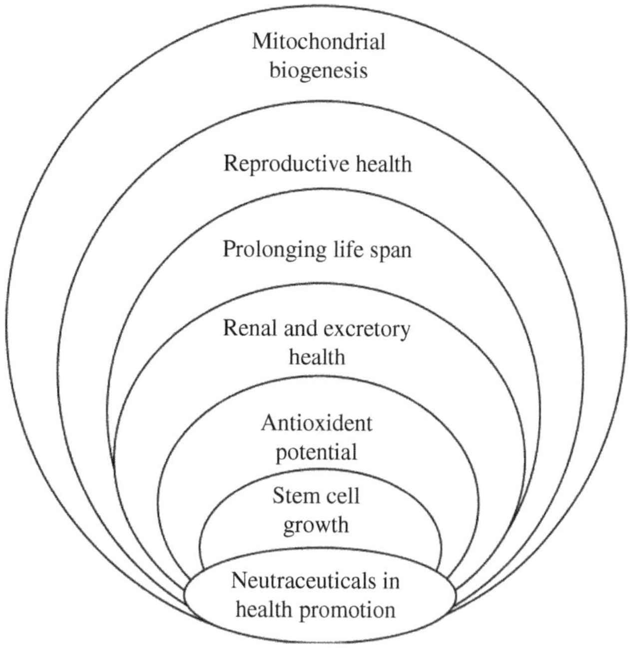

Figure 1.3 Role of nutraceuticals in disease prevention and health promotion.

1.6.3.1 Silibinin

Silibinin, a flavanolignan found in milk thistle seeds (*Silybum marianum*), exhibits potent anticarcinogenic properties against several tumors, including PCa. Silibinin demonstrates substantial anticancer effects in treating PCa in a preclinical animal model. Phase II clinical trials have assessed its bioavailability. More extensive clinical trials are required to validate silibinin's biological efficacy and effectiveness as a nutraceutical for managing advanced or localized forms of PCa.

1.6.3.2 Soy Isoflavones

Soy isoflavones are polyphenolic flavonoids mostly present in soybeans, red clover, and kudzu root and are commonly utilized in Asian and African cuisines. Clinical trials have demonstrated the benefits of soy isoflavones in combating PCa by significantly impacting inflammatory signaling and insulin levels. As reported, isoflavones have shown immuno-modulatory activities in the plasma of asymptomatic bio-chemically-recurrent PCa patients. An elevated level of genistein in the bloodstream was associated with a 69% decrease in the likelihood of developing PCa in Chinese individuals. Short-term administration of isoflavones for six weeks inhibits the cell cycle of prostate tumors and induces apoptotic signaling. However, it does not affect testosterone levels, PSA, free testosterone, or total cholesterol in patients with PCa. A six-month clinical investigation found that the use of soy protein did not impact the molecular markers associated with cell growth.

To assess apoptosis, the EGFR, Bax: Bcl-2, and Bax: PCNA ratios are examined in patients with high-risk and low-grade PCa. Conversely, soy protein (alcohol-washed) consumption reduced the levels of Bax and PCNA in tissues compared to patients who were treated with milk protein. Zhu and his team found that the cholesterol-reducing properties of soy protein isolate were improved when combined with soy soluble polysaccharide at greater levels of glycation.

1.6.4 Skincare Nutraceuticals

The skin is the body's biggest organ, protecting humans from pathogens, UV radiation, and toxins while contributing to sensitivity. Skin plays a significant role in safeguarding the body and may experience changes like immunological dysfunction, photo-aging, and inflammation, which can negatively impact human health. Nutraceuticals can help to postpone or reduce premature skin aging and alleviate skin-related illnesses. Nutraceuticals may include bioactive peptides, polysaccharides, botanical extracts, carotenoids, and more. These items have been shown to reduce symptoms of aging and provide protection against UV-induced aging in many human trials. CoQ10, a potent antioxidant, is used in antiwrinkle creams to delay aging signs (Iwaniak et al. 2019).

1.6.4.1 Peptides with Biological Activity

Peptide moieties are composed of two or more amino acids, possess a relatively tiny molecular weight of less than 3 kDa, and can manifest notable biological activity referred to as bioactive peptides. A wide variety of dietary proteins, derived from both plant and animal sources, have been subjected to extraction to get bioactive peptides. Eggs, milk (including casein and whey), and meat proteins are the primary animal protein sources. Bioactive peptides are frequently derived from several plant sources, including soy, oats, pulses (chickpeas, beans, peas, and lentils), canola, wheat, flaxseed, and hemp seed (Arya et al. 2022). Cosmetics often use peptides derived from collagen, which serve as nutraceutical formulations owing to their improved bioavailability and solubility. The effects of bioactive collagen peptide (BCP) have been demonstrated in a controlled study conducted on VERISOL®. The study encompassed the administration of VERISOL® and a placebo to a cohort of volunteers over eight weeks. The assessment of skin wrinkles was conducted both before and following the treatment period. After an eight-week treatment period, it was shown that BCP exhibited a significant reduction in eye wrinkle volume compared to the placebo. Furthermore, using BCP led to heightened concentrations of elastin and procollagen type 1, along with an augmentation in fibrillin content. The treatment reduced wrinkles and showed encouraging effects on skin matrix synthesis (Proksch et al. 2020). Nutraceuticals, namely Peptan F and Peptan P, are extracted from fish and include collagen peptides. Their purpose is to decelerate the aging process by effectively maintaining the moisture levels in the layers of the skin. According to a recent study, the nutraceutical product Celergen® has been shown to improve skin characteristics without inducing oxidative harm, demonstrating its efficacy and safety as a dietary supplement. The nutraceutical composition includes marine collagen peptides derived from deep sea fish, grape skin, coenzyme Q10, and lutein (Montero et al. 2022; Proksch et al. 2014).

1.6.4.2 Biologically Active Polysaccharides

Sugar-based polymers exhibit energy storage and structural functionality. They exist in various living forms like plants, fungi, mammals, and prokaryotic organisms, displaying various monosaccharide combinations. The most beneficial nutraceutical formulations are glycosaminoglycans derived from marine sources. These consist of an unbranched disaccharide unit of amino sugars *N*-acetylglucosamine or *N*-acetyl galactosamine, and a uronic acid called glucuronic or iduronic acid. A human trial was done using Imedeen® DermOne® formulations containing protein fractions and glycosaminoglycans, serving as dietary supplements for skin care. The supplement also included zinc gluconate and vitamin C, which are beneficial for skin health. During the trial, 10 women received Imedeen® for 90-days at a dosage of 0.5 mg. The parameters assessed were dryness, brittleness of hair and nails, wrinkles, and mottles. After 90 days, all these indications showed improvement, and observations confirmed that Imedeen® increased skin thickness and elasticity (Lassus et al. 1991).

1.6.4.3 Botanical Extracts with Biological Activity

The extracts are complex combinations of several chemicals with varied structures and origins. They have been used and reviewed for a long time (Sarma et al. 2023). Polyphenols are important natural substances used in cosmetics, characterized by a wide variety of families and structures. These are dietary plant-based micronutrients. These are advantageous as adjunctive therapy in disease prevention and can enhance illness prognosis. Polyphenols vary in bioavailability, with the most common ones in our diet having the highest concentration of active metabolites in target tissues. Their content and proportion differ based on the extraction process and families (Das et al. 2024; Shivangi et al. 2021). Pycnogenol® is a formulation high in catechins, flavonoids, procyanidins (B1, B2, B3, B7, C1, and C2), as well as phenolic acids including ferulic acid and caffeic acid. The plant has been proven to have several effects, including reducing cholesterol and providing cardiovascular benefits due to its antidiabetic, anti-inflammatory, and antioxidant qualities.

1.7 Current Status of Nutraceuticals

Several plant-based nutraceuticals have gained popularity in the recent past and have been used for the prevention and treatment of various ailments. Plant-based nutraceuticals are now used for skin care, wellness, health management, and treatment, and the stake of these plant-based nutraceuticals has emerged a high impact on the global market. Healthy people are consuming plant-based nutraceuticals to maintain their health. A new consumer market is established for those who are taking plant-based nutraceuticals for improving or maintaining their health, such as omega 3-fatty acid, antioxidants, flavonoids, polyphenols, or herbs like ginseng, green tea, aloe vera, algae, and wheatgrass (Jain et al. 2022). Approximately 40% of Chinese, 70% of Indians, and 90% of Africans use plant-based products for various diseases at the primary level (Calvani et al. 2020). The report suggests that the nutraceutical market in India will achieve a mark of USD 18 billion by 2025 (Malve and Bhalerao 2023). Among Association of Southeast Asian Nations (ASEAN) countries, China is the biggest exporter of nutraceuticals after Malaysia (Tripathi et al. 2020). After COVID-19, the nutraceutical market's immunity-boosting nutraceuticals boomed, and by 2027, the Asian market will touch the level of USD 140,178.6 million and the American market the level of USD 118.73 billion, respectively (Jadhav et al. 2023).

1.8 Challenges and Future Aspects

Nutraceuticals are gaining popularity worldwide, but their use has various challenges. No proper regulatory authorities have been established to control production, distribution, and efficacy testing. There is no identification system for identifying the raw or processed material, and the quality and purity assessment is also challenging. The clinical data for safety and efficacy are also lacking (Siddiqui and Moghadasian 2020). Now, the safety and efficacy of nutraceuticals are researchers' primary concerns, and various studies have been conducted to establish the safety data. In the future, more studies are required to assess nutraceuticals' bioavailability, efficacy, and toxicity. A comprehensive universal guideline for the processing and assessment of nutraceuticals is needed. Current technology, such as network pharmacology, metabolomics, and nutrigenomics, must be used to address these challenges (Da Costa 2017). Advanced drug delivery systems, such as nanoparticle microspheres, should be used to improve the bioavailability and efficacy of nutraceuticals (Chavan et al. 2024; Manocha et al. 2022).

1.9 Conclusion

Nutraceuticals are gaining popularity worldwide for their tremendous health benefits, and they are capturing a big stake in the world economy. Nutraceuticals are the main components of food that provide health benefits to the consumer. They are generally plant-based or phytoconstituents that improve or manage health conditions such as cancer, brain-related disorders, and metabolic disorders. But still, these are out-of-consumer approaches, and proper dosage forms are not available. They are now also cultivated using genetic engineering that will improve the economy of producers and ensure that consumers get it at cheaper cost with additional health benefits. Various traditional and nontraditional nutraceuticals have been practiced to date, and a variety of chemical constituents have been isolated, identified, and tested for claimed efficacy; however, the extracted compound has a complex system, and efficacy assessment is very difficult concerning the isolated or pure compound. They also have some other challenges, there are no proper techniques to identify the purity, quality, bioavailability, and efficacy of nutraceuticals. Some of the nutraceuticals have scientifically proven efficacy and known side effects, but to date, for the majority of nutraceuticals, we have a lack of information regarding the interactions, side effects, and benefit-risk ratio of nutraceuticals. However, researchers are striving to solve these questions. Hopefully, shortly we will get the answers. Academicians and pharmaceutical industries are working on various plant-based nutraceuticals using high-trough techniques such as network pharmacology, metabolomics, and nutrigenomics to explore the mechanism of action and efficiency of nutraceuticals with cost-effective nutraceuticals.

References

Ahmad, S.S., Khalid, M., Kamal, M.A., and Younis, K. (2021). Study of nutraceuticals and phytochemicals for the management of Alzheimer's disease: a review. *Current Neuropharmacology* 19 (11): 1884–1895. https://doi.org/10.2174/1570159X19666210215122333.

AlAli, M., Alqubaisy, M., and Aljaafari, M.N. (2021). Nutraceuticals: transformation of conventional foods into health promoters/disease preventers and safety considerations. *Molecules* 26 (9): 2540. https://doi.org/10.3390/molecules26092540.

Alkhatib, A., Tsang, C., Tiss, A. et al. (2017). Functional foods and lifestyle approaches for diabetes prevention and management. *Nutrients* 9 (12): 1310. https://doi.org/10.3390/nu9121310.

Anand, S. and Bharadvaja, N. (2022). Potential benefits of nutraceuticals for oxidative stress management. *Revista Brasileira de Farmacognosi* 32: 211–220. https://doi.org/10.1007/s43450-022-00246-w.

Andrew, R. and Izzo, A.A. (2017). Principles of pharmacological research of nutraceuticals. *British Journal of Pharmacology* 174 (11): 1177–1194. https://doi.org/10.1111/bph.13779.

Arai, S., Vattem, D.A., and Kumagai, H. (2016). *Functional Foods-History and Concepts*. DEStech Publications, Inc.

Arya, R.K.K., Kumar, P., Singh, A. et al. (2022). Chapter2.3—coenzyme Q: an endogenous antioxidant. In: *Antioxidants Effects in Health* (ed. S.M. Nabavi and A.S. Silva), 91–104. Elsevier. https://doi.org/10.1016/B978-0-12-819096-8.00023-9.

Barbieri, D., Gabriele, M., Summa, M. et al. (2020). Antioxidant, nutraceutical properties, and fluorescence spectral profiles of bee pollen samples from different botanical origins. *Antioxidants* 9 (10): 1001. https://doi.org/10.3390/antiox9101001.

Barry, A.R. and Dixon, D.L. (2021). Omega-3 fatty acids for the prevention of atherosclerotic cardiovascular disease. *Pharmacotherapy.* 41 (12): 1056–1065. https://doi.org/10.1002/phar.2615.

Bayat Mokhtari, R., Homayouni, T.S., Baluch, N. et al. (2017). Combination therapy in combating cancer. *Oncotarget* 8 (23): 38022–38043. https://doi.org/10.18632/oncotarget.16723.

Bhaskarachary, K. (2016). Traditional foods, functional foods and nutraceuticals. *Proceedings of the Indian National Science Academy* 2016 (82): 1565–1577.

Bhat, Z.F. and Bhat, H. (2011). Milk and dairy products as functional foods: a review. *International Journal of Dairy Science* 6: 1–12.

Bojarczuka, A., Skąpskaa, S., Khaneghaha, A.M., and Marszałek, K. (2022). Health benefits of starch: a review of literature. *Journal of Functional Foods* 93: 105094. https://doi.org/10.1016/j.jff.2022.105094.

Bommakanti, V., Ajikumar, A.P., Sivi, C.M. et al. (2023). An overview of herbal nutraceuticals, their extraction, formulation, therapeutic effects and potential toxicity. *Separations* 2023 10: 177. https://doi.org/10.3390/separations10030177.

Calvani, M., Pasha, A., and Favre, C. (2020). Nutraceutical boom in cancer: inside the labyrinth of reactive oxygen species. *International Journal of Molecular Sciences* 21 (6): 1936. https://doi.org/10.3390/ijms21061936.

Castro-Castaneda, C.R., Altamirano-Lamarque, F., Ortega-Macias, A.G. et al. (2022). Nutraceuticals: promising therapeutic approaching ophthalmology. *Nutrients* 14: 5014. https://doi.org/10.3390/nu14235014.

Chapman, C.M.C., Gibson, G.R., and Rowland, I. (2011). Health benefits of probiotics: are mixtures more effective than single strains? *European Journal of Nutrition* 50 (1): 1–17. https://doi.org/10.1007/s00394-010-0166-z.

Chavan, P., Singh, G.P., Aayush, K. et al. (2024). Development and characterization of pearl millet (AHB 1200) starch nanoparticle based edible films: a paradigm shift in food packaging. *Journal of Food Process Engineering* 47: e14575. https://doi.org/10.1111/jfpe.14575.

Chopra, A.S., Lordan, R., Horbanczuk, O.K. et al. (2022). The current use and evolving landscape of nutraceuticals. *Pharmacological Research* 175: 106001. https://doi.org/10.1016/j.phrs.2021.106001.

Da Costa, J.P. (2017). A current look at nutraceutical concepts and future prospects. *Trends in Food Science & Technology* 62: 68–78. https://doi.org/10.1016/j.tifs.2017.02.010.

Das, A.K., Gandhi, H., Panwar, V. et al. (2024). *14—Natural Product-Loaded Nanoparticles for Wound Healing, Nanotechnological Aspects for Next-Generation Wound Management*, 319–331. Academic Press https://doi.org/10.1016/B978-0-323-99165-0.00004-6.

Davani-Davari, D., Negahdaripour, M., Karimzadeh, I. et al. (2019). Prebiotics: definition, types, sources, mechanisms, and clinical applications. *Foods* 8 (3): 92. https://doi.org/10.3390/foods8030092.

De Araújo, F.F., De Paulo Farias, D., Neri-Numa, I.A., and Pastore, G.M. (2021). Polyphenols and their applications: an approach in food chemistry and innovation potential. *Food Chemistry* 338: 127535. https://doi.org/10.1016/j.foodchem.2020.127535.

De, A.K. and De, M. (2018). Functional and therapeutic applications of some important spices. *The Role of Functional Food Security in Global Health* 499–510. https://doi.org/10.1016/B978-0-12-813148-0.00029-3.

De Vrese, M. and Schrezenmeir, J. (2008). Probiotics, prebiotics, and synbiotics. *Advances in Biochemical Engineering/Biotechnology* 2008 (111): 1–66. https://doi.org/10.1007/10_2008_097.

Di Napoli, R., Balzano, N., Mascolo, A. et al. (2023). What is the role of nutraceutical products in cancer patients? A systematic review of randomized clinical trials. *Nutrients* 15 (14): 3249. https://doi.org/10.3390/nu15143249.

Dixit, S., Shukla, A., Singh, V., and Upadhyay, S.K. (2021). Bioprospecting of natural compounds for industrial and medical applications: current scenario and bottleneck. In: *Bioprospecting of Plant Biodiversity for Industrial Molecules*. Wiley; Portico. https://doi.org/10.1002/9781119718017.ch3.

Donno, D. (2020). Screening, identification, and quantification of nutritional components and phytochemicals in foodstuffs. *Foods* 10 (1): 125. https://doi.org/10.3390/foods10010125.

Drakoularakou, A., Mccartney, A., Rastall, R.A., and Gibson, G.R. (2004). Established and emerging prebiotics and their effects on the gut microflora. *Agro Food Industry Hi-Tech* 15: 18–20.

Durazzo, A., Lucarini, M., Durazzo, A. et al. (2020). Nutraceuticals in human. *Health*. https://doi.org/10.3390/books978-3-03936-458-9.

El-Beltagi, H.S., Mohamed, A.A., Mohamed, H.I.A. et al. (2022). Phytochemical and potential properties of seaweeds and their recent applications: a review. *Marine Drugs* 20 (6): 342. https://doi.org/10.3390/md20060342.

Emran, T.B., Shahriar, A., Mahmud, A.R. et al. (2022). Multidrug resistance in cancer: understanding molecular mechanisms, immunoprevention and therapeutic approaches. *Frontiers in Oncology* 12: 891652. https://doi.org/10.3389/fonc.2022.891652.

FAO/WHO (2001). Evaluation of Health and Nutritional Properties of Probiotics in Food Including Powder Milk With Live Acid Bacteria. Report of A Joint Fao/W Expert Consultation, Córdoba, Argentina. http://www.who.int/foodsafety/publications/fs_management/en/probiotics.pdf?ua=1.

Feng, M. et al. (2014). 1 H-NMR metabonomics study of the therapeutic mechanism of total alkaloids and Ajmalicine from Rauvolfia Verticillata in spontaneously hypertensive rats. *Analytical Methods* 6 (16): 6473–6482. https://doi.org/10.1039/C4AY00851K.

Gonçalves, R.F., Martins, J.T., Duarte, C.M. et al. (2018). Advances in nutraceutical delivery systems: from formulation design for bioavailability enhancement to efficacy and safety evaluation. *Trends in Food Science & Technology* 78: 270–291. https://doi.org/10.1016/j.tifs.2018.06.011.

Gupta, S., Chauhan, D., Mehla, K. et al. (2010). An overview of nutraceuticals: current scenario. *Journal of Basic and Clinical Pharmacy* 1 (2): 55–62.

He, Y., Wang, B., Wen, L. et al. (2022). Effects of dietary fiber on human health. *Food Science and Human Wellness* 11: 1–10.

Heghes, S.C., Vostinaru, O., Mogosan, C. et al. (2022). Safety profile of nutraceuticals rich in coumarins: an update. *Frontiers in Pharmacology* 13: 803338. https://doi.org/10.3389/fphar.2022.803338.

Helal, N.A., Eassa, H.A., Amer, A.M. et al. (2019). Nutraceuticals' novel formulations: the good, the bad, the unknown and patents involved. *Recent Patents on Drug Delivery & Formulation* 13 (2): 105–156. https://doi.org/10.2174/1872211313666190503112040.

Hsiao, Y.C., Peng, S.F., Lai, K.C. et al. (2019). Genistein induces apoptosis in vitro and has antitumor activity against human leukaemia HL-60 cancer cell xenograft growth in vivo. *Environmental Toxicology* 34 (4): 443–456. https://doi.org/10.1002/tox.22698.

Hutkins, R.W., Krumbeck, J.A., Bindels, L.B. et al. (2016). Prebiotics: why definitions matter. *Current Opinion in Biotechnology* 37: 1–7. https://doi.org/10.1016/j.copbio.2015.09.001.

Iwaniak, A., Darewicz, M., Mogut, D., and Minkiewicz, P. (2019). Elucidation of the role of in silico methodologies in approaches to studying bioactive peptides derived from foods. *Journal of Functional Foods* 61: 103486. https://doi.org/10.1016/j.jff.2019.103486.

Jadhav, H.B., Sablani, S., Gogate, P. et al. (2023). Factors governing consumers buying behavior concerning nutraceutical products. *Food Science & Nutrition* 11 (9): 4988–5003. https://doi.org/10.1002/fsn3.3518.

Jain, S., Purohit, A., Nema, P. et al. (2022). A Brief review on nutraceuticals and its application. *Asian Journal of Dental and Health Sciences* 2 (1): 7–13. https://doi.org/10.22270/ajdhs.v2i1.10.

Jalgaonkar, K., Mahawar, M.K., Bibwe, B. et al. (2019). Nutraceuticals and functional foods. *Trends and Prospects in Processing of Horticultural Crops* 231–250.

Jha, S.K., Roy, P., and Chakrabarty, S. (2021). Nutraceuticals with pharmaceutics: its importance and their applications. *International Journal of Drug Development and Research* 13. https://www.ijddr.in/.

Jones, K., Zhang, Y., Kong, Y. et al. (2021). Epigenetics in prostate cancer treatment. *Journal of Translational Genetics and Genomics* 5: 341–356. https://doi.org/10.20517/jtgg.2021.19.

Kaur, A.P., Bhardwaj, S., Dhanjal, D.S. et al. (2021). Plant prebiotics and their role in the Amelioration of diseases. *Biomolecules* 11 (3): Article 3. https://doi.org/10.3390/biom11030440.

Kausar, M., Arya, R.K.K., Joshi, T. et al. (2023). *Chapter 2—Bioactive Natural Products in Management of CVS Disorder Natural Products an Answer to Cardiovascular Health* (ed. A. Upalganlawar and T. Joshi), 33–56. Nova Publisher.

Kausar, M., Bisht, D., Keservani, R.K., Arya, R.K., 2023. Health benefits of herbal nutraceuticals: an overview hard Pages: Est. 354pp w/index.

Komala, M.G., Ong, S.G., Qadri, M.U. et al. (2022). Investigating the regulatory process, safety, efficacy and product transparency for nutraceuticals in the USA, Europe and Australia. *Foods* 12 (2): 427. https://doi.org/10.3390/foods12020427.

Kumar Paswan, V., Rose, H., Shekhar Singh, C. et al. (2021). Herbs and spices fortified functional dairy products. *Intech Open.* https://doi.org/10.5772/intechopen.98775.

Kumar, K., Kumar, D., Jindal, D. et al. (2016). Comparative antioxidant activity of roots and fruits of Piper sylvaticum Roxb. *Journal of Comprehensive Pharmacy* 3: 132–135.

Lassus, A., Jeskanen, L., Happonen, H. P., & Santalahti, J. (1991). Imedeen® for the Treatment of Degenerated Skin in Females. *Journal of International Medical Research*, 19(2), 147–152. https://doi.org/10.1177/030006059101900208

Lestari, L.A. (2021). Opportunities and challenges of functional foods and nutraceuticals development during Covid-19 pandemic. *BIO Web of Conferences* 41: 02009. https://doi.org/10.1051/bioconf/20214102009.

Li, Y., Ahmad, A., Kong, D. et al. (2014). Recent progress on nutraceutical research in prostate cancer. *Cancer Metastasis Reviews* 33 (2–3): 629–640. https://doi.org/10.1007/s10555-013-9478-9.

Liao, W., Sun, G., Xu, D. et al. (2021). The blood-pressure-lowering effect of food-protein-derived peptides: a meta-analysis of recent clinical trials. *Foods (Basel, Switzerland)* 10 (10): 2316. https://doi.org/10.3390/foods10102316.

Maiuolo, J., Gliozzi, M., Carresi, C. et al. (2021). Nutraceuticals and cancer: potential for natural polyphenols. *Nutrients* 13 (11): 3834.

Majeed, M., Ahmad, F., Mundkur, L., and Appian, S. (2022). Pharmacology of α-spinasterol, a phytosterol with nutraceutical values: a review. *Phytotherapy Research: PTR* 36 (10): 3681–3690. https://doi.org/10.1002/ptr.7560.

Malve, H. and Bhalerao, P. (2023). Past, present, and likely future of nutraceuticals in India: evolving role of pharmaceutical physicians. *Journal of Pharmacy & Bioallied Sciences* 15 (2): 68–74. https://doi.org/10.4103/jpbs.jpbs_96_23.

Mannucci, C., Casciaro, M., Sorbara, E.E. et al. (2021). Nutraceuticals against oxidative stress in autoimmune disorders. *Antioxidants (Basel, Switzerland)* 10 (2): 261. https://doi.org/10.3390/antiox10020261.

Manocha, S., Dhiman, S., Grewal, A.S., and Guarve, K. (2022). Nanotechnology: an approach to overcome bioavailability challenges of nutraceuticals. *Journal of Drug Delivery Science and Technology* 72: 103418. https://doi.org/10.1016/j.jddst.2022.103418.

de Melo, F.H.M., Oliveira, J.S., Sartorelli, V.O.B., and Montor, W.R. (2018). Cancer chemoprevention: classic and epigenetic mechanisms inhibiting tumorigenesis. What have we learned so far? *Frontiers in Oncology* 8: 644. https://doi.org/10.3389/fonc.2018.00644.

Mohammad, A. and Imran, M. (2019). Prospects of medicinal plants derived nutraceuticals: a re-emerging new era of medicine and health aid. *Progress in Chemical and Biochemical Research* 2 (4): 150–169.

Moller, N.P., Scholz-Ahrens, K.E., Roos, N., and Schrezenmeir, J. (2008). Bioactive peptides and proteins from foods: indication for health effects. *European Journal of Nutrition* 47 (4): 171–182. https://doi.org/10.1007/s00394-008-0710-2.

Monteiro, N.O., Monteiro, T.M., Nogueira, T.S.R. et al. (2022). Antihypertensive activity of the alkaloid aspidocarpine in normotensive wistar rats. *Molecules (Basel, Switzerland)* 27 (20): 6895. https://doi.org/10.3390/molecules27206895.

Montero, M., Acosta, Ó.G., and Bonilla, A.I. (2022). Membrane fractionation of gelatins extracted from skin of yellowfin tuna (*Thunnus albacares*): effect on molecular sizes and gelling properties of fractions. *CyTA—Journal of Food* 20 (1): 183–189. https://doi.org/10.1080/19476337.2022.2107707.

Mukherjee, P., Harwansh, R., Bahadur, S. et al. (2016). Factors to consider in the development of nutraceutical and dietary supplements. *Pharmacognosy* 653–661. https://doi.org/10.1016/B978-0-12-802104-0.00034-2.

Nwosu, O.K. and Ubaoji, K.I. (2020). Nutraceuticals: history, classification and market demand. In: *Functional Food and Nutraceuticals* (ed. C. Egbuna and G. Dable Tupas), 13–22. Cham: Springer https://doi.org/10.1007/978-3-030-42319-3_2.

Ohta, S., Asanoma, M., Irie, N. et al. (2022). Soy phospholipids exert a renoprotective effect by inhibiting the nuclear factor Kappa B pathway in macrophages. *Metabolites* 12 (4): 330. https://doi.org/10.3390/metabo1204033.

Poli, A., Marangoni, F., Corsini, A. et al. (2021). Phytosterols, cholesterol control, and cardiovascular disease. *Nutrients* 13 (8): 2810. https://doi.org/10.3390/nu13082810.

Proksch, E., Schunck, M., and Oesser, S. (2020). Positive effect of fish-derived bioactive collagen peptideson skin health. *Nutrafoods* 1: 127–133. https://doi.org/10.17470/NF-020-0018.

Proksch, E., Schunck, M., Zague, V. et al. (2014). Oral intake of specific bioactive collagen peptides reduces skin wrinkles and increases dermal matrix synthesis. *Skin Pharmacology and Physiology* 27 (3): 113–119. https://doi.org/10.1159/000355523.

Puri, V., Nagpal, M., Singh, I. et al. (2021). A comprehensive review on nutraceuticals: therapy support and formulation challenges. *Nutrients* 14 (21): 4637. https://doi.org/10.3390/nu14214637.

Puttasiddaiah, R., Lakshminarayana, R., Somashekar, N.L. et al. (2022). Advances in nanofabrication technology for nutraceuticals: new insights and future trends. *Bioengineering (Basel)* 9 (9): 478. https://doi.org/10.3390/bioengineering9090478.

Saini, R.K., Prasad, P., Sreedhar, R.V. et al. (2021). Omega-3 polyunsaturated fatty acids (PUFAs): emerging plant and microbial sources, oxidative stability, bioavailability, and health benefits-A review. *Antioxidants* 10 (10): 1627. https://doi.org/10.3390/antiox10101627.

Sarma, M.K., Saha, D., Das, B.K. et al. (2023). A delve into the pharmacological targets and biological mechanisms of *Paederia foetida* Linn.: a rather invaluable traditional medicinal plant. *Naunyn-Schmiedeberg's Archives of Pharmacology* 396: 2217–2240. https://doi.org/10.1007/s00210-023-02496-4.

Shahgholian, N. (2022). *Introduction to Nutraceuticals and Natural Products: Biological, Medicinal, and Nutritional Properties and Applications*, 1–14. Wiley. https://doi.org/10.1002/9781119746843.

Shahidi, F. (2012). Nutraceuticals, functional foods and dietary supplements in health and disease. *Journal of Food and Drug Analysis* 20 (1): 226–230.

Sharma, P., Bisht, D., Dwivedee, B.P. et al. (2022). Identification and quantification of biological active constituents of Amritarishta, a herbal formulation. *Indian Journal of Traditional Knowledge* 21 (4): 754–759. https://doi.org/10.56042/ijtk.v21i4.32514.

Shenoy, A., Buttar, H.S., Dicholkar, P. et al. (2022). Role of nutraceuticals, functional foods, and spices in the management of metabolic syndrome and related disorders. *Functional Foods and Nutraceuticals in Metabolic and Non-Communicable Diseases* 583–601.

Shivangi, S., Dorairaj, D., Negi, P.S., and Shetty, N.P. (2021). Development and characterization of a pectin-based edible film that contains mulberry leaf extract and its bio-active components. *Food Hydrocolloids* 121: 107046. https://doi.org/10.1016/j.foodhyd.2021.107046.

Siddiqui, R.A. and Moghadasian, M.H. (2020). Nutraceuticals and nutrition supplements: challenges and opportunities. *Nutrients* 12 (6): 1593. https://doi.org/10.3390/nu12061593.

Silva, R.F. and Pogačnik, L. (2019). Polyphenols from food and natural products: neuroprotection and safety. *Antioxidants* 9 (1): 61. https://doi.org/10.3390/antiox9010061.

Singh, J. and Sinha, S. (2012). Classification, regulatory acts and applications of nutraceuticals for health. *International Journal of Pharma and Bio Sciences* 2: 177–187.

Stahl, W. and Sies, H. (2003). Antioxidant activity of carotenoids. *Molecular Aspects of Medicine* 24 (6): 345–351. https://doi.org/10.1016/S0098-2997(03)00030-X.

Swennen, K., Courtin, C.M., and Delcour, J.A. (2006). Non-digestible oligosaccharides with prebiotic properties. *Critical Reviews in Food Science and Nutrition* 46 (6): 459–471. https://doi.org/10.1080/10408390500215746.

Thakur, M., Verma, R., Kumar, D. et al. (2024). Revisiting the ethnomedicinal, ethnopharmacological, phytoconstituents and phytoremediation of the plant *Solanum viarum* dunal. *Naunyn-Schmiedeberg's Archives of Pharmacology*. https://doi.org/10.1007/s00210-024-03034-6.

Tripathi, C., Girme, A., Champaneri, S. et al. (2020). Nutraceutical regulations: an opportunity in ASEAN countries. *Nutrition* 74: 110729. https://doi.org/10.1016/j.nut.2020.110729.

Tsiaka, T., Kritsi, E., Tsiantas, K. et al. (2022). Design and development of novel nutraceuticals: current trends and methodologies. *Nutraceuticals* 2: 71–90. https://doi.org/10.3390/nutraceuticals2020006.

Umadevi, M., Pushpa, R., Sampathkumar, K.P., and Bhowmik, D. (2012). Rice-traditional medicinal plant in India. *Journal of Pharmacognosy and Phytochemistry* 1 (1): 06–12.

Ummalyma, S.B., Sindhu, R., Binod, P. et al. (2019). *Production and Applications of Polyunsaturated Fatty Acids*, 109–125. EPFL Scientific Publications. https://doi.org/10.1002/9781119555384.ch6.

Upadhyay, S.K. (Éd.) (2021). *Genome Engineering for Crop Improvement*. Wiley. https://doi.org/10.1002/9781119672425.

Upadhyay, S.K. (Éd.) (2024). *Applications of Genome Engineering in Plants*. Wiley. https://doi.org/10.1002/9781394183913.

Upadhyay, S.K. and Singh, S.P. (Éds.) (2021). *Bioprospecting of Plant Biodiversity for Industrial Molecules*. Wiley. https://doi.org/10.1002/9781119718017.

Upadhyay, S.K. and Singh, S.P. (Éds.) (2023). *Plants as Bioreactors for Industrial Molecules*. Wiley. https://doi.org/10.1002/9781119875116.

Williamson, E.M., Liu, X., and Izzo, A.A. (2020). Trends in use, pharmacology, and clinical applications of emerging herbal nutraceuticals. *British Journal of Pharmacology* 177 (6): 1227–1240. https://doi.org/10.1111/bph.1494.

Zhang, X.-K., Lan, Y.-B., Huang, Y. et al. (2021). Targeted metabolomics of anthocyanin derivatives during prolonged wine aging: evolution, color contribution and aging prediction. *Food Chemistry 339*: 127795. https://doi.org/10.1016/j.foodchem.2020.127795.

2

Herbal Nutraceutical as Alternative Medicine

Alok Sharma[1], Madhu[2], Sapna Avinash Kondalkar[1], Ajay Kumar Meena[1], and Santosh Kumar Upadhyay[2]

[1] *Regional Ayurveda Research Institute, Gwalior, Madhya Pradesh, India*
[2] *Department of Botany, Panjab University, Chandigarh, India*

2.1 Introduction

Nutraceuticals are either food or food derivatives that provide health benefits and nutrition and help in disease prevention and treatment. Nutraceuticals include herbal products, dietary supplements, vitamins, and genetically modified foods (AlAli et al. 2021; Ivanišová et al. 2021). These products have various bioactive compounds that boost immunity and help in disease treatment. Herbal nutraceuticals, as their name suggests, are developed by plant parts like seeds, flowers, berries, roots, and oils and help with the acute and chronic illnesses caused by unhealthy habits by improving overall wellness. Traditionally, herbal products were used as medicine and supplements (Priya and Satheeshkumar 2020). The main countries known for their traditional practices of using plants as medicine are India and China. Moreover, in recent times, this traditional knowledge is not just limited to theory but has become a highly specialized area and is supported by clinical data generated by various recent research programs (Ivanišová et al. 2021). The data suggest that 80% of the population is using plant products as medicine (Atherton 1994; Lipman 2005). Examples include the use of purple foxglove, poppy, cinchona, willow, rauwolfia, and yew for pain, cough, fever, and hypertension (Lipman 2005).

In Ayurveda also, Rigvedic literature highlights the importance of spices, food, plants, and herbs in disease cure and prevention, which ultimately boosts health (Misra 2013; Patwardhan 2014). The phrase "let food be the medicine and medicine be the food" by Hippocrates (460–377 BC; father of medicines) further strengthens the role of plants and their products in health benefits and connects the traditional approaches from almost 2500 years ago to modern clinical approaches (Anand and Bharadvaja 2022; Das et al. 2012; Helal et al. 2019; Patwardhan 2014). Therefore, both these approaches build up the concept of nutraceuticals for the betterment of human health. Nutraceutical products are a blend of nutrition and pharmaceuticals but are not patentable like pharmaceutical compounds (Nasri et al. 2014). The ever-evolving nature of microbes and the side effects of clinical drugs compelled us to look beyond them, and nutraceuticals have emerged as potent candidates for the health benefits (AlAli et al. 2021). Herbal nutraceutical usage in daily life not only provides us with better health, immunity, and body functions but also saves us from the side effects of

Herbal Nutraceuticals: Products and Processes, First Edition. Edited by Santosh Kumar Upadhyay and Sudhir Pratap Singh.

pharmaceutical antibiotics, which arise due to their long-run usage. Although these herbal nutraceuticals are not meant as antibiotics, they help in the improvement of immune response and indirectly work as an antibiotic (Chanda et al. 2019). Herbal nutraceuticals are found to be involved in the improvement of digestion, vitamins and minerals absorption, detoxification, growth of beneficial biota, and hampering the growth of harmful bacteria and biochemical reactions (Anand and Bharadvaja 2022; Zhou et al. 2020). Moreover, these products also have a positive impact on health with properties like anticancerous, antiaging, antioxidant, immunity booster, and hypersensitivity and autoimmune reducer (AlAli et al. 2021; Anand and Bharadvaja 2022; Carr and Maggini 2017; Zhou et al. 2020). Additionally, herbal nutraceuticals are known to help in the control of diabetes, neurodegeneration, hypertension, cardiac diseases, oxidative stress, allergy, eye diseases, Alzheimer's, Parkinson's disease, inflammation, obesity, and liver diseases (Anand and Bharadvaja 2022; Das et al. 2012; Dixit et al. 2021a; Sachdeva et al. 2020). Although the active constituents of nutraceuticals can also have a toxic effect, various clinical studies have proved that these are safe if taken in a controlled manner and at a specific dosage. Most importantly, the things that are making nutraceuticals a growing and demanding approach are their easy availability, better tolerance, cost-effectiveness, fewer side effects, broader immune improvement, and higher efficacy (AlAli et al. 2021; Kim and Kim 2010). In recent times, herbal nutraceuticals attained worldwide popularity. Some of the important and known herbal nutraceuticals are green tea, ginseng, folic acid, Echinacea, omega-3, folic acid, lutein, turmeric, *Terminalia chebula*, *Terminalia bellerica*, *Emblica officinalis*, *Tinospora cordifolia*, *Paederia foetida*, *Allium sativum*, *Zingiber officinale*, *Medicago sativa*, *Foeniculum vulgare*, *Curcuma longa*, *Glycyrrhiza glabra* (AlAli et al. 2021; Petrovska 2012; Upadhyay and Singh 2021, Table 2.1).

2.2 Classification of Nutraceuticals

2.2.1 Traditional Nutraceuticals

These are raw products obtained from nature and taken in their natural form without any modifications. Some examples of traditional nutraceuticals are lycopene from tomatoes, saponins from soy, and sulforaphane from broccoli, which are used for various health benefits. Moreover, β-carotene and astaxanthin were found to reduce the level of reactive oxygen species (ROS), work as an inflammation mediator, and show antiinflammation properties (AlAli et al. 2021; Kang and Kim 2017; Lobo et al. 2010). Additionally, plants like rice, soybean, citrus, and lentils, are also examples of traditional nutraceuticals and part of the diet that provides various health benefits (Ruchi 2017; Sikand et al. 2015). The traditional nutraceutical can be further classified into three categories: (i) chemical constituents (ii) probiotic microorganisms, and (iii) nutraceutical enzymes.

In the case of chemical constituents, the primary metabolites like various amino acids, fatty acids, and vitamins are known for their positive impacts on the various biochemical functions of the body and help in the cure of cardiac, kidney, and respiratory-related diseases. Some of the examples include the benefits of natural products in brittle bone, hemoglobin level, and neuron transmission. Moreover, the combined dose of nutraceuticals and herbs are also found to cure various chronic diseases, for instance, menthol, a terpenoid obtained from *Mentha piperita,* and the terpenoid obtained from *Lavendula aungustifolia* has shown benefits in cold and asthma-induced diseases, respectively. Similarly, the bark of *Salix nigra* and *Petroselinum crispum* is used as an anti-inflammatory, analgesic, antiarthritic, antipyretic, astringent, carminative, and diuretic, respectively. Phytochemicals like flavonoids and phenolic acids are secondary metabolites of plants that are useful in the reduction of numerous chronic diseases like diabetes, cardiac-related diseases,

Table 2.1 List of some herbal nutraceuticals with their uses.

Part used	Botanical name	Vernacular name	Uses	References
Aerial parts	*Hypericum perforatum* L.	St. John's wort	Antidepressant, hepatitis	Chauhan et al. (2013)
	Sanguisorba minor Scop.	Burnet	Dysentery, diarrhea	Chauhan et al. (2013)
	Artemisia absinthium L.	Wormwood	Loss of appetite, hepatoprotective	Msaada et al. (2015)
	Clinopodium nepeta (L.) Kuntze	Calamint	Antimicrobial, antioxidant	Debbabi et al. (2020)
Bulb	*Allium sativum* L.	Garlic	Antibacterial, anti-inflammatory	Bayan et al. (2014)
	Allium cepa L.	Onion	Antibiotic, hypoglycemic	Chauhan et al. (2013)
Flower	*Crocus sativus* L.	Saffron	Memory enhancer, immunity booster	Rahaiee et al. (2015)
	Matricaria recutita L.	Chamomile	Ulcer, insomnia	Bommakanti et al. (2023)
	Achillea millefolium L.	Yarrow	Diarrhea, fever	Bommakanti et al. (2023)
Fruit	*Aegle marmelos* (L.) Corrêa	Bael	Appetizer, diarrhea	Chauhan et al. (2013)
	Myrica pensylvanica Mirb.	Bayberry	Nausea, ulcers	Zhang et al. (2015)
Gum-resin	*Ferula narthex* Boiss.	Asafoetida	Stimulant, carminative	Chauhan et al. (2013)
Leaves	*Aloe vera* (L.) Burm.f.	Aloe	Wound healing, anti-inflammatory	Chauhan et al. (2013)
	Senna alexandrina	Senna	Purgative	Chauhan et al. (2013)
	Ginkgo biloba L.	Gingko	Memory tonic, antioxidant	Chauhan et al. (2013)
	Hyssopus officinalis L.	Hyssop	Nervous dysfunction, digestive problem	Aćimović et al. (2021)
	Laurus nobilis L.	Bay	Antidiabetic, antidiarrheal	Alejo-Armijo et al. (2017)
	Cymbopogon citratus (DC.) Stapf	Lemon grass	Antioxidant, menstrual flow	Bommakanti et al. (2023)
	Mentha x *piperata* L.	Mentha	Digestive problem, antiseptic	Bommakanti et al. (2023)
	Monarda didyma L.	Bee Balm	Antispasmodic, carminative	Bommakanti et al. (2023)
	Ocimum tenuiflorum L.	Tulsi	Antimicrobial, anti-inflammatory	Yamani et al. (2016)
	Origanum vulgare L.	Oregano	Asthma, cough	Ocana-Fuentes et al. (2010)
	Petroselinum crispum (Mill.)	Parsley	Diuretic, blood pressure	Bommakanti et al. (2023)
	Salvia officinalis L.	Sage	Dizziness, cramps	Ghorbani and Esmaeilizadeh (2017)
	Salvia rosmarinus Spenn.	Rosemary	Muscular pain	Chauhan et al. (2013)
	Thymus vulgaris L.	Thyme	Cough, blood pressure	Chauhan et al. (2013)

(*Continued*)

Table 2.1 (Continued)

Part used	Botanical name	Vernacular name	Uses	References
Rhizome	*Curcuma longa* L.	Turmeric	Antiseptic, antiarthritic	Chauhan et al. (2013)
	Zingiber officinale Roscoe	Ginger	Cough, hypoglycemic	Chauhan et al. (2013)
Root	*Glycyrrhiza glabra* L.	Liquorice	Expectorant, anti-inflammatory	Chauhan et al. (2013)
	Hydrastis canadensis L.	Goldenseal	Astringent, antimicrobial	Chauhan et al. (2013)
	Panax ginseng C.A. Mey.	Ginseng	Immune booster, neuro tonic	Chauhan et al. (2013)
	Valeriana officinalis L.	Valeriana	Bronchial spasm, menstrual pain	Chauhan et al. (2013)
	Angelica archangelica L.	Angelica	Insomnia, arthritis	Zhao et al. (2018)
	Polyscias fruticose (L.) Harms	Aralia	Arthritis, hepatitis	Bommakanti et al. (2023)
Seed	*Anethum graveolens* L.	Dill	Stomach pain, bad breath	Bommakanti et al. (2023)
	Carum Carvi L.	Caraway	Stimulant, expectorant	Johri (2011)
	Foeniculum vulgare Mill.	Fennel	Bone development, blood sugar balance	Badgujar et al. (2014)
	Pimpinella anisum L.	Anise	Indigestion, migraine	Rebey et al. (2019)
	Silybum marianum (L.) Gaertn.	Milk thistle	Jaundice, indigestion	Aziz et al. (2021)
Stem	*Tinospora cordifolia* (Willd.) Hook. f. & Thomson	Tinospora	Hypoglycemic, fever	Saha and Ghosh (2012)
Whole plant	*Centella asiatica* (L.) Urb.	Brahmi	Spasmolytic, neurotonic	Chauhan et al. (2013)
	Echinacea purpurea (L.) Moench	Echinacea	Immunomodulator, anti-inflammatory	Chauhan et al. (2013)
	Tropaeolum majus L.	Nasturtium	Antifungal, bronchitis	Bommakanti et al. (2023)

cancer, and neurodegenerative diseases, and play a crucial role in maintaining the chemical balance of the brain (Ehrlich 2009; Kumar and Khanum 2012; Lobo et al. 2010). Curcumin is an example of one of the most common phytochemicals obtained in daily life through curcumin, which also has various health benefits.

2.2.1.1 Probiotic Microorganisms

These are the good microbes that maintain the balance of the digestive system by removing harmful flora from the body. The most common probiotic used is the different species of lactobacillus, which is found in the gut of humans and helps in the breakdown of lactose into simple sugar components; it plays a vital role in lactose intolerance (Kechagia et al. 2013; Pineiro and Stanton 2007; Singh and Upadhyay 2021).

2.2.1.2 Nutraceutical Enzymes

Another approach used in nutraceuticals is enzyme therapy. The enzymes are proteinaceous in nature and work as a biocatalyst and speed up the metabolism. Nowadays, enzyme therapy is used for several severe diseases like diabetes, Hunter syndrome, and Gaucher disease. However, the most preferred source for this is microbes rather than plants as that is more cost-friendly (Chanda et al. 2019).

2.2.2 Nonconventional Nutraceuticals

These are the types of nutraceuticals that are modified in nature through biotechnological and breeding approaches. These foods are modified to enhance their level of nutrients, which makes them a better food choice to improve human health. Examples include the β-carotene enriched rice, vitamins enriched broccoli, and minerals and vitamins enriched cereals (Upadhyay 2021). These nontraditional foods could be fortified foods or recombinant foods. In the case of fortified foods, modifications have been done at the level of breeding to add the desired and useful nutrition to the selected foods. The cholecalciferol fortified milk, calcium-fortified orange, and iron-fortified cereals are some examples (AlAli et al. 2021; Casey et al. 2010; Gupta et al. 2010; Ruchi 2017; Singh and Sinha 2012; Sui et al. 2016; Swaroopa and Srinath 2017; Upadhyay and Singh 2023). The second type, i.e., recombinant nutraceuticals, includes foods that are modified through recombination technology and not at the level of breeding (Drake et al. 2017; Singh and Sinha 2012). Golden rice and golden mustard, fortified with vitamin A, multivitamin corn fortified with three vitamins (vitamins A, B9, and C), and gold kiwifruit fortified with vitamin C are examples of recombinant nutraceuticals (AlAli et al. 2021; Alamgir 2017; Dixit et al. 2021b; Skinner et al. 2011; Stonehouse et al. 2013; Upadhyay 2024).

2.3 Types of Herbal Nutraceuticals Used as Alternative Medicine

Nutraceuticals are categorized into three classes based on their food availability, chemical nature, and mode of action, which can be used as medicine to cure many diseases.

2.3.1 Herbal Nutraceuticals Based on Food Availability

There are a number of herbal plants that we use in our daily life for food requirements due to their diverse types of primary metabolites such as amino acids, vitamins, minerals, fatty acids, and phytochemicals, which are important for a healthy lifestyle. These plants have a number of health benefits and after the extraction of their bioactive components, they can be used as a source of pharmaceuticals. Here, we are discussing some important herbal plants that are used as nutraceuticals and can also be taken as alternative medicine.

2.3.1.1 *Angelica*

In the Chinese book "Compendium of Materia Medica," *Angelica sinensis* has been described as a herbal medicine (Hou et al. 2021). It has been used as a tonic to treat disorders like mensuration and amenorrhea (Lv et al. 2016). In addition, the allopathic system of medicine considers *A. sinensis* as an antitumor, antioxidant, immunomodulatory, and hematopoietic. It is used to cure diseases like arthritis, heartburn, flatulence, anorexia, circulation problems, respiratory catarrh, insomnia, nervousness, and plague (Wei et al. 2016). The medicinal properties of this plant are due to its bioactive compounds like β-phellandrene, furocoumarins such as angelicin, bergapten, and xanthotoxin, umbelliprenin, and phenolic compounds (Hook 2014).

2.3.1.2 *Anise*

Anise is an aromatic plant, and its importance has been recognized in both conventional and pharmaceutical systems of medicine (Sun et al. 2019). It serves a dual purpose—one, it is consumed in food and the other, it is taken as a medicine (Sun et al. 2019). The active components of anise seeds are acetophenone, estragol, pinene, p-anisaldehyde, and limonene. The anise seeds consist of essential oil which has medicinal properties such as antioxidant, antibiotic, analgesic, anti-inflammatory, and antidiabetic (Sun et al. 2019).

2.3.1.3 *Allium sativum*

It is commonly called garlic, a member of the alliaceae family (Tesfaye 2021). *A. sativum* consists of sulfur components such as allicin, S-allylcysteine, and diallyl disulfide, which are important for their therapeutic properties (Houshmand et al. 2013). It has been reported that garlic has antibacterial, antifungal, anticancerous, and antidiabetic properties, and it also controls blood pressure, hyperlipidemia, and atherosclerosis (Bayan et al. 2014). Besides, it also consists of lectin proteins with numerous applications (Upadhyay et al. 2010a, b, 2011; Upadhyay and Singh 2012). The wide range of therapeutic effects makes garlic a potential alternative pharmaceutical.

2.3.1.4 *Artemisia annua*

It is an annual herbaceous plant of family asteraceae (Ekiert et al. 2021). In 2015, the Nobel Prize was awarded in medicine for the discovery of sesquiterpene lactone artemisinin in *A. annua* (Efferth et al. 2015). The study showed that this active compound in *A. annua* is effective against malaria (Efferth et al. 2015). The medicinal properties of this herb have also been discussed in ancient literature, where it has been reported that it is used for the cure of jaundice, dysentery, tuberculosis, and autoimmune diseases (Alesaeidi and Miraj 2016; Bora and Sharma 2011). Moreover, it has antiviral, antibacterial, antifungal, anti-inflammatory, antioxidant, analgesic, and nephroprotective properties (Ekiert et al. 2021).

2.3.1.5 *Boswellia serrata*

B. serrata is a member of the family burseraceae and is commonly known as Salai guggul (Siddiqui 2011). This plant exudate is called oleo gum-resin, which consists of resin, essential oils, and polysaccharides (Siddiqui 2011). In Ayurveda, it has been mentioned that *B. serrata* has anti-rheumatic properties and it is also effective against diseases like dysentery, fever, cardiovascular diseases, bronchitis, and jaundice (Sharma et al. 2004). The allopathic system of medicine described it as an antihyperlipidemic, analgesic, antiarthritic, and antiatherosclerotic (Sharma et al. 2004).

2.3.1.6 *Curcuma longa*

C. longa is a plant from the Zingiberaceae family, commonly called turmeric. It is commonly used as spice in Indian curries due to its color and flavor (Araujo and Leon 2001; Hewlings and Kalman 2017). In the Ayurvedic system of medicine, it is considered a medicine for wounds, respiratory, and liver problems (Goel et al. 2008). The bioactive compounds of turmeric are curcumin, bisdemethoxycurcumin, and demethoxycurcumin which are collectively called curcuminoids. These curcuminoids are considered potential therapeutics. They are effective against inflammation, diabetes, cancer, arthritis, and Alzheimer's disease (Gupta et al. 2013).

2.3.1.7 *Foeniculum vulgare*

Fennel is the common name of *F. vulgare* that has been used for a long time in traditional pharmacopoeia to treat respiratory, digestive, and reproductive disorders (Badgujar et al. 2014). The bioactive compounds of fennel are phenols, flavonoids, amino acids, and fatty acids (Barros et al. 2010). The active compounds of fennel have therapeutic properties such as anti-inflammatory, antibiotic, antipyretic, antitumor, hypoglycemic, chemomodulatory, and antispasmodic. Fennel is considered a potential herb in the traditional system of medicine and is a good alternative in the pharmaceutical industry for the development of new drugs (Neves et al. 2009).

2.3.1.8 *Glycyrrhiza glabra*

This plant has a strong history in ethnobotanical science as this has been used in traditional medicine in Eastern and European countries (Sharma et al. 2018). The active compounds of *G. glabra* include glycyrrhizin, alkaloids, flavonoids, saponins, glycyrrhetic acid, and glabridin (Hill 1952). These compounds are effective against diseases like cancer, inflammation, ulcers, and microbial infections (Sharma et al. 2018). Due to multifaceted activities, *G. glabra* is used in folk medicine against various diseases. Researchers have discovered the potential effects of bioactive constituents of *G. glabra* which could be of great importance in drug discovery in the present era (Sharma et al. 2018).

2.3.1.9 *Medicago sativa*

M. sativa belongs to the family fabaceae (Al-Snafi 2015). The nutritional and phytochemical constituents of this plant are carbohydrates, proteins, alkaloids, lignin, tannin, saponins, phenols, carotenoids, isoflavonoids, and glycosides (Karimi et al. 2013). In Ayurveda and homeopathy, it is used to treat diseases like arthritis and kidney disorders (Karimi et al. 2013). Furthermore, its active compounds are considered anticancerous, antiscorbutic, depurative, and cardiotonic (Rana et al. 2010). Moreover, it has immunological, neuroprotective, dermatological, antimicrobial, cytotoxic, antidiabetic, and cardioprotective effects (Avato et al. 2006; Gatouillat et al. 2014; Gomathi et al. 2014; Hong et al. 2009; Mehranjani et al. 2007; Zadeh et al. 2014). This herbal nutraceutical plant has innumerable curative health effects, which is why this should be the best alternative in the pharmacological system.

2.3.1.10 *Moringa oleifera*

M. oleifera is a native of India, is one of the most studied plants, and has immense nutritional and therapeutic properties (Milla et al. 2021). This plant is nutritionally very rich as it is a rich source of vitamins (A, C, and niacin), minerals (Ca, K, Zn, Mg, Fe, P, and Cu), proteins, carbohydrates, and fatty acids (El-Massry et al. 2013; Milla et al. 2021). Additionally, the secondary metabolites of *M. oleifera* are alkaloids, carotenoids, flavonoids, phenolic acids, glucosinolates, and tocopherols. The bioactive compounds of *M. oleifera* are considered anticancerous, anti-inflammatory, antibiotic, antioxidant, antidiabetic, hepatic, and cardioprotective (Bhattacharya et al. 2018; Koul and Chase 2015; Saini et al. 2016; Williams 2013). This tree has high nutraceutical and therapeutic properties and can be used as an alternative for medicine in allopathy.

2.3.1.11 *Zingiber officinale*

Ginger is rich in bioactive metabolites such as phenolic compounds, lipids, polysaccharides, terpenes, and organic acids (Mao et al. 2019). The major benefits of ginger are because of metabolites such as shogaols and gingerols (Dugasani et al. 2010). The major biological activities of ginger are antioxidants, anticancer, anti-inflammatory, antiobesity, antinausea, and antiemetic (Dhanik et al. 2017). These therapeutic properties of ginger make it a potential candidate for a new drug for the prevention of chronic disorders.

2.3.2 Herbal Nutraceuticals Based on Chemical Nature

Herbal nutraceuticals based on chemical nature are classified into different categories such as carbohydrates, lipids, fatty acids, amino acid-based substances, phenolic compounds, isoprenoid derivatives, and minerals (Chanda et al. 2019).

2.3.2.1 Carbohydrates-based Herbal Nutraceutical for Alternative Medicine

Carbohydrates are one of the main macronutrients which are essential for our health. It includes mainly starch, fibers, and sugars, which are important energy providers of our diet (Lamothe et al. 2019). The carbohydrates that have nutraceutical properties and could be used as alternative medicine are ascorbic acid, oligosaccharides, and polysaccharides (nonstarch) (Chanda et al. 2019). In the research field, it has been established that carbohydrate-based nutraceuticals could be used to treat diseases like diabetes, cancer, obesity, and gastric disorders (Naidu 2003). Among carbohydrates, ascorbic acid i.e., vitamin C, which is one of the most known nutraceuticals, is a derivative of glucose (Meščić Macan et al. 2019). Vitamin C is found naturally in many fruits such as citrus, kiwi, grapes, strawberries, pomegranate, and vegetables like cabbage, cauliflower, peppers, and broccoli (García-Closas et al. 2004). It is a powerful antioxidant which imparts various health and medical benefits. The daily nutritional requirement of vitamin C is 90 mg/day for men and 75 mg/day for women (Frei and Traber 2001). It has been reported that enough dietary intake of ascorbic acid reduces the chances of diseases like scurvy, common cold, breast cancer, autism, cardiovascular disease, microbial diseases, and COVID-19 (Carr and Frei 1999; Naidu 2003). Ascorbic acid, a potent nutraceutical, could be a good alternative to therapeutics in common cold infection as it stimulates the T-cell proliferation during infection, thus increasing the immune response (Hemilä and Chalker 2013). It also plays a vital role in wound healing as it stimulates the synthesis of collagen protein (Naidu 2003). Moreover, ascorbic acid is effective against atherosclerosis as it prevents the peroxidation of low-density lipoproteins (LDL) (Packer 1997). The main significance of vitamin C is reported by Nobel laureates Cameron and Pauling, who suggested that the intake of high dosages of vitamin C can cure cancer and sometimes complete revival (Naidu 2003). However, the mechanism of this anticancerous property of ascorbic acid is still not fully discovered. However, the proposed numerous health benefits of vitamin C make it a potential pharmaceutical candidate to cure various chronic diseases.

In the modern lifestyle, where people are eating different types of junk food that cause health complications and make them devoid of nutrition, people need to increase their consciousness toward their diet (Rahim et al. 2021). In this case, people should take dietary supplements to complete their nutritional requirements. Among these dietary supplements, two of them are oligosaccharide and polysaccharide nutraceuticals (Jagtap and Manohar 2021). These are present in the form of probiotics (Jagtap and Manohar 2021). They are found in different types of foods such as curd, milk, vegetables, fruits, honey, and marine algae (Jagtap and Manohar 2021). However, marine algae is a great source of probiotics and it can be easily cultivated. By using microbial enzymes, the extraction of oligosaccharides and polysaccharides from marine algae can be improved and used to improve health (de Borba Gurpilhares et al. 2019). The three main polysaccharides from brown algae, which are used as dietary fiber, are alginate, laminaran, and fucoidan (Gómez-Ordóñez et al. 2010). The nutritionally important polysaccharides of red algae are agar-agar, glucan, floridean starch, and carrageenan (Wells et al. 2017). The main polysaccharides of green algae are ulvans, glucose, xylose, glucuronic acid, and oligosaccharides galactose, mannose, and arabinose (Wells et al. 2017). Among different classes of algae, the products of green algae are mostly used. However, the important health benefits of different polysaccharides derived from diverse types of algae are not completely understood (de Borba Gurpilhares et al. 2019). Due to the innumerable health benefits of oligosaccharide nutraceuticals, they are in great demand in the nutrition industry (Martins et al. 2023). The aim of food scientists is to increase the nutritional efficiency of oligo and polysaccharides by enzymatic hydrolysis and transglycosylation (Martins et al. 2023).

2.3.2.2 Lipids and Fatty Acids Based Herbal Nutraceutical for Alternative Medicine

Lipids are natural compounds, and they have different categories which include vitamins that are fat soluble such as vitamins A, D, E, and K, fats, sterols, phospholipids, waxes, and carotenoids (Bernal et al. 2011). The main role of lipids is energy storage, they are the main structural components of cell membranes, and these are also key signaling molecules (Welte and Gould 2017). Plant-based lipids that we use in our diet are fatty acids, phytosterols, glycerolipids, squalene, terpenes, and different carotenoids (Yokoyama et al. 2017).

Fatty acids are vital nutritional compounds that have both nutraceutical and pharmaceutical properties (Benatti et al. 2004). Among fatty acids, long-chain polyunsaturated fatty acids (PUFA) such as omega-3, and α-linolenic acid have beneficial health effects; these fatty acids are known to prevent diseases like autoimmune disorders, cardiovascular diseases, high blood pressure, hyperglycemic, and inflammation (Benatti et al. 2004). The important second category of herbal nutraceuticals based on lipids is plant sterols (Schwartz et al. 2008). The main function of phytosterols is to block the cholesterol absorption sites, which decreases the plasma cholesterol related to LDL and thus, helps to maintain the cholesterol level in humans (Bakowska-Barczak et al. 2009; Nyam et al. 2009; Ramadan et al. 2006). Moreover, in some studies, these are found to be effective against cancer, inflammation, and thrombotic activities (Bernal et al. 2011). Due to their large number of both nutraceutical and therapeutical health benefits, the US FDA has approved phytosterols as a food additive (Bernal et al. 2011). Carotenoids are a large group of natural pigments that are present in plants and algae. Carotenoids are classified into two categories: carotenes and xanthophylls (Rmiki et al. 1999). These pigments have antioxidant and immunomodulation activities and are effective against prostate and gastrointestinal tract cancer, diabetes, and cardiovascular diseases (Milani et al. 2017).

2.3.2.3 Amino Acid-based Herbal Nutraceutical for Alternative Medicine

Amino acids, peptides, and proteins-based herbal nutraceuticals are important in the pharmaceutical industry (Bernal et al. 2011). For instance, different types of amino acids that are present in sprouts, alga, and legumes, affect the nervous system. These amino acids have antioxidant, anticancerous, and weight loss properties (Herrero et al. 2007; Kim et al. 2006; Mosca et al. 2008). In garlic, immunomodulatory proteins are present (Clement et al. 2010). The main nutraceutical properties have been studied in legume proteins (Carbonaro et al. 2015). It has been recommended that we consume legume seeds due to their beneficial effects on curing diseases such as diabetes, cancer, osteoporosis, high blood pressure, aging problems, and mental health (Bouchenak and Lamri-Senhadji 2013). Few proteins of legumes (soybeans, peas, chickpeas, and lentils) showed hormone-like activities (Roy et al. 2010). The soybean trypsin/chymotrypsin inhibitor proteins have weight loss, anti-inflammatory, and anticancerous properties (Dukariya et al. 2020). The jack bean hemagglutinins have immunostimulatory and anticancerous properties

(Zhang and Abbaspourrad 2018). The fermented soybean has antihypertensive properties (Li et al. 2022). Due to the innumerable functions of herbal proteins, the commercialization of these proteins is in demand in the market. People who are in sports take these proteins in supplement form (Kårlund et al. 2019). Moreover, doctors prescribe protein nutraceuticals to patients due to their pharmaceutical properties (Kårlund et al. 2019). It has been reported that the consumption of soy proteins significantly decreases the total cholesterol level, triglycerides, and LDL, and increases the level of high-density lipoproteins (HDL) (Pipe et al. 2009).

2.3.2.4 Phenolic Compound-based Herbal Nutraceutical for Alternative Medicine

Phenolic compounds are the products of secondary metabolism and are known to exert various therapeutic properties such as antioxidant, antibiotic, and anti-inflammatory, and also have nutritional values (Saibabu et al. 2015; Saltveit 2017). In the traditional system of Chinese medicine, it has been found that plants such as *Nepeta italica* subsp. Cadmea and *Teucrium sandrasicum* are rich in phenols and flavonoids and show antioxidant properties (Kaska et al. 2019). Moreover, plants such as orchid tree, clove, thyme, oregano, and rosemary have the highest number of phenolic compounds (Saricaoglu and Turhan 2018; Shahana et al. 2017). There are a number of plants such as *Bellis perennis*, *Althaea officinalis*, *Dendrobium densiflorum*, and *Pinus cembra*, which produce phenolic compounds that show antioxidant properties (Bouallag et al. 2022; Paudel et al. 2019; Wu et al. 2020). Therapeutic properties such as antifungal and antibacterial activities are shown by the phenolic compounds of plants *Scutellaria baicalensis*, *Albizia procera*, *Atalantia monophylla*, *Cassia fistula*, *Costus speciosus*, and *Cinnamomum verum* (Jayaprakasha et al. 2006; Khatoon et al. 2013; Lu et al. 2011; Luximon-Ramma et al. 2002; Thirugnanasampandan et al. 2015). *C. longa*'s bioactive phenolic compounds have been known for their anticancerous property against the B164A5 murine melanoma cell line (Soumya et al. 2023). The flavonoids gliricidia and quercetin of *Asplenium nidus* are chemopreventive against carcinoma HeLa cells in humans (Sun and Shahrajabian 2023). Flavones such as quercetin, genistein, and curcumin are found to be effective against pancreatic, prostate, and skin cancers (Batra and Sharma 2013). Phenolic compounds such as rutin, luteolin, kaempferol, and resveratrol have cardiotoxicity (Alotaibi et al. 2021). The flavonoids of *Phyllanthus acidus* have analgesic, anti-inflammatory, and antioxidant properties (Chakraborty et al. 2012). The phenolic compounds that are effective against COVID-19 are curcumin, myricetin, luteolin, resveratrol, quercetin, and scutellarein (Gligorijevic et al. 2021). The phenolic compounds of plants such as *G. glabra* and *Rheum palmatum* are effective against respiratory disorders (Setayesh et al. 2022). Therefore, due to the high therapeutic and nutraceutical potential of phenolic compounds in plants, these can be targeted for drug research.

2.3.2.5 Terpenes-based Herbal Nutraceutical for Alternative Medicine

Terpenoids are important secondary metabolites of plants that are used in food and pharmaceutical industries (Fan et al. 2023). Terpenes have various types of medicinal properties which include anticancerous, antioxidant, anti-inflammatory, hypoglycemic, antibiotic, and prevention of cardiovascular diseases (Fan et al. 2023). The terpenoids such as perillyl alcohol, paclitaxel, and geraniol possess antitumor properties (Chen et al. 2021; Chopra et al. 2021; Yang et al. 2020). Paeoniflorin terpene, which is extracted from the root of *Paeonia lactiflora*, showed anti-inflammatory activity (Zhao et al. 2016). Similarly, a sesquiterpene from the plant *Inula japonica* has anti-inflammatory properties (Wu et al. 2016). Ginsenoside, a terpenoid component of the ginseng plant, is an anti-inflammatory agent and also exhibits neuroprotective properties (Yang et al. 2023). Monoterpenes such as 1,8- eucalyptus, limonene, geranialdehyde, sabinene, menthol, sabinol, and carvone have antibacterial properties (Mohammadi-Cheraghabadi and Hazrati 2023). Diterpenes such as artemisinin, andrographolide, and triterpene like oleanolic acid also have antibacterial properties (Mohammadi-Cheraghabadi and Hazrati 2023). Artemisinin is a sesquiterpene found to be the most effective drug against malaria (Martínez et al. 2012). Tanshinone IIA, the active component of *Salvia miltiorrhiza*, is effective against atherosclerosis (Guo et al. 2020). The diterpenoid stevioside of *Stevia rebaudiana* has an antihyperglycemic effect (López et al. 2016).

2.3.2.6 Mineral-based Herbal Nutraceutical for Alternative Medicine

Minerals are an important part of our food as they help in proper functioning of muscles and nerves, regulate water balance, facilitate bone development, and are part of enzymes and hormones (Miller 2017; Soetan et al. 2010). Minerals have an important role in maintaining the optimum functioning of the immune system (Gombart et al. 2020). In general, a balanced diet is enough to fulfill the daily requirement of minerals. But older people have chronic illnesses and pregnant women face mineral deficiencies. To fulfill these deficiencies, these people are advised to take mineral supplements. These supplements complete the daily requirements of minerals and protect individuals from various types of lifestyle diseases (Lukaski 2004). For instance, magnesium (Mg) has a role in immune responses, and it helps to cure infectious diseases (Ashique et al. 2023). The plant sources of Mg are bananas, oats, and cooked spinach (Bohn 2008). Another important mineral is zinc (Zn) which has both nutraceutical and therapeutic properties (Duan et al. 2023). Zn is also important in regulating the immune system (Duan et al. 2023). Similarly, copper, iron, and selenium have a role in immunity and are important in infectious disease prevention (Munoz et al. 2007). Calcium is an important mineral for our bone health, digestive system, heart function, and blood regulation (Brini et al. 2013). Potassium is important for heart and muscle functioning (Haddy et al. 2006; Sejersted and Sjøgaard 2000). Cobalt is found to be significant in vitamin B12 synthesis (Osman et al. 2021). The deficiency of any mineral could lead to chronic diseases. So, it is important in today's lifestyle that we should take minerals and nutraceuticals in our diet as they serve both nutritional and medicinal purposes.

2.3.3 Herbal Nutraceuticals Based on Mode of Action

Nutraceuticals based on their therapeutic properties are classified into different categories and a few of them are discussed in this section.

2.3.3.1 Nutraceutical with Antioxidant Therapeutic Properties

Oxidative stress in our body causes different types of chronic diseases like hypertension, cancer, diabetes, autoimmune, and atherosclerosis (Sharifi-Rad et al. 2020). Oxidative stress occurs due to the production of excessive amounts of ROS. Therefore, it is important to maintain the balance of ROS, and, for that, we need antioxidants (Lushchak 2014; Sharma et al. 2012). For instance, vitamin C, E, phenolic, and terpenoid components of plants are the main antioxidants (Asif 2015). Vitamin C is a very strong antioxidant. It protects the cell from oxidative damage. The main function of vitamin C is protection from lipid peroxidation (Poljsak and Ionescu 2009). Lipid peroxidation in cells causes chronic diseases such as cardiovascular diseases, diabetes, and neurodegenerative diseases (Muthuraman et al. 2020). In such cases, antioxidant therapy has been suggested. In one study, vitamin C supplementation was found to reduce the lipid peroxidation product by 20–30% in urine, which indicates that antioxidant therapy by vitamin C supplementation reduces oxidative stress (Taghiyar et al. 2013). In another study, patients with cardiovascular disorders, when treated with antioxidant supplementation which includes vitamin C, E, coenzyme Q10, and minerals such as selenium, showed positive results (Myung et al. 2013). Vitamin E is another potent antioxidant that is required to cure hemolytic anemia and peripheral neuropathy (Torquato et al. 2020). It has been studied that people who smoke cigarettes have low levels of vitamin E and increased amounts of lipid peroxidation. When these people were supplemented with vitamin C, the vitamin E disappearance in plasma was restored (Bruno et al. 2006). A phenolic component like quercetin possesses antioxidant properties (Dueñas et al. 2011). The demand for antioxidants can be fulfilled by different types of fruits, vegetables, and nuts in the diet (Landete 2013). But if we have deficiency symptoms of these antioxidants, we need to seek help from a medical practitioner who can advise to take vitamin supplementation (Landete 2013). Therefore, health research is nowadays focused on utilizing these nutraceuticals as an alternative medicine to cure chronic diseases that are associated with a deficiency of antioxidants (Adefegha 2018).

2.3.3.2 Nutraceutical with Antimicrobial Therapeutic Properties

The plant's secondary metabolites possess nutraceutical properties. A number of such kinds of nutraceuticals have antimicrobial properties that could be utilized in therapeutic industries (Hayat et al. 2017). The main antimicrobial secondary metabolites are polyphenols and terpenes. They inhibit the growth of different types of microbes (Gokhale and Wadhwani 2015). In the case of SARS-COV-2, tannin inhibits the binding of spike protein to the receptors (Haddad et al. 2022). Other antimicrobial secondary metabolites are phenylpropanoids for instance, eugenol, terpenoids, tannins, benzylisothiocyanate, and allicin (Jubair et al. 2021). Allicin is found to be effective against both gram-positive and gram-negative bacteria (Derbe and Kidanemariam 2015). Essential oil of *Rosmarinus officinalis* consists of terpenes such as borneol, α-pinene, β-pinene, camphor, myrcene, verbenone, and polyphenol rosmarinic acid, which shows powerful antimicrobial activity (Ali et al. 2019). Similarly, the essential oil of the oregano plant is effective against *Staphylococcus aureus* and *Pseudomonas aeruginosa* (Özkalp et al. 2010). The polyphenolic content of olive oil has a toxic effect on microbes such as *Moraxella catarrhalis, S. aureus, Haemophilus influenza Salmonella spp.*, and *Vibrio parahaemolyticus* (Aissaoui and Boukhari 2021). Epicatechin gallate, a flavonoid present in green tea, is effective against *S. aureus* (Taylor et al. 2005). The effective response of this antimicrobial plant nutraceutical is a good candidate for pharmaceuticals.

2.3.3.3 Nutraceutical with Anti-inflammatory Therapeutic Properties

Different types of chronic diseases such as cardiovascular diseases, cancer, diabetes, and respiratory diseases are associated with inflammation problems (Khansari et al. 2009). In such cases, various types of nutraceuticals showed anti-inflammatory properties (Zhu et al. 2018). Curcumin, an anti-inflammatory agent, showed a good response against inflammation cytokines such as cyclooxygenase-2 (COX-2), TNF-α, and interleukins (Lee et al. 2020). Curcumin is also recommended for the treatment of periodontitis (Li et al. 2021). Another important nutraceutical is PUFAs, which are effective anti-inflammatory agents for diseases like Duchenne muscular dystrophy (Rodríguez-Cruz et al. 2018). Lycopene, a component of nutraceutical, has a good potential to reduce cardiovascular diseases like myocarditis and atherosclerosis (Nikhra 2018). Some of the probiotics e.g. *Bifidobacterium lactis* and *Bifidobacterium longum* show anti-inflammatory activity against proinflammatory cytokine (Sichetti et al. 2018). Plants such as ginger, peppermint, and cinnamon also have anti-inflammatory properties (Kokate et al. 2023). In one study, when rats with necrotizing enterocolitis were orally administrated with ginger, they showed a significant reduction in inflammatory cytokines such as TNF-α, interleukin-1β, and interleukin-6 (Cakir et al. 2018). These anti-inflammatory responses of herbal nutraceuticals are important to study their therapeutic response.

2.4 Conclusions

Natural resources have been recognized since ancient times for their health benefits. However, over the past few decades, the market for herbal nutraceuticals has expanded globally, including in India, as a result of their putative medicinal benefits and the general public's growing interest in athletics, body toning, and muscular building. Moreover, the various research studies validated their significant contribution to several chronic and acute diseases, and in improving the metabolism and functioning of the body. Furthermore, the broader spectrum of therapeutic usage and lower chances of side effects are the main reasons for the recent rise in interest in herbal nutraceuticals. This chapter comprises the multidimensional medicinal usage of various herbal nutraceuticals like turmeric, ginger, Tinospora, garlic, fennel, licorice, and alfalfa in cardiac diseases, gastro-related diseases, neurodegenerative diseases, obesity, diabetes, and digestion-related diseases, and also discusses the different health benefiting nutritional resources. Collectively, the chapter enlightens that these herbal nutraceuticals can be potent and effective alternatives for medicines and could be a source of various nutritional values as well.

Acknowledgments

The authors are very grateful to the Director General of CCRAS, Ministry of AYUSH, New Delhi, India and Panjab University, Chandigarh, India for the encouragement and facilities provided for this work. M is thankful to UGC for the senior research fellowship.

References

Aćimović, M., Pezo, L., Zeremski, T. et al. (2021). Weather conditions influence on hyssop essential oil quality. *Processes* 9 (7): 1152.

Adefegha, S.A. (2018). Functional foods and nutraceuticals as dietary intervention in chronic diseases; novel perspectives for health promotion and disease prevention. *Journal of Dietary Supplements* 15 (6): 977–1009.

Aissaoui, Y. and Boukhari, Y. (2021). Antioxidant, antibacterial and antifungal effects of phenolic extracts of extra virgin olive oil from two western regions of Algeria: a comparative study. *Phytothérapie* 19 (2): 93–99.

AlAli, M., Alqubaisy, M., Aljaafari, M.N. et al. (2021). Nutraceuticals: transformation of conventional foods into health promoters/disease preventers and safety considerations. *Molecules* 26 (9): 2540.

Alamgir, A.N.M. (2017). *Therapeutic Use of Medicinal Plants and Their Extracts*, vol. 1. Springer International Publishing AG.

Alejo-Armijo, A., Altarejos, J., and Salido, S. (2017). Phytochemicals and biological activities of laurel tree (*Laurus nobilis*). *Natural Product Communications* 12 (5): 1934578X1701200519.

Alesaeidi, S. and Miraj, S. (2016). A systematic review of anti-malarial properties, immunosuppressive properties, anti-inflammatory properties, and anti-cancer properties of Artemisia annua. *Electronic Physician* 8 (10): 3150.

Ali, A., Chua, B.L., and Chow, Y.H. (2019). An insight into the extraction & fractionation technologies of the essential oils and bioactive compounds in *Rosmarinus officinalis* L: past, present and future. *TrAC Trends in Analytical Chemistry* 118: 338–351.

Alotaibi, B.S., Ijaz, M., Buabeid, M. et al. (2021). Therapeutic effects and safe uses of plant-derived polyphenolic compounds in cardiovascular diseases: a review. *Drug Design, Development and Therapy* 4713–4732.

Al-Snafi, A.E. (2015). *Encyclopedia of the Constituents and Pharmacological Effects of Iraqi Medicinal Plants*. Rigi Publication.

Anand, S. and Bharadvaja, N. (2022). Potential benefits of nutraceuticals for oxidative stress management. *Revista Brasileira de Farmacognosia* 32 (2): 211–220.

Araujo, C.A.C. and Leon, L.L. (2001). Biological activities of *Curcuma longa* L. *Memórias do Instituto Oswaldo Cruz* 96: 723–728.

Ashique, S., Kumar, S., Hussain, A. et al. (2023). A narrative review on the role of magnesium in immune regulation, inflammation, infectious diseases, and cancer. *Journal of Health, Population and Nutrition* 42 (1): 74.

Asif, M. (2015). Chemistry and antioxidant activity of plants containing some phenolic compounds. *Chemistry International* 1 (1): 35–52.

Atherton, D.J. (1994). Towards the safer use of traditional remedies. *BMJ* 308 (6930): 673–674.

Avato, P., Bucci, R., Tava, A. et al. (2006). Antimicrobial activity of saponins from *Medicago sp.*: structure-activity relationship. *Phytotherapy Research* 20 (6): 454–457.

Aziz, M., Saeed, F., Ahmad, N. et al. (2021). Biochemical profile of milk thistle (*Silybum marianum* L.) with special reference to silymarin content. *Food Science & Nutrition* 9 (1): 244–250.

Badgujar, S.B., Patel, V.V., and Bandivdekar, A.H. (2014). *Foeniculum vulgare* mill: a review of its botany, phytochemistry, pharmacology, contemporary application, and toxicology. *BioMed Research International* 2014: 842674.

Bakowska-Barczak, A.M., Schieber, A., and Kolodziejczyk, P. (2009). Characterization of Canadian black currant (*Ribes nigrum* L.) seed oils and residues. *Journal of Agricultural and Food Chemistry* 57 (24): 11528–11536.

Barros, L., Carvalho, A.M., and Ferreira, I.C. (2010). The nutritional composition of fennel (*Foeniculum vulgare*): shoots, leaves, stems and inflorescences. *LWT-Food Science and Technology* 43 (5): 814–818.

Batra, P. and Sharma, A.K. (2013). Anti-cancer potential of flavonoids: recent trends and future perspectives. *3 Biotech* 3: 439–459.

Bayan, L., Koulivand, P.H., and Gorji, A. (2014). Garlic: a review of potential therapeutic effects. *Avicenna Journal of Phytomedicine* 4 (1): 1.

Benatti, P., Peluso, G., Nicolai, R., and Calvani, M. (2004). Polyunsaturated fatty acids: biochemical, nutritional and epigenetic properties. *Journal of the American College of Nutrition* 23 (4): 281–302.

Bernal, J., Mendiola, J.A., Ibáñez, E., and Cifuentes, A. (2011). Advanced analysis of nutraceuticals. *Journal of Pharmaceutical and Biomedical Analysis* 55 (4): 758–774.

Bhattacharya, A., Tiwari, P., Sahu, P.K., and Kumar, S. (2018). A review of the phytochemical and pharmacological characteristics of *Moringa oleifera*. *Journal of Pharmacy & Bioallied Sciences* 10 (4): 181.

Bohn, T. (2008). Dietary factors influencing magnesium absorption in humans. *Current Nutrition & Food Science* 4 (1): 53–72.

Bommakanti, V., Puthenparambil, A., Sivi, C.M. et al. (2023). An overview of herbal nutraceuticals, their extraction, formulation. *Therapeutic Effects and Potential Toxicity. Separations* 10 (3): 177.

Bora, K.S. and Sharma, A. (2011). The genus *Artemisia*: a comprehensive review. *Pharmaceutical Biology* 49 (1): 101–109.

de Borba Gurpilhares, D., Cinelli, L.P., Simas, N.K. et al. (2019). Marine prebiotics: polysaccharides and oligosaccharides obtained by using microbial enzymes. *Food Chemistry* 280: 175–186.

Bouallag, O., Zeghichi-Hamri, S., Banethumr-Bey, M. et al. (2022). Optimization of phenolic compounds extraction from *Bellis perennis* flowers and assessment for antioxidant properties. *The Annals of the University Dunarea de Jos of Galati.* Fascicle VI-Food Technology 46 (1): 125–139.

Bouchenak, M. and Lamri-Senhadji, M. (2013). Nutritional quality of legumes, and their role in cardiometabolic risk prevention: a review. *Journal of Medicinal Food* 16 (3): 185–198.

Brini, M., Ottolini, D., Calì, T., and Carafoli, E. (2013). Calcium in health and disease. *Metal Ions in Life Sciences* 13: 81–137.

Bruno, R.S., Leonard, S.W., Atkinson, J. et al. (2006). Faster plasma vitamin E disappearance in smokers is normalized by vitamin C supplementation. *Free Radical Biology and Medicine* 40 (4): 689–697.

Cakir, U., Tayman, C., Serkant, U. et al. (2018). Ginger (*Zingiber officinale* Roscoe) for the treatment and prevention of necrotizing enterocolitis. *Journal of Ethnopharmacology* 225: 297–308.

Carbonaro, M., Maselli, P., and Nucara, A. (2015). Structural aspects of legume proteins and nutraceutical properties. *Food Research International* 76: 19–30.

Carr, A. and Frei, B. (1999). Does vitamin C act as a pro-oxidant under physiological conditions? *The FASEB Journal* 13 (9): 1007–1024.

Carr, A.C. and Maggini, S. (2017). Vitamin C and immune function. *Nutrients* 9 (11): 1211.

Casey, C.F., Slawson, D.C., and Neal, L.R. (2010). Vitamin D supplementation in infants, children, and adolescents. *American Family Physician* 81 (6): 745–748.

Chakraborty, R., Biplab, D., Devanna, N., and Sen, S. (2012). Antiinflammatory, antinociceptive and antioxidant activities of *Phyllanthus acidus* L. extracts. *Asian Pacific Journal of Tropical Biomedicine* 2 (2): S953–S961.

Chanda, S., Tiwari, R.K., Kumar, A., and Singh, K. (2019). Nutraceuticals inspiring the current therapy for lifestyle diseases. *Advances in Pharmacological and Pharmaceutical Sciences* 2019: 6908716.

Chauhan, B., Kumar, G., Kalam, N., and Ansari, S.H. (2013). Current concepts and prospects of herbal nutraceutical: a review. *Journal of Advanced Pharmaceutical Technology & Research* 4 (1): 4.

Chen, T.C., da Fonseca, C.O., Levin, D., and Schönthal, A.H. (2021). The monoterpenoid perillyl alcohol: anticancer agent and medium to overcome biological barriers. *Pharmaceutics* 13 (12): 2167.

Chopra, B., Dhingra, A.K., Dhar, K.L., and Nepali, K. (2021). Emerging role of terpenoids for the treatment of cancer: a review. *Mini Reviews in Medicinal Chemistry* 21 (16): 2300–2336.

Clement, F., Pramod, S.N., and Venkatesh, Y.P. (2010). Identity of the immunomodulatory proteins from garlic (*Allium sativum*) with the major garlic lectins or agglutinins. *International Immunopharmacology* 10 (3): 316–324.

Das, L., Bhaumik, E., Raychaudhuri, U., and Chakraborty, R. (2012). Role of nutraceuticals in human health. *Journal of Food Science and Technology* 49: 173–183.

Debbabi, H., El Mokni, R., Chaieb, I. et al. (2020). Chemical composition, antifungal and insecticidal activities of the essential oils from Tunisian *Clinopodium nepeta* subsp. *nepeta* and *Clinopodium nepeta* subsp. *glandulosum*. *Molecules* 25 (9): 2137.

Derbe, T. and Kidanemariam, B. (2015). Antibacterial activity of garlic extracts against gram-negative and gram-positive bacteria. *European Journal of Biomedical and Pharmaceutical Sciences* 2 (1): 335–341.

Dhanik, J., Arya, N., and Nand, V. (2017). A review on *Zingiber officinale*. *Journal of Pharmacognosy and Phytochemistry* 6 (3): 174–184.

Dixit, S., Shukla, A., Singh, V., and Upadhyay, S.K. (2021a). *Bioprospecting of Natural Compounds for Industrial and Medical Applications.* Wiley https://doi.org/10.1002/9781119718017.ch3.

Dixit, S., Shukla, A., Singh, V., and Upadhyay, S.K. (2021b). *Engineering of Plant Metabolic Pathway for Nutritional Improvement.* Wiley doi: 10.1002/9781119672425.ch20.

Drake, P.M., Szeto, T.H., Paul, M.J. et al. (2017). Recombinant biologic products versus nutraceuticals from plants–a regulatory choice? *British Journal of Clinical Pharmacology* 83 (1): 82–87.

Duan, M., Li, T., Liu, B. et al. (2023). Zinc nutrition and dietary zinc supplements. *Critical Reviews in Food Science and Nutrition* 63 (9): 1277–1292.

Dueñas, M., Surco-Laos, F., González-Manzano, S. et al. (2011). Antioxidant properties of major metabolites of quercetin. *European Food Research and Technology* 232: 103–111.

Dugasani, S., Pichika, M.R., Nadarajah, V.D. et al. (2010). Comparative antioxidant and anti-inflammatory effects of [6]-gingerol,[8]-gingerol,[10]-gingerol and [6]-shogaol. *Journal of Ethnopharmacology* 127 (2): 515–520.

Dukariya, G., Shah, S., Singh, G., and Kumar, A. (2020). Soybean and its products: nutritional and health benefits. *Journal of Nutrition Science and Health Diet* 1 (2): 22–29.

Efferth, T., Zacchino, S., Georgiev, M.I. et al. (2015). Nobel prize for artemisinin brings phytotherapy into the spotlight. *Phytomedicine* 22 (13): A1–A3.

Ehrlich, S.D. (2009). *Peppermint (Mentha piperita), Private Practice Specializing in Complementary and Alternative Medicine*. Phoenix: VeriMed Healthcare Network.

Ekiert, H., Świątkowska, J., Klin, P. et al. (2021). Artemisia annua–importance in traditional medicine and current state of knowledge on the chemistry, biological activity and possible applications. *Planta Medica* 87 (08): 584–599.

El-Massry, F.H., Mossa, M.E., and Youssef, S.M. (2013). *Moringa oleifera* plant. *Egyptian Journal of Agricultural Research* 91 (4): 1597–1909.

Fan, M., Yuan, S., Li, L. et al. (2023). Application of terpenoid compounds in food and pharmaceutical products. *Fermentation* 9 (2): 119.

Frei, B. and Traber, M.G. (2001). The new US dietary reference intakes for vitamins C and E. *Redox report* 6 (1): 5–9.

García-Closas, R., Berenguer, A., Tormo, M.J. et al. (2004). Dietary sources of vitamin C, vitamin E and specific carotenoids in Spain. *British Journal of Nutrition* 91 (6): 1005–1011.

Gatouillat, G., Alabdul Magid, A., Bertin, E. et al. (2014). Cytotoxicity and apoptosis induced by alfalfa (*Medicago sativa*) leaf extracts in sensitive and multidrug-resistant tumor cells. *Nutrition and Cancer* 66 (3): 483–491.

Ghorbani, A. and Esmaeilizadeh, M. (2017). Pharmacological properties of *Salvia officinalis* and its components. *Journal of Traditional and Complementary Medicine* 7 (4): 433–440.

Gligorijevic, N., Radomirovic, M., Nedic, O. et al. (2021). Molecular mechanisms of possible action of phenolic compounds in COVID-19 protection and prevention. *International Journal of Molecular Sciences* 22 (22): 12385.

Goel, A., Kunnumakkara, A.B., and Aggarwal, B.B. (2008). Curcumin as "Curecumin": from kitchen to clinic. *Biochemical Pharmacology* 75 (4): 787–809.

Gokhale, M. and Wadhwani, M. (2015). Antimicrobial activity of secondary metabolites from plants-a review. *International Journal of Pharmacognosy* 22: 23.

Gomathi, R., Vijipriya, M., and Usha, K. (2014). Cardioprotective effect of ethanolic extract of *Medicago sativa* stem on isoproterenol induced myocardial infarction in Wistar Albino rats. *International Journal of Pharmacy and Pharmaceutical Sciences* 6 (2): 839–842.

Gombart, A.F., Pierre, A., and Maggini, S. (2020). A review of micronutrients and the immune system working in harmony to reduce the risk of infection. *Nutrients* 12 (1): 236.

Gómez-Ordóñez, E., Jiménez-Escrig, A., and Rupérez, P. (2010). Dietary fibre and physicochemical properties of several edible seaweeds from the northwestern Spanish coast. *Food Research International* 43 (9): 2289–2294.

Guo, R., Li, L., Su, J. et al. (2020). Pharmacological activity and mechanism of tanshinone IIA in related diseases. *Drug Design, Development and Therapy* 14: 4735–4748.

Gupta, S., Chauhan, D., Mehla, K. et al. (2010). An overview of nutraceuticals: current scenario. *Journal of Basic and Clinical Pharmacy* 1 (2): 55.

Gupta, S.C., Patchva, S., and Aggarwal, B.B. (2013). Therapeutic roles of curcumin: lessons learned from clinical trials. *The AAPS Journal* 15: 195–218.

Haddad, M., Gaudreault, R., Sasseville, G. et al. (2022). Molecular interactions of tannic acid with proteins associated with SARS-CoV-2 infectivity. *International Journal of Molecular Sciences* 23 (5): 2643.

Haddy, F.J., Vanhoutte, P.M., and Feletou, M. (2006). Role of potassium in regulating blood flow and blood pressure. *American Journal of Physiology-Regulatory, Integrative and Comparative Physiology* 290 (3): R546–R552.

Hayat, M., Abbas, M., Munir, F. et al. (2017). Potential of plant flavonoids in pharmaceutics and nutraceutics. *Journal of Biomolecules and Biochemistry* 1 (1): 12–17.

Helal, N.A., Eassa, H.A., Amer, A.M. et al. (2019). Nutraceuticals' novel formulations: the good, the bad, the unknown and patents involved. *Recent Patents on Drug Delivery & Formulation* 13 (2): 105–156.

Hemilä, H. and Chalker, E. (2013). Vitamin C for preventing and treating the common cold. *Cochrane Database of Systematic Reviews* 2013 (1): CD000980.

Herrero, M., Ibáñez, E., Fanali, S., and Cifuentes, A. (2007). Quantitation of chiral amino acids from microalgae by MEKC and LIF detection. *Electrophoresis* 28 (15): 2701–2709.

Hewlings, S.J. and Kalman, D.S. (2017). Curcumin: a review of its effects on human health. *Foods* 6 (10): 92.

Hill, A.F. (1952). *Economic Botany: A Textbook of Useful Plants and Plant Products*, 2nde. McGraw-Hill.

Hong, Y.H., Huang, C.J., Wang, S.C., and Lin, B.F. (2009). The ethyl acetate extract of alfalfa sprout ameliorates disease severity of autoimmune-prone MRL-lpr/lpr mice. *Lupus* 18 (3): 206–215.

Hook, I.L. (2014). Danggui to *Angelica sinensis* root: are potential benefits to European women lost in translation? A review. *Journal of Ethnopharmacology* 152 (1): 1–13.

Hou, C., Yin, M., Lan, P. et al. (2021). Recent progress in the research of *Angelica sinensis* (Oliv.) Diels polysaccharides: extraction, purification, structure and bioactivities. *Chemical and Biological Technologies in Agriculture* 8: 1–14.

Houshmand, B., Mahjour, F., and Dianat, O. (2013). Antibacterial effect of different concentrations of garlic (*Allium sativum*) extract on dental plaque bacteria. *Indian Journal of Dental Research* 24 (1): 71–75.

Ivanišová, E., Kačániová, M., Savitskaya, T.A., and Grinshpan, D.D. (2021). *Medicinal Herbs: Important Source of Bioactive Compounds for Food Industry*. Herbs and Spices—New Processing Technologies.

Jagtap, A.S. and Manohar, C.S. (2021). Overview on microbial enzymatic production of algal oligosaccharides for nutraceutical applications. *Marine Biotechnology* 23: 159–176.

Jayaprakasha, G.K., Ohnishi-Kameyama, M., Ono, H. et al. (2006). Phenolic constituents in the fruits of *Cinnamomum zeylanicum* and their antioxidant activity. *Journal of Agricultural and Food Chemistry* 54 (5): 1672–1679.

Johri, R.K. (2011). *Cuminum cyminum* and *Carum carvi*: an update. *Pharmacognosy Reviews* 5 (9): 63.

Jubair, N., Rajagopal, M., Chinnappan, S. et al. (2021). Review on the antibacterial mechanism of plant-derived compounds against multidrug-resistant bacteria (MDR). *Evidence-Based Complementary and Alternative Medicine* 2021: 3663315.

Kang, H. and Kim, H. (2017). Astaxanthin and-carotene in helicobacter pylori-induced gastric inflammation: a mini-review on action mechanisms. *Journal of Cancer Prevention* 22: 57–61.

Karimi, E., Oskoueian, E., Oskoueian, A. et al. (2013). Insight into the functional and medicinal properties of *Medicago sativa* (Alfalfa) leaves extract. *Journal of Medicinal Plants Research* 7 (7): 290–297.

Kårlund, A., Gómez-Gallego, C., Turpeinen, A.M. et al. (2019). Protein supplements and their relation with nutrition, microbiota composition and health: is more protein always better for sportspeople? *Nutrients* 11 (4): 829.

Kaska, A., Çiçek, M., and Mammadov, R. (2019). Biological activities, phenolic constituents and mineral element analysis of two endemic medicinal plants from Turkey: *Nepeta italica* and *Teucrium sandrasicum*. *South African Journal of Botany* 124: 63–70.

Kechagia, M., Basoulis, D., Konstantopoulou, S. et al. (2013). Health benefits of probiotics: a review. *International Scholarly Research Notices* 2013: 481651.

Khansari, N., Shakiba, Y., and Mahmoudi, M. (2009). Chronic inflammation and oxidative stress as a major cause of age-related diseases and cancer. *Recent Patents on Inflammation & Allergy Drug Discovery* 3 (1): 73–80.

Khatoon, M., Islam, E., Islam, R. et al. (2013). Estimation of total phenol and in vitro antioxidant activity of *Albizia procera* leaves. *BMC Research Notes* 6: 1–7.

Kim, M. and Kim, Y. (2010). Hypocholesterolemic effects of curcumin via up-regulation of cholesterol 7a-hydroxylase in rats fed a high fat diet. *Nutrition Research and Practice* 4 (3): 191–195.

Kim, S.J., Kawaharada, C., Suzuki, T. et al. (2006). Effect of natural light periods on rutin, free amino acid and vitamin C contents in the sprouts of common (*Fagopyrum esculentum* Moench) and tartary (F. tataricum Gaertn.) buckwheats. *Food Science and Technology Research* 12 (3): 199–205.

Kokate, A.A., Kumavat, S.R., Vakhariya, R.R., and Mohite, S.K. (2023). Review on multidimensional therapeutic strategies and potential antiinflammatory medicinal plants. *Asian Journal of Research in Pharmaceutical Science* https://doi.org/10.52711/2231-5659.2023.00022.

Koul, B. and Chase, N. (2015). *Moringa oleifera* Lam.: Panacea to several maladies. *Journal of Chemical and Pharmaceutical Research* 7 (6): 687–707.

Kumar, G.P. and Khanum, F. (2012). Neuroprotective potential of phytochemicals. *Pharmacognosy Reviews* 6 (12): 81.

Lamothe, L.M., Lê, K.A., Samra, R.A. et al. (2019). The scientific basis for healthful carbohydrate profile. *Critical Reviews in Food Science and Nutrition* 59 (7): 1058–1070.

Landete, J.M. (2013). Dietary intake of natural antioxidants: vitamins and polyphenols. *Critical Reviews in Food Science and Nutrition* 53 (7): 706–721.

Lee, S.Y., Cho, S.S., Li, Y. et al. (2020). Anti-inflammatory effect of *Curcuma longa* and *Allium hookeri* co-treatment via NF-κB and COX-2 pathways. *Scientific Reports* 10 (1): 5718.

Li, Y., Jiao, J., Qi, Y. et al. (2021). Curcumin: a review of experimental studies and mechanisms related to periodontitis treatment. *Journal of Periodontal Research* 56 (5): 837–847.

Li, T., Zhang, X., Ren, Y. et al. (2022). Antihypertensive effect of soybean bioactive peptides: a review. *Current Opinion in Pharmacology* 62: 74–81.

Lipman, T.O. (2005). Herbal supplements. *Current Gastroenterology Reports* 7 (4): 302–307.

Lobo, V., Patil, A., Phatak, A., and Chandra, N. (2010). Free radicals, antioxidants and functional foods: impact on human health. *Pharmacognosy Reviews* 4 (8): 118.

López, V., Pérez, S., Vinuesa, A. et al. (2016). *Stevia rebaudiana* ethanolic extract exerts better antioxidant properties and antiproliferative effects in tumour cells than its diterpene glycoside stevioside. *Food & Function* 7 (4): 2107–2113.

Lu, Y., Joerger, R., and Wu, C. (2011). Study of the chemical composition and antimicrobial activities of ethanolic extracts from roots of *Scutellaria baicalensis* Georgi. *Journal of Agricultural and Food Chemistry* 59 (20): 10934–10942.

Lukaski, H.C. (2004). Vitamin and mineral status: effects on physical performance. *Nutrition* 20 (7-8): 632–644.

Lushchak, V.I. (2014). Free radicals, reactive oxygen species, oxidative stress and its classification. *Chemico-Biological Interactions* 224: 164–175.

Luximon-Ramma, A., Bahorun, T., Soobrattee, M.A., and Aruoma, O.I. (2002). Antioxidant activities of phenolic, proanthocyanidin, and flavonoid components in extracts of *Cassia fistula*. *Journal of Agricultural and Food Chemistry* 50 (18): 5042–5047.

Lv, J.L., Chen, H.L., Duan, J.A., and Liu, J.W. (2016). Research progress of structures and pharmacological activities of phthalides from *Angelica sinensis*. *China Journal of Chinese Materia Medica* 41 (2): 167–176.

Mao, Q.Q., Xu, X.Y., Cao, S.Y. et al. (2019). Bioactive compounds and bioactivities of ginger (*Zingiber officinale*). *Foods* 8 (6): 185.

Martínez, M.J.A., Del Olmo, L.M.B., Ticona, L.A., and Benito, P.B. (2012). The *Artemisia L.* genus: a review of bioactive sesquiterpene lactones. *Studies in Natural Products Chemistry* 37: 43–65.

Martins, M., Ávila, P.F., Poletto, P., and Goldbeck, R. (2023). Polysaccharide degradation for oligosaccharide production with nutraceutical potential for the food industry. In: *Polysaccharide Degrading Biocatalysts*, 335–363. Academic Press.

Mehranjani, M., Shariatzadeh, M., Desfulian, A. et al. (2007). Effects of *Medicago sativa* on nephropathy in diabetic rats. *Indian Journal of Pharmaceutical Sciences* 69 (6): 768.

Meščić Macan, A., Gazivoda Kraljević, T., and Raić-Malić, S. (2019). Therapeutic perspective of vitamin C and its derivatives. *Antioxidants* 8 (8): 247.

Milani, A., Basirnejad, M., Shahbazi, S., and Bolhassani, A. (2017). Carotenoids: biochemistry, pharmacology and treatment. *British Journal of Pharmacology* 174 (11): 1290–1324.

Milla, P.G., Peñalver, R., and Nieto, G. (2021). Health benefits of uses and applications of *Moringa oleifera* in bakery products. *Plants* 10 (2): 318.

Miller, D.D. (2017). Minerals. In: *Fennema's Food Chemistry*, 627–679. CRC Press.

Misra, L.N. (2013). Traditional phytomedicinal systems, scientific validations and current popularity as nutraceuticals. *International Journal of Traditional and Natural Medicines* 2: 27–75.

Mohammadi-Cheraghabadi, M. and Hazrati, S. (2023). Terpenoids, steroids, and phenolic compounds of medicinal plants. In: *Phytochemicals in Medicinal Plants, Biodiversity, Bioactivity and Drug Discovery*, 105. De Gruyter.

Mosca, M., Boniglia, C., Carratù, B. et al. (2008). Determination of α-amylase inhibitor activity of phaseolamin from kidney bean (Phaseolus vulgaris) in dietary supplements by HPAEC-PAD. *Analytica Chimica Acta* 617 (1–2): 192–195.

Msaada, K., Salem, N., Bachrouch, O. et al. (2015). Chemical composition and antioxidant and antimicrobial activities of wormwood (*Artemisia absinthium* L.) essential oils and phenolics. *Journal of Chemistry* 2015 (2): https://doi.org/10.1155/2015/804658.

Munoz, C., Rios, E., Olivos, J. et al. (2007). Iron, copper and immunocompetence. *British Journal of Nutrition* 98 (S1): S24–S28.

Muthuraman, A., Rishitha, N., Paramakrishnan, N. et al. (2020). Role of lipid peroxidation process in neurodegenerative disorders. In: *Lipid Peroxidation Research*, vol. 1, 13. London: IntechOpen.

Myung, S.K., Ju, W., Cho, B. et al. (2013). Efficacy of vitamin and antioxidant supplements in prevention of cardiovascular disease: systematic review and meta-analysis of randomised controlled trials. *BMJ* 346: f10.

Naidu, K.A. (2003). Vitamin C in human health and disease is still a mystery? An overview. *Nutrition Journal* 2: 1–10.

Nasri, H., Baradaran, A., Shirzad, H., and Rafieian-Kopaei, M. (2014). New concepts in nutraceuticals as alternative for pharmaceuticals. *International Journal of Preventive Medicine* 5 (12): 1487.

Neves, J.M., Matos, C., Moutinho, C. et al. (2009). Ethnopharmacological notes about ancient uses of medicinal plants in Trás-os-Montes (northern of Portugal). *Journal of Ethnopharmacology* 124 (2): 270–283.

Nikhra, V. (2018). Nutraceuticals for improving cardiovascular health and prognosis in cardiovascular disease. *Journal of Nutrition and Diet Probiotics* 1 (2): 180009.

Nyam, K.L., Tan, C.P., Lai, O.M. et al. (2009). Physicochemical properties and bioactive compounds of selected seed oils. *LWT-Food Science and Technology* 42 (8): 1396–1403.

Ocana-Fuentes, A., Arranz-Gutierrez, E., Senorans, F.J., and Reglero, G. (2010). Supercritical fluid extraction of oregano (*Origanum vulgare*) essentials oils: anti-inflammatory properties based on cytokine response on THP-1 macrophages. *Food and Chemical Toxicology* 48 (6): 1568–1575.

Osman, D., Cooke, A., Young, T.R. et al. (2021). The requirement for cobalt in vitamin B12: a paradigm for protein metalation. *Biochimica et Biophysica Acta (BBA)-Molecular Cell Research* 1868 (1): 118896.

Özkalp, B., Sevgi, F., Özcan, M., and Özcan, M.M. (2010). The antibacterial activity of essential oil of oregano (*Origanum vulgare* L.). *Journal of Food Agriculture and Environment* 8 (2): 6–8.

Packer, L. (1997). *Vitamin C in Health and Disease*, vol. 4. CRC Press.

Patwardhan, B. (2014). Bridging ayurveda with evidence-based scientific approaches in medicine. *EPMA Journal* 5: 1–7.

Paudel, M.R., Chand, M.B., Pant, B., and Pant, B. (2019). Assessment of antioxidant and cytotoxic activities of extracts of *Dendrobium crepidatum*. *Biomolecules* 9 (9): 478.

Petrovska, B.B. (2012). Historical review of medicinal plants' usage. *Pharmacognosy Reviews* 6 (11): 1.

Pineiro, M. and Stanton, C. (2007). Probiotic bacteria: legislative framework—requirements to evidence basis. *The Journal of Nutrition* 137 (3): 850S–853S.

Pipe, E.A., Gobert, C.P., Capes, S.E. et al. (2009). Soy protein reduces serum LDL cholesterol and the LDL cholesterol: HDL cholesterol and apolipoprotein B: apolipoprotein AI ratios in adults with type 2 diabetes. *The Journal of Nutrition* 139 (9): 1700–1706.

Poljsak, B. and Ionescu, J.G. (2009). Pro-oxidant versus antioxidant effects of vitamin C. In: *Handbook of Vitamin C Research: Daily Requirements, Dietary Sources and Adverse Effects*, vol. 153. New York: Nova Biomedical Books.

Priya, S. and Satheeshkumar, P.K. (2020). Natural products from plants. In: *Functional and Preservative Properties of Phytochemicals*, 145–163. Amsterdam: Elsevier.

Rahaiee, S., Moini, S., Hashemi, M., and Shojaosadati, S.A. (2015). Evaluation of antioxidant activities of bioactive compounds and various extracts obtained from saffron (*Crocus sativus* L.): a review. *Journal of Food Science and Technology* 52: 1881–1888.

Rahim, M.A., Saeed, F., Khalid, W. et al. (2021). Functional and nutraceutical properties of fructo-oligosaccharides derivatives: a review. *International Journal of Food Properties* 24 (1): 1588–1602.

Ramadan, M.F., Sharanabasappa, G., Seetharam, Y.N. et al. (2006). Characterisation of fatty acids and bioactive compounds of kachnar (*Bauhinia purpurea* L.) seed oil. *Food Chemistry* 98 (2): 359–365.

Rana, M.G., Katbamna, R.V., Padhya, A.A. et al. (2010). In vitro antioxidant and free radical scavenging studies of alcoholic extract of *Medicago sativa* L. *Romanian Journal of Biology-Plant Biology* 55 (1): 15–22.

Rebey, I.B., Wannes, W.A., Kaab, S.B. et al. (2019). Bioactive compounds and antioxidant activity of *Pimpinella anisum* L. accessions at different ripening stages. *Scientia horticulturae* 246: 453–461.

Rmiki, N.E., Lemoine, Y.V.E.S., and Schoeff, B. (1999). Carotenoids and stress in higher plants and algae. In: *Handbook of Plant and Crop Stress*, 2nde, 465–482. New York: Dekker.

Rodríguez-Cruz, M., del Rocío Cruz-Guzmán, O., Almeida-Becerril, T. et al. (2018). Potential therapeutic impact of omega-3 long chain-polyunsaturated fatty acids on inflammation markers in Duchenne muscular dystrophy: a double-blind, controlled randomized trial. *Clinical Nutrition* 37 (6): 1840–1851.

Roy, F., Boye, J.I., and Simpson, B.K. (2010). Bioactive proteins and peptides in pulse crops: pea, chickpea and lentil. *Food Research International* 43 (2): 432–442.

Ruchi, S. (2017). Role of nutraceuticals in health care: a review. *International Journal of Green Pharmacy (IJGP)* 11 (03).

Sachdeva, V., Roy, A., and Bharadvaja, N. (2020). Current prospects of nutraceuticals: a review. *Current Pharmaceutical Biotechnology* 21 (10): 884–896.

Saha, S. and Ghosh, S. (2012). *Tinospora cordifolia*: one plant, many roles. *Ancient Science of Life* 31 (4): 151.
Saibabu, V., Fatima, Z., Khan, L.A., and Hameed, S. (2015). Therapeutic potential of dietary phenolic acids. *Advances in Pharmacological and Pharmaceutical Sciences* 2015: 823539.
Saini, R.K., Sivanesan, I., and Keum, Y.S. (2016). Phytochemicals of *Moringa oleifera*: a review of their nutritional, therapeutic and industrial significance. *3 Biotech* 6: 1–14.
Saltveit, M.E. (2017). Synthesis and metabolism of phenolic compounds. In: *Fruit and Vegetable Phytochemicals: Chemistry and Human Health*, 2nde, 115–124. Wiley.
Saricaoglu, F.T. and Turhan, S. (2018). Antimicrobial activity and antioxidant capacity of thyme, rosemary and clove essential oils and their mixtures. *Journal of Innovative Science and Engineering* 2 (1): 25–33.
Schwartz, H., Ollilainen, V., Piironen, V., and Lampi, A.M. (2008). Tocopherol, tocotrienol and plant sterol contents of vegetable oils and industrial fats. *Journal of Food Composition and Analysis* 21 (2): 152–161.
Sejersted, O.M. and Sjøgaard, G. (2000). Dynamics and consequences of potassium shifts in skeletal muscle and heart during exercise. *Physiological Reviews* 80 (4): 1411–1481.
Setayesh, M., Karimi, M., Zargaran, A. et al. (2022). Efficacy of a Persian herbal medicine compound on coronavirus disease 2019 (COVID-19): a randomized clinical trial. *Integrative Medicine Research* 11 (3): 100869.
Shahana, S., Nikalje, A.P.G., and Nikalje, G. (2017). A brief review on Bauhinia variegata: phytochemistry, antidiabetic and antioxidant potential. *American Journal of PharmTech Research* 7: 186–197.
Sharifi-Rad, M., Anil Kumar, N.V., Zucca, P. et al. (2020). Lifestyle, oxidative stress, and antioxidants: back and forth in the pathophysiology of chronic diseases. *Frontiers in Physiology* 11: 694.
Sharma, P., Jha, A.B., Dubey, R.S., and Pessarakli, M. (2012). Reactive oxygen species, oxidative damage, and antioxidative defense mechanism in plants under stressful conditions. *Journal of Botany* 2012: 217037.
Sharma, V., Katiyar, A., and Agrawal, R.C. (2018). *Glycyrrhiza glabra*: chemistry and pharmacological activity. *Sweeteners* 87–100. https://doi.org/10.1007/978-3-319-27027-2_21.
Sharma, S., Thawani, V., Hingorani, L. et al. (2004). Pharmacokinetic study of 11-keto β-Boswellic acid. *Phytomedicine* 11 (2-3): 255–260.
Sichetti, M., De Marco, S., Pagiotti, R. et al. (2018). Anti-inflammatory effect of multistrain probiotic formulation (*L. rhamnosus*, *B. lactis*, and *B. longum*). *Nutrition* 53: 95–102.
Siddiqui, M.Z. (2011). *Boswellia serrata*, a potential antiinflammatory agent: an overview. *Indian Journal of Pharmaceutical Sciences* 73 (3): 255.
Sikand, G., Kris-Etherton, P., and Boulos, N.M. (2015). Impact of functional foods on prevention of cardiovascular disease and diabetes. *Current Cardiology Reports* 17: 1–16.
Singh, J. and Sinha, S. (2012). Classification, regulatory acts and applications of nutraceuticals for health. *International Journal of Pharma and Bio Sciences* 2 (1): 177–187.
Singh, S.P. and Upadhyay, S.K. (2021). *Bioprospecting of Microorganism based Industrial Molecules*. Wiley https://doi.org/10.1002/9781119717317.
Skinner, M.A., Loh, J.M., Hunter, D.C., and Zhang, J. (2011). Gold kiwifruit (*Actinidia chinensis*) for immune support. *The Proceedings of the Nutrition Society* 70 (2): 276–280.
Soetan, K.O., Olaiya, C.O., and Oyewole, O.E. (2010). The importance of mineral elements for humans, domestic animals and plants: a review. *African Journal of Food Science* 4 (5): 200–222.
Soumya, T., Jayasree, P.R., and Manish Kumar, P.R. (2023). Zingiberaceae plants: a cornucopia of promising chemotherapeuticals for cancer cure. In: *Bioprospecting of Tropical Medicinal Plants*, 427–462. Springer.

Stonehouse, W., Gammon, C.S., Beck, K.L. et al. (2013). Kiwifruit: our daily prescription for health. *Canadian Journal of Physiology and Pharmacology* 91 (6): 442–447.

Sui, X., Zhang, Y., and Zhou, W. (2016). Bread fortified with anthocyanin-rich extract from black rice as nutraceutical sources: its quality attributes and in vitro digestibility. *Food Chemistry* 196: 910–916.

Sun, W. and Shahrajabian, M.H. (2023). Therapeutic potential of phenolic compounds in medicinal plants—natural health products for human health. *Molecules* 28 (4): 1845.

Sun, W., Shahrajabian, M.H., and Cheng, Q. (2019). Anise (*Pimpinella anisum* L.), a dominant spice and traditional medicinal herb for both food and medicinal purposes. *Cogent Biology* 5 (1): 1673688.

Swaroopa, G. and Srinath, D. (2017). Nutraceuticals and their health benefits. *International Journal of Pure and Applied Bioscience* 5 (4): 1151–1155.

Taghiyar, M., Darvishi, L., Askari, G. et al. (2013). The effect of vitamin C and e supplementation on muscle damage and oxidative stress in female athletes: a clinical trial. *International Journal of Preventive Medicine* 4 (Suppl 1): S16.

Taylor, P.W., Hamilton-Miller, J.M., and Stapleton, P.D. (2005). Antimicrobial properties of green tea catechins. *Food Science and Technology Bulletin* 2: 71.

Tesfaye, A. (2021). Revealing the therapeutic uses of garlic (*Allium sativum*) and its potential for drug discovery. *The Scientific World Journal* 2021: 8817288.

Thirugnanasampandan, R., Gunasekar, R., and Gogulramnath, M. (2015). Chemical composition analysis, antioxidant and antibacterial activity evaluation of essential oil of *Atalantia monophylla* Correa. *Pharmacognosy Research* 7 (Suppl 1): S52.

Torquato, P., Marinelli, R., Bartolini, D., and Galli, F. (2020). Vitamin E: nutritional aspects. In: *Molecular Nutrition*, 447–485. Academic Press.

Upadhyay, S.K. (2021). *Genome Engineering for Crop Improvement*. Wiley https://doi.org/10.1002/9781119672425.

Upadhyay, S.K. (2024). *Applications of Genome Engineering in Plants*. Wiley https://doi.org/10.1002/9781394183913.

Upadhyay, S.K., Mishra, M., Singh, H. et al. (2010b). Interaction of *Allium sativum* leaf agglutinin (ASAL) with midgut BBMV proteins and its stability in *Helicoverpa armigera*. *Proteomics* 10: 4431–4440.

Upadhyay, S.K., Saurabh, S., Rai, P. et al. (2010a). SUMO fusion facilitates expression and purification of garlic lectin but modifies some of its properties. *Journal of Biotechnology* 146: 1–8.

Upadhyay, S.K., Saurabh, S., Singh, R. et al. (2011). Purification and characterization of a lectin with high hemagglutination property isolated from *Allium altaicum*. *The Protein Journal* 30: 374–383.

Upadhyay, S.K. and Singh, P.K. (2012). Receptors of garlic (*Allium sativum*) lectins and their role in insecticidal action. *The Protein Journal* 31 (6): 439–446.

Upadhyay, S.K. and Singh, S.P. (2021). *Bioprospecting of Plant Biodiversity for Industrial Molecules*. Wiley doi: 10.1002/9781119718017.

Upadhyay, S.K. and Singh, S.P. (2023). *Plants as Bioreactors for Industrial Molecules*. Wiley https://doi.org/10.1002/9781119875116.

Wei, W.L., Zeng, R., Gu, C.M. et al. (2016). *Angelica sinensis* in China-A review of botanical profile, ethnopharmacology, phytochemistry and chemical analysis. *Journal of Ethnopharmacology* 190: 116–141.

Wells, M.L., Potin, P., Craigie, J.S. et al. (2017). Algae as nutritional and functional food sources: revisiting our understanding. *Journal of Applied Phycology* 29: 949–982.

Welte, M.A. and Gould, A.P. (2017). Lipid droplet functions beyond energy storage. *Biochimica et Biophysica Acta (BBA)-Molecular and Cell Biology of Lipids* 1862 (10): 1260–1272.

Williams, L.L. (2013). *Moringa olefiera*: could this be an answer to our need for an alternative to fighting drug-resistance and chronic infections. *Medicinal and Aromatic Plants* 2 (743): 2167–0412.

Wu, X.D., Ding, L.F., Tu, W.C. et al. (2016). Bioactive sesquiterpenoids from the flowers of *Inula japonica*. *Phytochemistry* 129: 68–76.

Wu, D.T., Nie, X.R., Shen, D.D. et al. (2020). Phenolic compounds, antioxidant activities, and inhibitory effects on digestive enzymes of different cultivars of okra (*Abelmoschus esculentus*). *Molecules* 25 (6): 1276.

Yamani, H.A., Pang, E.C., Mantri, N., and Deighton, M.A. (2016). Antimicrobial activity of Tulsi (*Ocimum tenuiflorum*) essential oil and their major constituents against three species of bacteria. *Frontiers in Microbiology* 7: 681.

Yang, W., Chen, X., Li, Y. et al. (2020). Advances in pharmacological activities of terpenoids. *Natural Product Communications* 15 (3): 1934578X20903555.

Yang, Y., Xu, S., Yang, K. et al. (2023). Characterization and in vitro antioxidant and anti-inflammatory activities of ginsenosides extracted from forest-grown wild *Panax quinquefolius L. Foods* 12 (23): 4316.

Yokoyama, Y., Levin, S.M., and Barnard, N.D. (2017). Association between plant-based diets and plasma lipids: a systematic review and meta-analysis. *Nutrition Reviews* 75 (9): 683–698.

Zadeh, M.Z., Mirzaei, A., and Fazilati, M. (2014). Effects of hydro-alcoholic extract of *Medicago sativa* on the chemical burn injuries in rats. *World Journal of Pharmacy and Pharmaceutical Sciences* 3 (2): 910–918.

Zhang, Z. and Abbaspourrad, A. (2018). How much bean hemagglutinin is safe for human consumption? *Journal of Agricultural and Food Chemistry* 66 (27): 6937–6939.

Zhang, X., Huang, H., Zhang, Q. et al. (2015). Phytochemical characterization of Chinese bayberry (Myrica rubra Sieb. et Zucc.) of 17 cultivars and their antioxidant properties. *International Journal of Molecular Sciences* 16 (6): 12467–12481.

Zhao, C., Jia, Y., and Lu, F. (2018). Angelica stem: a potential low-cost source of bioactive phthalides and phytosterols. *Molecules* 23 (12): 3065.

Zhao, D.D., Jiang, L.L., Li, H.Y. et al. (2016). Chemical components and pharmacological activities of terpene natural products from the genus *Paeonia*. *Molecules* 21 (10): 1362.

Zhou, W., Cheng, Y., Zhu, P. et al. (2020). Implication of gut microbiota in cardiovascular diseases. *Oxidative Medicine and Cellular Longevity* 2020: 5394096.

Zhu, F., Du, B., and Xu, B. (2018). Anti-inflammatory effects of phytochemicals from fruits, vegetables, and food legumes: a review. *Critical Reviews in Food Science and Nutrition* 58 (8): 1260–1270.

3

Significance of Nutraceuticals in Modern Health Maintenance and Disease Prevention

Mahra Mohammad Moosa Ali Alblooshi, Kok-Song Lai, Swee-Hua Erin Lim, and Shamshul Ansari

Health Sciences Division, Abu Dhabi Women's College, Higher Colleges of Technology, Abu Dhabi, UAE

3.1 Introduction

Long before the advent of modern pharmacology, natural medications were sought after due to the presence of bioactive compounds that exhibited tangible improvements in human health. Nutraceuticals, derived from food sources, offer additional health benefits and contribute significantly to overall well-being. These bioactive compounds, including vitamins and antioxidants, play a pivotal role in promoting health and preventing age-related chronic diseases (El-Sohaimy 2012).

Historical records from China, Egypt, Greece, and India illustrate a profound appreciation for plant-based foods, not just as dietary staples but also for their therapeutic properties. These unprocessed and pure ingredients are packed with essential nutrients, such as omega-3 fatty acids found in cold-water fish and lycopene in tomatoes.

3.1.1 Addressing Nutritional Gaps

As dietary preferences shift toward ketogenic, vegetarian, and vegan lifestyles, a noticeable nutritional gap has emerged due to the absence of certain nutrients. Nutraceuticals have risen to the challenge by providing essential vitamins and minerals. For instance, beans and lentils are excellent sources of protein, while alternatives to vitamin B12, such as mushrooms and kimchi, cater to individuals abstaining from meat consumption. Moreover, the health of the gut is paramount, influencing overall health by fostering beneficial bacteria that aid in digestion and bolster immune functions. Plant-based functional foods, with their anti-inflammatory and antioxidant properties, play a pivotal role in this regard (Nwosu and Ubaoji 2020).

In the traditions of Chinese medicine, plant-based treatments are revered for their role in enhancing immunity, mitigating inflammation, and facilitating the body's natural recovery mechanisms. The ayurvedic system of India is also renowned for its extensive use of herbal remedies designed to fortify the immune system, particularly in treating respiratory ailments (Nwosu and Ubaoji 2020).

Herbal Nutraceuticals: Products and Processes, First Edition. Edited by Santosh Kumar Upadhyay and Sudhir Pratap Singh.

3.1.2 Confluence of Nutrition and Medicine

The evolution of nutraceuticals represents a convergence of nutrition and medicine. This diverse range of health-enhancing substances, encompassing vital nutrients and beneficial probiotics, has become integral to dietary regimes focused on health maintenance and disease prevention. The age-old adage by Hippocrates, advocating for food as a conduit to health, has regained prominence, emphasizing the significance of a nutrient-dense diet in disease management and prevention.

Continuing studies delve into the functional aspects of various foods and their capacity to counteract health issues. These investigations spotlight the substantial benefits inherent in a diet rich in naturally occurring health-promoting ingredients.

The exploration and implementation of functional foods continue to progress, revealing the enduring importance and untapped potential of plant-based diets in enhancing overall health and longevity (Halliwell 2024). As society continues to embrace alternative dietary choices, the role of nutraceuticals in filling nutritional gaps and promoting well-being becomes ever more pronounced. The integration of traditional wisdom with modern science underscores the profound impact of these natural compounds on human health and the prevention of chronic diseases.

3.2 Plant-based Nutraceuticals

The global interest in plant-based nutraceuticals has intensified, driven by an increasing awareness of health and wellness alongside a preference for natural products. This trend is observable across various regions, including the European Union (EU), North America, and Australia, where certain plant-based nutraceuticals have gained prominence due to their health-enhancing properties. Despite regional variations in consumer preferences and regulatory frameworks governing these nutraceuticals, a core list of plant-based compounds has been recognized for their therapeutic benefits (Dixit et al. 2021; Upadhyay and Singh 2021, 2023). The following sections enumerate five notable plant-based nutraceuticals, detailing their associated health benefits and the bioactive constituents responsible for these effects, while acknowledging the disparate landscape of popularity and regulatory endorsements across these regions.

Turmeric (*Curcuma longa*) is extensively studied for its anti-inflammatory and antioxidant effects, beneficial for joint, cardiovascular, and cognitive health. Its primary bioactive component, curcumin, is instrumental in these health benefits. Research focuses on curcumin's role in addressing chronic inflammation, preventing cancer, and enhancing brain function (Musial et al. 2020).

Green tea (*Camellia sinensis*) is acknowledged within the scientific community for its beneficial effects on heart health, its role in weight management, and its potent antioxidative properties, which contribute to reduced cancer risk. The primary bioactive molecules responsible for these benefits are catechins, with epigallocatechin gallate (EGCG) being the most significant compound. Extensive research has been conducted on green tea's influence on metabolic health, its antioxidative capacity, and its protective effects against certain cancers (Bayan et al. 2014; Rodriguez-Leyva et al. 2010).

Flaxseed is recognized as another significant plant-based nutraceutical, valued for its rich content of dietary fiber, omega-3 fatty acids, particularly alpha-linolenic acid (ALA), and lignans. These components collectively contribute to improved cardiovascular health, enhanced digestive well-being, and a potential reduction in cancer risk. Consequently, scientific investigations have frequently centered on flaxseed's cardiovascular advantages, its utility in diabetes management, and its anticancer properties (Yu et al. 2019).

Garlic (*Allium sativum*) is noted for its cardiovascular benefits, such as reducing blood pressure and cholesterol levels, in addition to possessing antimicrobial and anti-inflammatory effects. These health benefits are primarily attributed to allicin and other sulfur-containing compounds, stimulating research into garlic's impact on cardiovascular health, its antimicrobial properties, and its role in cancer prevention (Bayan et al. 2014).

Ginseng is recognized for its ability to enhance energy levels, mitigate stress, improve brain function, and bolster immune health, with notable anti-inflammatory effects. Ginsenosides, the major active components, underlie most of ginseng's medicinal properties. This has led to studies into ginseng's effects on cognitive function, its potential in managing erectile dysfunction, and its impact on immune health, thereby highlighting its multifaceted role in promoting wellness (Kim et al. 2017; Yu et al. 2019).

3.3 Plant-based Nutraceuticals in the Middle East

The last decades have witnessed a pivot toward functional food and nutraceutical consumer trends in the western world, attributed to the global recognition of the deep nutritional value of plant-based foods (Subramoniam 2014), supported by studies that demonstrated the role and beneficial functions of nutraceuticals in decreasing oxidation and inflammation caused by pharyngitis (Chopra et al. 2022).

With the emergence of awareness regarding medicinal plants over the last one and a half decade, there has been an increase in the establishment of new industry departments in India focused on plant and herbal science research. This development is attributed to the integral role of traditional medicine in Indian culture, with more than 46,000 recorded plant species (El Khoury et al. 2012). The side effects of modern medicine, such as chemotherapy, have necessitated global research to mitigate side effects and enhance treatment effectiveness with the use of plant and herbal treatments (Kumar 2017). Epirubicin, a chemotherapy drug causing cardiac alterations and arrhythmias, can be treated with high doses of doxorubicin (Greenwell and Rahman 2015), a chemotherapeutic drug that may result in significant toxicity, such as cardiac toxicity induced by high doses of therapeutic drugs (Williamson et al. 2019).

The Middle East's diverse flora has attracted significant attention in the field of botanical nutraceuticals, merging ancient herbal knowledge with modern scientific inquiry. This region's unique vegetation is meticulously researched for its potential in addressing a range of chronic health conditions. Central to this botanical treasure is the pomegranate, indigenous to Middle Eastern areas like Iran. Celebrated for its array of bioactive substances, particularly potent antioxidants, the pomegranate plays a pivotal role in counteracting cellular damage from environmental factors. These antioxidants are crucial in DNA protection, contributing to a reduced risk of cancer and positively influencing cholesterol levels by enhancing HDL and managing LDL (de Oliveira et al. 2020). The fruit's significant health benefits are attributed to its rich phenolic content, placing it at the forefront of nutraceutical research.

The date palm (*Phoenix dactylifera*), thriving in the warm climates of nations like Saudi Arabia and the UAE, has a long history of medicinal use. Dates are packed with beneficial compounds including phenolics, tannins, flavonoids, phytosterols, and carotenoids (Bhatt et al. 2023), known for their effectiveness in preventing cardiovascular diseases and other chronic illnesses. Their dense nutritional profile, rich in vitamins and minerals, has made dates a focal point of studies investigating their impact on tumor growth inhibition, diabetes management, and cholesterol reduction, partly due to the bioactive β-glucan content (Maqsood et al. 2020). Dates are also recognized for enhancing gut health.

Extra virgin olive (*Olea europaea*) oil, a staple in the Middle Eastern diet, is rich in polyphenols, powerful antioxidants effective in cellular protection against oxidative damage. Regular consumption of this oil is linked to improved HDL functionality (Visioli et al. 2018) and is associated with a decreased risk of diseases like breast cancer and Alzheimer's, alongside aiding in managing type 2 diabetes.

Black seed (*Nigella sativa*), with its longstanding use in traditional remedies, offers a wide range of pharmacological properties. It comprises essential proteins, carbohydrates, and fats crucial for various bodily functions, including energy production. Recent findings underscore its role in enhancing gastrointestinal, cardiovascular, and immune system health, and in providing pain relief in various conditions (Tavakkoli et al. 2017).

Aniseed (*Pimpinella anisum*), known for its strong antioxidant properties and effectiveness against multidrug-resistant bacteria, owes its abilities to its phenolic content. It also contains monoterpenes such as estragole, *cis*-anethole, and *trans*-anethole (AlBalawi et al. 2023), which contribute to its distinctive aroma and antimicrobial efficacy.

In conclusion, the Middle East's diverse plant-based nutraceuticals, from pomegranates and dates to olive oil, black seed, and aniseed, represent a successful convergence of ancient traditions and modern science. These botanicals are not just part of cultural heritage but also hold significant promise for health promotion and preventive care in the contemporary world (Berman et al. 2017).

3.4 Modified Versus Organic Plant-based Nutraceuticals

Recent trends among consumers have shown a marked increase in the preference for plant-based foods that are cultivated using environmentally sustainable practices, with a particular emphasis on organic agriculture. As defined by the Council of the European Union, organic agriculture is characterized by a stringent prohibition on the use of genetically modified organisms (GMOs) and ionizing radiation, and it imposes significant restrictions on the employment of synthetic agrochemicals. Despite the widespread adoption of organic farming practices globally, consumer perceptions regarding the nutritional superiority of organic products as opposed to nonorganic variants are varied. At the heart of this discussion lies the concentration of phenolic compounds in plant-based foods, which exhibit considerable variability across different species and are subject to influence by agricultural practices. These compounds have been recognized for their pivotal role in the prevention of chronic diseases. Nevertheless, the scientific community has not yet achieved a consensus regarding the assertion that organic farming categorically enhances the nutritional value of these foods in comparison to conventional farming methods. The body of ongoing research, illustrated by investigations such as those conducted in Tarragona, Spain, seeks to shed light on the differential impacts of organic versus nonorganic farming practices on the antioxidant properties and phenolic composition of plant-based foods (Cruz-Carrión et al. 2023; Kotha and Luthria 2019).

3.5 The Preventive Potency of Plant-based Nutraceuticals

Plant-based nutraceuticals and plant sources, due to the bioactive compounds they contain, provide a wide range of benefits to the body, decreasing the chances of inflammation, cardiovascular diseases, and more. One of the major compounds present is omega-3 fatty acids, which contribute to anti-inflammatory and heart health effects. Plant sterols are another example of bioactive compounds, as shown in in (Table 3.1).

Overlapped risk factors for cancer and cardiovascular diseases are often due to poor nutrition. Foods containing high fats could cause cells to be moved to high fat cells due to the high amount that is found in this part of body, this causes inflammation that contributes to an increase in cell division (Duan et al. 2018; Pati et al. 2023).

Table 3.1 Bioactive compounds and their health benefits.

Compound	Source(s)	Health benefits	References
Probiotic cultures	Fermented foods	Enhances gastrointestinal and immune health, supports gut flora for digestion and nutrient absorption.	Dixit et al. (2023)
Free radical scavengers	Berries, plant-based foods	Combat oxidative stress, reduce risk of chronic diseases, protect cells from damage.	Halliwell (2024)
Curcuminoids	Turmeric	Anti-inflammatory, antioxidant, may improve cognitive function, treat arthritis and gastrointestinal disorders.	Halliwell (2024) and Liu et al. (2022)
Campesterol	Bananas, peppers, plant-based foods	Prevents LDL cholesterol absorption, anti-inflammatory, beneficial for cardiovascular health and arthritis.	Panknin et al. (2023)
Anthocyanidins	Fruits	Anti-inflammatory, antioxidative, manage diabetic cardiovascular diseases.	Cersosimo and Hong (1986)
Flavonols and flavones	Fruits, vegetables	Anti-inflammatory, antioxidant, antibacterial, help in chronic disease prevention.	Sapian et al. (2022) and Ullah et al. (2020)
Kaempferol	Various plants	Antiatherogenic, may prevent heart disease.	Chagas et al. (2022)
Resveratrol	Grapes	Cardioprotective, anti-aging, potential in treating diabetes, obesity, and certain cancers.	National Center for Biotechnology Information (2024) and Singh et al. (2019)
Omega-3 lipids	Fatty fish	Benefits cardiovascular and brain health, reduces heart disease risk, may improve cognitive function.	Dixit et al. (2023)
Lycopene	Tomatoes	Antioxidant, associated with reduced risk of certain cancers and heart disease.	Kong et al. (2013)
Allicin	Garlic	Antimicrobial, cardioprotective, reduces blood pressure and cholesterol levels.	Story et al. (2010)
Catechins	Green tea	Antioxidant, anti-inflammatory, aids reducing heart disease risk.	Banerjee and Maulik (2002)
Capsaicinoids	Chili peppers	Pain relief, metabolism boosting, may aid in weight loss and pain management for arthritis.	Babu and Liu (2008)
Isoflavonoids	Soy products	Reduce menopause symptoms, risk of osteoporosis, reduces risk of bladder cancers.	Zheng et al. (2017) and Chen and Chen (2021)
Beta-glucans	Oats, barley	Lower cholesterol levels, improve heart health, manage blood sugar, enhance immune response.	Jaime and Santoyo (2021)

A study by the American College of Cardiology has stated that a plant-based diet intake of bioactive compounds like fiber and antioxidants found in vegetables, fruits, and wholegrain, has shown lower risk of cardiovascular disease (Sachdeva et al. 2020); plant-based diets are proven to reduce IHD related mortality, the plants are filled with a protein called glutamic acid which contributes to being a source of lowering blood pressure (Satija and Hu 2018). Additionally, glutamic acid helps with blood flow stimulation and fluidity (Durante 2019). When it comes to cancer, it is known that there is no specific reason for how it occurs. However, specialists have observed that a diet lacking essential nutrients can impact disease development. Studies have shown that bioactive compounds, such as phytochemicals found in fruits, vegetables, and whole grains, contain anti-carcinogenic properties that can help prevent the advancement and progression of certain cancer varieties (Pandey and Rizvi 2009).

Plant-derived nutraceuticals have showcased their effectiveness in addressing a spectrum of diseases and disorders, including neurodegenerative conditions, gastrointestinal issues, various cancers (Peña-Jorquera et al. 2023), immune-associated disorders, and hypertension (AlAli et al. 2021). Despite concerns about the potential risks associated with plant-based diets, it is emphasized that these diets are rich in micronutrients, which have demonstrated substantial support in preventing chronic diseases. Additionally, a study involving 76,364 women indicated that the consumption of tree nuts and peanuts was correlated with a 13–19% reduction in cardiovascular risk and a 15–23% lower risk of coronary heart diseases (Chen et al. 2019).

Bioactive compounds which are migrated and mainly found in plants have shown the effects of prevention in some of the most severe chronic diseases like obesity and heart issues. These bioactive compounds contain probiotics, peptides, and omega-3 fatty acids. Plant metabolites are categorized into primary and secondary types. The primary constituents include carbohydrates, amino acids, proteins associated with nucleic acids, chlorophyll, purines, and pyrimidines. The secondary constituents comprise a variety of substances such as alkaloids, terpenes, flavonoids, lignans, steroids, curcumin, saponins, glucosides, and diverse phenolic compounds. It is moreover very rare to find all these components in a single plant; they are, on the other hand, found in several different plants with a very wide variety as shown in Figure 3.1 (Soumya et al. 2021).

Primary metabolites are deemed indispensable nutrients sourced from plants. They comprise carbohydrates, pivotal as the main energy substrate, crucial for cerebral functions. They also

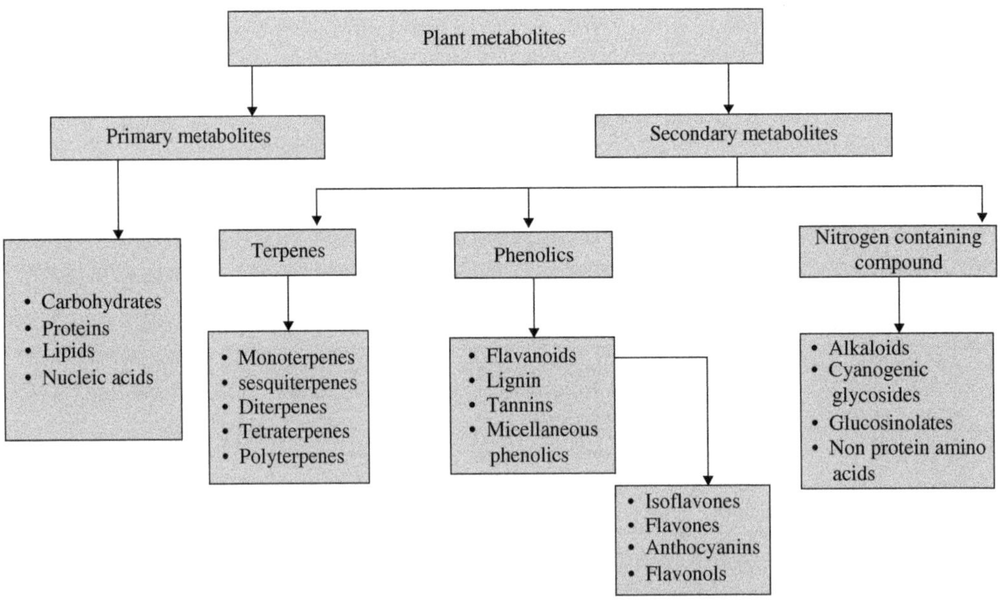

Figure 3.1 Metabolites found in plant-based nutraceuticals. *Source:* Soumya et al. 2021/OpenJournal Systems.com/CC BY 4.0.

encompass amino acids, imperative for tissue restoration and enzymatic reactions, as well as nucleic acids that are fundamental for cellular replication and operation.

Conversely, secondary metabolites are distinguished by their medicinal attributes. While not crucial for the basic sustenance of plant life, they possess compounds with health benefits. These substances enhance human health by providing a broad spectrum of advantages.

3.6 Prospective Developments in Plant-based Functional Foods: Global and Middle Eastern Perspectives

The domain of plant-based nutraceuticals is experiencing a notable ascent, characterized by a confluence of innovation, scientific advancement, and evolving consumer preferences. At the vanguard of this evolution is personalized nutrition, propelled by genomics and biotechnological innovations, facilitating bespoke nutrition solutions attuned to individual health requisites. Simultaneously, advancements in technology, such as nanotechnology and encapsulation techniques, are enhancing the bioavailability of phytonutrients, ensuring optimized nutrient assimilation.

Sustainability emerges as a pivotal trend within this narrative, with an escalating demand for ethically harvested and produced constituents. Agricultural innovations, including vertical farming and hydroponics, are becoming indispensable in the realm of sustainable agricultural practices. Mirroring a shift toward greater product transparency, the market is increasingly gravitating toward clean label products, devoid of additives and synthetic constituents.

In the Middle Eastern context, the fusion of contemporary scientific methodologies with traditional knowledge is paving the way for the inception of novel functional food products. Anticipated is the influence of the region's storied history with medicinal flora on the identification of innovative functional ingredients, promising to broaden the functional food market. The emphasis on endemic plants and herbs portends the emergence of regional superfoods, while the ascendancy of nutraceuticals mirrors dietary predilections distinct to the Middle East.

The imperative for collaborative research endeavors stands pronounced, as regional and global entities unite to explore the nutritive value of indigenous plant life. This collective effort is deemed essential amidst the growing adoption of plant-based diets across diverse cultural spectrums, heralding a paradigm shift toward the incorporation of functional foods within health and wellness frameworks. These diets, abundant in bioactive compounds, constitute an integral component of a holistic health approach.

Future projections posit plant-based functional foods at the epicenter of the global health dialogue. Their capacity to mitigate chronic disease, augment dietary quality, and advocate for sustainable food systems earmarks them as pivotal contributors to forthcoming health initiatives. The sector navigates challenges, including the imperative for coherent regulatory frameworks to guarantee product integrity and safety, alongside the need to enlighten consumers about the virtues of plant-based nutrition. Tackling noncommunicable diseases presents a significant avenue, with these foods poised to play a substantial role in both prevention and management, thereby alleviating healthcare expenditures.

As the sector matures, the synthesis of global trends with local exigencies will prove critical. Plant-based nutraceuticals stand on the threshold of becoming indispensable to health and wellness discourses, particularly as the Middle East harnesses its rich botanical heritage and burgeoning scientific acumen to contribute meaningfully to this global health movement.

3.7 Conclusions

The exploration of plant-based nutraceuticals reveals a promising avenue for the enhancement of human health and the prevention of chronic diseases. By bridging traditional wisdom with contemporary scientific research, these natural compounds stand at the forefront of nutritional science and medicine. The increasing global interest in sustainable, health-promoting dietary options underscores the potential of plant-based nutraceuticals to fill nutritional gaps and offer therapeutic benefits. The future of functional foods lies in leveraging advancements in biotechnology, sustainable agriculture, and personalized nutrition to optimize the delivery and efficacy of bioactive compounds. As the world continues to embrace plant-based diets, the integration of nutraceuticals into everyday health and wellness regimens becomes increasingly critical. With ongoing research and development, plant-based nutraceuticals are poised to play a pivotal role in global health initiatives, offering a sustainable and effective approach to disease prevention and health maintenance. This review underscores the importance of a collaborative effort in research and education to further elucidate the benefits of plant-based nutrition and to promote its adoption for a healthier future.

References

AlAli, M., Alqubaisy, M., Aljaafari, M.N. et al. (2021). Nutraceuticals: transformation of conventional foods into health promoters/disease preventers and safety considerations. *Molecules* 26 (9): 2540. https://doi.org/10.3390/molecules26092540.

AlBalawi, A.N., Elmetwalli, A., Baraka, D.M. et al. (2023). Chemical constituents, antioxidant potential, and antimicrobial efficacy of *Pimpinella anisum* extracts against multidrug-resistant bacteria. *Microorganisms* 11: 1024. https://doi.org/10.3390/microorganisms11041024.

Babu, P.V. and Liu, D. (2008). Green tea catechins and cardiovascular health: an update. *Current Medicinal Chemistry* 15 (18): 1840–1850. https://doi.org/10.2174/092986708785132979.

Banerjee, S.K. and Maulik, S.K. (2002). Effect of garlic on cardiovascular disorders: a review. *Nutrition Journal* 1: 4. https://doi.org/10.1186/1475-2891-1-4.

Bayan, L., Koulivand, P.H., and Gorji, A. (2014). Garlic: a review of potential therapeutic effects. *Avicenna Journal of Phytomedicine* 4 (1): 1–14.

Berman, A.Y., Motechin, R.A., Wiesenfeld, M.Y. et al. (2017). The therapeutic potential of resveratrol: a review of clinical trials. *NPJ Precision Oncology* 1: 35. https://doi.org/10.1038/s41698-017-0038-6.

Bhatt, N., Al-Subhi, L., Hilles, A.R., and Waly, M.I. (2023). *Health Benefits and Nutraceutical Properties of Dates*. Date Palm. https://doi.org/10.1079/9781800620209.0016.

Cersosimo, R.J. and Hong, W.K. (1986). Epirubicin: a review of the pharmacology, clinical activity, and adverse effects of an adriamycin analogue. *Journal of Clinical Oncology* 4 (3): 425–439. https://doi.org/10.1200/JCO.1986.4.3.425.

Chagas, M.D.S.S., Behrens, M.D., Moragas-Tellis, C.J. et al. (2022). Flavonols and flavones as potential anti-inflammatory, antioxidant, and antibacterial compounds. *Oxidative Medicine and Cellular Longevity* 2022: 9966750. https://doi.org/10.1155/2022/9966750.

Chen, L.R. and Chen, K.H. (2021). Utilization of isoflavones in soybeans for women with menopausal syndrome: an overview. *International Journal of Molecular Sciences* 22 (6): 3212. https://doi.org/10.3390/ijms22063212.

Chen, L.R., Ko, N.Y., and Chen, K.H. (2019). Isoflavone supplements for menopausal women: a systematic review. *Nutrients* 11 (11): 2649. https://doi.org/10.3390/nu11112649.

Chopra, A.S., Lordan, R., Horbańczuk, O.K. et al. (2022). The current use and evolving landscape of nutraceuticals. *Pharmacological Research* 175: 106001. https://doi.org/10.1016/j.phrs.2021.106001.

Cruz-Carrión, Á., Ruiz de Azua, M.J., Muguerza, B. et al. (2023). Organic versus non-organic plant-based foods-A comparative study on phenolic content and antioxidant capacity. *Plants* 12 (1): 183. https://doi.org/10.3390/plants12010183.

Dixit, V., Joseph Kamal, S.W., Bajrang Chole, P. et al. (2023). Functional foods: exploring the health benefits of bioactive compounds from plant and animal sources. *Journal of Food Quality* 2023: 5546753. https://doi.org/10.1155/2023/5546753.

Dixit, S., Shukla, A., Singh, V., and Upadhyay, S.K. (2021). Bioprospecting of Natural Compounds for Industrial and Medical Applications: Current Scenario and Bottleneck. In: *Bioprospecting of Plant Biodiversity for Industrial Molecules*. Wiley; Portico. https://doi.org/10.1002/9781119718017.ch3.

Duan, Y., Zeng, L., Zheng, C. et al. (2018). Inflammatory links between high fat diets and diseases. *Frontiers in Immunology* 9: 2649. https://doi.org/10.3389/fimmu.2018.02649.

Durante, W. (2019). The emerging role of l-glutamine in cardiovascular health and disease. *Nutrients* 11 (9): 2092. https://doi.org/10.3390/nu11092092.

El Khoury, D., Cuda, C., Luhovyy, B.L., and Anderson, G.H. (2012). Beta glucan: health benefits in obesity and metabolic syndrome. *Journal of Nutrition and Metabolism* 2012: 851362. https://doi.org/10.1155/2012/851362.

El-Sohaimy, S.A. (2012). Functional foods and nutraceuticals-modern approach to food science. *World Applied Sciences Journal* 20 (5): 691–708. https://doi.org/10.5829/idosi.wasj.2012.20.05.66119.

Greenwell, M. and Rahman, P.K. (2015). Medicinal plants: their use in anticancer treatment. *International Journal of Pharmaceutical Sciences and Research* 6 (10): 4103–4112. https://doi.org/10.13040/IJPSR.0975-8232.6(10).4103-12.

Halliwell, B. (2024). Understanding mechanisms of antioxidant action in health and disease. *Nature Reviews Molecular Cell Biology* 25: 13–33. https://doi.org/10.1038/s41580-023-00645-4.

Jaime, L. and Santoyo, S. (2021). The health benefits of the bioactive compounds in foods. *Food* 10: 325. https://doi.org/10.3390/foods10020325.

Kim, J.H., Yi, Y.S., Kim, M.Y., and Cho, J.Y. (2017). Role of ginsenosides, the main active components of Panax ginseng, in inflammatory responses and diseases. *Journal of Ginseng Research* 41 (4): 435–443. https://doi.org/10.1016/j.jgr.2016.08.004.

Kong, L., Luo, C., Li, X. et al. (2013). The anti-inflammatory effect of kaempferol on early atherosclerosis in high cholesterol fed rabbits. *Lipids in Health and Disease* 12: 115. https://doi.org/10.1186/1476-511X-12-115.

Kotha, R.R. and Luthria, D.L. (2019). Curcumin: biological, pharmaceutical, nutraceutical, and analytical aspects. *Molecules* 24: 2930. https://doi.org/10.3390/molecules24162930.

Kumar, A. (2017). Peer review #3 of "Phytochemical-rich medicinal plant extracts suppress bacterial antigens-induced inflammation in human tonsil epithelial cells (v0.2)." *Peer J.* https://doi.org/10.7287/peerj.3469v0.2/reviews/3.

Liu, S., Liu, J., He, L. et al. (2022). A comprehensive review on the benefits and problems of curcumin with respect to human health. *Molecules* 27 (14): 4400. https://doi.org/10.3390/molecules27144400.

Maqsood, S., Adiamo, O., Ahmad, M., and Mudgil, P. (2020). Bioactive compounds from date fruit and seed as potential nutraceutical and functional food ingredients. *Food Chemistry* 308: https://doi.org/10.1016/j.foodchem.2019.125522.

Musial, C., Kuban-Jankowska, A., and Gorska-Ponikowska, M. (2020). Beneficial properties of green tea catechins. *International Journal of Molecular Sciences* 21 (5): 1744. https://doi.org/10.3390/ijms21051744.

National Center for Biotechnology Information. (2024). PubChem compoaund summary for CID 173183, campesterol. https://pubchem.ncbi.nlm.nih.gov/compound/Campesterol (Accessed 18 January 2024).

Nwosu, O. and Ubaoji, K. (2020). *Nutraceuticals: History, Classification and Market Demand*. Springer. https://doi.org/10.1007/978-3-030-42319-3_2.

de Oliveira, F.L., Arruda, T.Y.P., da Silva Lima, R. et al. (2020). Pomegranate as a natural source of phenolic antioxidants: a review. *Journal of Food Bioactives* 9: 10–22. https://doi.org/10.31665/JFB.2020.9214.

Pandey, K.B. and Rizvi, S.I. (2009). Plant polyphenols as dietary antioxidants in human health and disease. *Oxidative Medicine and Cellular Longevity* 2 (5): 270–278. https://doi.org/10.4161/oxim.2.5.9498.

Panknin, T.M., Howe, C.L., Hauer, M. et al. (2023). Curcumin supplementation and human disease: a scoping review of clinical trials. *International Journal of Molecular Sciences* 24 (5): 4476. https://doi.org/10.3390/ijms24054476.

Pati, S., Irfan, W., Jameel, A. et al. (2023). Obesity and cancer: a current overview of epidemiology, pathogenesis, outcomes, and management. *Cancers* 15 (2): 485. https://doi.org/10.3390/cancers15020485.

Peña-Jorquera, H., Cid-Jofré, V., Landaeta-Díaz, L. et al. (2023). Plant-based nutrition: exploring health benefits for atherosclerosis, chronic diseases, and metabolic syndrome—a comprehensive review. *Nutrients* 15 (15): 3244. https://doi.org/10.3390/nu15143244.

Rodriguez-Leyva, D., Dupasquier, C.M., McCullough, R., and Pierce, G.N. (2010). The cardiovascular effects of flaxseed and its omega-3 fatty acid, alpha-linolenic acid. *Canadian Journal of Cardiology* 26 (9): 489–496. https://doi.org/10.1016/s0828-282x(10)70455-4.

Sachdeva, V., Roy, A., and Bharadvaja, N. (2020). Current prospects of nutraceuticals: a review. *Current Pharmaceutical Biotechnology* 21 (10): 884–896. https://doi.org/10.2174/1389201021666200130113441.

Sapian, S., Taib, I.S., Katas, H. et al. (2022). The role of anthocyanin in modulating diabetic cardiovascular disease and its potential to be developed as a nutraceutical. *Pharmaceuticals (Basel)* 15 (11): 1344. https://doi.org/10.3390/ph15111344.

Satija, A. and Hu, F.B. (2018). Plant-based diets and cardiovascular health. *Trends in Cardiovascular Medicine* 28 (7): 437–441. https://doi.org/10.1016/j.tcm.2018.02.004.

Singh, A.P., Singh, R., Verma, S.S. et al. (2019). Health benefits of resveratrol: evidence from clinical studies. *Medicinal Research Reviews* 39 (5): 1851–1891. https://doi.org/10.1002/med.21565.

Soumya, N.P., Mini, S., Sivan, S., and Mondal, S. (2021). Bioactive compounds in functional foods and their role as therapeutics. *Bioactive Compounds in Health and Disease* 4 (3): 24–39. https://doi.org/10.31989/bchd.v4i3.786.

Story, E.N., Kopec, R.E., Schwartz, S.J., and Harris, G.K. (2010). An update on the health effects of tomato lycopene. *Annual Review of Food Science and Technology* 1: 189–210. https://doi.org/10.1146/annurev.food.102308.124120.

Subramoniam, A. (2014). Present scenario, challenges and future perspectives in plant-based medicine development. *Annals of Phytomedicine* 3 (1): 31–36.

Tavakkoli, A., Mahdian, V., Razavi, B.M., and Hosseinzadeh, H. (2017). Review on clinical trials of black seed (*Nigella sativa*) and its active constituent, thymoquinone. *Journal of Pharmacopuncture* 20 (3): 179–193. https://doi.org/10.3831/KPI.2017.20.021.

Ullah, A., Munir, S., Badshah, S.L. et al. (2020). Important flavonoids and their role as a therapeutic agent. *Molecules* 25 (22): 5243. https://doi.org/10.3390/molecules25225243.

Upadhyay, S.K. and Singh, S.P. (Éds.) (2021). *Bioprospecting of Plant Biodiversity for Industrial Molecules*. Wiley. https://doi.org/10.1002/9781119718017.

Upadhyay, S.K. and Singh, S.P. (Éds.) (2023). *Plants as Bioreactors for Industrial Molecules*. Wiley. https://doi.org/10.1002/9781119875116.

Visioli, F., Franco, M., Toledo, E. et al. (2018). Olive oil and prevention of chronic diseases: summary of an international conference. *Nutrition, Metabolism, and Cardiovascular Diseases* 28 (7): 649–656.

Williamson, E.M., Liu, X., and Izzo, A.A. (2019). Trends in use, pharmacology, and clinical applications of emerging herbal nutraceuticals. *British Journal of Pharmacology* 177 (6): 1227–1240. https://doi.org/10.1111/bph.14943.

Yu, S.E., Mwesige, B., Yi, Y.S., and Yoo, B.C. (2019). Ginsenosides: the need to move forward from bench to clinical trials. *Journal of Ginseng Research* 43 (3): 361–367. https://doi.org/10.1016/j.jgr.2018.09.001.

Zheng, J., Zheng, S., Feng, Q. et al. (2017). Dietary capsaicin and its anti-obesity potency: from mechanism to clinical implications. *Bioscience Reports* 37 (3): BSR20170286. https://doi.org/10.1042/BSR20170286.

4

Applications of Spices in Nutraceuticals

Ashwini Prabhu

Division of Cancer Research and Therapeutics (CaRT), Yenepoya Research Centre, Yenepoya (Deemed to be University), Mangalore, Karnataka, India

4.1 Introduction

Indian cuisine is well-known worldwide for its unique flavor, color, and aroma, which accounts for the variety of spices and condiments found there. India currently leads the world's major spice-producing nations and provides 70% of the world's total spice production. Spices were used in many different ways in antiquity and have a long traditional history. In the past, spices were more often employed for rendering health benefits than for cooking (Sachan et al. 2018; Shukla and Yadav 2018; Upadhyay and Singh 2021, 2023).

Due to their diverse origins, spices are known around the world for their use and flavor. Spices are well distributed across the globe throughout various countries, which include the Asian subcontinent, Mediterranean countries, and South Europe. Many nations currently grow spices, including India, China, Nepal, Brazil, Pakistan, Indonesia, Germany, Turkey, Vietnam, Madagascar, Bangladesh, Japan, Sri Lanka, Ethiopia, and Guatemala. The Indian subcontinent is referred to as "land of spices" since it produces the most spices of all of these nations, accounting for 70% of global production and keeping a sizable portion of its supply for domestic use (Jaffee 2005; Lal 2018; Sahu and Mishra 2013).

Spices are utilized in traditional medicinal systems in addition to being traditionally eaten as cuisine adjuncts to enhance the sensory aspects of food ingredients. Spices, since times immemorial, are known to provide long-term health benefits. They serve as a major source of bioactives by acting as functional foods. Spices help lower daily sodium intake and achieve a healthy fat intake by adding flavor to food without adding salt or fat. Numerous health benefits of these popular dietary adjuncts have been experimentally shown in recent years through a number of animal studies and clinical trials. The bioactive chemicals found in spices possess antioxidant qualities and hence, they are essential in preventing oxidative stress, which is a factor in the development of diseases like cancer, neurological diseases, cardiovascular disease, and inflammatory diseases. Therefore, spices should be seen as an essential and natural part of our daily diet, in addition to their function as food flavor enhancers (Dixit et al. 2021; Srinivasan 2011).

Herbal Nutraceuticals: Products and Processes, First Edition. Edited by Santosh Kumar Upadhyay and Sudhir Pratap Singh.

4.2 Benefits of Spices as Nutraceuticals

Over the past three decades, numerous positive health impacts of spices have been empirically proven, indicating their possible uses as food adjuncts beyond imparting flavor and taste. Research conducted in preclinical scenarios, including cell lines and animal studies, as well as a few clinical trials, have demonstrated the potential health advantages of spices (Srinivasan 2011). Spices are known to be palatability enhancers, rather than contributing to nutritional aspects. The major pharmacological effects of spices are described below (Table 4.1).

4.2.1 Anticancer Potential of Spices

Numerous studies have indicated the potential of spices to possess anti-inflammatory, antioxidant, and immunomodulatory properties. Many cancers, including those of the stomach, prostate, cervix, lung, breast, liver, and cervix, can be prevented and treated with the help of these pharmacological effects of spices. *Curcuma longa* (turmeric), *Nigella sativa* (black cumin), *Zingiber officinale* (ginger), *Piper nigrum* (black pepper), *Capsicum annum* (chili pepper), and *Crocus sativus* (saffron) are among the various spices that contain significant functional components, such as thymoquinone, piperine, capsaicin, and curcumin. Inducing programmed cell death, preventing tumor cell proliferation, invasion, and expatriation, and rendering tumors highly susceptible to radiotherapy and chemotherapy are the primary modes of action of these bioactive compounds from spices.

Curcumin, extracted from turmeric rhizomes, is a secondary metabolite containing polyphenolic molecules. It modulates several molecular regulators to demonstrate a variety of pharmacological activities, including anticancer efficacy (Staege et al. 2013). The extracellular regulated protein kinase (ERK)-1/2 signaling pathway in mice xenografts was affected by curcumin treatment, in a translational study, which could prevent the proliferation of nasopharyngeal carcinoma (NPC) cells (Xie et al. 2014). Curcumin had inhibitory effects on NPC in another study by upregulating the tumor protein 53 (TP53) gene expression and suppressing miR-125a-5p expression (Gao et al. 2014). Treatment of human lung cancer cells with curcumin was linked to promote DNA damage and suppress the activity of proteins linked to DNA damage repair pathways, including 14-3-3 protein σ, BRCA1, mediator of DNA damage checkpoint 1 (MDC1), and O6-methylguanine-DNA methyltransferase (MGMT) (Ting et al. 2015). In small-cell lung cancer, curcumin promoted an enhancement in the expression of Bax while decreasing the expression of B-cell lymphoma-xL (Bcl-xL) and B-cell lymphoma-2 (Bcl-2). This led to an elevation of intracellular reactive oxygen species (ROS) and apoptosis. Extensive activation of Caspase-3 and caspase-9 was noted (Yang et al. 2012). In squamous cell lung carcinoma (SCC) tumor cell lines, a study reported the inhibition of protein inhibitor of activated STAT3 (PIAS3), which served as an endogenous inhibitor of STAT3. Curcumin therapy enhanced the expression of endogenous PIAS3, while significantly reducing cell growth and survival (Abbas et al. 2015). Curcumin significantly reduced the viability and boosted apoptotic cell death in nonsmall-cell lung carcinoma cells by reducing ERK1/2 activity, p65 expression, and COX-2. It also inhibited the development of orthotopic human NSCLC xenografts and enhanced the lifespan of athymic mice administered with curcumin (Lev-Ari et al. 2014; Ranjan et al. 2016). It also inhibited the JAK2 activity. Consequently, curcumin effectively suppressed the development of lung carcinomas in xenografted mouse models (Wu et al. 2015).

Triple-negative breast cancer is indicated by a poor prognosis and is an aggressive subtype of the disease. Curcumin effectively inhibited the multiplication of MDA-MB 231 triple-negative breast cancer cells, by blocking the epidermal growth factor receptor (EGFR) signaling pathway

Table 4.1 Commonly used spices and their nutraceutical effects.

Spice	Nutraceutical phytocomponents	Culinary applications	Pharmacological activities	References
Nutmeg	Elemicin, Myristicin, Sabinene, Eugenol, Caryophyllene, β-myrcene, and α-pinene	It is used in powdered form for masala and garnishing. It is utilized in the preparation of shampoos, fragrances, and soaps.	Used in the treatment ofrheumatism, cholera, hemorrhoids, psychosis, nausea, anxiety, and stomach cramps	Gupta and Rajpurohit (2011) and Rabade et al. (2022)
Turmeric	Curcumin, Alpha curcumene, Curcumol, Curdione, Dehydrocurdione, Zingiberene	It is beneficial to impart flavor and aroma to salad dressings, pickles, smoothies, and vegetable curries.	Serves as anticancer, anti-inflammatory, antioxidant, antimutagenic, immunostimulant, and hepatoprotective	Alok et al. (2015)
Fennel	Kampferol-3-O-glucuronide, isoquercetin, isorhamnetin-3-O-glucoside, kaempferol-3-Orutinoside, kaempferol-3-O-glucoside, limonene and quercetin-3-O-glucuronide	Fennel seeds are steeped into teas, pickles, alcoholic beverages, and sweet and savory meals.	Helps with digestion and relief of colic pain, used in the treatment of bronchitis, hyperglycemia, renal calculi, chronic fever, and persistent cough; possesses antimicrobial, anti-inflammatory, antioxidant, cardioprotective and anti-hyperlipidemic properties	Dheebisha and Vishwanath (2020), Noreen et al. (2023), Olaru and Popa (2019), and Riska (2020)
Ginger	Gingerol, Zingerone, Quercetin, Shogaol, β-bisabolene, α-curcumene, Zingiberene, α-farnesene, β-sesquiphellandrene, Gingerenone-A	Most of the times, ginger is used in sweet dishes like ginger ale, gingerbread, ginger snaps, parkin, a classic English cake, and speculaas, which are spiced shortbread biscuits from the Netherlands. Ginger tea is considered a tasty beverage in Asia.	Possesses antioxidant, antimicrobial, anti-inflammatory, and Anticancer properties	Yeh et al. (2014), Sharma et al. (2022), and Mao et al. (2019)
Carom	Para-cymene, γ-terpenine, α- and β-pinenes, Dipentene, α-terpinene	The tadkas of pickles, curries, and dal are enhanced with carom seeds.	Exerts antimicrobial, digestive stimulant, antihypertensive, diuretic, anti-inflammatory, gastroprotective, antiseptic and ameliorative. Also possesses colic pain relieving properties	Ahsan et al.(1990), Anilakumar et al. (2009), Bonjar (2004), Choudhury et al. (1998), Dubey et al. (2013), and Ramaswamy et al. (2010)

(*Continued*)

Table 4.1 (Continued)

Spice	Nutraceutical phytocomponents	Culinary applications	Pharmacological activities	References
Pepper	Piperine, Nerolidol,Piperamide, β-caryophyllene, Limonene, β-pinene, Copaene, Napthalenol, β-myrcene	It is a common component of salad dressings, rubs, and peppercorn blends.	Possesses antihypertensive, antiasthmatic, antipyretic, anti-inflammatory, antiapoptotic, antispasmodic, and antioxidant activities	Aziz et al. (2012), Gu et al. (2018), Manoharan et al. (2009), Saleem et al. (2022), and Taqvi et al. (2008)
Tulsi	Eugenol, Myretenal, Rosmarinic acid,Luteolin, Apigenin, β-sitosterol, Carnosic acid	Basil leaves are utilized as garnishing material for salads, meat, cheese, egg dishes,and pesto. Decoction of the leaves is made in the form of tea as well.	Exerts antimicrobial, hepatoprotective, antioxidant, anti-inflammatory, chemopreventive, cardioprotective, antidiabetic, antiulcer, memory enhancement, and anticoagulant activities	Mahajan et al. (2013), Mohan et al. (2011), Mondal et al. (2009), and Pattanayak et al. (2010)
Fenugreek	Choline, Gentianine, Carpaine, Trigonelline, Apigenin, Orientin, Luteolin, Quercetin, Vitexin, Isovitexin, n-alkanes and Sesquiterpenes	Fenugreek leaves are added to soups, sauces, curries, and vegetable dishes, both fresh and dried. Fenugreek seeds are used in ground form or left whole to prepare spice mixtures like panch phoran (Indian five spice) and garam masala, or they can be used as dry rubs for meat.	Exerts hypoglycemic, hypocholesterolemic, immunomodulatory, anticancer, and neuroprotective efficacy	Abdelwahab et al. (2021), Alrumaihi et al. (2021), Alsemari et al. (2014), Foltynie and Kahan (2013), Heshmat-Ghahdarijani et al. (2020), Manjunatha et al. (2020), Singh and Garg (2006), and Wani and Kumar (2018)
Rosemary	Carnosic acid, Carnosol,Rosmarinic acid, Ursolic acid and Betulinic acid	A wide range of foods, including soups, casseroles, salads, and stews, employ rosemary as a spice.	Possesses antidepressant, memory-boosting, neuroprotective, antihyperglycemic, and antihyperalgesic properties	Balu and Lucki (2009), Borrás-Linares et al. (2014), Farr et al. (2016), Gonul et al. (2005), Nemeroff (2002),Park et al. (2010), and Rasoulian et al. (2019)
Asafoetida	Umbelliprenin, Tadshiferin, Galbanic acid, Conferol, Gummosin, Asafoetidin, Polyanthinin, Kamolonol, Vanillin, Picealactone C, Taraxacin, β-Sitosterol, Falcarinolone	It is a basic component in Indian cookery and is frequently combined with turmeric in a range of vegetable dishes and lentil meals like dal.	Exerts antiulcer, antiasthma, and antihypertensive activity	Amalraj and Gopi (2017) and Singh et al. (2022)

Table 4.1 (Continued)

Spice	Nutraceutical phytocomponents	Culinary applications	Pharmacological activities	References
Garlic	Allin, Allicin, Diallyl disulfide, Diallyl trisulfide, Thiacremanone, Ajoene, Lectins		Possesses anticancer, antidiabetic, antiobesity, antithrombotic, antimicrobial, pesticidal and anti-inflammatory activities	Bhatwalkar et al. (2021), Faroughi et al. (2018), Kwak et al. (2014), Upadhyay et al. (2010a,b, 2011, 2012), and Okoro et al. (2023)
Coriander	Ferulic acid, Gallic acid,Caffeic acid, Esculetin, Esculin, Scopoletin, 4-Hydroxycoumarin, Umbelliferone, Dicoumarin, Hyperoside, Rutin, Hesperidin, Vicenin, Diosmin, Luteolin, Apigenin, Orientine, Dihydroquercetin, Catechin, and Arbutin	Asian, Oriental, and Middle Eastern cuisine frequently uses fresh, zesty coriander leaves as a garnish for rice dishes, curries, soups, and stews. Coriander seeds when crushed or ground into spicy curries, or when roasted, give a meaty depth to meat rubs and marinades.	Exerts antidiabetic, anticancer, antioxidant, and antihypercholesterolemic effects	Kousar et al. (2011), Mahleyuddin et al. (2022), and Oganesyan et al. (2007)

(Sun et al. 2012). In addition, 5-fluorouracil (5-FU and) curcumin together shielded surrounding normal cells from undergoing apoptosis and allowed for greater doses or prolonged therapeutic durations of 5-FU, enhancing the antimetabolite's cytotoxic side effects during chemotherapy (Ferguson and Orlando 2015). By altering the regulation of m1/m2 macrophages in the tumor milieu, curcumin was shown to prevent metastatic breast cancer in mice (Shiri et al. 2015). Curcumin-induced downregulation of platelet endothelial cell adhesion molecule-1 (PECAM-1), p65, and cyclin D1 is associated with the prevention of tumor development and neovascularization in a preclinical breast cancer system (Bimonte et al. 2015). Breast cancer resistance protein was inhibited *in vivo* by curcumin (Kusuhara et al. 2012). Curcumin was prescribed in a clinical trial to treat metastatic breast cancer at a dose of 6000 mg/day for seven days every three weeks in addition to docetaxel (Bayet-Robert et al. 2010).

Curcumin inhibited human endometrial cancer cell growth and proliferation by downregulating expression of AR in these cells, facilitated via the signaling pathway of Wnt (Feng et al. 2014). Cervical cancer cells with HPV positivity were specifically eradicated, and expressions of EGFR and transformation of antigen E6 were suppressed, with the concurrent stimulation of p53 following treatment with a cervical cream containing curcumin as an active ingredient. The study demonstrated that applying a vaginal lotion containing curcumin to the epithelial layer of the vagina of healthy mice eliminated HPV+ cancer cells while leaving noncancerous tissue unaffected (Debata et al. 2013).

It has been demonstrated that the main bioactive component extracted from *N. sativa* black seeds, thymoquinone, exhibits antitumor properties against a variety of neoplasms (Attoub et al. 2013). It was found that extracts and oil isolated from *N. sativa* seeds decreased the growth and proliferation of carcinoma cells from human lungs, along with a change in their morphology, and this trend was found to be dose-dependent (Al-Sheddi et al. 2014). Thymoquinone and cisplatin in combination were found to be safe with a high tolerance and significantly decreased volume and weight of tumor mass in a mouse xenograft model without any toxicity to the vital organs (Jafri et al. 2010).

Both in cell line and animal preclinical models, thymoquinone combined with gemcitabine synergistically inhibited the growth of tumors and increased apoptosis in pancreatic cancer cells by modulating multiple molecular signaling targets, including downregulating phosphatase and tensin homolog deleted on chromosome ten (PTEN), upregulating Notch1 and Notch intracellular domain (NICD), and inactivating Akt/mTOR/S6 signaling pathways. The combination treatment boosted the release of cytochrome c and downregulated the expression of pro-apoptotic molecules including Bax, caspase-3, and caspase-9 as well as antiapoptotic proteins like XIAP, Bcl-xL, and Bcl-2. Pretreatment with thymoquinone after administration of gemcitabine led to a synergistic increase in the suppression of tumor mass and apoptosis in both orthotopic and xenograft *in vivo* models via injecting PANC-1 cells and pancreatic cancer cells *in vitro* (Mu et al. 2015).

Thymoquinone prevented the oxidative stress induced by the administration of 1,2-dimethylhydrazine (DMH) at its onset and prevented tumor formation both before and after treatment. Furthermore, thymoquinone promoted cell death in human colon carcinoma cells by inhibiting Janus kinase (JAK) 2 and Src-mediated activation of EGFR tyrosine kinase, which in turn hindered STAT3 signaling. Additionally, in a colon cancer xenograft model using HCT116 cells, thymoquinone significantly prevented the proliferation of cancer cells (Gali-Muhtasib et al. 2008).

Widely used as a spice in food and drink, ginger (*Z. officinale*) is high in a number of phenolic compounds which are biologically active, including some aromatic spicy nonvolatile components such as shogaols, paradols, and gingerols (Karna et al. 2012). In women affected with breast cancer and obesity, ginger supplementation was observed to lower Malondialdehyde (MDA) and raise adiponectin, NO, and GPx (Karimi and Roshan 2013). The effects exerted by ginger on colorectal carcinomas were studied in a number of clinical trials. In a pilot investigation, ginger promoted apoptosis and differentiation and decreased proliferation in the colonic epithelium with a normal healthy appearance in individuals at increased risk of colorectal cancer. In another study, individuals at risk of colorectal cancer showed a considerable drop in the levels of arachidonic acid following administration of medication containing ginger. Ginger reduced COX, lowering PGE-2 concentrations as well as the incidence of adenomas (Citronberg et al. 2013; Zick et al. 2015).

Natural organosulfur compounds, isolated from garlic, such as S-allyl mercaptocysteine, allicin, diallyl tetrasulfide, diallyl trisulfide, diallyl disulfide, and diallyl sulfide are widely recognized for their antioxidant and chemo-sensitization qualities (Czepukojc et al. 2014). In human breast cancer cells, diallyl trisulfide induced adecrease in the function of estrogen receptor-α (ER-α). When diallyl trisulfide was added to MCF-7 and T47D cells, the ER-α protein was downregulated and its nuclear levels decreased. However, this effect did not occur when 17β-estradiol was present (Hahm and Singh 2014). Garlic cloves were used to isolate N-benzyl-N-methyldecan-1-amine (NBNMA). NBNMA promoted arrest of cell cycle at G2/M phase and cell death in U937 cells. The expression of the Cdk inhibitor p21WAF1/CIP1 was upregulated whereas the expression of CDC2 and cyclin-dependent kinase (Cdk) 2, the regulator genes of G2/M phase progression, was repressed (Jeong et al. 2014). A substance derived from garlic called z-ajoene was thought to be a viable treatment option for glioblastoma multiforme due to its effects on the cancer stem cells (Jung et al. 2014).

Ocimum sanctum (OS), often known as holy basil or tulsi, is an aromatic herb belonging to the Lamiaceae family. It is native to the old-world tropical regions and is widely grown as both an escaped weed and a cultivated plant. About 71% of the volatile oil in the leaf of OS is composed of eugenol and methyl eugenol content is found to be 20%. Sesquiterpine hydrocarbon caryophyllene and carvacrol are also present in the oil (Indian Drug Manufacturers Association 2002). It was inferred that the extract of OS leaves prepared in alcohol modulates the metabolism of carcinogens by acting on the enzymes involved, including cytochrome B5, cytochrome P-450, glutathione-S-transferase (GST), and aryl hydrocarbon hydroxylase. These enzymes are critical in the elimination of mutagens and carcinogens from the body (Madhuri and Pandey 2010).

As saffron contains bioactive chemicals including crocin and crocetin, it has been shown to have anticancer properties. In adenocarcinoma cells of the stomach, crocetin promoted apoptosis, increased the Bax expression and alleviated the expression of Bcl-2, both *in vitro* and in the rat model of stomach cancer. Additionally, lactate dehydrogenase and serum antioxidant activity alterations brought on by 1-methyl-3-nitro-1-nitrosoguanidine were reversed by crocetin (Bathaie et al. 2013). In highly aggressive prostate cancer cell lines, cellular growth and proliferation were inhibited by saffron extract and its primary component, crocin. Crocin and saffron extract elevated the expression of Bax while decreasing the expression of Bcl-2 (D'Alessandro et al. 2013). Crocetin significantly reduced both the invasion and proliferation of MDA-MB-231 cells, a highly invasive cell line. Pro-MMP-2/MMP-2 protein levels and Pro-MMP-9 activity were also considerably reduced by crocetin, which also strongly repressed pro-MT2-MMP and pro-MT1-MMP gene and protein expression (Chryssanthi et al. 2011). K-562 human chronic myeloid leukemia cells showed cytotoxic effects that were regulated by crocin and safranal (Geromichalos et al. 2014).

By initiating caspase-3 and PARP cleavage, piperine from black pepper suppressed growth and triggered apoptosis. Piperine inhibited the production of MMP-9 produced by EGF via interfering with p38 MAPK and ERK1/2, which decreased the cell migration. Piperine increased the sensitivity towards paclitaxel treatment in human EGFR (HER-2)-overexpressing breast cancer cells (Do et al. 2013). Piperine suppressed the proliferation of prostate cancer cells PC-3, LNCaP, DU-145, and 22RV1. When prostate tumor cells were xenotransplanted into nude mouse models, the development of both androgen-independent and -dependent tumors was significantly inhibited by piperine administration (Samykutty et al. 2013). Piperine had a cytostatic effect on HRT-18 human rectal cancer cells by inhibiting their metabolic activity. Piperine promotes the arrest of cell cycle and induces cell death (Yaffe et al. 2013).

The most fragrant component of red chili peppers, capsaicin, has a strong anticancer effect on a variety of human cancers. Capsaicin specifically inhibited angiogenesis in nonsmall-cell lung cancer cells, wherein it downregulated the vascular endothelial growth factor (VEGF) expression. Capsaicin demonstrated anti-proliferative effects on human small-cell lung cancer in preclinical models (Brown et al. 2010; Chakraborty et al. 2014). Capsaicin reduced phosphorylated ERK1 and 2, cleaved caspase-3, Bcl-2, p38 MAPK, or JNK expression, and prevented the growth of human gastric cancer cells (AGS cells) by inducing programmed cell death (Park et al. 2014). Using a transgenic mouse model of prostate adenocarcinoma, the anticancer efficacy of capsaicin against prostate carcinoma cells was studied. Comparing the capsaicin-treated mice to the control group, the former displayed a tendency toward lesser disease burden with highly developed adenocarcinoma. *In vitro*, PC3 cells' ability to invade and migrate was inhibited, and the metastatic rate was decreased (Venier et al. 2015).

Chinese traditional medicine has employed clove (*Syzygium aromaticum*), an unopened flower bud, which is sun-dried, of the plant, as a popular spice. Clove and a few other spices including bay leaves, cinnamon, and basil all contain a significant amount of eugenol. Clove extract contains oleanolic acid, which is also linked to its anticancer properties. Eugenol increased the levels of polyamines and glycolytic metabolites and decreased oxidative stress and adenosine triphosphate (ATP) utilization in oral squamous cell carcinoma cells (Koh et al. 2013). Eugenol demonstrated anti-inflammatory and proapoptotic effects in human cervical carcinoma cells (Hussain et al. 2011).

One of the Apiaceae family's common culinary spices and medicinal herbs is coriander (*Coriandrum sativum*). The coriander root ethyl acetate extract prevented DNA damage, reduced MCF-7 cell migration brought on by H_2O_2 and had antiproliferative effects on MCF-7 cells.

Furthermore, one of the key ingredients in coriander that contributes to its anticancer properties is linalool. Linalool had a pro-oxidant effect on tumor tissue and suppressed tumor growth *in vivo* (Jana et al. 2014; Tang et al. 2013).

4.2.2 Spices and Antidiabetic Activity

Aqueous extracts of *Z. officinale* rhizomes were used to test the antihyperglycemic effect on Sprague Dawley rats, which is triggered by streptozotocin. Aqueous extract at a dose of 60 mg/kg demonstrated inhibitory effects on hyperglycemia along with weight loss. *Z. officinale* was studied for its antihyperglycemic effects in alloxan-induced and insulin-resistant rats. According to this study, ginger improved insulin sensitivity and decreased blood sugar in both diabetic rats (Hassan et al. 2016; Iranloye et al. 2011).

A popular seasoning all around the world, garlic is mostly found in India, Sri Lanka, and Northeastern Iran in Central Asia. The alpha-amylase inhibitory activity of *Allium sativum* bulb aqueous extracts was found to have increased dormant antidiabetic potential *in vitro*. This inhibition aids in reducing blood glucose release. Bulb extract of *A. sativum* was used to achieve the antihyperglycemic effect. Both the lipid profile and blood sugar level were found to be declining. Therefore, the treatment of diabetes mellitus with garlic bulbs is quite effective (Andleeb et al. 2020; Sureshkumar et al. 2019).

Cumin is an abundantly available spice in India and the Middle East. The antidiabetic potential of ethanolic extracts from cumin seeds has been demonstrated in both normoglycemic rats and type-2 diabetes mellitus models induced using streptozotocin. Tests on oral glucose tolerance, glucose absorption, alpha-glucosidase, and aldose reductase activity were performed. The findings from this investigation inferred that cumin seeds may be utilized for improving the lipid profile and reducing the hyperglycemia (Mohamed et al. 2018).

C. longa rhizome's antidiabetic efficacy was investigated in both normoglycemic and hypoglycemic rats administered with alloxan. In diabetic rats treatment with 200 mg/kg concentration showed a considerable reduction in cholesterol, Low-density lipoprotein (LDL), and blood glucose levels and a rise in high-density lipoprotein (HDL) and total protein. Therefore, *C. longa* aqueous extract is useful in the treatment of diabetes mellitus. *C. longa* root methanolic extracts exhibited superior hypoglycemic efficacy in diabetic rats compared to aqueous extracts (Mohammed et al. 2017; Olatunde et al. 2014).

Essential oils from fennel seeds had significant hypoglycemic effects in rats with diabetes induced by streptozotocin. Consumption of fennel oil reduced hyperglycemia and assisted in the regeneration of diabetic mellitus-related damaged tissues. The study investigated the *in vitro* hyperglycemic activity of *Foeniculum vulgare* seeds using three different solvents: benzene, n-butanol, and ethyl acetate. Benzene extracts from the fennel seeds inhibited the enzymatic activities of α-glucosidase and α-amylase (El-Soud et al. 2011; Godavari et al. 2018).

Aqueous extracts of *Allium cepa* were evaluated for their hypoglycemic potential in diabetic rats induced by alloxan. Serum cholesterol, serum lipids, and blood glucose levels were decreased in onion extract treated rats (Jevas 2011).

Clove is primarily from the Indonesian Maluku islands, but it may also be found in tropical regions of Asia, Africa, and the ocean. Alpha-amylase enzyme activity was significantly inhibited by the essential oils of *S. aromaticum* at 100 μg/ml concentration, as indicated by the results of *in vitro* tests. Thus, by blocking the activity of certain enzymes, the essential oils derived from cloves can effectively treat diabetes (Tahir et al. 2015)

Cinnamon, also known as *Cinnamomum zeylanicum*, is commonly found in South Asia, Sri Lanka, and India. Studies conducted with bark extracts on α-amylase and α-glucosidase inhibitory assays have demonstrated the remarkable efficacy of cinnamon in treating diabetes by inhibiting the activities of both enzymes. In alloxan-induced rats, 200 mg/kg of *Cinnamomum verum* bark extracts showed latent action with a decrease in Triglycerides (TG), blood glucose levels, LDL, and total cholesterol. In addition, an improvement in body weight gain (El-Dosoky et al. 2012; Gulcin et al. 2019).

4.2.3 Spices in the Prevention of Cardiovascular Disorders

Spices do serve as the potential providers of antioxidants. They also act as powerful inhibitors of inflammation and tissue damage induced by elevated blood sugar and circulating lipid levels. Spices are a good source of free-radical scavengers and other potentially bioactive elements for the diet because of their low-calorie content and affordability (Muthulakshmi et al. 2009). Spices are considered to be significant players in the management of heart disorders since they can be used as preventive and therapeutic agents.

In an obesity model consisting of natural mutants, at four weeks, diet incorporated with turmeric at concentrations of 1% and 5%, significantly increased HDL cholesterol and decreased triglyceride and cholesterol concentrations. Turmeric administration reduces blood glucose, kidney damage, and LDL oxidation in diabetics. Furthermore, studies have shown that it diminishes endothelial dysfunction, cyclooxygenase, thromboxane, platelet aggregation, and smooth muscle cell proliferation. Because of their antioxidant and anti-inflammatory properties, curcumin and turmeric have been shown to be efficient in treating several conditions, such as myocardial infarctions, pancreatitis, inflammatory bowel diseases, neurodegenerative illnesses, lung and hepatic damage, muscle injuries, and cystic fibrosis (Krishnaswamy 2006). Research indicates that curcumin has several different targets for its preventive effects on heart disease. Cardiotoxicity caused by adriamycin can be suppressed by the free-radical scavenging properties of curcumin (Venkatesan 1998). The compound curcumin, found in turmeric, inhibits the transcription factor NF-κB by inhibiting IKK, a kinase required for the activation of NF-kB. It also improves the shooting-up of pro-inflammatory cytokines during cardiopulmonary bypass (CBP) and reduces the apoptosis in cardiomyocytes following reperfusion injury/global cardiac ischemia (Soni and Kuttan 1992).

It has been suggested that garlic helps prevent heart disease (Mykolas 2009). The use of garlic and the advancement of cardiovascular disease are inversely correlated, according to epidemiologic research. Increased levels of total serum cholesterol, elevated LDL and its oxidation, elevated aggregation of platelets,smoking, and hypertension are among the several variables linked to cardiovascular disease (Rahman and Lowe 2006). Allicin, a garlic component, dramatically reduced the formation of aortic sinus fatty streaks (Abramovitz et al. 1999). A study demonstrated that the water-soluble organosulfur molecule S-allyl cysteine, which is found in aged garlic extract, is a strong preventer of cholesterol synthesis and deposition. Garlic lowers blood pressure, lowers cholesterol, prevents platelet aggregation, and boosts antioxidant status. Consumption of garlic (900 mg/day) on average reduced the total blood cholesterol levels by about 9%. In preclinical models (both *in vitro* and *in vivo*), it has been demonstrated that ingestion of garlic has been associated with antiatherosclerotic ability to elevate the levels of HDL, which may aid in the removal of excess cholesterol from artery tissue. It has been shown to have antiplatelet, antioxidant, and fibrinolytic properties in addition to decreasing blood pressure and cholesterol (Ebadi 2002; Tapsell et al. 2006; Yu-Yan and Liu 2001).

As a member of the Lauraceae family, the genus *Cinnamomum* includes cinnamon, which is native to Australia, China, India, Egypt, and Sri Lanka. Bark and leaves of cinnamon are widely utilized as culinary spices and to make essential oils (Jayaprakasha et al. 2003). Without altering the bile composition, dietary cinnamon boosts the amount of phospholipids as well as cholesterol that are secreted by the bile. At a dosage of 250 mg/kg body weight (one-time), a 50% alcoholic extract of cinnamon was found to have a strong antihypercholesterolemic impact and lower the serum triglyceride levels in rats. In another study, hyperlipidemia induced in rats using triton WR-1339 was utilized as a model. Cinnamon stimulated NO levels, a powerful vasodilator, and activated PPARγ, which improves insulin resistance and lowers fasted LDL-c, controlling hyperlipidemia linked to obesity. Due to its possession of higher concentrations of polyphenols, potent reducing capacity, and superoxide radical scavenging activity, cinnamon has shown a considerable ability to suppress the progression of lipid peroxidation (Kim et al. 1999; Srinivasan and Sambaiah 1991; Yadav and Bhatnagar 2007).

Coriander seeds exhibit a noteworthy hypolipidemic effect. In the tissues of the experimental group of rats that received coriander seeds, there was a notable alleviation in the levels of total cholesterol and triglycerides. The experimental group showed notable increases in plasma lecithin cholesterol acyl transferase activity and -hydroxy, -methyl glutaryl CoA reductase. When comparing the experimental group with that of the control group, the amount of LDL + Very-low density lipoprotein (VLDL) cholesterol dropped, and the level of HDL cholesterol increased. Its hypocholesterolemic impact appears to be caused by increased breakdown of cholesterol to neutral sterols and fecal bile acids, which enhanced the activity of plasma Lecithin-cholesterol acyltransferase (LCAT), which promoted hepatic bile acid production (Chithra and Leelamma 1997). Water extracts of curry leaf and coriander leaf were administered to human platelets at different concentrations. Both of these extracts inhibited the aggregation of platelets. A preclinical study showed a considerable drop-down in the levels of Total count (TC), Triacylglycerols (TAG), TG, and low-density lipoprotein cholesterol (LDL-c) in plasma, and a notable elevation in the levels of high-density lipoprotein cholesterol (HDL-c) was found in rats fed with cholesterol-rich (1%) basal diet and subsequently treated with coriander seed oil (Ramadan et al. 2008).

Piperine, an active ingredient of black pepper, was able to prevent oxidative damage by blocking the formation of ROS and free radicals. Black pepper or piperine administration also lowered lipid peroxidation in animal models and promoted the synthesis of antioxidant enzymes, cellular thiols, and antioxidant molecules (Srinivasan 2007). *Piper longum* methanolic extract was found to be effective in offering significant protection against cardiotoxicity induced by the administration of adriamycin, by means of its free-radical scavenging property. Black pepper influences the metabolism of fats, majorly by the fatty acid mobilization. Supplementation of black pepper in rats increased the HDL-c concentrations and brought down the LDL-c concentrations and plasma very-low-density lipoprotein cholesterol (VLDL-c) in comparison with the levels in high fat fed rats without the intervention of any supplementation. Since black pepper is rich in vanadium, it promotes functional recovery of heart health in myocardial infarction and pressure overload–induced hypertrophy (Shenuarin and Fukunaga 2009; Vijayakumar et al. 2002).

4.3 Conclusions

Spices have taken on a significant role in our daily lives. With their flavor, color, taste, and scent, spices enhance the appearance and palatability of food. Spices do possess bioactives that function as nutraceuticals and have been shown to be beneficial for certain medical conditions. Spices such as pepper, ginger, nutmeg, fenugreek, bishop's weed, cumin, star anise, garlic, coriander, and cinnamon have long been used in culinary applications and they do exert health benefits as well. Not only will these spices enhance the aroma, taste, and presentability of our cuisine, but their appropriate use in our day-to-day diet will promote health. The nutraceutical sector has grown at a rate much greater than the food and medicine sectors. The future demand for nutraceuticals will be determined by their medicinal and curative properties. Regulatory toxicologists, health professionals, and nutritionists should strategically collaborate to plan appropriate regulations for nutraceuticals, despite their great potential to promote human health and prevent disease. This will ensure that mankind benefits from these products to the fullest capacity in terms of healthcare applications.

References

Abbas, R., McColl, K.S., Kresak, A. et al. (2015). PIAS3 expression in squamous cell lung cancer is low and predicts overall survival. *Cancer Medicine* 4: 325–332.

Abdelwahab, N.S., Morsi, A., Ahmed, Y.M. et al. (2021). Ecological HPLC method for analyzing an antidiabetic drug in real rat plasma samples and studying the effects of concurrently administered fenugreek extract on its pharmacokinetics. *RSC Advances* 11 (8): 4740–4750.

Abramovitz, D., Gavri, S., Harats, D. et al. (1999). Allicin-induced decrease in formation of fatty streaks (atherosclerosis) in mice fed a cholesterol-rich diet. *Coronary Artery Disease* 10: 515–519.

Ahsan, S.K., Shah, A.H., Tanira, M.O. et al. (1990). Studies on some herbal drugs used against kidney stones in Saudi folk medicine. *Fitoterapia* 61 (5): 435–438.

Alok, A., Singh, I.D., Singh, S. et al. (2015). Curcumin–pharmacological actions and its role in oral submucous fibrosis: a review. *Journal of Clinical and Diagnostic Research* 9 (10): ZE01.

Alrumaihi, F.A., Khan, M.A., Allemailem, K.S. et al. (2021). Methanolic fenugreek seed extract induces p53-dependent mitotic catastrophe in breast cancer cells, leading to apoptosis. *Journal of Inflammation Research* 16: 1511–1535.

Alsemari, A., Alkhodairy, F., Aldakan, A. et al. (2014). The selective cytotoxic anti-cancer properties and proteomic analysis of *Trigonella foenum-graecum*. *BMC Complementary and Alternative Medicine* 14: 1–9.

Al-Sheddi, E.S., Farshori, N.N., Al-Oqail, M.M. et al. (2014). Cytotoxicity of *Nigella sativa* seed oil and extract against human lung cancer cell line. *Asian Pacific Journal of Cancer Prevention* 15: 983–987.

Amalraj, A. and Gopi, S. (2017). Biological activities and medicinal properties of Asafoetida: a review. *Journal of Traditional and Complementary Medicine* 7 (3): 347–359.

Andleeb, S., Tariq, F., Muneer, A. et al. (2020). *In vitro* bactericidal, antidiabetic, cytotoxic, anticoagulant, and hemolytic effect of green-synthesised silver nanoparticles using *Allium sativum* clove extract incubated at various temperatures. *Green Processing and Synthesis* 9: 538–553.

Anilakumar, K.R., Saritha, V., Khanum, F., and Bawa, A.S. (2009). Ameliorative effect of ajwain extract on hexachlorocyclohexane-induced lipid peroxidation in rat liver. *Food and Chemical Toxicology* 47 (2): 279–282.

Attoub, S., Sperandio, O., Raza, H. et al. (2013). Thymoquinone as an anticancer agent: evidence from inhibition of cancer cells viability and invasion *in vitro* and tumor growth *in vivo*. *Fundamentals and Clinical Pharmacology* 27: 557–569.

Aziz, S., Naher, S., Abukawsar, M.D., and Roy, S.K. (2012). Comparative studies on physicochemical properties and GC-MS analysis of essential oil of the two varieties of the black pepper (*Piper nigrum* Linn.). *International Journal of Pharmaceutical and Phytopharmacological Research* 2 (2): 67–70.

Balu, D.T. and Lucki, I. (2009). Adult hippocampal neurogenesis: regulation, functional implications, and contribution to disease pathology. *Neuroscience & Biobehavioral Reviews* 33 (3): 232–252.

Bathaie, S.Z., Hoshyar, R., Miri, H., and Sadeghizadeh, M. (2013). Anticancer effects of crocetin in both human adenocarcinoma gastric cancer cells and rat model of gastric cancer. *Biochemistry and Cell Biology* 91: 397–403.

Bayet-Robert, M., Kwiatkowski, F., Leheurteur, M. et al. (2010). Phase I dose escalation trial of docetaxel plus curcumin in patients with advanced and metastatic breast cancer. *Cancer Biology and Therapy* 9: 8–14.

Bhatwalkar, S.B., Mondal, R., Krishna, S.B. et al. (2021). Antibacterial properties of organosulfur compounds of garlic (*Allium sativum*). *Frontiers in Microbiology* 12: 613077.

Bimonte, S., Barbieri, A., Palma, G. et al. (2015). Dissecting the role of curcumin in tumour growth and angiogenesis in the mouse model of human breast cancer. *Biomed Research International* 20 (15): 1–7.

Bonjar, G.S. (2004). Anti yeast activity of some plants used in traditional herbal-medicine of Iran. *Journal of Biological Sciences* 4: 212–215.

Borrás-Linares, I., Stojanović, Z., Quirantes-Piné, R. et al. (2014). *Rosmarinus officinalis* leaves as a natural source of bioactive compounds. *International Journal of Molecular Sciences* 15 (11): 20585–20606.

Brown, K.C., Witte, T.R., Hardman, W.E. et al. (2010). Capsaicin displays anti-proliferative activity against human small cell lung cancer in cell culture and nude mice models via the E2F pathway. *PLoS ONE* 5: 495.

Chakraborty, S., Adhikary, A., Mazumdar, M. et al. (2014). Capsaicin-induced activation of p53-SMAR1 auto-regulatory loop down-regulates VEGF in non-small cell lung cancer to restrain angiogenesis. *PLoS ONE* 9: 495.

Chithra, V. and Leelamma, S. (1997). Hypolipidemic effect of coriander seeds (*Coriandrum sativum*), mechanism of action. *Plant Foods in Human Nutrition* 51: 167–172.

Choudhury, S., Ahmed, R., Kanjilal, P.B., and Leclercq, P.A. (1998). Composition of the seed oil of *Trachyspermum ammi* (L.) Sprague from Northeast India. *Journal of Essential Oil Research* 10 (5): 588–590.

Chryssanthi, D.G., Dedes, P.G., Karamanos, N.K. et al. (2011). Crocetin inhibits invasiveness of MDA-MB-231 breast cancer cells via downregulation of matrix metalloproteinases. *Planta Medica* 77: 146–151.

Citronberg, J., Bostick, R., Ahearn, T. et al. (2013). Effects of ginger supplementation on cell-cycle biomarkers in the normal-appearing colonic mucosa of patients at increased risk for colorectal cancer: results from a pilot, randomized, and controlled trial. *Cancer Prevention Research* 6: 271–281.

Czepukojc, B., Baltes, A.K., Cerella, C. et al. (2014). Synthetic polysulfane derivatives induce cell cycle arrest and apoptotic cell death in human hematopoietic cancer cells. *Food and Chemical Toxicology* 64: 249–257.

D'Alessandro, A.M., Mancini, A., Lizzi, A.R. et al. (2013). *Crocus sativus* stigma extract and its major constituent crocin possess significant antiproliferative properties against human prostate cancer. *Nutrition and Cancer* 65: 930–942.

Debata, P.R., Castellanos, M.R., Fata, J.E. et al. (2013). A novel curcumin-based vaginal cream Vacurin selectively eliminates apposed human cervical cancer cells. *Gynecologic Oncology* 129: 145–153.

Dheebisha, C. and Vishwanath, Y.C. (2020). Advances in cultivation of fennel. *Journal of Pharmacognosy and Phytochemistry* 9 (2): 1295–1300.

Dixit, S., Shukla, A., Singh, V., Santosh Kumar, Upadhyay (2021) *Bioprospecting of Natural Compounds for Industrial and Medical Applications*. Wiley. https://doi.org/10.1002/9781119718017.ch3

Do, M.T., Kim, H.G., Choi, J.H. et al. (2013). Antitumor efficacy of piperine in the treatment of human HER2-overexpressing breast cancer cells. *Food Chemistry* 141: 2591–2599.

Dubey, S., Maity, S., Singh, M. et al. (2013). Phytochemistry, pharmacology and toxicology of *Spilanthes acmella*: a review. *Advances in Pharmacological and Pharmaceutical Sciences* 2013.

Ebadi, M. (2002). *The Pharmacodynamic Basis of Herbal Medicine*. BocaRaton: CRC Press.

El-Dosoky, G.E., Mourad, A.M., Aboul, S., and Al-numair, K.S. (2012). Antidiabetic and hyperlipidemic effects of Ceylon cinnamon. *Journal of Medicinal Plant Research* 6: 1685–1691.

El-Soud, N.A., El-Laithy, N., El-Saeed, M.S. et al. (2011). Antidiabetic activities of *Foreniculum vulgare* mill. Essential oil in streptozotocin-induced diabetic rats. *Macedonian Journal of Medical Sciences* 4: 139–146.

Faroughi, F., Charandabi, S.M., Javadzadeh, Y., and Mirghafourvand, M. (2018). Effects of garlic pill on blood glucose level in borderline gestational diabetes mellitus: a triple blind, randomized clinical trial. *Iranian Red Crescent Medical Journal* 20 (7).

Farr, S.A., Niehoff, M.L., Ceddia, M.A. et al. (2016). Effect of botanical extracts containing carnosic acid or rosmarinic acid on learning and memory in SAMP8 mice. *Physiology & Behavior* 165: 328–338.

Feng, W., Yang, C.X., Zhang, L. et al. (2014). Curcumin promotes the apoptosis of human endometrial carcinoma cells by downregulating the expression of androgen receptor through Wnt signal pathway. *European Journal of Gynaecological Oncology* 35: 718–723.

Ferguson, J.E. and Orlando, R.A. (2015). Curcumin reduces cytotoxicity of 5-Fluorouracil treatment in human breast cancer cells. *Journal of Medicinal Food* 18: 497–502.

Foltynie, T. and Kahan, J. (2013). Parkinson's disease: an update on pathogenesis and treatment. *Journal of Neurology* 260: 1433–1440.

Gali-Muhtasib, H., Ocker, M., Kuester, D. et al. (2008). Thymoquinone reduces mouse colon tumor cell invasion and inhibits tumor growth in murine colon cancer models. *Journal of Cellular and Molecular Medicine* 12: 330–342.

Gao, W., Chan, J.Y., and Wong, T. (2014). Curcumin exerts inhibitory effects on undifferentiated nasopharyngeal carcinoma by inhibiting the expression ofmiR-125a-5p. *Clinical Science* 127: 571–579.

Geromichalos, G.D., Papadopoulos, T., Sahpazidou, D., and Sinakos, Z. (2014). Safranal, a *Crocus sativusl* constituent suppresses the growth of K-562 cells of chronic myelogenous leukemia: in silico and *in vitro* study. *Food and Chemical Toxicology* 74: 45–50.

Godavari, A., Amudha, K., and Manicka, M.N. (2018). *In-vitro* hypoglycemic effect of *Foeniculum vulgare* mill. Seeds on the carbohydrate hydrolyzing enzymes. α-amylase and α-glucosidase. *International Journal of Pharmaceutical Science Research* 9: 4441–4445.

Gonul, A.S., Akdeniz, F., Taneli, F. et al. (2005). Effect of treatment on serum brain–derived neurotrophic factor levels in depressed patients. *European Archives of Psychiatry and Clinical Neuroscience* 255: 381–386.

Gu, F., Huang, F., Wu, G., and Zhu, H. (2018). Contribution of polyphenol oxidation, chlorophyll and vitamin C degradation to the blackening of *Piper nigrum* L. *Molecules* 23 (2): 370.

Gulcin, I., Kaya, R., Goren, A.C. et al. (2019). Anticholinergic, antidiabetic and antioxidant activities of cinnamon (*Cinnamomum verum*) bark extracts:polyphenol contents analysis by LC-MS/MS. *International Journal of Food Properties* 22: 1511–1526.

Gupta, A.D. and Rajpurohit, D. (2011). Antioxidant and antimicrobial activity of nutmeg (*Myristica fragrans*). In: *Nuts and Seeds in Health and Disease Prevention* (ed. V.R. Preedy, R.R. Watson, and V.B. Patel), 831–839. Academic Press.

Hahm, E.R. and Singh, S.V. (2014). Diallyl trisulfide inhibits estrogen receptor-alpha activity in human breast cancer cells. *Breast Cancer Research and Treatment* 144: 47–57.

Hassan, N.A., Karunakaran, R., Sankar, U.A., and Aye, K.M. (2016). Antidiabetic effect of *Zingiber officinale* on sprague dawley rats. *International Journal of Pharmacogonosy and Phytochemistry* 8: 1940–1943.

Heshmat-Ghahdarijani, K., Mashayekhiasl, N., Amerizadeh, A. et al. (2020). Effect of fenugreek consumption on serum lipid profile: a systematic review and meta-analysis. *Phytotherapy Research* 34 (9): 2230–2245.

Hussain, A., Brahmbhatt, K., Priyani, A. et al. (2011). Eugenol enhances the chemotherapeutic potential of gemcitabine and induces anticarcinogenic and anti-inflammatory activity in human cervical cancer cells. *Cancer Biotherapy and Radiopharmaceuticals* 26: 519–527.

Indian Drug Manufacturers Association (2002). *Indian Herbal Pharmacopoeia*, 272. Mumbai: Indian drug Manufacturers' Association.
Iranloye, B.O., Arikawe, A.P., Rotimi, G., and Sogbade, A.O. (2011). Antidiabetic and anti-oxidant effects of *Zingiber officinale* on alloxan-induced and insulin-resistant diabetic male rats. *Nigerian Journal of Physiological Sciences* 26: 89–96.
Jaffee, S. (2005). Delivering and taking the heat: Indian spices and evolving product and process standards. In: *The International Bank for Reconstruction and Development*, 1–42. The World Bank.
Jafri, S.H., Glass, J., Shi, R. et al. (2010). Thymoquinone and cisplatin as a therapeutic combination in lung cancer: *in vitro* and *in vivo*. *Journal of Experimental and Clinical Cancer Research* 29: 87.
Jana, S., Patra, K., Sarkar, S. et al. (2014). Antitumorigenic potential of linalool is accompanied by modulation of oxidative stress: an *in vivo* study in sarcoma-180 solid tumor model. *Nutrition and Cancer* 66: 835–848.
Jayaprakasha, G.K., Rao, L.J., and Sakaraiah, K.K. (2003). Volatile constituents from *Cinnamomum zeylanicum* fruit stalks and their antioxidant activities. *Journal of Agricultural and Food Chemistry* 51: 4344–4348.
Jeong, J.W., Park, S., Park, C. et al. (2014). *N*-benzyl-*N*-methyldecan-1-amine, a phenylamine derivative isolated from garlic cloves, induces G2/M phase arrest and apoptosis in U937 human leukemia cells. *Oncology Reports* 32: 373–381.
Jevas, C. (2011). Anti-diabetic effects of *Allium cepa* (onions) aqueous extracts on alloxan-induced diabetic *Rattus novergicus*. *Journal of Medicinal Plant Research* 5: 1134–1139.
Jung, Y., Park, H., Zhao, H.Y. et al. (2014). Systemic approaches identify a garlic-derived chemical, Z-ajoene, as a glioblastoma multiforme cancer stem cell-specific targeting agent. *Molecular Cell* 37: 547–553.
Karimi, N. and Roshan, V.D. (2013). Change in adiponectin and oxidative stress after modifiable lifestyle interventions in breast cancer cases. *Asian Pacific Journal of Cancer Prevention* 14: 2845–2850.
Karna, P., Chagani, S., Gundala, S.R. et al. (2012). Benefits of whole ginger extract in prostate cancer. *British Journal of Nutrition* 107: 473–484.
Kim, N.J., Jung, E.A., Kim, D.H., and Lee, S. (1999). Studies on the development of antihyperlipidemic drugs from oriental herbal medicine. *Korean Journal of Pharmacognosy* 30: 368–374.
Koh, T., Murakami, Y., Tanaka, S. et al. (2013). Changes of metabolic profiles in an oral squamous cell carcinoma cell line induced by eugenol. *In Vivo* 27: 233–243.
Kousar, S., Jahan, N., Khalil-ur-Rehman, K.U., and Nosheen, S. (2011). Antilipidemic activity of *Coriandrum sativum*. *Bioscience Research* 8 (1): 8–14.
Krishnaswamy, K. (2006). *Turmeric-The Salt of the Orient is the Spice of Life*. New Delhi: Allied Publishers Pvt. Ltd.
Kusuhara, H., Furuie, H., Inano, A. et al. (2012). Pharmacokinetic interaction study of sulphasalazine in healthy subjects and the impact of curcumin as an *in vivo* inhibitor of BCRP. *British Journal of Pharmacology* 166: 1793–1803.
Kwak, J.S., Kim, J.Y., Paek, J.E. et al. (2014). Garlic powder intake and cardiovascular risk factors: a meta-analysis of randomized controlled clinical trials. *Nutrition Research and Practice* 8 (6): 644.
Lal, G. (2018). Scenario, importance and prospects of seed spices: a review. *Current Investigations in Agriculture and Current Research* 4 (2): 491–498.
Lev-Ari, S., Starr, A., Katzburg, S. et al. (2014). Curcumin induces apoptosis and inhibits growth of orthotopic human non-small cell lung cancer xenografts. *The Journal of Nutritional Biochemistry* 25: 843–850.
Madhuri, S. and Pandey, G. (2010). Effect of ProImmu, a herbal drug on estrogen caused uterine and ovarian cytotoxicity. *Journal of Biomedical Science* 5 (1): 57–62.
Mahajan, N., Rawal, S., Verma, M. et al. (2013). A phytopharmacological overview on Ocimum species with special emphasis on *Ocimum sanctum*. *Biomedicine & Preventive Nutrition* 3 (2): 185–192.

Mahleyuddin, N.N., Moshawih, S., Ming, L.C. et al. (2022). *Coriandrum sativum* L.: a review on ethnopharmacology, phytochemistry, and cardiovascular benefits. *Molecules* 27 (1): 209.

Manjunatha, K.P., Rao, S., Sathyanarayana, M.L. et al. (2020). Evaluation of immunomodulatory cells CD4+ and CD8+ and their ratio using alcoholic seed extract of *Trigonella foenum graecum* and alcoholic leaves extract of *Coccinia indica* by flow cytometry in streptozotocin-induced diabetic rats. *Journal of Pharmacognosy and Phytochemistry* 9 (5): 2943–2947.

Manoharan, S., Balakrishnan, S., Menon, V.P. et al. (2009). Chemopreventive efficacy of curcumin and piperine during 7, 12-dimethylbenz (a) anthracene-induced hamster buccal pouch carcinogenesis. *Singapore Medical Journal* 50 (2): 139.

Mao, Q.Q., Xu, X.Y., Cao, S.Y. et al. (2019). Bioactive compounds and bioactivities of ginger (*Zingiber officinale* Roscoe). *Foods* 8 (6): 185.

Mohamed, D.A., Hamed, I.M., and Fouda, K.A. (2018). Antioxidant and anti- diabetic effects of cumin seeds crude ethanol extract. *Research Journal of Biological Sciences* 18: 251–259.

Mohammed, A., Wudil, A.M., Alhasaan, A.J. et al. (2017). Hypoglycemic activity of *Curcuma longa* linn root extracts on alloxan-induced diabetic rats. *Haya: The Saudi Journal of Life Sciences* 2: 43–49.

Mohan, L., Amberkar, M.V., and Kumari, M. (2011). *Ocimum sanctum* Linn (Tulsi)-an overview. *International Journal of Pharmaceutical Sciences Review and Research* 7 (1): 51–53.

Mondal, S., Mirdha, B.R., and Mahapatra, S.C. (2009). The science behind sacredness of Tulsi (*Ocimum sanctum* Linn.). *Indian Journal ofPhysiology and Pharmacology* 53 (4): 291–306.

Mu, G.G., Zhang, L.L., Li, H.Y. et al. (2015). Thymoquinone pretreatment overcomes the insensitivity and potentiates the antitumor effect of gemcitabine through abrogation of notch1, PI3K/Akt/mTOR regulated signaling pathways in pancreatic cancer. *Digestive Diseases and Sciences* 60: 1067–1080.

Muthulakshmi, V., Vijayakumar, V., Vasanthkumar, M., and Vasanthi, H.R. (2009). *Functional Foods for Chronic Diseases*, vol. 4 (ed. D.M. Martirosyan), 274–317. D & A Inc/FF Publishing.

Mykolas, A. (2009). *Functional Foods for Chronic Diseases*, vol. 4 (ed. D.M. Martirosyan), 234–241. D & A Inc/FF Publishing.

Nemeroff, C.B. (2002). Recent advances in the neurobiology of depression. *Psychopharmacology Bulletin* 36: 6–23.

Noreen, S., Tufail, T., Bader Ul Ain, H. et al. (2023). Antioxidant activity and phytochemical analysis of fennel seeds and flaxseed. *Food Science & Nutrition* 11 (3): 1309–1317.

Oganesyan, E.T., Nersesyan, Z.M., and Parkhomenko, A.Y. (2007). Chemical composition of the above-ground part of *Coriandrum sativum*. *Pharmaceutical Chemistry Journal* 41 (3): 149–153.

Okoro, B.C., Dokunmu, T.M., Okafor, E. et al. (2023). The ethnobotanical, bioactive compounds, pharmacological activities and toxicological evaluation of garlic (*Allium sativum*): a review. *Pharmacological Research-Modern Chinese Medicine* 8: 100273.

Olaru, G. and Popa, E.M. (2019). *In vitro* research on the inhibitory effects of fennel, sage and seabuckthorn essential oils on some food spoilage fungi. *Scientific Bulletin Series F: Biotechnologies* 23: 87–90.

Olatunde, A., Joel, E.B., Tijjani, H. et al. (2014). Anti-diabetic activity of aqueous extract of *Curcuma longa* (Linn) rhizome in normal and alloxan-induced diabetic rats. *Researcher* 6: 58–65.

Park, S.Y., Kim, J.Y., Lee, S.M. et al. (2014). Capsaicin induces apoptosis and modulates MAPK signaling in human gastric cancer cells. *Molecular Medicine Reports* 9: 499–502.

Park, D.H., Park, S.J., Kim, J.M. et al. (2010). Subchronic administration of rosmarinic acid, a natural prolyl oligopeptidase inhibitor, enhances cognitive performances. *Fitoterapia* 81 (6): 644–648.

Pattanayak, P., Behera, P., Das, D., and Panda, S.K. (2010). *Ocimum sanctum* Linn.a reservoir plant for therapeutic applications: an overview. *Pharmacognosy Reviews* 4 (7): 95.

Rabade, V., Ingole, A., Gangane, P. et al. (2022). Tasty spices as nutraceuticals. In: *Herbs and Spices-New Advances*. IntechOpen.

Rahman, K. and Lowe, G.M. (2006). Garlic and cardiovascular disease, a critical review. *The Journal of Nutrition* 136: 736–740.

Ramadan, F.M., Amer, M.M.A., Awad, A., and El-Sayed, S. (2008). Coriander (*Coriandrum sativum* L.) seed oil improves plasma lipid profile in rats fed a diet containing cholesterol. *European Food Research and Technology* 21: 319–328.

Ramaswamy, S., Sengottuvelu, S., Sherief, S.H. et al. (2010). Trachyspermum ammi fruit. *International Journal of Pharma and Bio Sciences* 1 (1).

Ranjan, A.P., Mukerjee, A., Gdowski, A. et al. (2016). Curcumin-ER prolonged subcutaneous delivery for the treatment of non-small cell lung cancer. *Journal of Biomedical Nanotechnology* 12: 679–688.

Rasoulian, B., Hajializadeh, Z., Esmaeili-Mahani, S. et al. (2019). Neuroprotective and antinociceptive effects of rosemary (*Rosmarinus officinalis* L.) extract in rats with painful diabetic neuropathy. *The Journal of Physiological Sciences* 69: 57–64.

Riska, S.R. (2020). Overview of traditional use, phytochemical and pharmacological activities of fennel (*Foeniculum vulgare*). *International Journal of Modern Pharmaceutical Research* 5 (1): 1–9.

Sachan, A.K., Kumar, S., Kumari, K., and Singh, D. (2018). Medicinal uses of spices used in our traditional culture: worldwide. *Journal of Medicinal Plants Studies* 6: 116–122.

Sahu, P.K. and Mishra, P. (2013). Modelling and forecasting production behaviour and import-export of total spices in two most populous countries of the world. *Journal of Agricultural Research* 51 (1): 81–97.

Saleem, A., Naureen, I., Naeem, M. et al. (2022). Therapeutic role of *Piper nigrum* L (Black Pepper) and pharmacological activities. *Scholars International Journal of Biochemistry* 5 (1): 15–21.

Samykutty, A., Shetty, A.V., Dakshinamoorthy, G. et al. (2013). Piperine, a bioactive component of pepper spice exerts therapeutic effects on androgen dependent and androgen independent prostate cancer cells. *PLoS ONE* 8: 495.

Sharma, S., Shukla, M.K., Sharma, K.C. et al. (2022). Revisiting the therapeutic potential of gingerols against different pharmacological activities. *Naunyn-Schmiedeberg's Archives of Pharmacology* 396 (4): 633–647. https://doi.org/10.1007/s00210-022-02372-7.

Shenuarin, B. and Fukunaga, K. (2009). Cardioprotection by vanadium compounds targeting Akt-mediated signaling. *Journal of Pharmacological Sciences* 110: 1–113.

Shiri, S., Alizadeh, A.M., Baradaran, B. et al. (2015). Dendrosomal curcumin suppresses metastatic breast cancer in mice by changing m1/m2 macrophage balance in the tumor microenvironment. *Asian Pacific Journal of Cancer Prevention* 16: 3917–3922.

Shukla, A. and Yadav, N. (2018). Role of Indian spices in Indian history. *International Journal of Management Research and Reviews* 8: 1–6.

Singh, S., Agrawal, N., and Upadhyay, P.K. (2022). *Chemistry, Biological Activities, and Uses of Asafetida. InGums, Resins and Latexes of Plant Origin: Chemistry, Biological Activities and Uses*, 629–656. Cham: Springer International Publishing.

Singh, V. and Garg, A.N. (2006). Availability of essential trace elements in Indian cereals, vegetables and spices using INAA and the contribution of spices to daily dietary intake. *Food Chemistry* 94 (1): 81–89.

Soni, K.B. and Kuttan, R. (1992). Effect of administration on oral curcumin serum peroxides and cholesterol levels in human volunteers. *Indian Journal of Physiology and Pharmacology* 36: 273–275.

Srinivasan, K. (2007). Black pepper and its pungent principle—piperine, a review of diverse physiological effects. *Critical Reviews in Food Science and Nutrition* 47: 735–748.

Srinivasan, K. (2011). Spices fortaste and flavour: nutraceuticals for human health. In: *Spices: The Elixir of Life*, 43–62. London: Originals (An Imprint of Low Price Publications).

Srinivasan, K. and Sambaiah, K. (1991). The effect of spices on cholesterol 7 alpha-hydroxylase activity and on serum and hepatic cholesterol levels in the rat. *International Journal for Vitamin and Nutrition Research* 61: 363–369.

Staege, M., Kewitz, S., and Volkmer, I. (2013). Curcuma contra cancer? Curcumin and hodgkin's lymphoma. *Cancer Growth Metastasis* 6: 35–52.

Sun, X.D., Liu, X.E., and Huang, D.S. (2012). Curcumin induces apoptosis of triple-negative breast cancer cells by inhibition of EGFR expression. *Molecular Medicine Reports* 6: 1267–1270.

Sureshkumar, S., Senthilkumar, B., Rajeshkumar, S., and Nangalingam, M. (2019). Antihyperglycemic and antioxidant effect of *Allium sativum* bulb extract against alloxan-induced diabetic male albino rats. *Drug Invention Today* 12: 965–1968.

Tahir, H.U., Sarfraz, R.A., Ashraf, A., and Shazia, A. (2015). Chemical composition and anti-diabetic activity of essential oils obtained from two species (*Syzygium aromaticum* and *Cuminum cyminum*). *International Journal of Food Properties* 19: 2156–2164.

Tang, E.L., Rajarajeswaran, J., Fung, S.Y., and Kanthimathi, M.S. (2013). Antioxidant activity of *Coriandrum sativum* and protection against DNA damage and cancer cell migration. *BMC Complementary and Alternative Medicine* 13: 495.

Tapsell, L.C., Hemphill, I., Cobiac, L. et al. (2006). Health benefits of herbs and spices, the past, the present, the future. *The Medical Journal of Australia* 185: 4–24.

Taqvi, S.I., Shah, A.J., and Gilani, A.H. (2008). Blood pressure lowering and vasomodulator effects of piperine. *Journal of Cardiovascular Pharmacology* 52 (5): 452–458.

Ting, C.Y., Wang, H.E., Yu, C.C. et al. (2015). Curcumin triggers DNA damage and inhibits expression of DNA repair proteins in human lung cancer cells. *Anticancer Research* 35: 3867–3873.

Upadhyay, S.K. and Singh, P.K. (2012). Receptors of garlic (Allium sativum) lectins and their role in insecticidal action. *The Protein Journal* 31 (6): 439–446.

Upadhyay, S.K. and Singh, S.P. (2021). *Bioprospecting of Plant Biodiversity for Industrial Molecules*. Wiley. https://doi.org/10.1002/9781119718017.

Upadhyay, S.K. and Singh, S.P. (2023). *Plants as Bioreactors for Industrial Molecules*. Wiley. https://doi.org/10.1002/9781119875116.

Upadhyay, S.K., Mishra, M., Singh, H. et al. (2010a). Interaction of Allium sativum leaf agglutinin (ASAL) with midgut BBMV proteins and its stability in Helicoverpa armigera. *Proteomics* 10: 4431–4440.

Upadhyay, S.K., Saurabh, S., Rai, P. et al. (2010b). SUMO fusion facilitates expression and purification of garlic lectin but modifies some of its properties. *Journal of Biotechnology* 146: 1–8.

Upadhyay, S.K., Saurabh, S., Singh, R. et al. (2011). Purification and characterization of a lectin with high hemagglutination property isolated from Allium altaicum. *The Protein Journal* 30: 374–383.

Venier, N.A., Yamamoto, T., Sugar, L.M. et al. (2015). Capsaicin reduces the metastatic burden in the transgenic adenocarcinoma of the mouse prostate model. *Prostate* 75: 1300–1311.

Venkatesan, N. (1998). Curcumin attenuation of acute adriamycin myocardial toxicity in rats. *British Journal of Pharmacology* 124: 425–427.

Vijayakumar, R.S., Surya, D., Senthilkumar, R., and Nalini, N. (2002). Hypolipidemic effect of black pepper (*Piper nigrum* Linn.) in rats fed high fat diet. *Journal of Clinical Biochemistry and Nutrition* 32: 31–42.

Wani, S.A. and Kumar, P. (2018). Fenugreek: a review on its nutraceutical properties and utilization in various food products. *Journal of the Saudi Society of Agricultural Sciences* 17 (2): 97–106.

Wu, L., Guo, L., Liang, Y. et al. (2015). Curcumin suppresses stem-like traits of lung cancer cells via inhibiting the JAK2/STAT3 signaling pathway. *Oncology Reports* 34: 3311–3317.

Xie, Y.Q., Wu, X.B., and Tang, S.Q. (2014). Curcumin treatment alters ERK-1/2 signaling *in vitro* and inhibits nasopharyngeal carcinoma proliferation in mouse xenografts. *International Journal of Clinical and Experimental Medicine* 7: 108–114.

Yadav, A.S. and Bhatnagar, D. (2007). Modulatory effect of spice extracts on iron-induced lipid peroxidation in rat liver. *Biofactors* 29: 147–157.

Yaffe, P.B., Doucette, C.D., Walsh, M., and Hoskin, D.W. (2013). Piperine impairs cell cycle progression and causes reactive oxygen species-dependent apoptosis in rectal cancer cells. *Experimental and Molecular Pathology* 94: 109–114.

Yang, C.L., Ma, Y.G., Xue, Y.X. et al. (2012). Curcumin induces small cell lung cancer NCI-H446 cell apoptosis via the reactive oxygen species-mediated mitochondrial pathway and not the cell death receptor pathway. *DNA and Cell Biology* 31: 139–150.

Yeh, H.Y., Chuang, C.H., Chen, H.C. et al. (2014). Bioactive components analysis of two various gingers (*Zingiber officinale* Roscoe) and antioxidant effect of ginger extracts. *LWT-Food Science and Technology* 55 (1): 329–334.

Yu-Yan, Y. and Liu, L. (2001). Cholesterol lowering effect of garlic extracts and organosulfur compounds, human and animal studies. *The Journal of Nutrition* 131: 989–993.

Zick, S.M., Turgeon, D.K., Ren, J. et al. (2015). Pilot clinical study of the effects of ginger root extract on eicosanoids in colonic mucosa of subjects at increased risk for colorectal cancer. *Molecular Carcinogenesis* 54: 908–915.

5

Antioxidant Properties of Food Plants, Spices, Beverages, and Nutraceuticals—Health and Anti-aging Potentials

Michael Wink

Institute of Pharmacy and Molecular Biotechnology, Heidelberg University, Heidelberg, Germany

5.1 Introduction

About 400,000 plant species have been recognized. Their biodiversity is matched by the diversity of natural products, also termed specialized or secondary metabolites (PSM). Presently, over 100,000 PSM have been identified from 20% to 30% of plants; it is likely that many more PSM will be discovered when finally, all plants have been analyzed phytochemically (Harborne and Baxter 1993; Hartmann 2007; Seigler 1998; Wink 1988, 2003, 2008a, 2008b, 2010a, 2010b, 2020; Upadhyay and Singh 2021, 2023; Zenk and Juenger 2007). All plants produce and sequester hundreds of structurally different PSMs (Table 5.1), which consist of complex mixtures. The composition of these mixtures can differ in the different parts of a plant, i.e. a root stores a unique set of PSM, which can be distinguished from that of leaves, flowers, and seeds. In addition, PSM patterns can differ in individuals in a population or between populations and between developmental stages (Wink 2008a, b, 2010a, 2020). Some types of PSM are produced in nearly all plant families, such as terpenoids and phenolics (flavonoids, tannins), whereas others, such as alkaloids, cyanogenic glucosides, glucosinolates, nonprotein amino acids, polyacetylenes, and anthraquinones mostly occur in phylogenetically related taxa. However, there are interesting exceptions which could be due to convergent evolution or old phylogenetically shared pathway genes (Wink 2003, 2010a, 2013, 2019, 2020).

The biosynthetic pathways leading to PSM all derive from primary metabolites, such as glucose, acetyl-CoA, or amino acids (Seigler 1998; Wink 2010a). The sites of synthesis can be the cytosol or even organelles, such as chloroplasts. PSMs are either stored in vacuoles (water-soluble PSM), laticifers, resin ducts or oil cells, specialized glandular hairs, and trichomes (lipophilic PSM), where their concentrations are usually quite high (a requirement for their ecological functions). In some instances, the site of biosynthesis is not identical to the site of storage. Several alkaloids, such as pyrrolizidine alkaloids and nicotine are synthesized in roots and are transported to aerial parts with help of the xylem. Many PSM are made in the aerial organs, such as the leaf (quinolizidine alkaloids, glucosinolates); they are often translocated to the fruits and other plant parts via the phloem (Maeshima 2001; Rehman et al. 2016; Wink 1997, 2010a, 2019).

It is a trivial statement that plants are sessile organisms that lack an active adaptive immune system (as mammals have) and the capability to run away in case of danger. Thus, plants cannot escape when attacked by an herbivore. Plants had to fight against herbivores and microorganisms,

Herbal Nutraceuticals: Products and Processes, First Edition. Edited by Santosh Kumar Upadhyay and Sudhir Pratap Singh.

Table 5.1 Chemical diversity of plant specialized metabolites (PSM).

Class of PSM	Estimated numbers of specialized metabolites	Degree of antioxidant activities
With nitrogen		
Alkaloids	27,000	Moderate
Nonprotein amino acids	900	Low
Amines	100	Low
Cyanogenic glucosides	60	Low
Glucosinolates/isothiocyanates	130	Active
Alkamides	290	Active
Peptides	2500	Low
Without nitrogen		
Monoterpenes (including iridoids)	2500	Low
Sesquiterpenes	5000	Low
Diterpenes	2500	Low
Triterpenes, steroids, saponins	5000	Low
Tetra- and polyterpenes	700	Active
Phenylpropanoids, phenolic acids, coumarins, lignans, lignin	2000	Very active
Flavonoids, anthocyanins, xanthones, and tannins	6000	Very active
Polyacetylenes, fatty acids, waxes	1000	Active
Polyketides, quinones	1500	Active
Carbohydrates	200	Low
Organic acids	100	Low

Harborne and Baxter (1993); Seigler (1998); van Wyk and Wink (2015), (2017); Wink and van Wyk (2008) and their potential antioxidant activities (graded in low, moderate, active, and very active). This table is an extension of an earlier version (van Wyk and Wink (2015); Wink (2022)).

such as bacteria, fungi, and also viruses, already when the first land plants occurred on the planet more than 400 million years ago. Like other sessile organisms, plants evolved a multitude of bioreactive small organic molecules, which can modulate the activity of molecular targets (such as proteins, biomembranes, DNA/RNA) in animals and microbes (van Wyk and Wink 2015; Wink 2007, 2008b, 2015, 2022). The main function of PSM is defense against animals, microbes, and competing plants. In addition, several of them serve as signal compounds important for plant–animal and plant–microbe interactions (Figure 5.1) (Antil et al. 2023; Hartmann 2004, 2007; Kaur et al. 2021; Krauss and Nies 2014; Madhu et al. 2024; Seigler 1998; Shumayla and Upadhyay 2023; Upadhyay 2021, 2024; Upadhyay and Shumayla 2022). The angiosperms (flowering plants) started their development around 140 million years ago and diversified during the tertiary (Wink 2020). Angiosperms need insects and other animals for pollination or seed dispersal. To fulfill these functions, some of the defense PSM have a further function as a signal molecule to attract animals. Another ecological function of PSM derives from the need of plants for protection against ROS and UV light (Figure 5.1); as discussed below, some of the defense compounds also exhibit substantial antioxidant activity, which will be the main focus of this chapter.

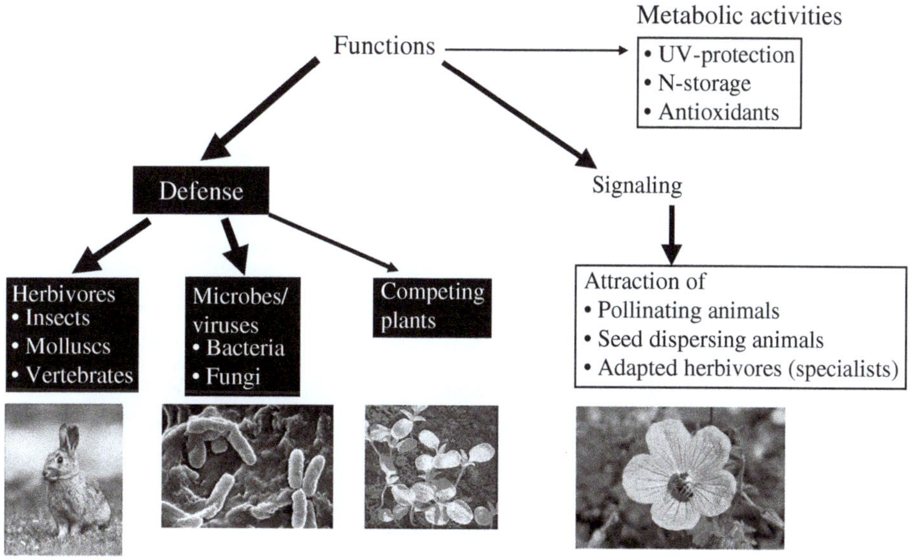

Figure 5.1 The main ecological functions of PSM as defense and signal compounds and, in addition, metabolic activities. *Source:* JJ Harrison/Wikimedia Commons/CC BY-SA 3.0; PIXNIO/Wikimedia Commons/ CC BY SA 4.0 (photo of Herbivores and Microbes/viruses).

As a consequence of over 400 million years of natural selection, all plants that exist today produce a broad library of PSM with interesting pharmacological properties, some of them useful for humans as spices, perfumes, or drugs to treat health conditions and infections (Van Wyk and Wink 2015, 2017).

In this chapter, the main groups of PSM with antioxidant activities will be summarized with regard to potential applications to treat certain health conditions related to oxidative stress. As a laboratory model system, we used the nematode *Caenorhabditis elegans* to determine which PSM can serve as antioxidants *in vivo*. ROS can counteract oxidative stress, enhance survival and longevity, and retard the development of neurodegenerative diseases, including M. Parkinson, Morbus Alzheimer, or M. Huntington and longevity (Li et al. 2024; Mudd and Liceaga 2022; Roxo and Wink 2022; Wink 2022).

5.2 Reactive Oxygen Species and Health Implications

It is well established that our cells can generate ROS from O_2 and H_2O_2, including superoxide, hydroxyl radical, or singlet oxygen $^1O_2$ (Alberts et al. 2015) (Figure 5.2). ROS helps cells, tissues, and organs fight bacterial infections, virus-infected cells, and even cancer cells; they serve for cell signaling, immune response, inflammation, and cell differentiation (Shields et al. 2021; Sies and Jones 2020). ROS are reactive molecules that can oxidize DNA, lipids, and proteins, and thus they can kill pathogens or cancer cells. Furthermore, ROS can regulate essential processes (among them cell signaling and cell differentiation) (Alberts et al. 2015; Shields et al. 2021; Sies and Jones 2020).

As mentioned above, ROS are generated during mitochondrial respiration, but their production can be stimulated by external factors (infection, inflammation, stress, pain, radiation, UV light, air pollution, and smoking; Figure 5.2). The overproduction of ROS can lead to oxidative

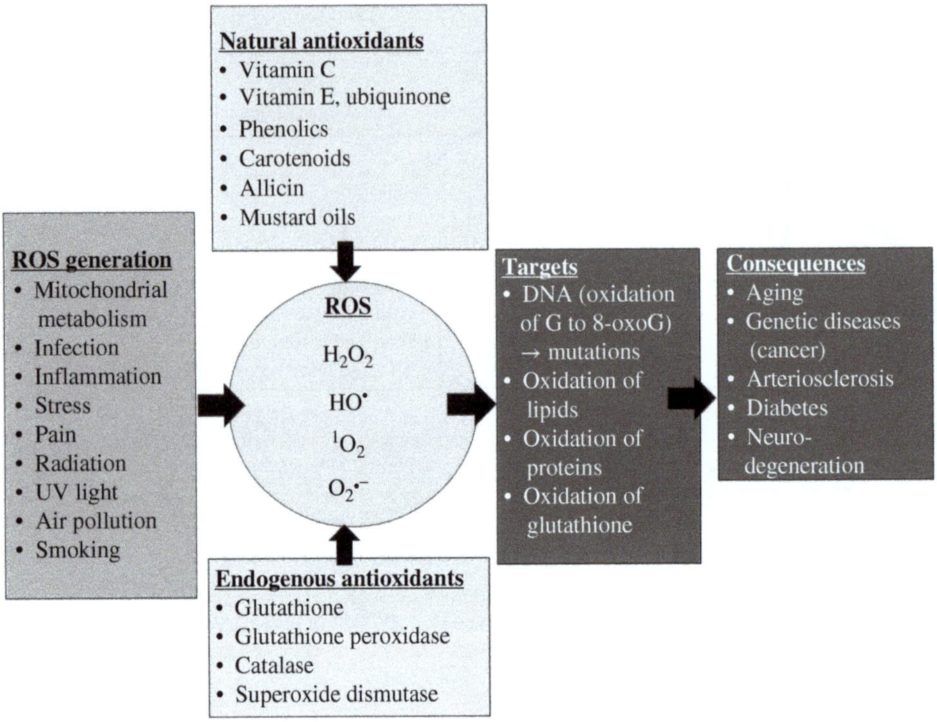

Figure 5.2 Generation of ROS, negative effects of oxidative stress, protective function of endogenous and exogenous antioxidants.

stress, albeit cells have their own antioxidant strategies, including glutathione, and enzymes which can inactivate ROS [glutathione peroxidase, catalase, superoxide dismutase (SOD-3)]. Oxidative stress can be detrimental to essential macromolecules in our cells and body (proteins, lipids, DNA; Figure 5.2), which can lead to several health conditions, including arteriosclerosis (especially after oxidation of phospholipids in Low Density Proteins (LDL), cardiovascular diseases, diabetes, and even cancer and neurodegeneration (Roxo and Wink 2022; Shields et al. 2021; Wink 2022). Oxidative stress appears to be relevant for aging and aging-associated health conditions (Franceschi et al. 2018; Shields et al. 2021; Sies and Jones 2020). Importantly, ROS is known to oxidize the DNA base guanosine (G). As a result, 8-deoxoguanosine (8-oxoG) is generated. Since 8-oxoG cannot pair with C (as G would do) but with A, point mutations are generated (Alberts et al. 2015). Mutations in protein-coding genes and in regulatory DNA sequences (promotor, enhancer) can lead to functionless proteins, or they can modulate gene expression; these changes can be the cause of several genetic diseases, including cancer, arteriosclerosis, and cardiovascular diseases (Roxo and Wink 2022; Scalbert et al. 2005; Sies and Jones 2020; Wink 2022).

Antioxidants are known to reduce or inhibit oxidative stress. Epidemiological studies have revealed that the regular consumption of antioxidant-rich food (vegetables, fruits, spices) or drinking beverages (tea, coffee, fruit juices) (Tables 5.2 and 5.3) can help to prevent the development of cancer, arteriosclerosis, cardiovascular conditions, and diabetes; furthermore, they appear to have a positive effect on the aging process (Devore et al. 2012; Kotecha et al. 2016; Mursu et al. 2014; Scalbert et al. 2005; Wang et al. 2011).

Table 5.2 Extracts and PSM from plants (ordered with regards to their families) exhibiting antioxidant properties and promoting longevity and neuroprotection in *Caenorhabditis elegans*.

Plant species/ common name	Extracts/ compounds	Main PSM	Bioactivity	Reference
Amaranthaceae				
Beta vulgaris Beet	Roots	Betalain	1, 2, 3, 4, 6	Guerrero-Rubio et al. (2020); Imamura et al. (2022)
Anacardiaceae				
Anacardium occidentale Cashew	Aerial parts	α-linolenic acid, and β-caryophyllene	1, 2, 3, 4, 5	Duangjan et al. (2019a, 2021c)
Aquifoliaceae				
Ilex paraguariensis Mate	Leaves	Polyphenols, methylxanthines (caffeine)	1, 2, 4	Lima et al. (2014)
Arecaceae				
Euterpe precatoria; *E. oleracea* Acai palm	Fruits	Anthocyanins	1, 2, 3, 5, 6, 7	de Freitas Bonomo et al. (2014); Peixoto et al. (2016 a, b)
Asteraceae				
Arctium lappa Burdock	Roots	Lignans	1, 2, 3	Su and Wink (2015)
Carlina acaulis Carline thistle	Leaves	Carlina oxide	1, 2, 3, 6	Link et al. (2016)
Tagetes erecta Aztek marigold	Flowers	Phenolics: digallic acid, laricitrin	1, 2, 3, 4, 6	Moliner et al. (2018)
Bixaceae				
Bixa orellana Achiote	Seeds	Carotenoids, tocotrienol	1, 2, 3, 4	Gómez-Linton et al. (2021)
Brassicaceae				
Brassica oleracea Broccoli	Leaves	Mustard oils: sulphoraphane	1, 2, 3, 4, 5, 6	Qi et al. (2021)
Red cabbage	Leaves	Anthocyanins	1, 2, 3, 4	Zhang et al. (2021)
Caryocaraceae				
Caryocar villosum	Fruit shells	Phenolic acids, tannins, triterpenoid saponins	1, 2, 4	Roxo et al. (2020)
Ebenaceae				
Euclea divinorum Diamond leaf	Leaves	Tannins	1, 2, 3	Bogari et al. (2022)
Ericaceae				
Vaccinium myrtillus Blueberry	Fruits	Anthocyanins, tannins	1, 2, 4	Wang et al. (2018)
Vaccinium vitis-idaea Lowbush cranberry	Fruits	Anthocyanins and other flavonoids	1, 4	Scerbak et al. (2018)

(*Continued*)

Table 5.2 (Continued)

Plant species/ common name	Extracts/ compounds	Main PSM	Bioactivity	Reference
Fabaceae				
Aspalathus linearis Rooibos	Leaves	Aspalathin and other flavonoids	1, 2, 3	Chen et al. (2013a)
Caesalpinia mimosoides	Leaves and young twigs	3-O-methylgallate, 4- theogallin, gallic acid	1, 2, 3, 5	Rangsinth et al. (2019)
Cassia abbreviata Long-tail cassia	Roots	Procyanidins	1, 2, 3	Sobeh et al. (2018d)
Cassia fistula Golden shower tree	Leaf and stems	Flavonoids, procyanidins anthraquinones	1, 2, 3, 6, 7	Thabit et al. (2018)
Glycyrrhiza glabra Liquorice	Fruits	Isoliquiritigenin	1, 2, 3, 5, 6	Link et al. (2015); Link and Wink (2019)
Schotia brachypetala Weeping boer-bean	Leaves	Flavonoids, procyanidins, anthocyanins	1, 2, 3	Sobeh et al. (2016)
Styphnolobium japonicum Pagoda tree	Fruits	Rutin, tannins	1, 2, 3, 6	Thabit et al. (2019, 2023)
Geraniaceae				
Pelargonium sidoides African geranium (Umckaloabo)	Roots	Polyphenols, coumarins (EPs7630®)	1, 2, 3, 4	Rezaizadehnajafi and Wink (2014)
Humiriaceae				
Endopleura uchi	Bark	Polyphenols	1, 2, 3, 4	Peixoto et al. (2019)
Lamiaceae				
Clerodendrum speciosum Glorybower	Leaves	Flavonoids and other polyphenols		Youssef et al. (2022)
Salvia rosmarinus; (formerly *Rosmarinus officinalis*) Rosemary	Leaves	Flavonoids, rosmarinic acid and other phenolic acids	1, 2, 4	Zamberlan et al. (2016)
Sideritis scardica Greek mountain tea	Aerial parts	Caffeoylquinic acids, flavonoids, acteoside	1, 3, 4, 6, 7	Heiner et al. (2018); Nishioka et al. (2020)
Malvaceae				
Hibiscus sabdariffa Hibiscus	Flowers	Flavonoids, anthocyanins	1, 5, 6	Koch et al. (2020)
Myrtaceae				
Eugenia uniflora Purple pitanga	Fruits	Flavonoids, anthocyanins	1, 2, 3, 4	Tambara et al. (2018)
Syzygium nervosum Daly river satinash	Fruit pulp	Anthocyanins	2, 3, 4, 5, 6	Prasanth et al. (2019)

Table 5.2 (Continued)

Plant species/ common name	Extracts/ compounds	Main PSM	Bioactivity	Reference
Olacaceae				
Ximenia americana Tallow wood	Leaves	Flavonoids, tannins, quinones	1, 2, 3	Bakrim et al. (2022)
Oleaceae				
Olea europaea Olive tree	Leaves	Polyphenols	2, 3	Luo et al. (2019)
Poaceae				
Triticum aestivum Purple wheat	Grains	Anthocyanins	1, 3, 4	Chen et al. (2013c)
Rosaceae				
Agrimonia procera, Agrimony	Aerial parts	Phenolic acids, flavonoids, ellagitannins, procyanidins	1, 2, 4	Saier et al. (2018)
Rubiaceae				
Calycophyllum spruceanum Tree of youth	Stem bark	Flavonoids, coumarins (gardenoside, cyanidin, taxifolin)	1, 2, 3, 4, 5	Peixoto et al. (2018)
Coffea arabica Coffee tree	Seeds	Caffeine, polyphenols	1, 3, 4	Li et al. (2019)
Uncaria tomentosa Cat's claw	Leaves	Alkaloids: mitraphylline,	1, 2, 3	Azevedo et al. (2019)
Rutaceae				
Citrus sinensis Red orange	Fruits	Carotenoids: lycopene, carotene	1, 2, 3, 4, 6	de Oliveira Caland et al. (2019)
Phyllanthaceae				
Glochidion zeylanicum Umbrella cheese tree	Leaves	Gallic acid, catechins, tannins	1, 2, 3, 4, 5	Duangjan et al. (2019b, 2021a)
Salicaceae				
Salix subserrata Willow	Leaves, bark	Flavonoids, salicylates	1, 2, 3	Tawfeek et al. (2019)
Sapindaceae				
Paullinia cupana Guarana	Seeds	Caffeine and other methylxanthines; catechins	1, 2, 3, 4, 5, 6, 7	Boasquívis et al. (2018); Peixoto et al. (2017)
Solanaceae				
Lycium barbarum Goji	Fruits	Flavonoids, carotenoids	Toxic and pro-oxidant activities	de Freitas Rodrigues et al. (2021)

(*Continued*)

Table 5.2 (Continued)

Plant species/ common name	Extracts/ compounds	Main PSM	Bioactivity	Reference
Theaceae				
Camelia sinensis L. Tea	Green tea, EGCG	EGCG and other catechins	1, 2, 3, 4, 6	Abbas and Wink (2009, 2010, 2014); Wink and Abbas (2012); Zhang et al. (2020, 2022)
Vitaceae				
Vitis vinifera Grape	Leaves	Flavonoids, tannins	1, 2, 3, 6	Duangjan et al. (2021b)
		Resveratrol	1, 2, 3	Chen et al. (2013b)

Reported bioactivities: 1 = acute oxidative stress resistance; 2 = reduction of ROS levels; 3 = *hsp-16.2, gst-4,* and *sod-3* expression; 4 = lifespan extension; 5 = anti-aging effects (lipofuscin accumulation; pharyngeal pumping rate); 6 = retardation of Aβ-induced paralysis in M. Alzheimer model; 7 = reduction in the number of polyQ40 aggregates in M. Parkinson model.
This table is an extension of an earlier version (details in Roxo and Wink 2022; Wink 2022).

Table 5.3 Commonly used food plants, spices, and fruits with antioxidant and additional antimicrobial, anti-inflammatory, carminative, and diuretic properties (van Wyk 2014; van Wyk and Wink 2017).

Plant species	Plant part	Antioxidants and other PSM	Additional biochemical activities
Amaranthaceae			
Beta vulgaris (red beet)	Roots	Betalains	
Amaryllidaceae			
Allium sativum (garlic); *A. cepa* (onion); *A. ursinum* and other *Allium* species	Bulb, leaves	Allicin, other sulfur-containing PSM; alliin is converted to pungent and bioactive allicin	2, 4, 7
Tulbaghia violacea (society garlic)	Leaves	Alliin is converted to bioactive allicin and other sulfur-containing PSM	2, 4, 7
Adoxaceae			
Sambucus nigra (elderberry)	Fruits	Anthocyanins, flavonoids, triterpenes	3
Anacardiaceae			
Mangifera indica (mango)	Fruits	Hydrolysable tannins	3
Aquifoliaceae			
Ilex paraguariensis (maté)	Leaves	Caffeine, simple phenolic acids, polyphenols, triterpene saponins	5, 6
Apiaceae			
Anethum graveolens (dill)	Fruits	Monoterpenes: (*R* (−) carvone, limonene, dill ether), flavonoids, coumarins	2, 4, 7
Angelica archangelica (Archangel, garden angelica)	Roots, leaves, fruits	Furanocoumarins, coumarins; phellandrene, pinene	2, 4, 7

Table 5.3 (Continued)

Plant species	Plant part	Antioxidants and other PSM	Additional biochemical activities
Apium graveolens (celery)	Stems, tubers	Monoterpenes (limonene), apiole, butylphthalides, methylphthalides, furanocoumarins, flavonoids	2, 4, 5, 7
Carum carvi (caraway)	Fruits	Monoterpenes (*D* (+)carvone, limonene, carveol), phenylpropanoids, flavonoids	2, 4, 7
Coriandrum sativum (coriander)	Fruits	Monoterpenes (linalool), phenylpropanoids, triterpenes	2, 4, 7
Daucus carota (carrot)	Roots	Carotene	Provitamin A
Foeniculum vulgare (fennel)	Fruits, stems	Trans-anethole and other phenylpropanoids, monoterpenes, flavonoids	2, 3, 4, 7
Levisticum officinale (lovage)	Rhizome, roots, fruits, leaves	Alkylphthalides, mono- and sesquiterpenes, coumarins, furanocoumarins, falcarindiol	2, 3, 4, 7
Petroselinum crispum (parsley)	Aerial parts, roots	Phenylpropanoids (apiol, myristicin), flavonoids, furanocoumarins	5, 7
Pimpinella anisum (anise), *P. major* (saxifrage)	Fruits, roots	Phenylpropanoids (trans-anethole), sesquiterpenes	2, 4, 7
Arecaceae			
Euterpe oleracea (açai palm)	Fruits	Anthocyanins, flavonoids, procyanidins	
Asteraceae			
Artemisia dracunculus (estragon, tarragon)	Leaves	Estragol, phellandrene, ocimene	4, 7, 9
Calendula officinalis (marigold)	Flower heads	Flavonoids, saponins, triterpenes	3, 4, 7
Cichorium intybus (chicory)	Roots, leaves	Inulin, sesquiterpene lactones, phenolic acids, flavonoids	1, 2, 3
Cynara cardunculus (artichoke)	Leaves, flower heads	Phenolic acids, flavonoids, sesquiterpene lactones	1, 3
Matricaria chamomilla (chamomile)	Flowers	Terpenoids, flavonoids, polyacetylenes	2, 3, 4, 7
Silybum marianum (milk-thistle)	Fruits	Flavonolignans (silymarin)	3
Berberidaceae			
Berberis vulgaris (barberry) and related species	Fruits, bark	Berberine and other isoquinoline alkaloids, carotenoids	1, 2, 3
Brassicaceae			
Armoracia rusticana (horseradish)	Roots	Glucosinolates are converted into bioactive and pungent isothiocyanates	2, 4, 7

(Continued)

Table 5.3 (Continued)

Plant species	Plant part	Antioxidants and other PSM	Additional biochemical activities
Brassica nigra; B. alba (mustard)	Seeds	Glucosinolates are converted into bioactive and pungent isothiocyanates	2, 7
Brassica oleracea (broccoli and other cabbages)	Aerial parts	Chlorophyll; mustard oils (sulforaphane)	2, 7
Eutrema japonicum (Wasabi)	Roots	Glucosinolates are converted into bioactive and pungent isothiocyanates	2, 7
Sinapis alba (White mustard)	Seeds	Glucosinolates which are converted to pungent allyl isothiocyanates	2, 4, 7
Tropaeolum majus (nasturtium)	Leaves	Glucosinolates are converted into bioactive and pungent isothiocyanates	2, 5, 7
Cannabaceae			
Humulus lupulus (hop plant)	Hops	Lupulone and other phloroglucinols, 2-methyl-3-buten-2-ol, flavonoids, procyanidins	1, 9
Cucurbitaceae			
Cucurbita pepo (pumpkin)	Fruits; seeds	Carotenoids; tocopherol; linoleic acid	3
Cupressaceae			
Juniperus communis (juniper)	Cone (berry)	Monoterpenes (pinene, terpinen-4-ol), phenolics	2, 5, 7
Eleagnaceae			
Hippophae rhamnoides (sea buckthorn)	Fruits	Ascorbic acid, flavonoids, carotenoids	
Ericaceae			
Arctostaphylos uva-ursi (bearberry)	Fruits, leaves	Arbutin, gallotannins, ellagitannins	1, 2
Gaultheria procumbens (wintergreen)	Leaves	Methyl salicylate, arbutin, tannins	2, 3, 7
Vaccinium erythrocarpum, Vaccinium macrocarpon, Vaccinium microcarpum, Vaccinium oxycoccos (cranberry)	Fruits	Resveratrol, anthocyanins, condensed tannins	2, 3
Vaccinium myrtillus (Blueberry; bilberry)	Fruits	Anthocyanins, tannins, flavonoids, resveratrol	2, 3
Fabaceae			
Aspalathus linearis (rooibos tea plant)	Leaves, twigs	Flavonoids, dihydrochalcones	4
Cyclopia intermedia (honeybush tea)	Leaves and twigs	Mangiferin and other phenolics, flavonoids, isoflavones	5

Table 5.3 (Continued)

Plant species	Plant part	Antioxidants and other PSM	Additional biochemical activities
Glycine max (soybean)	Seeds, sprouts	Isoflavonoids, saponins	8
Glycyrrhiza glabra (licorice) and related species	Rhizomes	Liquiritin and other flavonoids and isoflavonoids, triterpene saponins (glycyrrhizic acid)	2, 3, 4
Medicago sativa (alfalfa)	Seeds, sprouts	Organic acids, nonprotein amino acids, isoflavonoids, saponins	3, 8
Trifolium pratense (red clover)	Aerial parts	Genistein and other isoflavones	3, 8
Trigonella foenum-graecum (fenugreek)	Seeds	Alkaloids, flavonoids, saponins	3, 8
Grossulariaceae			
Ribes nigrum (Blackcurrant)	Fruits	Phenolic acids, flavonoids, anthocyanins, tannins, procyanidins	2, 3
Lamiaceae			
Hyssopus officinalis (hyssop)	Aerial parts	Pinocamphone; marrubiin, rosmarinic acid	2, 3, 4, 7
Lavandula angustifolia (lavender)	Flowers, leaves	Linaloyl acetate, linalool, phenolics	2, 4, 7, 9
Melissa officinalis (lemon balm)	Leaves	Terpenoids (citronellal, citral A and B), rosmarinic acid, flavonoids	2, 3, 7
Mentha x *piperita* (peppermint) and other mints	Leaves	Menthol and other monoterpenes	2, 4, 7
Ocimum basilicum (basil); *Ocimum tenuiflorum* (Holy basil)	Leaves	Monoterpenes, caryophyllene, (methylchavicol, linalool), eugenol, flavonoids	2, 4, 7
Origanum vulgare (oregano); *Origanum majorana* (sweet majoram)	Aerial parts	Monoterpenes (carvacrol, *p*-cymene, sabinene, terpinene, terpinen-4-ol, 1,8-cineole), flavonoids	2, 3, 4, 7
Orthosiphon aristatus (orthosiphon)	Leaves	Borneol, limonene, thymol; diterpenes, flavonoids	2, 3, 4, 7
Salvia rosmarinus (formerly *Rosmarinus officinalis*) (rosemary)	Aerial parts	Monoterpenes (1,8-cineole, pinene, camphor), rosmarinic acid, carnosic acid, flavonoids, di- and triterpenes	2, 3, 4, 7
Salvia officinalis (sage) and related species	Aerial parts	Monoterpenes, rosmarinic acid, carnosic acid, flavonoids, di- and triterpenes	2, 3, 4, 7
Satureja montana (winter savory)	Flowering tops	Monoterpenes (carvacrol, p-cymol, terpinen-4-ol, thymol), rosmarinic acid	2, 3, 4, 7
Sideritis scardica (mountain tea)	Aerial parts	Mono- and diterpenes, triterpenes, phenolic acids, polyphenols	2, 3, 7

(Continued)

Table 5.3 (Continued)

Plant species	Plant part	Antioxidants and other PSM	Additional biochemical activities
Thymus vulgaris (thyme)	Aerial parts	Thymol, carvacrol and other terpenoids, rosmarinic acid, flavonoids, triterpenes	2, 3, 4, 7
Lauraceae			
Cinnamomum verum (cinnamon) and related species	Bark	Cinnamaldehyde, eugenol, procyanidins	2, 4, 7
Laurus nobilis (laurel)	Leaves	1,8-Cineole, linalool and other monoterpenes, eugenol, flavonoids, sesquiterpene lactones, alkaloids	2, 3, 4, 7
Lythraceae			
Punica granatum (pomegranate)	Fruits	Gallotannins, ellagitannins (punicalin), piperidine alkaloids (pelletierine)	1, 2, 3, 4
Malpighiaceae			
Malpighia glabra (Acerola)	Fruits	Vitamin C; carotenoids	
Malvaceae			
Cola acuminata (cola nut tree)	Seeds	Caffeine, simple phenolic acids, polyphenols	5, 6
Hibiscus sabdariffa (hibiscus)	Flowers	Anthocyanins, organic acids, mucilage	
Malva sylvestris (mallow)	Leaves, flowers	Mucilage, anthocyanins, tannins	3
Theobroma cacao (Cacao)	Seeds	Caffeine, simple phenolic acids, polyphenols, catechins	5, 6
Myristicaceae			
Myristica fragrans (nutmeg)	Seeds	Monoterpenes (sabinene, pinene), phenylpropanoids (myristicin, elemicin, eugenol), lignans	2, 3, 4, 6, 7
Myrtaceae			
Eucalyptus globulus (bluegum) and related species	Leaf oil	1,8-Cineole and other monoterpenes; sesquiterpenes, flavonoids	2, 3, 4, 7
Pimenta dioica (allspice)	Fruits	Eugenol, 1,8-cineole, caryophyllene	2, 3, 7
Syzygium aromaticum (clove tree)	Flower buds	Eugenol, flavonoids, gallotannins, triterpenes	2, 3, 7
Orchidaceae			
Vanilla planifolia (vanilla)	Fruits	Vanillin (formed from the glycoside vanilloside)	3
Oleaceae			
Olea europaea (olive tree)	Leaves, fruits	Unsaturated fatty acids, secoiridoids, flavonoids	3

Table 5.3 (Continued)

Plant species	Plant part	Antioxidants and other PSM	Additional biochemical activities
Passifloraceae			
Passiflora edulis (passion fruit)	Fruits	Resveratrol, flavonoids, anthocyanins	9
Piperaceae			
Piper nigrum (pepper), *Piper cubeba, Piper longum* and related species	Fruits	Terpenoids (sabinene, pinene, phellandrene, linalool), piperine and related alkaloids	2, 4, 7
Polygonaceae			
Fagopyrum esculentum (buckwheat)	Fruits, leaves	Rutin and other flavonoids	3
Ranunculaceae			
Nigella sativa (black cumin)	Seeds	Thymoquinone and other monoterpenes, saponins, flavonoids	2, 3, 4, 7
Rosaceae			
Aronia melanocarpa (black chokeberry)	Fruits	Phenolic acids, anthocyanins, condensed tannins, vitamins (B, C, E)	3
Crataegus monogyna (hawthorn) and related species	Fruits	Flavonoids, oligomeric procyanidins	Cardiotonic
Fragaria x ananassa (strawberry)	Fruits	Phenolic acids, flavonoids, ellagitannins, anthocyanins	3
Malus domestica (apple)	Fruits	Phenolic acids, flavonoids, cyanogenic glucosides	3
Prunus cerasus (cherry) and other *Prunus* species	Fruits	Phenolic acids, anthocyanins	
Pyrus communis (pear)	Fruits	Phenolic acids	
Rosa canina (rose)	Fruits, seeds	Carotenoids; polyphenols	3 (seeds)
Rubus fruticosus (Blackberry)	Fruits	Phenolic acids, anthocyanins, flavonoids, ellagitannins	3
Rubus idaeus (raspberry)	Fruits	Phenolic acids, flavonoids, anthocyanins, tannins	3
Sorbus aucuparia (mountain ash)	Fruits	Sorbitol, ascorbic acid, carotenoids	
Rubiaceae			
Coffea arabica (Coffee tree)	Seeds	Caffeine, simple phenolic acids, polyphenols	5, 6
Rutaceae			
Citrus limon (lemon); *C. aurantiifolia* (lime); *C. sinensis* (sweet orange); *C. aurantium* (bitter orange); *Citrus x paradisi* (grapefruit)	Fruits, peel	Vitamin C, naringin and other flavonoids, triterpenes, carotenoids	1, 2, 3, 7

(*Continued*)

Table 5.3 (Continued)

Plant species	Plant part	Antioxidants and other PSM	Additional biochemical activities
Sapindaceae			
Paullinia cupana (guaraná)	Seeds	Caffeine, simple phenolic acids, polyphenols, procyanidins	5, 6
Schisandraceae			
Schisandra chinensis (schisandra)	Fruits	Lignans, flavonoids	3
Solanaceae			
Capsicum frutescens; C. annuum (pepper, chili)	Fruits	Capsaicinoids; carotenoids, vitamin C	3, 7
Lycium barbarum (goji); *L. chinense* (Chinese wolfberry)	Fruits	Carotenoids, terpenoids	3
Solanum esculentum (tomato)	Mature fruits	Lycopene and other carotenoids	3
Theaceae			
Camellia sinensis (tea plant)	Leaves	Caffeine; EGCG and other catechins	2, 5, 6
Verbenaceae			
Aloysia citridora (vervain)	Leaves	Citral, limonene; flavonoids	7, 9
Verbena officinalis (vervain)	Aerial parts	Iridoid glucosides, phenolic acids	1, 2, 3
Vitaceae			
Vitis vinifera (grapes)	Fruits, seeds	Anthocyanins (red wines); resveratrol, catechins, procyanidins, tannins	3
Zingiberaceae			
Alpinia officinarum (galangal)	Rhizomes	Mono- and sesquiterpenes, gingerols	2, 3, 4, 7
Curcuma longa (turmeric); *C. xanthorrhiza; C. zedoria*	Rhizomes	Curcumin and other curcuminoids, terpenoids	2, 3, 7
Elettaria cardamomum (cardamom)	Fruits, seeds	1,8-Cineole and other monoterpenes	2, 4, 7
Zingiber officinale (ginger)	Rhizomes	Terpenoids (camphene, phellandrene; zingiberene), gingerols	2, 3, 4, 7

Species are ordered systematically. This table is an extension of an earlier version (Wink 2022). Other sensorial and pharmacological properties include: 1 = bitter taste; 2 = antimicrobial, 3 = anti-inflammatory, 4 = carminative, antispasmodic, 5 = diuretic, 6 = stimulant, 7 = aromatic or spicy, 8 = phytoestrogen; 9 = sedative.

5.3 Antioxidant Molecules from Nature

As discussed in Section 0, PSM evolved as defense and signal compounds (Figure 5.1). An additional evolutionary challenge can be related to the metabolism of plants itself, especially the need to counteract oxidative stress (Figure 5.2). ROS also cause problems in plants themselves and they probably evolved defensive PSM which could also function as antioxidants. Thus, evolution apparently favored several bioactivities of the same PSM molecule. Antioxidant or radicle scavenging activities are common properties of vitamins C and E, phenolics (flavonoids, stilbenes, lignans, anthocyanins, catechins, tannins, and anthraquinones), carotenoids, and related terpenoids. Also sulfur-containing PSM, such as allicin (derived from the prodrug alliin after hydrolysis by alliinase), isothiocyanates, and even the common chlorophyll exhibit antioxidant activities (Table 5.1) (Roxo and Wink 2022; Wink 2022).

5.4 Antioxidant Activities in the Model System *Caenorhabditis elegans*

Do we have experimental evidence (besides epidemiological data) that antioxidative PSM is effective against ROS? In the laboratory, several chemical assay systems can be employed to investigate the antioxidant activities of compounds, such as CUPRAC, GOR, PHENYL, trolox equivalent antioxidant capacity (TEAC), ABTS, and DPPH (2,2-diphenyl-1-picrylhydrazyl) (Leopoldini et al. 2011; Perron and Brumaghim 2009).

More important are, however, *in vivo* studies with laboratory animals and human intervention studies. In the literature, a large number of studies involving cell cultures, and laboratory animals (fruit flies, rats, mice, rabbits) have been reported, indicating that many antioxidant PSMs are bioavailable and can scavenge ROS inside the cells or body (Barardo et al. 2017; Kenyon 2011; Leonov et al. 2015; Li et al. 2024; Link 2006; Mudd and Liceaga 2022).

In my laboratory, we have used the nematode *C. elegans* as a model system to study the *in vivo* activities of PSM on oxidative stress, gene expression, longevity, and neurodegeneration (summarized in Roxo and Wink 2022; Wink 2022). Therefore, this review has a bias toward our own research.

C. elegans has many advantages as a model system, as it can easily be cultivated in the lab in large numbers and its life expectancy is less than three weeks. Thus, experiments can be carried out with larger sets of animals, which even allows longevity studies in short-term experiments. *C. elegans* shares about 41% of its genes with vertebrates, thus many signal pathways are homologous to those in humans, and findings in *C. elegans* can be relevant for us (Kim et al. 2018; Markaki and Tavernarakis 2010). In addition, scientists have developed many *C. elegans* mutants which help study the metabolic, pharmacological, and toxicological properties of exogenous metabolites (Giunti et al. 2021; Mudd and Liceaga 2022). Furthermore, many hundred mutants (including gene knockouts) have been developed to investigate several diseases (Giunti et al. 2021; Mudd and Liceaga 2022), which are freely available from WormBase. Especially useful are mutants in which fluorescent proteins [such as green fluorescent protein (GFP)] are coupled to disease-related proteins or to gene promotors; their activities can be analyzed by fluorescent microscopy or spectroscopy and quantified by qPCR (quantitative polymerase chain reaction). In our context, it is important to note that several nematode strains have been engineered—GFP has been coupled with genes that are activated by stress, and the transcription factors DAF-16 and SKN-1 have been linked with (Link 2006; Link et al. 1999, 2003). We can distinguish between transcriptional and translational reporters. In transcriptional reporters, the genes of interest, such as HSP-16.2, SOD-3, and GST-4p, have been replaced with GFP; their promoters orchestrate the expression of GFP. In translational reporters (including SKN-1, DAF-16, and HSF-1), GFP is directly connected with the gene of interest and expressed as a fusion protein. The activity of transcriptional and translational reporters can be investigated by fluorescence microscopy (Link et al. 1999, 2003; Link and Johnson 2002; Roxo and Wink 2022).

The ROS levels in nematodes can be investigated with the help of the fluorescing H_2DCFDA (2′, 7′-dichlorofluoresceindiacetate) (Labuschagne and Brenkman 2013). ROS are generated spontaneously; their level in *C. elegans* can be reduced by 20–30% when antioxidative PSM (polyphenols, carotenoids, mustard oils, or chlorophyll), added to the nematode growth medium, had been adsorbed by the worms (Roxo and Wink 2022; Wink 2022). When the pro-oxidant juglone (a naphthoquinone from *Juglans regia*) is added to the growth medium, a burst of toxic ROS is generated. Pre-treatment of the nematodes with polyphenols, carotenoids, or mustard oils can reduce the elevated ROS concentration. Juglone concentrations exceeding 80 μM will kill 70–80% of the nematodes within two days. When the nematodes were pre-treated with antioxidative PSM, their survival rate was substantially higher. These experiments already indicate that even polar PSM, such as phenolics, are adsorbed; they are thus bioavailable and can serve as antioxidants in the body and inside the cells (Roxo and Wink 2022; Wink 2022).

What happens when antioxidants are adsorbed? A central pathway of *C. elegans* to cope with oxidative stress is the insulin/IGF-1 pathway (ILS pathway). The pathway is modulated by insulin-like peptides (ILPs), ligands of the insulin-like transmembrane receptor DAF-2 (homologous to IGFR in mammals). Upon activation, DAF-2 is phosphorylated and activates a series of proteins and enzymes (mostly by phosphorylation) in the signal pathway such as AGE-1 (homologous to PI3K in mammals), phosphatidylinositol 4,5-biphosphate (PIP_2), 3-phosphoinositide-dependent kinase 1 (PDK-1), AKT-1 and AKT-2, and SGK-1 (Hertweck et al. 2004). An important member of this signal pathway is DAF-16 (a transcription factor homologous to Human FOXO3 and FOXO4). The phosphorylated DAF-16 is localized in the cytosol, thus preventing the expression of DAF-16 associated genes. When DAF-16 is not activated, it can migrate to the nucleus and turn on genes related to stress response and aging. SKN-1 (a homologous transcription factor to the mammalian nuclear factor erythroid-2-related factor Nrf) is retained in the cytosol after phosphorylation through the AKT-1/AKT-2/SGK-1 complex similar to DAF-16 (Lee et al. 2003; Lee and Johnson 2004; Link et al. 1999, 2003; Link and Johnson 2002; Roxo and Wink 2022; Tullet et al. 2008).

When the nematodes are challenged by heat stress, ROS, or starvation, the key proteins of the ILS pathway such as PIP_2, AKT-1, and DAF-16/SKN-1 are not phosphorylated. As mentioned previously, only the un-phosphorylated transcription factors DAF-16 and SKN-1 are translocated to the nucleus where they regulate associated genes. Furthermore, other important pathways (mTOR and SIRT) modulate DAF-16 activity (Robida-Stubbs et al. 2012). The transcription factor DAF-16 is important for the transcriptional regulation of several genes encoding proteins important to establish resistance toward stress and the promotion of longevity: Targets are SOD-3, heat shock proteins (HSP-16.2), catalase (CTL-1, CTL-2), and metallothionein-1 (MTL-1) (Landis and Murphy 2010; Lee et al. 2003; Murphy et al. 2003). Genes related to detoxification (e.g. glutathione S-transferase; GST-4) and proteostasis are under the transcriptional control of SKN-1 (Blackwell et al. 2015; Tullet et al. 2008).

As mentioned previously, PSM, such as polyphenols (including tannins), flavonoids, phenylpropanoids, and other phenolics, allicin, mustard oils (sulforaphane), carotenoids, betalains, and even the ubiquitously occurring chlorophylls of vegetables function as active antioxidants in the *C. elegans* system (Table 5.2) (Camboim Rockett et al. 2020; Fu et al. 2021; Gómez-Linton et al. 2021; Guerrero-Rubio et al. 2020; Kampkötter et al. 2007; Kirchweger et al. 2023; Llopis et al. 2019; Mudd and Liceaga 2022; Okoro et al. 2021; Peixoto et al. 2016a, b, 2017, 2018, 2019; Qi et al. 2021; Wang et al. 2018; Wang et al. 2022; Wang and Wink 2016). These antioxidants reduce the ROS load and can also ameliorate the toxic effects of commonly used pro-oxidants (paraquat, tert-butyl hydroperoxide, juglone) (Senchuk et al. 2017) or of heat and osmotic stress. At higher concentrations, several phenolics have the opposite effect and function as pro-oxidants (Roxo and Wink 2022). The antioxidants not only scavenge ROS but can directly modulate the expression of genes necessary to combat ROS. In addition, phenolics and polyphenols can directly influence the activity of the proteins in the ILS pathway (Fraga et al. 2010; Ozdal et al. 2013; Wink 2008a, b, 2015). Phenolics, especially polyphenols, are characterized by several hydroxyl groups. The phenolic hydroxyl groups can dissociate *in vivo* to negatively charged phenolate ions under physiological conditions. The multiple hydroxyl groups can form hydrogen and ionic bonds with proteins that carry positively charged amino groups. As a consequence, the 3D structure of proteins and their bioactivity also becomes compromised (Wink 2008a, b, 2015). Polyphenols not only interact with proteins or DNA (Chanphai and Tajmir-Riahi 2019; Choi et al. 2007; N'Soukpoe-Kossi et al. 2015); they have multiple other activities: they can modulate lipid peroxidation and can influence the fluidity and stability of biomembranes (Fadel et al. 2011; Oteiza et al. 2005; Sirk et al. 2009).

Several antioxidant flavonoids, resveratrol, anthocyanins, catechins, and other polyphenols can inhibit the formation of protein aggregates in nematodes designed as disease models for neurodegenerative diseases such as Morbus Parkinson, Alzheimer's, and Huntington, which are also linked to oxidative stress (Anwar et al. 2021; Brignull et al. 2006; Giunti et al. 2021; Lin et al. 2023; Link 2006; Morley et al. 2002; Regitz et al. 2016; Wu et al. 2006).

In Table 5.2, I have summarized some findings from our lab and other research groups, which demonstrate the bioactivities of PSM (phenolics, carotenoids, mustard oils) in *C. elegans*, and confirm that these PSM are taken up into the body, function inside cells against oxidative stress, and trigger the gene expression in DAF-2, DAF-16, and SKN-1 pathways and other pathways relevant for survival and longevity. Several phenolics, mustard oils, and carotenoids, in addition to caloric restriction, can prolong the lifespan of *C. elegans* by 10–40% (Abbas and Wink 2009; Bene and Salmon 2023; Ding and Zhao 2022; Fu et al. 2021; Gómez-Linton et al. 2021; Kirchweger et al. 2023; Lin et al. 2023; Mudd and Liceaga 2022; Song and Zhang 2023; Wang et al. 2022; Yazaki et al. 2011) (Table 5.2).

In addition to *C. elegans,* we also used cell cultures (Sobeh et al. 2019a), rats, and mice to study the antioxidative potential of PSM in the liver and pancreas. Liver damage, which can lead to cirrhosis, can be induced in rats and mice by treating them with ethanol or toxic chemicals (diethylnitrosamine, cyclophosphamide, galactosamine, chloroform). ROS plays an important role in these health conditions. In our experiments, we could ameliorate ROS-mediated liver damage by treating the animals with extracts from plants containing elevated concentrations of polyphenols, especially of gallotannins, ellagitannins, and catechins (Amen et al. 2020; El-Saied et al. 2018; El-Domiaty et al. 2009; Emam et al. 2019; Ghareeb et al. 2019; Hamdan et al. 2014; Mahmoud et al. 2013; Osman et al. 2016; Sherif et al. 2023; Sobeh et al. 2017a, b, c, 2018a, b, c, d, e, 2019 b,c, 2020; Youssef et al. 2016, 2019).

In rats and mice with artificially induced diabetes (via the application of streptozotocin), the effect could be decreased by treating the animals with extracts from medicinal plants rich in polyphenols, especially with curcumin, gallotannins, ellagitannins, and catechins (Nurcahyanti et al. 2021; Sobeh et al. 2017c, 2019c; Youssef et al. 2017).

5.5 Potential Health Benefits of Herbal Products with Antioxidant Properties

Nutraceuticals often consist of minerals, vitamins, essential amino acids, essential fatty acids, and probiotics but also of PSM with antioxidant, anti-inflammatory, and antimicrobial properties (Wink 2022). Normally, nutraceuticals do not bear a health claim. However, the *European Food Safety Authority* (EFSA) has granted a health claim for several established compounds such as vitamins. Sometimes these compounds are mixed with PSM, such as antioxidants; these nutraceuticals can use the health claim of the vitamins. Although antioxidative PSM may be useful, they are not included in the EFSA list. In the following, we will list some of the positive health aspects of natural antioxidants.

The *in vitro* and *in vivo* studies clearly indicate that some PSM (phenolics, curcumins, gingerols, carotenoids, allicin, mustard oils, and chlorophyll), besides vitamins C and E, have prominent antioxidant activities. Human intervention studies and epidemiological analyses implicate that plant products with antioxidants (spices, medicinal herbs, phytotherapeutics, and nutraceuticals) show positive effects on human health in preventing cancer, arteriosclerosis, cardiovascular diseases, diabetes, and neurodegeneration (overviews in Scalbert et al. 2005; Wang et al. 2011; Devore et al. 2012; Mursu et al. 2014; Kotecha et al. 2016; Dixit et al. 2021a, Nurcahyanti et al. 2022).

Therefore, health and food agencies often advise that our daily diet should contain antioxidants. They can come in the form of vegetables, spices, fruits, fruit juices, nuts, and beverages (tea, coffee, wine). The food industry also promotes the intake of antioxidants in the form of nutraceuticals. It needs to be shown that nutraceuticals are as potent as PSM in a plant matrix. As mentioned in 0, plants produce mixtures of PSM. Some compounds in a mixture, such as saponins, can facilitate the uptake of polar compounds, such as phenolics (Hebestreit and Melzig 2003). This would create a synergistic effect that would not occur when an antioxidant PSM is given alone (as in most nutraceuticals) (Wagner and Ulrich-Merzenich 2009; Wink 2008b, 2015).

In Table 5.3, common food plants, spices, fruits, and plants used in beverages (tea, coffee, mate) are listed, which contain PSM with antioxidant activities and might thus promote health and healthy aging. Several PSMs in these plants also exhibit further pharmacological properties, such as antimicrobial, anti-inflammatory, sedative, carminative, stimulant, estrogenic, and antispasmodic activities. These properties are used in medicinal plants (van Wyk and Wink 2015, 2017) but may also be interesting in the context of food plants.

5.6 Conclusions

Natural selection over more than 400 million years of land plant evolution is responsible for the impressive diversity of PSM. Apparently, PSM evolved for defense against plant-eating animals and microorganisms but also as signal compounds for seed-dispersing and pollinating animals. Plants also use some PSM to store nitrogen or carbon, as antioxidants and UV light protectives. As a consequence of this evolutionary background, all PSMs display one or several biochemical and pharmacological activities. Because PSM exhibit substantial bioactivities, they can be useful for medicine, nutrition, health, and the promotion of longevity (Bene and Salmon 2023; Hartmann 2004; Kirchweger et al. 2023; Krauss and Nies 2014; Li et al. 2024; Varela-López et al. 2023; Wink 1988, 2003, 2008a, 2010a, b, 2015, 2020, 2022).

Plants synthesize and store mixtures of structurally different PSM, which can interfere with a multitude of molecular targets in cells, tissues, and organs (Wink 2000, 2007, 2015, 2022; Wink and Schimmer 2010). PSM in these mixtures not only show additive activities, but some of them also exhibit synergistic interactions (Wagner and Ulrich-Merzenich 2009; Wink 2008b, 2015), e.g. amphiphilic saponins interact with biomembranes and can synergistically facilitate the uptake of polar PSM (Hebestreit and Melzig 2003), such as phenolics.

Many food plants, spices, and fruits (Tables 5.2 and 5.3) contain PSM with antioxidant and other medicinally interesting activities (anti-inflammatory, antispasmodic, analgesic, or antimicrobial properties). These plants and derived nutraceuticals appear to be beneficial for health and longevity. Although a large body of studies has been published, we need controlled clinical or intervention studies to understand not only the benefits but also possible limitations of antioxidants to fight oxidative stress and related health conditions and to promote longevity (Gruber and Halliwell 2017; Kirchweger et al. 2023; Li et al. 2024; Okoro et al. 2021; Song and Zhang 2023). It should be mentioned that many medicinal drugs induce ROS as an important mechanism, for example in the chemotherapy of cancer. The concomitant use of antioxidant food, species, or nutraceuticals would be counterproductive.

Acknowledgments

Several Masters and PhD students have collaborated with me on the phytochemistry and pharmacology of toxic and medicinal plants in my group at the IPMB of Heidelberg University over the years. I would like to thank Mariana Roxo, Shaoxiong Zhang, Erjia Wang, Jehad Dumireih, Hanmei Li, Cornelius Domhan, Junxian Zhou, Sara Shody Thabit, Markus Braun, Felix Heiner, Clarissa Dziggel, Eva Arnold, Tamer Mohamed, Sonja Krstin, Herbenya Peixoto, Mansour Sobeh, Vera Herbel, Pille Link, Farukh Sharopov, Agustina Nurcahyanti, Razan Hamoud, Ikhwan Resmala Sudji, Xiaobo Fan, Shan Su, Rasha Abou Aleinein, Leila Rezaizadehnajafi, Wei Chen, Safaa Eid, Ahmad Tahrani, Dorothea Kaufmann, Florian Herrmann, Mahmoud Zaki El Readi, Sri Mulyaningsih, Lei Zhao, Sami Abbas, Endalkachew Nibret Semegn, Mohamed Ashour, Dalia Hamdan, Tiwatt Kuljanabhagavad, Bernhard Wetterauer, Geeta Pakalapati, Melanie Distl, Yonggang Ma, Maren Möller, Vera Rosenkranz, Pablo Ibieta Hillerns, Marit-Saskia Wahrendorf, Ramon Eloy Aguiar Escobar, Girma Moges Woldemichael, Andreas Tei, Thorsten Schmeller, Assem Mohamed El Shazly, and Birgit Latz-Brüning for their contributions. Our research was supported by grants from the German Science Foundation (DFG), German Academic Exchange Service (DAAD), Chinese Scholarship Council (CSC), Cusanuswerk, CNPq, Coordenação de Aperfeiçoamento de Pessoal de Nível Superior (CAPES), National Council of Science and Technology (CONACYT), Science and Technology Development Fund (STDF), EU, BMFT, Land Baden-Württemberg, and the Egyptian Government.

References

Abbas, S. and Wink, M. (2009). Epigallocatechin gallate (EGCG) from green tea (*Camellia sinensis*) increases lifespan and stress resistance in *Caenorhabditis elegans*. *Planta Medica* 75: 216–221.

Abbas, S. and Wink, M. (2010). Epigallocatechin gallate inhibits beta-amyloid oligomerization in *Caenorhabditis elegans* and affects the daf-2/insulin-like signalling pathway. *Phytomedicine* 17: 902–909.

Abbas, S. and Wink, M. (2014). Green tea extract induces the resistance of *Caenorhabditis elegans* against oxidative stress. *Antioxidants* 3: 129–143.

Alberts, B., Johnson, A., Lewis, J. et al. (2015). *Molecular Biology of the Cell*, 6e. Garland Science: New York, NY.

Amen, Y., Sherif, A.E., Shawky, N.M. et al. (2020). Grape leaf extract attenuates alcohol-induced liver injury *via* interference with NF-κB signalling pathway. *Biomolecules* 2020 (10): 558. https://doi.org/10.3390/biom10040558.

Antil, D., Shumayla, and Upadhyay, S.K. (2023). Evolutionary history of plant receptor-like kinases. In: *Plant Receptor-Like Kinases*, 25–37. Elsevier. https://doi.org/10.1016/b978-0-323-90594-7.00010-7.

Anwar, H.M., Georgy, G.S., Hamad, S.R. et al. (2021). A leaf extract of *Harrisonia abyssinica* ameliorates neurobehavioral, histological and biochemical changes in the hippocampus of rats with aluminum chloride-induced Alzheimer's disease. *Antioxidants (Basel)* 2021 (10): 947. https://doi.org/10.3390/antiox10060947.

Azevedo, B., Pereira, A.M.S., Roxo, M. et al. (2019). Antioxidant activity of an aqueous leaf extract from *Uncaria tomentosa* and its major alkaloids mitraphylline and isomitraphylline in *Caenorhabditis elegans*. *Molecules* 2019 (24): 3299. https://doi.org/10.3390/molecules24183299.

Bakrim, W.B., Nurcahyanti, A.D.R., Dmirieh, M. et al. (2022). Phytochemical profiling of the leaf extract of *Ximenia americana* var. *caffra* and its antioxidant, antibacterial, and anti-ageing activities *in vitro* and in *Caenorhabditis elegans*: a cosmeceutical and dermatological approach. *Oxidative Medicine and Cellular Longevity* 2022: 3486257. doi:10.1155/2022/3486257.

Barardo, D., Thornton, D., Thoppil, H. et al. (2017). The DrugAge database of ageing—related drugs. *Ageing Cell* 16: 594–597.

Bene, M. and Salmon, A.B. (2023). Testing the evidence that lifespan-extending compound interventions are conserved across laboratory animal model species. *Geroscience* 45 (3): 1401–1409. https://doi.org/10.1007/s11357-022-00722-0.

Blackwell, T.K., Steinbaugh, M.J., Hourihan, J.M. et al. (2015). SKN-1/Nrf, stress responses, and ageing in *Caenorhabditis elegans*. *Free Radical Biology & Medicine* 88: 290–301.

Boasquívis, P.F., Silva, G.M.M., Paiva, F.A. et al. (2018). Guarana (*Paullinia cupana*) extract protects *Caenorhabditis elegans* models for Alzheimer disease and Huntington disease through activation of antioxidant and protein degradation pathways. *Oxidative Medicine and Cellular Longevity* 2018: 9241308. https://doi.org/10.1155/2018/9241308.

Bogari, H.A., Rashied, R.M.H., Abdelfattah, M.A.O. et al. (2022). *Euclea divinorum* Hiern: chemical profiling of the leaf extract and its antioxidant activity in silico, in vitro and in *Caenorhabditis elegans* model. *Metabolites* 12: 1031. https://doi.org/10.3390/metabo12111031.

Brignull, H.R., Moore, F.E., Tang, S.J., and Morimoto, R.I. (2006). Polyglutamine proteins at the pathogenic threshold display neuron-specific aggregation in a pan-neuronal *Caenorhabditis elegans* model. *The Journal of Neuroscience* 26: 7597–7606.

Camboim Rockett, F., de Oliveira, S.H., Schmidt, L. et al. (2020). Phenolic compounds and antioxidant activity *in vitro* and *in vivo* of Butia and Opuntia fruits. *Food Research International* 137: 109740. https://doi.org/10.1016/j.foodres.2020.109740.

Chanphai, P. and Tajmir-Riahi, H. (2019). Structural dynamics of DNA binding to tea catechins. *International Journal of Biological Macromolecules* 125: 238–243.

Chen, W., Joubert, E., van Wyk, B.E., and Wink, M. (2013a). Ameliorative effect of aspalathin from rooibos (*Aspalathus linearis*) on acute oxidative stress in *Caenorhabditis elegans*. *Phytomedicine* 20: 380–386.

Chen, W., Müller, D., Richling, E., and Wink, M. (2013c). Anthocyanin-rich purple wheat prolongs the life span of *Caenorhabditis elegans* probably by activating the DAF-16/FOXO transcription factor. *Journal of Agricultural Food Chemistry* 61: 3047–3053.

Chen, W., Rezaizadehnajafi, L., and Wink, M. (2013b). Influence of resveratrol on oxidative stress resistance and life span in *Caenorhabditis elegans*. *The Journal of Pharmacy and Pharmacology* 65: 682–688.

Choi, S.Y., Hwang, J.H., Ko, H.C. et al. (2007). Nobiletin from citrus fruit peel inhibits the DNA-binding activity of NF-kappaB and ROS production in LPS-activated RAW 264.7 cells. *Journal of Ethnopharmacology* 113: 149–155.

Devore, E.E., Kang, J.H., Breteler, M.M., and Grodstein, F. (2012). Dietary intakes of berries and flavonoids in relation to cognitive decline. *Annals of Neurology* 72: 135–143.

Ding, F. and Zhao, Y. (2022). Astaxanthin induces transcriptomic responses associated with lifespan extension in *Caenorhabditis elegans*. *Antioxidants (Basel)* 11 (11): 2115. https://doi.org/10.3390/antiox11112115.

Dixit, S., Shukla, A., Singh, V., and Upadhyay, S.K. (2021a). *Bioprospecting of Natural Compounds for Industrial and Medical Applications*. Wiley https://doi.org/10.1002/9781119718017.ch3.

Duangjan, C., Rangsinth, P., Gu, X. et al. (2019a). Lifespan extending and oxidative stress resistance properties of a leaf extracts from *Anacardium occidentale* L. in *Caenorhabditis elegans*. *Oxidative Medicine and Cellular Longevity* 2019: 9012396. https://doi.org/10.1155/2019/9012396.

Duangjan, C., Rangsinth, P., Gu, X. et al. (2021c, 2021). *Anacardium occidentale* L. leaf extracts protect against glutamate/H_2O_2-induced oxidative toxicity and induce neurite outgrowth in cultured neuronal cells. the involvement of SIRT1/Nrf2 signalling pathway and teneurin 4 transmembrane protein. *Frontiers in Pharmacology* 12: 627738.

Duangjan, C., Rangsinth, P., Gu, X. et al. (2019b). *Glochidion zeylanicum* leaf extracts exhibit lifespan extending and oxidative stress resistance properties in *Caenorhabditis elegans* via DAF-16/FoxO and SKN-1/Nrf-2 signalling pathways. *Phytomedicine* 64: 153061.

Duangjan, C., Rangsinth, P., Gu, X. et al. (2021b). *Vitis vinifera* leaf extract protects against glutamate-induced oxidative toxicity in HT22 hippocampal neuronal cells and increases stress resistance in *Caenorhabditis elegans*. *Frontiers in Nutrition* 8: 634100. https://doi.org/10.3389/fnut.2021.634100.

Duangjan, C., Rangsinth, P., Zhang, S. et al. (2021a, 2021). Neuroprotective effects of *Glochidion zeylanicum* leaf extract against H2O2/glutamate-induced toxicity in cultured neuronal cells and A_-induced toxicity in *Caenorhabditis elegans*. *Biology* 10: 800. https://doi.org/10.3390/biology10080800.

El-Domiaty, M.M., Wink, M., Abdel Aal, M.M. et al. (2009). Antihepatotoxic activity and chemical constituents of *Buddleja asiatica Lour*. *Journal of Biosciences* 64c: 11–19.

El-Saied, M., Sobeh, M., Youssif, L.T. et al. (2018). *Rheum palmatum* root extract inhibits hepatocellular carcinoma in rats treated with diethylnitrosamine. *Journal of Pharmacy and Pharmacology* 70: 821–829. https://doi.org/10.1111/jphp.12899.

Emam, M., El Raey, M.A., El-Haddad, A.E. et al. (2019). A new polyoxygenated flavonol gossypetin-3-*O*-β-D-robinobioside from *Caesalpinia gilliesii* (Hook.)D. Dietr. and *in vivo* hepatoprotective, anti-inflammatory and anti-ulcer activities of the leaf methanol extract. *Molecules* 24: 138. doi:10.3390/molecules24010138.

Fadel, O., El Kirat, K., and Morandat, S. (2011). The natural antioxidant rosmarinic acid spontaneously penetrates membranes to inhibit lipid peroxidation in situ. *Biochimica et biophysica Acta (BBA)-Biomembranes* 1808: 2973–2980.

Fraga, C.G., Galleano, M., Verstraeten, S.V., and Oteiza, P.I. (2010). Basic biochemical mechanisms behind the health benefits of polyphenols. *Molecular Aspects of Medicine* 31: 435–445.

Franceschi, C., Garagnani, P., Morsiani, C. et al. (2018). The continuum of ageing and age-related diseases: common mechanisms but different rates. *Frontiers in Medicine* 2018 (5): 61.

de Freitas Bonomo, L., Silva, D.N., Boasquivis, P.F. et al. (2014). Açaí (*Euterpe oleracea* Mart.) modulates oxidative stress resistance in *Caenorhabditis elegans* by direct and indirect mechanisms. *PloS One* 9: e89933.

de Freitas Rodrigues, C., Ramos Boldori, J., Valandro Soares, M. et al. (2021 Jun). Goji berry (*Lycium barbarum* L.) juice reduces lifespan and premature ageing of *Caenorhabditis elegans*: is it safe to consume it? *Food Research International* 144: 110297. https://doi.org/10.1016/j.foodres.2021.110297.

Fu, M., Zhang, X., Zhang, X. et al. (2021). Autophagy plays a role in the prolongation of the life span of *Caenorhabditis elegans* by astaxanthin. *Rejuvenation Research* 24 (3): 198–205. https://doi.org/10.1089/rej.2020.2355.

Ghareeb, M.A., Sobeh, M., El-Maadawy, W.H. et al. (2019). Chemical profiling of polyphenolics in *Eucalyptus globulus* and evaluation of its hepato-renal potential against cyclophosphamide induced toxicity in mice. *Antioxidants* 8: 415.

Giunti, S., Andersen, N., Rayes, D., and De Rosa, M.J. (2021). Drug discovery: insights from the invertebrate *Caenorhabditis elegans*. *Pharmacology Research & Perspectives* 9: e00721.

Gómez-Linton, D.R., Alavez, S., Navarro-Ocaña, A. et al. (2021). Achiote (*Bixa orellana*) lipophilic extract, bixin, and δ-tocotrienol effects on lifespan and stress resistance in *Caenorhabditis elegans*. *Planta Medica* 87 (5): 368–374. https://doi.org/10.1055/a-1266-6674.

Gruber, J. and Halliwell, B. (2017). Approaches for extending human healthspan: from antioxidants to healthspan pharmacology. *Essays in Biochemistry* 61 (3): 389–399. https://doi.org/10.1042/EBC20160091.

Guerrero-Rubio, M.A., Hernández-García, S., Escribano, J. et al. (2020). Betalain health-promoting effects after ingestion in *Caenorhabditis elegans* are mediated by DAF-16/FOXO and SKN-1/Nrf2 transcription factors. *Food Chemistry* 330: 127228.

Hamdan, D.I., Wink, M., and El-Shazly, A.M. (2014). Secondary metabolites isolated from the dichloromethane extract of rough lemon stem and hepatoprotective effect of limonianin. *British Journal of Pharmaceutical Research* 4: 1963–1975.

Harborne, J.B. and Baxter, H. (1993). *Phytochemical Dictionary. A Handbook of Bioactive Compounds from Plants*. London: Taylor & Francis.

Hartmann, T. (2004). Plant-derived secondary metabolites as defensive chemicals in herbivorous insects: a case study in chemical ecology. *Planta* 219: 1–4.

Hartmann, T. (2007). From waste products to ecochemicals: fifty years research of plant secondary metabolism. *Phytochemistry* 68: 2831–2846.

Hebestreit, P. and Melzig, M.F. (2003). Cytotoxic activity of the seeds from *Agrostemma githago* var. *githago*. *Planta Medica* 69: 921–925. https://doi.org/10.1055/s-2003-45101.

Heiner, F., Feistel, B., and Wink, M. (2018). *Sideritis scardica* extracts inhibit aggregation and toxicity of amyloid-? in *Caenorhabditis elegans* used as a model for Alzheimer's disease. *PeerJ* 6: e4683.

Hertweck, M., Göbel, C., and Baumeister, R. (2004). *C. elegans* SGK-1 is the critical component in the Akt/PKB kinase complex to control stress response and life span. *Developmental Cell* 6: 577–588.

Imamura, T., Isozumi, N., Higashimura, Y. et al. (2022). Red-beet betalain pigments inhibit amyloid-β aggregation and toxicity in amyloid-β expressing *Caenorhabditis elegans*. *Plant Foods for Human Nutrition* 77 (1): 90–97. https://doi.org/10.1007/s11130-022-00951-w.

Kampkötter, A., Nkwonkam, C.G., Zurawski, R.F. et al. (2007). Effects of the flavonoids kaempferol and fisetin on thermotolerance, oxidative stress and FoxO transcription factor DAF-16 in the model organism *Caenorhabditis elegans*. *Archives of Toxicology* 81: 849–858.

Kaur, A., Madhu, Taneja, M., and Upadhyay, S.K. (2021). Applications of calcium transport elements in plant improvement. In: *Calcium Transport Elements in Plants*, 427–445. Elsevier. https://doi.org/10.1016/b978-0-12-821792-4.00022-9.

Kenyon, C. (2011). The first long-lived mutants: discovery of the insulin/IGF-1 pathway for ageing. *Philosophical Transactions of the Royal Society of London. Series B, Biological Sciences* 366: 9–16.

Kim, W., Underwood, R.S., Greenwald, I., and Shaye, D.D. (2018). OrthoList 2: a new comparative genomic analysis of human and *Caenorhabditis elegans* genes. *Genetics* 210: 445–461.

Kirchweger, B., Zwirchmayr, J., Grienke, U., and Rollinger, J.M. (2023). The role of *Caenorhabditis elegans* in the discovery of natural products for healthy ageing. *Natural Product Reports* 40 (12): 1849–1873. https://doi.org/10.1039/d3np00021d.

Koch, K., Weldle, N., Baier, S. et al. (2020, 2020). *Hibiscus sabdariffa* L. extract prolongs lifespan and protects against amyloid-beta toxicity in *Caenorhabditis elegans*: involvement of the FoxO and Nrf2 orthologues DAF-16 and SKN-1. *European Journal of Nutrition* 59: 137–150.

Kotecha, R., Takami, A., and Espinoza, J.L. (2016). Dietary phytochemicals and cancer chemoprevention: a review of the clinical evidence. *Oncotarget* 7: 52517–52529.

Krauss, G.J. and Nies, D.H. (2014). *Ecological Biochemistry: Environmental and Interspecific interactions*. Weinheim: Wiley-VCH.

Labuschagne, C.F. and Brenkman, A.B. (2013). Current methods in quantifying ROS and oxidative damage in *Caenorhabditis elegans* and other model organism of ageing. *Ageing Research Reviews* 12: 918–930.

Landis, J.N. and Murphy, C.T. (2010). Integration of diverse inputs in the regulation of *C. elegans* DAF-16/FOXO. *Developmental Dynamics* 239: 1405–1412.

Lee, J.M. and Johnson, J.A. (2004). An important role of Nrf2-ARE pathway in the cellular defense mechanism. *BMB Reports* 37: 139–143.

Lee, S.S., Kennedy, S., Tolonen, A.C., and Ruvkun, G. (2003). DAF-16 target genes that control *C. elegans* life-span and metabolism. *Science* 300: 644–647.

Leonov, A., Arlia-Ciommo, A., Piano, A. et al. (2015). Longevity extension by phytochemicals. *Molecules* 20: 6544–6572.

Leopoldini, M., Russo, N., and Toscano, M. (2011). The molecular basis of working mechanism of natural polyphenolic antioxidants. *Food Chemistry* 125: 288–306.

Li, Y., Li, P., Zhang, W. et al. (2024). New wine in old bottle: *Caenorhabditis elegans* in food science. *Food Reviews International* 40 (1): 533–573. doi:10.1080/87559129.2023.2172429.

Li, H., Roxo, M., Cheng, X. et al. (2019). Pro-oxidant and lifespan extension effects of caffeine and related methylxanthines in *Caenorhabditis elegans. Food Chemistry* https://doi.org/10.1016/j.fochx.2019.100005.

Lima, M.E., Colpo, A.C., Salgueiro, W.G. et al. (2014). *Ilex paraguariensis* extract increases lifespan and protects against the toxic effects caused by paraquat in *Caenorhabditis elegans. International Journal of Environmental Research and Public Health* 11: 10091–10104.

Lin, Y., Lin, C., Cao, Y., and Chen, Y. (2023 Nov). *Caenorhabditis elegans* as an in vivo model for the identification of natural antioxidants with anti-ageing actions. *Biomedicine & Pharmacotherapy* 167: 115594. https://doi.org/10.1016/j.biopha.2023.115594.

Link, C.D. (2006). *C. elegans* models of age-associated neurodegenerative diseases: lessons from transgenic worm models of Alzheimer's disease. *Experimental Gerontology* 41: 1007–10013.

Link, C.D., Cypser, J.R., Johnson, C.J., and Johnson, T.E. (1999). Direct observation of stress response in *Caenorhabditis elegans* using a reporter transgene. *Cell Stress & Chaperones* 4: 235.

Link, P., Fu, Y., Wetterauer, B., and Wink, M. (2015). Extracts of *Glycyrrhiza uralensis* and isoliquiritigenin, one of its compounds, counteract amyloid-beta toxicity in *Caenorhabditis elegans. Planta Medica* 81: 357–362.

Link, C.D. and Johnson, C.J. (2002). Reporter transgenes for study of oxidant stress in *Caenorhabditis elegans. Methods in Enzymology* 353: 497–505.

Link, P., Roth, K., Sporer, F., and Wink, M. (2016). *Carlina acaulis* and carlina oxide exhibit antioxidant activity and counteracts Aβ toxicity in *Caenorhabditis elegans. Molecules* 21: 871. https://doi.org/10.3390/molecules21070871.

Link, C.D., Taft, A., Kapulkin, V. et al. (2003). Gene expression analysis in a transgenic *Caenorhabditis elegans* Alzheimer's disease model. *Neurobiol Ageing* 24: 397–413.

Link, P. and Wink, M. (2019). Isoliquiritigenin exerts antioxidant activity in *Caenorhabditis elegans* via insulin-like signalling pathway and SKN-1. *Phytomedicine* 55: 119–124.

Llopis, S.P., Rodrigo, M.J., González, N. et al. (2019). β-Cryptoxanthin reduces body fat and increases oxidative stress response in *Caenorhabditis elegans* model. *Nutrients* 11: 232.

Luo, S., Jiang, X., Jia, L. et al. (2019). *In vivo* and *in vitro* antioxidant activities of methanol extracts from olive leaves on *Caenorhabditis elegans. Molecules* 2019 (24): 704.

Madhu, Kaur, A., Shekhawat, J., and Upadhyay, S.K. (2024). An introduction to the defense-related proteins in plants. In: *Defense-Related Proteins in Plants*, 1–35. Elsevier. https://doi.org/10.1016/b978-0-443-13236-0.00001-4.

Maeshima, M. (2001). Tonoplast transporters: Organization and Function. *Annual Review of Plant Physiology and Plant Molecular Biology* 52: 469–497. https://doi.org/10.1146/annurev.arplant.52.1.469.

Mahmoud, M.F., Hamdan, D.I., Wink, M., and El-Shazly, A.M. (2013). Hepatoprotective effect of limonin a natural limonoid from the seed of *Citrus auratium* var. *bigaradia* on D-galactosamine-induced liver injury in rats. *Naunyn-Schmiedeberg's Archives of Pharmacology* https://doi.org/10.1007/s00210-013-0937-1.

Markaki, M. and Tavernarakis, N. (2010). Modeling human diseases in *Caenorhabditis elegans*. *Biotechnology Journal* 5: 1261–1276.

Moliner, C., Barros, L., Dias, M.I. et al. (2018). Edible flowers of *Tagetes erecta* L. as functional ingredients: phenolic composition, antioxidant and protective effects on *Caenorhabditis elegans*. *Nutrients* 10 (12): 2002. https://doi.org/10.3390/nu10122002.

Morley, J.F., Brignull, H.R., Weyers, J.J., and Morimoto, R.I. (2002). The threshold for polyglutamine-expansion protein aggregation and cellular toxicity is dynamic and influenced by ageing in *Caenorhabditis elegans*. *PNAS* 99: 10417–10422.

Mudd, N. and Liceaga, A.M. (2022). *Caenorhabditis elegans* as an *in vivo* model for food bioactives: a review. *Current Research in Food Science* 13 (5): 845–856. https://doi.org/10.1016/j.crfs.2022.05.001.

Murphy, C.T., McCarroll, S.A., Bargmann, C.I. et al. (2003). Genes that act downstream of DAF-16 to influence the lifespan of *Caenorhabditis elegans*. *Nature* 424: 277–283.

Mursu, J., Virtanen, J.K., Tuomainen, T.-P. et al. (2014). Intake of fruit, berries, and vegetables and risk of type 2 diabetes in Finnish men: the Kuopio Ischaemic Heart Disease Risk Factor Study. *The American Journal of Clinical Nutrition* 99: 328–333.

Nishioka, Y., Nishikawa, S., and Shibata, T. (2020). A hot water extract of *Sideritis scardica* prolongs life span and enhances heat shock resistance in *Caenorhabditis elegans*. *Natural Product Communications* 2020: 15.

N'Soukpoe-Kossi, C., Bourassa, P., Mandeville, J. et al. (2015). Structural modeling for DNA binding to antioxidants resveratrol, genistein and curcumin. *Journal of Photochemistry and Photobiology B: Biology* 151: 69–75.

Nurcahyanti, A.D.R., Cokro, F., Wulanjati, M.P. et al. (2022). Curcuminoid for metabolic syndrome: meta-analysis evidences toward personalized prevention and treatment management. *Frontiers in Nutrition* 9: 891339. https://doi.org/10.3389/fnut.2022.89133.

Nurcahyanti, A.D.R., Jap, A., Lady, J. et al. (2021, 2021). Function of selected natural antidiabetic compounds with potential against cancer via modulation of the PI3K/AKT/mTOR cascade. *Biomedicine & Pharmacotherapy* 144: 112138.

Okoro, N.O., Odiba, A.S., Osadebe, P.O. et al. (2021). Bioactive phytochemicals with anti-ageing and lifespan extending potentials in *Caenorhabditis elegans*. *Molecules* 26 (23): 7323. https://doi.org/10.3390/molecules26237323.

de Oliveira Caland, R.B., Cadavid, C.O.M., Carmona, L. et al. (2019). Pasteurized orange juice rich in carotenoids protects *Caenorhabditis elegans* against oxidative stress and β-amyloid toxicity through direct and indirect mechanisms. *Oxidative Medicine and Cellular Longevity* 18: 5046280. https://doi.org/10.1155/2019/5046280.

Osman, S.M., El-Haddad, A.E., El-Raey, M. et al. (2016). A new octadecenoic acid derivative from *Caesalpinia gilliesii* flowers with potent hepatoprotective activity. *Pharmacognosy Magazine* 12: S332–S336.

Oteiza, P.I., Erlejman, A.G., Verstraeten, S.V. et al. (2005). Flavonoid-membrane interactions: a protective role of flavonoids at the membrane surface? *Clinical and Developmental Immunology* 12: 19–25.

Ozdal, T., Capanoglu, E., and Altay, F. (2013). A review on protein–phenolic interactions and associated changes. *Foodservice Research International* 51: 954–970.

Peixoto, H., Roxo, M., Koolen, H. et al. (2018). *Calycophyllum spruceanum* (Benth.), the Amazonian "tree of youth" prolongs longevity and enhances stress resistance in *Caenorhabditis elegans*. *Molecules* 23: 534. https://doi.org/10.3390/molecules23030534.

Peixoto, H., Roxo, M., Krstin, S. et al. (2016b). An anthocyanin-rich extract of Açaí (*Euterpe precatoria* Mart.) increases stress resistance and retards ageing related markers in *Caenorhabditis elegans*. *Journal of Agricultural and Food Chemistry* 64: 1283–1290.

Peixoto, H., Roxo, M., Krstin, S., and Wink, M. (2016a). Anthocyanin-rich extracts of Acai (*Euterpe precatoria* Mart.) mediate neuroprotective activities *in Caenorhabditis elegans*. *Journal of Functional Foods* 26: 385–393.

Peixoto, H., Roxo, M., Röhrig, T. et al. (2017). Anti-ageing and antioxidant potential of *Paullinia cupana* var. sorbilis: findings in *Caenorhabditis elegans* indicate a new utilization for roasted seeds of guarana. *Medicines* 4: 61.

Peixoto, H., Roxo, M., Silva, E. et al. (2019). Bark extract of the amazonian tree *Endopleura uchi* (Humiriaceae) extends lifespan and enhances stress resistance in *Caenorhabditis elegans*. *Molecules* 2019 (24): 915.

Perron, N.R. and Brumaghim, J.L. (2009). A review of the antioxidant mechanisms of polyphenol compounds related to iron binding. *Cell Biochemistry and Biophysics* 53: 75–100.

Prasanth, M.I., Brimson, J.M., Chuchawankul, S. et al. (2019). Antiageing, stress resistance, and neuroprotective efficacies of *Cleistocalyx nervosum* var. paniala fruit extracts using *Caenorhabditis elegans* model. *Oxidative Medicine and Cellular Longevity* 2019: 7024785.

Qi, Z., Ji, H., Le, M. et al. (2021). Sulforaphane promotes *C. elegans* longevity and healthspan by DAF-16/DAF-2 insulin/IGF-1 signaling. *Aging* 13: 1649–1670.

Rangsinth, P., Prasansuklab, A., Duangjan, C. et al. (2019). Leaf extract of *Caesalpinia mimosoides* enhances oxidative stress resistance and prolongs lifespan in *Caenorhabditis elegans*. *BMC Complementary and Alternative Medicine* 19: 1–13.

Regitz, C., Fitzenberger, E., Mahn, F.L. et al. (2016). Resveratrol reduces amyloid-beta (A β 1–42)-induced paralysis through targeting proteostasis in an Alzheimer model of *Caenorhabditis elegans*. *European Journal of Nutrition* 55: 741–747.

Rehman, R., Hanif, M.A., Mushtaq, Z. et al. (2016). Biosynthetic factories of essential oils: the aromatic plants. *Natural Products Chemistry & Research* 4: 227. https://doi.org/10.4172/2329-6836.1000227.

Rezaizadehnajafi, L. and Wink, M. (2014). EPs7630® from *Pelargonium sidoides* increases stress resistance in *Caenorhabditis elegans* and enhances life span probably via the DAF-16/FOXO pathway. *Phytomedicine* 21: 547–550.

Robida-Stubbs, S., Glover-Cutter, K., Lamming, D.W. et al. (2012). TOR signaling and rapamycin influence longevity by regulating SKN-1/Nrf and DAF-16/FoxO. *Cell Metabolism* 15: 713–724.

Roxo, M., Peixoto, H., Wetterauer, P. et al. (2020). Piquiá shells (*Caryocar villosum*): a fruit by-product with antioxidant and antiageing properties in Caenorhabditis elegans. *Oxidative Medicine and Cellular Longevity* 2020: 7590707.

Roxo, M. and Wink, M. (2022). The use of the nematode *Caenorhabditis elegans* to study antioxidant and longevity promoting plant secondary metabolites. In: *New Findings from Natural Products*, vol. 2022 (ed. E. Ovidi, T. Karpinski, and A. Tiezzi), 133–163. Bentham Books.

Saier, C., Gommlich, I., Hiemann, V. et al. (2018). *Agrimonia procera* Wallr. Extract increases stress resistance and prolongs life span in *Caenorhabditis elegans* via transcription factor DAF-16 (FoxO Orthologue). *Antioxidants* 7: 192.

Scalbert, A., Manach, C., Morand, C. et al. (2005). Dietary polyphenols and the prevention of diseases. *Critical Reviews in Food Science and Nutrition* 2005 (45): 287–306.

Scerbak, C., Vayndorf, E., Hernandez, A. et al. (2018). Lowbush cranberry acts through DAF-16/FOXO signaling to promote increased lifespan and axon branching in ageing posterior touch receptor neurons. *Geroscience* 40: 151–162.

Seigler, D.S. (1998). *Plant Secondary Metabolism*. Dordrecht, London, Boston: Kluwer Academic Publishers.

Senchuk, M.M., Dues, D., and Van Raamsdonk, J.M. (2017). Measuring oxidative stress in *Caenorhabditis elegans*: paraquat and juglone sensitivity assays. *Bio-Protocol* 7: e2086.

Sherif, A.E., Amen, Y., Wink, M. et al. (2023). Psidium guajava L.: chemical composition and protective effects of a leaf extract against ethanol-induced cardiotoxicity. *South African Journal of Botany* 162: 334341.

Shields, H.J., Traa, A., and Van Raamsdonk, J.M. (2021). Beneficial and detrimental effects of reactive oxygen species on lifespan: a comprehensive review of comparative and experimental studies. *Frontiers in Cell and Development Biology* 11 (9): 628157. https://doi.org/10.3389/fcell.2021.628157.

Shumayla and Upadhyay, S.K. (2023). An overview of receptor-like kinases in plants. In: *Plant Receptor-Like Kinases*, 1–23. Elsevier. https://doi.org/10.1016/b978-0-323-90594-7.00005-3.

Sies, H. and Jones, D.P. (2020). Reactive oxygen species (ROS) as pleiotropic physiological signalling agents. *Nature Reviews Molecular Cell Biology* 2020 (21): 363–383.

Sirk, T.W., Brown, E.F., Friedman, M., and Sum, A.K. (2009). Molecular binding of catechins to biomembranes: relationship to biological activity. *Journal of Agricultural and Food Chemistry* 57: 6720–6728.

Sobeh, M., ElHawary, E., Peixoto, H. et al. (2016). Identification of phenolic secondary metabolites from *Schotia brachypetala* Sond. (Fabaceae) and demonstration of their antioxidant activities in *Caenorhabditis elegans*. *PeerJ* 4: e2404.

Sobeh, M., El-Raey, M., Rezq, S. et al. (2019b). Chemical profiling of secondary metabolites of *Eugenia uniflora* and their antioxidant, anti-inflammatory, pain killing and anti-diabetic activities: a comprehensive approach. *Journal of Ethnopharmacology* 240: 111939. https://doi.org/10.1016/j.jep.2019.111939.

Sobeh, M., Esmat, A., Petruk, G. et al. (2018e). Phenolic compounds from *Syzygium jambos* (Myrtaceae) exhibit distinct antioxidant and hepatoprotective activities *in vivo*. *Journal of Functional Foods* 41: 223–231.

Sobeh, M., Hamza, M.S., Ashour, M.L. et al. (2020). A polyphenol rich fraction from *Eugenia uniflora* exhibits antioxidant and hepatoprotective activities *in vivo*. *Pharmaceuticals* 2020 (13): 84. https://doi.org/10.3390/ph13050084.

Sobeh, M., Mahmoud, M., Abdelfattah, M.A.O. et al. (2017c). Hepatoprotective and hypoglycemic effects of a tannin rich extract from *Ximenia americana* var. *caffra* root. *Phytomedicine* 33: 36–42.

Sobeh, M., Mahmoud, M.F., Abdelfattah, M.A. et al. (2018d). A proanthocyanidin-rich extract from *Cassia abbreviata* exhibits antioxidant and hepatoprotective activities *in vivo*. *Journal of Ethnopharmacology* 213: 38–47.

Sobeh, M., Mahmoud, M.F., Abdelfattah, M.A.O. et al. (2017a). *Albizia harveyi:*: phytochemical profiling, antioxidant, antidiabetic and hepatoprotective activities of the bark extract. *Medicinal Chemistry Research* 26: 3091–3105. https://doi.org/10.1007/s00044-017-2005-8.

Sobeh, M., Mahmoud, M.F., Hasan, R.A. et al. (2018b). Tannin-rich extracts from *Lannea stuhlmannii* and *Lannea humilis* (Anacardiaceae) exhibit hepatoprotective activities *in vivo via* enhancement of the anti-apoptotic protein Bcl-2. *Scientific Reports* 8: 9343. https://doi.org/10.1038/s41598-018-27452-8.

Sobeh, M., Mahmoud, M.F., Hasan, R.A. et al. (2017b). *Senna singueana*: antioxidant, hepatoprotective, antiapoptotic properties and phytochemical profiling of a methanol bark extract. *Molecules* 2017 (22): 1502. doi:10.3390/molecules22091502.

Sobeh, M., Mahmoud, M.F., Hasan, R.A. et al. (2019c). Chemical composition, antioxidant and hepatoprotective activities of methanol extracts from leaves of *Terminalia bellirica* and *Terminalia sericea* (Combretataceae). *PeerJ* 7: e6322. https://doi.org/10.7717/peerj6322.

Sobeh, M., Mahmoud, M.F., Petruk, G. et al. (2018c). *Syzygium aqueum*: a polyphenol rich leaf extract exhibits antioxidant, hepatoprotective, pain-killing and anti-inflammatory activities in animal models. *Frontiers in Pharmacology* 9: 566. doi:10.3389/fphar.2018.00566.

Sobeh, M., Petruk, G., Osman, S. et al. (2019a). Isolation of myricitrin and 3,5-di-O-methylgossypetin from *Syzygium samarangense* and evaluation of their involvement to protect keratinocytes against oxidative stress via activation of the Nrf-2 pathway. *Molecules* 24 (9): 1839.

Sobeh, M., Youssef, F.S., Esmat, A. et al. (2018a). High resolution UPLC-MS/MS profiling of the polyphenolics in the methanol extract from *Syzygium samarangense* leaves and its hepatoprotective activity in rats with CCl_4-induced hepatic damage. *Food and Chemical Toxicology* 113: 145–153.

Song, L. and Zhang, S. (2023). Anti-ageing activity and modes of action of compounds from natural food sources. *Biomolecules* 13 (11): 1600. https://doi.org/10.3390/biom13111600.

Su, S. and Wink, M. (2015). Natural lignans from *Arctium lappa* as antiageing agents in *Caenorhabditis elegans*. *Phytochemistry* 117: 340–350.

Tambara, A.L., Moraes, L.L.S., Dal Forno, A.H. et al. (2018). Purple pitanga fruit (*Eugenia uniflora* L.) protects against oxidative stress and increase the lifespan in *Caenorhabditis elegans* via the DAF-16/FOXO pathway. *Food and Chemical Toxicology* 120: 639–650.

Tawfeek, N., Sobeh, M., Hamdan, D.I. et al. (2019). Phenolic compounds from *Populus alba* L. and *Salix subserrata* Willd. (Salicaceae) counteract oxidative stress in *Caenorhabditis elegans*. *Molecules* 24: 1999. doi:10.3390/molecules241019.

Thabit, S., Handoussa, H., El Sayed, N.S. et al. (2023). A fruit extract of *Styphnolobium japonicum* (L.) counteracts oxidative stress and mediates neuroprotection in *Caenorhabditis elegans* possibly through the SKN-1/Nrf-2 pathway. *BMC Complementary Medicine and Therapies* 23: 330.

Thabit, S., Handoussa, H., Roxo, M. et al. (2019). *Styphnolobium japonicum* (L.) Schott fruits increase stress resistance and exert antioxidant properties in *Caenorhabditis elegans* and mouse models. *Molecules* 24 (14): 2633.

Thabit, S., Handoussa, H., Roxo, M. et al. (2018). Evaluation of antioxidant and neuroprotective activities of *Cassia fistula* (L.) using the *Caenorhabditis elegans* model. *PeerJ* 6: e5159.

Tullet, J.M., Hertweck, M., An, J.H. et al. (2008). Direct inhibition of the longevity-promoting factor SKN-1 by insulin-like signalling in C. elegans. *Cell* 132: 1025–1038.

Upadhyay, S.K. (2021). *Calcium Transport Elements in Plants*. Elsevier. https://doi.org/10.1016/c2019-0-04172-6.

Upadhyay, S.K. (2024). *Defense-Related Proteins in Plants*. Elsevier. https://doi.org/10.1016/c2022-0-02023-7.

Upadhyay, S.K. and Shumayla (2022). *Plant Receptor-Like Kinases*. Elsevier. https://doi.org/10.1016/c2020-0-03988-5.

Upadhyay, S.K. and Singh, S.P. (2021). *Bioprospecting of Plant Biodiversity for Industrial Molecules*. Wiley https://doi.org/10.1002/9781119718017.

Upadhyay, S.K. and Singh, S.P. (2023). *Plants as Bioreactors for Industrial Molecules*. Wiley https://doi.org/10.1002/9781119875116.

Van Wyk, B.E. (2014). *Culinary Herbs & Spices of the World*. Pretoria: BRIZA.

Van Wyk, B.E. and Wink, M. (2015). *Phytomedicines, Herbal Drugs and Poisons*. London: Kew Publishing.

Van Wyk, B.E. and Wink, M. (2017). *Medicinal Plants of the World*, 2ee. Wallingford: CABI.

Varela-López, A., Romero-Márquez, J.M., Navarro-Hortal, M.D. et al. (2023). Dietary antioxidants and lifespan: relevance of environmental conditions, diet, and genotype of experimental models. *Experimental Gerontology* 178: 112221. https://doi.org/10.1016/j.exger.2023.112221.

Wagner, H. and Ulrich-Merzenich, G. (2009). Synergy research: approaching a new generation of phytopharmaceuticals. *Phytomedicine* 16: 97–110. https://doi.org/10.1016/j.phymed.2008.12.018.

Wang, H., Liu, J., Li, T., and Liu, R.H. (2018). Blueberry extract promotes longevity and stress tolerance via DAF-16 in *Caenorhabditis elegans*. *Food & Function* 9: 5273–5282.

Wang, S., Melnyk, J.P., Tsao, R., and Marcone, M.F. (2011). How natural dietary antioxidants in fruits, vegetables and legumes promote vascular health. *Foodservice Research International* 44: 14–22.

Wang, B., Tang, X., Mao, B. et al. (2022). Anti-ageing effects and mechanisms of anthocyanins and their intestinal microflora metabolites. *Critical Reviews in Food Science and Nutrition* 21: 1–17. https://doi.org/10.1080/10408398.2022.2123444.

Wang, E. and Wink, M. (2016). Chlorophyll enhances oxidative stress tolerance in *Caenorhabditis elegans* and extends its lifespan. *PeerJ* 4: e1879. https://doi.org/10.7717/peerj.1879.

Wink, M. (1988). Plant breeding: importance of plant secondary metabolites for protection against pathogens and herbivores. *Theoretical and Applied Genetics* 75: 225–233.

Wink, M. (1997). Compartmentation of secondary metabolites and xenobiotics in plant vacuoles. *Advances in Botanical Research* 25: 141–169.

Wink, M. (2000). Interference of alkaloids with neuroreceptors and ion channels. *Bioactive Natural Products* 21: 3–129.

Wink, M. (2003). Evolution of secondary metabolites from an ecological and molecular phylogenetic perspective. *Phytochemistry* 64: 3–19.

Wink, M. (2007). Molecular modes of action of cytotoxic alkaloids—from DNA intercalation, spindle poisoning, topoisomerase inhibition to apoptosis and multiple drug resistance. *The Alkaloids Chemistry and Biology* 64: 1–48.

Wink, M. (2008a). Plant secondary metabolism: diversity, function and its evolution. *Natural Product Communications* 3: 1205–1216.

Wink, M. (2008b). Evolutionary advantage and molecular modes of action of multi-component mixtures used in phytomedicine. *Current Drug Metabolism* 9: 996–1009.

Wink, M. (2010a). *Biochemistry of Plant Secondary Metabolism. Annual Plant Reviews*, vol. 40. Chichester: Wiley-Blackwell.

Wink, M. (2010b). *Function of Plant Secondary Metabolites and Their Exploitation in Biotechnology. Annual Plant Reviews*, vol. 39. Chichester: Wiley-Blackwell.

Wink, M. (2013). Evolution of secondary metabolites in legumes (Fabaceae). *South African Journal of Botany* 89: 164–175.

Wink, M. (2015). Modes of action of herbal medicines and plant secondary metabolites. *Medicine* 2: 251–286.

Wink, M. (2019). Quinolizidine and pyrrolizidine alkaloid chemical ecology—a mini-review on their similarities and differences. *Journal of Chemical Ecology* 45: 109–115. https://doi.org/10.1007/s10886-018-1005-6.

Wink, M. (2020). Evolution of the angiosperms and co-evolution of secondary metabolites, especially of alkaloids. In: *Co-Evolution of Secondary Metabolites. Reference Series in Phytochemistry* (ed. J.-M. Mérillon and K.G. Ramawat), 1–24. Springer https://doi.org/10.1007/978-3-319-76887-8_22-1.

Wink, M. (2022, 2022). Current understanding of modes of action of multicomponent bioactive phytochemicals: potential for nutraceuticals and antimicrobials. *Annual Review of Food Science and Technology* 13: 337–359.

Wink, M. and Abbas, S. (2012). Epigallocatchin gallate (EGCG) from green Tea (*Camellia sinensis*) and other natural products mediate stess resistance and slow down ageing processes in *Caenorhabditis elegans*. In: *Tea in Health and Disease Prevention*, 1005–1115. Elsevier.

Wink, M. and Schimmer, O. (2010). Molecular modes of action of defensive secondary metabolites. In: *Functions and Biotechnology of Plant Secondary Metabolites*, Annual Plant Reviews, vol. 39 (ed. M. Wink), 21–161. Wiley.

Wink, M. and Van Wyk, B.E. (2008). *Mind-Altering and Poisonous Plants of the World*. Timber Press.

Wu, Y., Wu, Z., Butko, P. et al. (2006). Amyloid-?-induced pathological behaviors are suppressed by Ginkgo biloba extract EGb 761 and ginkgolides in transgenic *Caenorhabditis elegans*. *The Journal of Neuroscience* 26: 13102–13113.

Yazaki, K., Yoshikoshi, C., Oshiro, S., and Yanase, S. (2011). Supplemental cellular protection by a carotenoid extends lifespan via Ins/IGF-1 signalling in *Caenorhabditis elegans*. *Oxidative Medicine and Cellular Longevity* 2011: 596240. https://doi.org/10.1155/2011/596240.

Youssef, F., Ashour, M.L., Ebada, S.S. et al. (2017). Anti-hyperglycaemic activity of the methanol extract from leaves of *Eremophila maculata* (Scrophulariaceae) in streptozotocin-induced diabetic rats. *Journal of Pharmacy and Pharmacology* 69: 733–742.

Youssef, F., Ashour, M., Elbeshbishy, H. et al. (2019, 2019). Metabolic profiling of *Buddleia indica* leaves using LC/MS and evidence of their antioxidant and hepatoprotective activity using different in vitro and in vivo experimental models. *Antioxidants* 8: 412. doi:10.3390/antiox8090412.

Youssef, F.S., Ashour, L., Sobeh, M. et al. (2016). *Eremophila maculata*—isolation of a rare naturally-occurring lignan glycoside and the hepatoprotective activity of the leaf extract. *Phytomedicine* 23: 1484–1493. https://doi.org/10.1016/j.phymed.2016.08.006.

Youssef, F.S., Sobeh, M., Dmirieh, M. et al. (2022). Metabolomics based profiling of *Clerodendrum specisosum* (Lamiaceae) leaves using LC/ESI/MS-MS and in vivo evaluation of its antioxidant activity using *Caenorhabditis elegans* model. *Antioxidants* 2022 (11): 330. https://doi.org/10.3390/antiox11020330.

Zamberlan, D., Amaral, G., Arantes, L. et al. (2016). *Rosmarinus officinalis* L. increases *Caenorhabditis elegans* stress resistance and longevity in a DAF-16, HSF-1 and SKN-1-dependent manner. *Brazilian Journal of Medical and Biological Research* 2016: 49.

Zenk, M.H. and Juenger, M. (2007). Evolution and current status of the phytochemistry of nitrogenous compounds. *Phytochemistry* 65: 2757–2772.

Zhang, S., Duangjan, C., Tencomnao, T. et al. (2020). Neuroprotective effects of oolong tea extracts against glutamate-induced toxicity in cultured neuronal cells and Aβ-induced toxicity in *Caenorhabditis elegans*. *Food & Function* 2020 (11): 8179–8192. https://doi.org/10.1039/D0FO01072C.

Zhang, S., Duangjan, C., Tencomnao, T. et al. (2022). Oolonghomobisflavans exert neuroprotective activities in cultured neuronal cells and anti-ageing effects in *Caenorhabditis elegans*. *Frontiers in Aging Neuroscience* 14: 967316. https://doi.org/10.3389/fnagi.2022.967316.

Zhang, N., Jiao, S., and Jing, P. (2021). Red cabbage rather than green cabbage increases stress resistance and extends the lifespan of *Caenorhabditis elegans*. *Antioxidants (Basel)*. 10 (6): 930. https://doi.org/10.3390/antiox10060930.

6

Nutraceuticals with Antidiabetic Potentials from Plants

Roman Lysiuk[1], Serhii Oliinyk[2], Andrii Boiko[3], Iryna Lozynska[4], Petro Oliinyk[5], and Andrian Boiko[6]

[1] *Department of Pharmacognosy and Botany, Danylo Halytsky Lviv National Medical University, Lviv, Ukraine*
[2] *Nobel – Ukraine Pharmaceutical Company, Kyiv, Ukraine*
[3] *Department of Management and Economy of Pharmacy and Drug Technology of Postgraduate Faculty, Danylo Halytsky Lviv National Medical University, Lviv, Ukraine*
[4] *Department of Biochemistry, Danylo Halytsky Lviv National Medical University, Lviv, Ukraine*
[5] *Department of Disaster Medicine and Military Medicine, Danylo Halytsky Lviv National Medical University, Lviv, Ukraine*
[6] *Department of Therapeutic Dentistry, Periodontology and Dentistry FPDE, Danylo Halytsky Lviv National Medical University, Lviv, Ukraine*

6.1 Introduction

Diabetes mellitus (DM) is a metabolic disease described by an inappropriate rise in blood glucose concentrations. It is known that the term was first used by Apollonius of Memphis about 250–300 BC. Ancient Greek, Indian, and Egyptian cultures detected the sweet character of urine in such conditions; therefore, the name DM was accepted (Sapra and Bhandari 2023).

DM is a significant health item worldwide (Al-Lawati 2017; Al-Yaarubi et al. 2014). Over the past 30 years, the number of diabetics has quadrupled. DM is currently the ninth leading cause of death, with approximately 1 in 11 adult persons with DM (Zheng et al. 2018); 537 million people were diagnosed with DM in 2021, and this quantity is projected to reach 643 million by 2030, and 783 million by 2045 (International Diabetes Federation 2019; Yahaya and Ufuoma 2020).

Along with cardiovascular disease, tumors, and respiratory ailments, diabetic states account for more than 80% of all premature deaths from noncommunicable diseases. People with DM have a 2–3 times higher risk of all-cause mortality. DM is related to elevated mortality from infectious and cardiovascular diseases, stroke, chronic kidney and liver diseases, and oncopathology. Furthermore, despite progress in improving population health and increasing life expectancy, DM is the second largest negative overall contributor to global health-adjusted life expectancy (Lin et al. 2020).

Despite the entire set of doctors' means to fight this disease, including effective drugs, modern technologies, and constant improvement of educational and preventive methods, humanity is gradually losing the fight against diabetes, which leads to disability due to its complications as well as puts patients' lives at risk (Tkachenko 2014).

Herbal Nutraceuticals: Products and Processes, First Edition. Edited by Santosh Kumar Upadhyay and Sudhir Pratap Singh.

6.2 Types of Diabetes

Different types of diabetes occur: type 1 diabetes, type 2 diabetes, maturity-onset diabetes of the young (MODY), gestational diabetes, and neonatal diabetes. The major types of diabetes comprise type 1 diabetes (T1DM) and type 2 diabetes (T2DM), which are caused by impaired insulin production and/or activity. T1DM occurs in children or adolescents; thus, T2DM is considered to affect middle-aged and older people with long-term hyperglycemia caused by poor lifestyle and nutrition. Chronic duration of DM can lead to significant aggravations, including autonomic and peripheral neuropathy, retinopathy, renal damage, atherosclerotic cardiovascular disease, peripheral artery disease, cerebrovascular ailments, and decreased resistance to infections. The pathogenesis of T1DM and T2DM is fundamentally distinct, so each type has a different etiology, manifestations, and treatment (Sapra and Bhandari 2023).

Gestational diabetes is hyperglycemia with blood sugar concentration exceeding normal but lower than the diagnostic indicators of DM that develop during pregnancy and usually disappear after childbirth. It can happen at any stage of pregnancy but is more common in the second or third trimester. Women with gestational diabetes have an elevated risk of aggravation while carrying a child and during childbirth. They and possibly their offspring are also susceptible to T2DM subsequently (Banday et al., 2020; Saravanan et al. 2020; Ye et al. 2022).

MODY is represented by various conditions accompanied by abnormally high plasma glucose concentration. Such types of DM usually begin under the age of 30 and thus might develop later. In MODY, increased level of blood glucose occurs due to decreased insulin production. Different types of MODY have congenital origins. The most common forms comprise HNF1A-MODY (MODY3), which accounts for 50–70% of cases, and GCK-MODY (MODY2), which accounts for 30–50%. Less common forms are represented by HNF4A-MODY (MODY1) and renal cysts and diabetes (RCAD) syndrome (HNF1B-MODY, or MODY5), each accounting for 5–10 %. Over ten other forms have been detected, all of which are very rare (Kim 2015b). An autosomal dominant mutation in particular genes implicated in insulin biosynthesis and metabolism can develop MODY. Such type of DM has to distinguish between pathogenetic and clinical manifestations. Therefore, its therapy might need a different approach than other DM forms. Thus, it is recommended for primary care providers to develop MODY medications and therapies, considering the established mechanism of activity and the phenotypic manifestations of its subforms (Yahaya and Ufuoma 2020).

Neonatal diabetes results from a single gene mutation. It mostly develops after the first four weeks of life up to the age of six months, but might occasionally occur under the age of one year. DM is the result of a complex interaction between environmental and genetic influences. Success in molecular genetics during recent years has accelerated the recognition that early onset DM is most often the outcome of underlying monogenic deficiency—disruptions related to alteration(s) in a single gene. Neonatal (or congenital) diabetes mellitus (NDM) occurs in around 1 in 90,000–160,000 live births. More than 20 genetic factors of NDM are known (Lemelman et al. 2018; Letourneau et al. 2017).

Type 1 diabetes, also recognized as non-insulin-dependent DM, is reported in about 5–10% of all diabetes patients. Symptoms comprise excessive urination (polyuria), thirst (polydipsia), permanent hunger, weight loss, vision impairment, and fatigue. Such manifestations may develop suddenly. T1DM is an autoimmune disease, representing T-cell-mediated damage of the pancreatic islets of Langerhans, leading to insulin insufficiency and eventually hyperglycemia. The pathogenetic mechanisms of this autoimmunity have not yet been fully realized. The speed of its development in infants and children (juvenile onset) is mostly fast, but in adults, it might develop gradually (late onset) (Sapra and Bhandari 2023).

Type 2 diabetes, also classified as non-insulin-dependent diabetes, occurs in about 90–95% of all diabetics. This kind of DM is denoted by two major insulin-associated abnormalities: insulin resistance and β-cell upset. Insulin resistance is the result of disruption of different cellular pathways, resulting in lowered response or sensitivity of cells to insulin in peripheral organs, including adipocytes and cells of muscular and hepatic tissues (Muoio and Newgard 2008).

6.3 Flavonoids in Ameliorating of Diabetes Mellitus Complications

Chronic hyperglycemia that results from diabetes causes long-lasting medical impediments such as deterioration and decline of many organs, including the nerves, heart, blood vessels, kidneys, and eyes as well. Among the used approaches aimed at reduction of hyperglycemia is the inhibition of the enzyme glucosidase, which hydrolyzes carbohydrates in the bowels and plays a significant role in accelerating glucose absorption by hydrolysis of carbohydrates to monosaccharides before absorption. Alpha-glucosidase inhibitors also make a significant contribution to reduction of insulin and glycemic peaks due to prolongation of carbohydrate absorption. However, it has been established that long-term application of antidiabetic medicines can lead to different side effects (Yen et al. 2021).

Flavonoids are phenolics that occur naturally and are often available as secondary metabolites in plants and fungi. They contain 15 carbon atoms in 2 aromatic nuclei, joined by three-carbon chain. These polyphenols are broadly divided into six main subtypes: flavonols, flavones, flavanones, isoflavones, flavanols, and anthocyanidins. Due to the presence of reactive phenol radicals and carbonyl groups in the molecule, they participate in various metabolic processes that determine their antioxidant, anti-inflammatory, anticancer, and anti-allergic effects. All natural flavonoids are low-toxic compounds, with a wide spectrum of biological activity, which makes them attractive in the development of phytopreparations (Chirumbolo et al. 2018; Gasmi et al. 2022a; Lysiuk and Hudz 2017). *In vivo* and *in vitro* experiments have revealed their hypoglycemic impact. Flavonoids are also able to regulate glucose metabolism, lipid profile, the activity of liver enzymes, protein kinase inhibitors, peroxime proliferator-activated receptors (PPAR), and adenosine monophosphate protein kinase (AMPK) with NF-κB (Nrf2/HO1), as well as regulate insulin secretion using many signaling pathways, which are relatively significant as a therapeutic tool for diabetes (Al Duhaidahawi et al. 2021; Al-Ishaq et al. 2019; Gasmi et al. 2022b; Serafini et al. 2010; Ullah et al. 2020).

Given the relationship between the development of diabetes and inflammation (Duncan et al. 2003; Pickup 2004; Vinayagam and Xu 2015) and the potential of these polyphenolics in protection of the body against free radicals and other prooxidant substances (Bjørklund et al. 2017; Rice-Evans et al. 1996), it is clear that uses of flavonoids or foods rich in flavonoids may lower the risk of DM (Pandey and Rizvi 2009). Accordingly, new concepts in the treatment of diabetes have emerged, such as nutraceuticals, dietary therapy, phytonutrients, and herbal medicine. These types of foods and herbal preparations make a positive contribution to maintaining plasma glucose concentration, glucose assimilation, and insulin production, and modulating immune mechanisms to prevent DM (Barchuk et al. 2017; Hajiaghaalipour et al. 2015; Hanhineva et al. 2010).

Strong evidence from cellular and animal experimental studies has shown that flavonoids or flavonoid-rich extracts are able to alleviate diabetic complications. Edible plant flavonoids decrease blood sugar levels by preserving pancreatic islets of Langerhans, stimulating the production of insulin, impeding glycogenolysis, gluconeogenesis, and enzymes involved in digestion and carbohydrates' metabolism. Many flavonoids alleviate complications of diabetes, such as retino-, nephro-, neuro-, and cardiopathy, vascular illness, and kidney insufficiency, by modulating the gut flora, altering the inflammatory response, and inhibiting the processes of oxidation (Lysiuk and Mansuetusmboya 2020; Xiao 2022).

6.4 Diabetic Cardiomyopathy

Diabetic cardiomyopathy is one of the main risk factors for mortality in patients with diabetes. It has been established that the majority of cardiac structural and functional changes in the state of diabetic cardiomyopathy occur as an outcome of hyperglycemia-induced persistent oxidative stress (OS) in the heart, which leads to maladaptive inflammatory reactions and apoptosis (Jubaidi et al. 2021). Genistein was found to be able to reduce oxidative damage to the heart as reflected by a decrease in malondialdehyde (MDA) levels in the heart compared to untreated diabetic rats (Jia et al. 2019). The isoflavone puerarin, commonly found in legumes and soybeans, inhibits NF-κB signaling, thereby inhibiting the cardiac inflammatory response both *in vivo* and *in vitro* (Yin et al. 2019).

Macrovascular manifestations of diabetes include coronary and peripheral artery disease. Early macrovascular disease is related to atherosclerotic plaques in the veins that provide the supply of blood to the heart, brain, limbs, and other body parts, while late phases of the macrovascular disease include total arterial occlusion with an elevated risk of myocardial infarction and stroke (Caro-Ordieres et al. 2020; Fowler 2011). Blueberry (*Vaccinium myrtillus* L.) anthocyanins reduced endothelial dysfunction of human aortic endothelial cells (HAEC), treated with palmitic acid, due to inhibition of Nox-mediated reactive oxygen species (ROS) formation (Bharat et al. 2018; Caro-Ordieres et al. 2020). The flavonoid rutin also arrested endothelial dysfunction under conditions of hyperglycemia both *in vitro* (human umbilical vein endothelial cells) and *in vivo* (carbohydrate-fed rats) by impeding Nox4-mediated ROS production and NLRP3 inflammasome stimulation (Caro-Ordieres et al. 2020; Wang et al. 2017).

Flavonoids liquiritigenin, apigenin, and stilbene resveratrol reduce diabetic damage to the cardiovascular system. Liquiritigenin, a flavanone isolated from *Radix Glycyrrhizae*, attenuated cardiac damage through its antifibrotic and anti-inflammatory effects by inhibiting the NF-κB signaling. Apigenin is preserved against cardiovascular injury by prohibiting cardiomyocyte hypertrophy and fibrosis and lessening oxidative and inflammatory states in DMT1 model. Resveratrol reduces ROS formation in HAEC by amending autophagy as an outcome of regulation of the AMPK-mTOR pathway and demonstrated cardioprotective impact *in vivo* while replacing destruction and thickening of muscle fibers in a diabetic rat model of myocardial ischemia. Furthermore, its combination with glibenclamide decreased arrhythmias due to ischemia/reperfusion in an experimental DM model in rats (Caro-Ordieres et al. 2020).

Resveratrol (3,4′,5-trihydroxy-trans-stilbene) is a plant polyphenolic substance that belongs to the group of stilbenoids. The resveratrol molecule exists in the form of *cis*- and *trans*-stereoisomers. Only the trans-isoform is biologically active. Resveratrol is metabolized by cytochrome P450 (CYP1B1) to its monohydroxylated form, piceathanol. This metabolite is more active than resveratrol due to the occurrence of more hydroxy groups, which increases the inhibition of the activity of the nuclear factor NF-κB, which induces inflammatory processes in the body (Oyenihi et al. 2016; Sokolova et al. 2020). In the body, resveratrol is rapidly absorbed, with peak plasma concentrations 30 minutes after oral administration, and about 70–75% of absorption occurs by transepithelial diffusion. Although resveratrol is readily absorbed, its systemic bioavailability is less than 1%. Glucuronidation and sulfation in the intestine and in the liver are the main metabolic pathways of resveratrol that limit its availability (Intagliata et al. 2019; Smoliga and Blanchard 2014; Sokolova et al. 2020).

Resveratrol exhibits anti-inflammatory, antioxidant, antihyperglycemic, detoxifying, antiviral, antithrombotic, cancerostatic, and neuroprotective effects; it enhances vasodilation and clearance of β-amyloid peptides (Mishra et al. 2022; Oyenihi et al. 2016). The high safety profile of resveratrol combined with its positive effects makes it an attractive candidate for protecting the body against diabetic cardiovascular damage (Sokolova et al. 2020).

Atherosclerosis causes the narrowing of the lumen of arterial vessels and the formation of blood clots, which leads to cardiovascular diseases. Atherosclerosis mainly affects the intima of the arteries' walls followed by the deposition of extracellular lipids, augmentation and migration of muscle fiber cells, as well as persistent inflammation. Low-density lipoproteins (LDLs) are involved in the atherosclerotic process (Sokolova et al. 2020). Preclinical studies have shown that resveratrol can improve the lipid profile by reducing plasma levels of triglycerides and LDL cholesterol, oxidized oxLDL, and apolipoprotein B-100 (ApoB), as well as increasing the content of HDL-cholesterol (Dyck et al. 2019). In addition, resveratrol can enhance the suppression of LDL-receptors (LDL-R) in parenchymal cells of the liver, contributing to the further lowering of LDL cholesterol in the blood. Due to its antioxidant properties, resveratrol inhibits the oxidation of LDL, which directly concerns atherogenesis, stimulating various internal antioxidant systems (Berrougui et al. 2009) and anti-inflammatory processes (Ramprasath and Jones 2010).

In general, the effect of resveratrol leads to inhibition of apoptosis, activation of endothelium and blood vessels in the process of inflammation, and improvement endothelial function (Oyenihi et al. 2016). In addition, resveratrol decreases the suppression of adhesion agents (ICAM-1 and VCAM-1), which contribute to the binding of leukocytes to the endothelium, by inhibiting the triggering of NF-κB (Deng et al. 2011; Dyck et al. 2019).

6.5 Diabetic Nephropathy

Diabetic nephropathy (DN) is significant diabetic microvascular impediment, and OS and inflammation might remarkably contribute to its incorporation and development. Due to the action of metabolic and hemodynamic factors, as well as cytokines and mediators of inflammation, changes develop in the basal membrane of the kidneys, which leads to a reduction in its negative charge and a growth in the diameter of the pores. At the same time, as a result of hyperglycemia, intraglomerular pressure increases due to enhanced renal blood flow and spasms of glomerular outflow vessels. The clinical symptoms of these pathological changes comprise increased albumin filtration, initially in the form of albuminuria of 30–300 mg/1 g of creatinine, and later in the form of proteinuria. Over time, glomerular hyalinosis, interstitial tissue fibrosis, and renal failure develop (Svintsitskyi and Haievski 2018).

Flavonoids have been found to possess therapeutic significance in DN by modulating OS and inflammatory pathways (Gasmi et al. 2022a; Hu et al. 2021). Common in many plants, quercetin, apigenin, baicalin, luteolin, hesperidin, genistein, proanthocyanidin, and eriodictyol are capable of mitigating OS. These natural polyphenols boost podocyte autophagy and suppress RAAS hyperactivity by inhibiting OS and inflammatory pathways, eventually ameliorating DN (Hu et al. 2021). Quercetin plays an important role in the treatment of DN by reducing OS, inflammation, and scavenging free radicals. The combination of quercetin and dasatinib eliminates senescent cells *in vitro*, improves physical function, and increases lifespan in mice *in vivo*. Administration of dasatinib with quercetin in patients with diabetic kidney disease and idiopathic lung disease has been shown to effectively reduce the expression of the senescence markers p16 and SA-β-gal (Fan et al. 2022; Hu et al. 2021).

Guo et al. investigated the protective impact of flavonoids from *Coreopsis tinctoria* Nutt on DN. Previous studies have shown that its extract, which contains 19 flavonoids, has anti-inflammatory and antioxidant effects. Treatment with the extract, especially at doses of 150 mg/kg and the same dose with metformin, remarkably reduced impaired glycolipid metabolism, kidney activity, and pathological kidney damage in db/db mice in *in vivo* and *in vitro* models. The extract decreased the impact of factors associated with renal OS, cytokines, and fibrotic agents. The authors consider that *C. tinctoria* extract may protect against diabetic kidney damage by suppressing renal oxidation and inflammatory and fibrotic processes via RhoA/ROCK and NF-κB/TGF-β/SMAD signaling (Guo et al. 2022).

The total extract of *Echinops spinosus* L. and the high-flavonoid fraction, containing kaempferol, quercetin, and rutin, demonstrated antidiabetic properties in streptozotocin-induced diabetic models in rats, as evidenced by a reduction in glucose and an elevation in insulin, the rate of insulin receptor expression, and glycogen synthesis. In addition, *E. spinosus* total extract and high-flavonoid fraction alleviated diabetic renal complications by reducing OS, modulating inflammatory mediators, and inhibiting the apoptotic cascade along with the correction of diabetic dyslipidemia (Khedher et al. 2021; Othman et al. 2022). Therefore, it can be concluded that extracts of *E. spinosus* L. and *C. tinctoria* Nutt might remarkably contribute to the treatment of T2DM and diabetes-related lesions, as well as oxidative damage to liver and kidney tissues.

Licorice (*Glycyrrhiza glabra* L.) flavonoids have demonstrated notable antidiabetic impacts. More than 400 substances have been detected in its roots, together with triterpene saponins, flavonoids, coumarins, phenolic substances, and pterocarpane. Among them, over 300 licorice-derived flavonoids with a C6-C3-C6 backbone have been identified, as well as flavanones, flavones, flavonols, chalcones, isoflavones, isoflavanones, isoflavans, and isoflavenes, which exhibit substantial structural diversity and possess anti-inflammatory, antioxidant, anticancer, antibacterial, antiviral, gastroprotective, and other types of impact (Wu et al. 2022a). The ethanol extract of licorice is able to reduce long-term hyperglycemic conditions and DN, and *Glycyrrhiza inflata* L. extract is a potent agent against DN and diabetes-related vascular obstacles and endothelial ailment (Wu et al. 2022a; Yang et al. 2020).

Licorice flavonoids, including glycibenzofuran, lycocoumarone, glycyrrhizoflavone, glycoflavone, isoliqiritigenin, lycoflavone A, apigenin, glycycoumarin, and isoglycicoumarin (Li et al. 2010a), inhibited the activity of protein tyrosine phosphatase (PTP1B), the main transcriptional repressor of the insulin signaling cascade, and activated the insulin signaling pathway (Li et al. 2013). Isoliquiritigenin impeded mesangial enlargement and matrix deposition in DN by deactivating TGF-β RI and TGF-β RII and hindering the SMAD signaling. Glucose-induced mesangial expansion and matrix dilation were reduced by type IV collagen accumulation, and diabetes-related kidney inflammation was diminished by reducing inflammatory ICAM-1 suppression and monocyte chemotactic protein-1 (MCP-1) production in the mesangial cells. Isoliqiritigenin 1–20 μM matured with HRMC for three days averted the development of mesangial fibrosis and glomerulosclerosis in kidney deficiency and end-stage renal disease by reducing the high-level accumulation of glucose-associated mesangial matrix (Li et al. 2010b).

Some studies describe the antioxidant effects of several flavonoids on DN. Thus, formonetin decreased OS and lessened renal injury in a rat model with type 2 diabetes by augmenting the action of antioxidants SOD and CAT in addition to the activation of Sirtuin-1 protein (Oza and Kulkarni 2019). Diosmetin, a flavonoid found in the leaves of *Olea europaea*, exhibits potent antioxidant properties, restores the lowered concentration of SOD and NO noticed in the kidneys of diabetic rats, and reduces the level of MDA, a marker of lipid peroxidation (Jiang et al. 2018). The flavanone naringin also diminishes ROS formation in diabetic renal tissues resulting in decrease of Nox4 suppression and an increase of the antioxidant enzymes SOD and GPx (Zhang et al. 2017).

6.6 Diabetic Neuropathy

Diabetic neuropathy is a peripheral nerve disease that acts on body parts, innervated by the autonomic nervous system, such as the heart, eye, digestive tract, kidney, and hepatic tissues. The ailment impacts more than half of all patients with diabetes and is usually related to a loss of perception in the feet, causing difficulties walking and footdrops. OS and inflammatory processes are of key importance in neuronal degeneration. OS conducts the stimulation of inflammatory and apoptotic mechanisms, leading to insufficient neurochemical growth agents, neurovascular disorder, neuronal demyelination, and enlarged autoimmune injury. Thus, inhibition of the cascades of OS and inflammation may be used as a tool to anticipate the spread of diabetic neuropathy (Caro-Ordieres et al. 2020).

Flavonoids impede several pathways related to the development of diabetic neuropathy, such as reduction of OS, lycogenolysis, and the generation of advanced glycation end-products (AGEs), increasing glucose utilization, and inhibiting the enzyme α-glucosidase. Some instances of the protective effect of flavonoids against diabetic neuropathy have been reported. Catechin in the streptozotocin-induced diabetes rat model diminished neuronal injury by elevating SOD and CAT, as well as decreasing MDA and lymphocytic infiltration in neural tissues (Addepalli and Suryavanshi 2018; Caro-Ordieres et al. 2020). Deguelin alleviates diabetic neuropathy, attenuates OS and neuroinflammation through initiation of the NRF2 signaling pathway, and restores to a certain extent the neuronal transmission velocity in diabetic rats by expanding (Na^+-K^+) ATPase action (Chen et al. 2018). A similar neuroprotective impact against OS and inflammation has been demonstrated for phloretin, which, in a single or combined form with the neuropathic pharmaceutical duloxetine, elevated nerve fiber regeneration in rat model of DM caused by administration of streptozotocin (Balaha et al. 2018; Sood et al. 2020).

6.7 Diabetic Retinopathy

Diabetic retinopathy (DR) is an impediment of DM that affects the eyes and can lead to severe visual impairment and blindness. A high level of sugar is related to injury of retinal vessels. Diabetic macular edema results from DR, which originates from swelling in an area of the retina known as macula (Caro-Ordieres et al. 2020; Semeraro et al. 2019). Investigations evaluating the impact of flavonoids on eye damage in diabetes have demonstrated their antioxidant, anti-inflammatory, and antiangiogenic effects in the eye, accompanied by a reduction in the symptoms of DR. For instance, catechin deactivated the NF-κB signaling pathway and lowered the suppression of target genes (e.g. IL-1β, IL-6, and TNF-α) in the rats' retinas under conditions of experimental DM (Wang et al. 2018). The flavanone naringenin lessened levels of markers of oxidation, apoptosis, angiogenesis, and neurodegeneration, while galangin (Al-Dosari et al. 2017; Zhang et al. 2019) attenuated inflammation and breakdown of the rats' blood–retinal barrier under diabetic condition. Myricetin, eriodictyol, and galangin have demonstrated anti-inflammatory and antioxidant effects in the retinal epithelial, ganglion, and microglial cells, respectively (Arumugam et al. 2019; Lv et al. 2019; Zhang et al. 2019). Baicalin exhibits a protective effect on human retinal pigment epithelial cells under hyperglycemic conditions by lowering proliferative, antiapoptotic, and anti-inflammatory impacts by upregulating miR-145 and subsequently inhibiting NF-κB and p38 MAPK pathways (Dai et al. 2019). Chrysin improves the functioning of the retinoid visual cycle by elevating the levels of retinoid-binding proteins (RPE65, LRAT, RDH5, and rhodopsin), reducing a cellular response to angiogenic agents, and arresting AGE-related endoplasmic reticulum stress in glucose-activated RPE cells and the db/db mouse retina (Kang et al. 2018).

6.8 Diabetic Skin Ulcer

A diabetic skin ulcer is a frequently occurring impediment in diabetics. Lu et al. (2022) reported that xanthohumol, a dietary flavonoid derived from hops, diminishes oxidative injury and accelerates diabetic wound healing via Nrf2 stimulation. *In vitro*, xanthohumol impaired hydrogen peroxide-induced cytotoxicity, ROS formation, cell apoptosis, and high glucose-induced cell injury. Some research also revealed that xanthohumol is able to stabilize Nrf2 and boost its nuclear displacement, which is related to AMPKα stimulation and covalent mitigation of Keap1. In natural conditions, xanthohumol expanded the suppression of Nrf2 and intensified the healing of diabetic wounds, being a promising substance for the therapy of skin ulcers in diabetics (Lu et al. 2022).

6.9 Improvement of Bioavailability of Flavonoids

As evidenced by the results of many studies, the therapeutic effects of flavonoids are hindered due to their fast metabolism, poor assimilation, and low solubility in water. In addition, individual flavonoids have a short half-life and require frequent dosage to sustain a constant plasma level to achieve a therapeutic outcome. The bioavailability of flavonoids can be improved using different nano-delivery systems. Nanocarriers exhibit enormous potential to improve the bioavailability of flavonoids. Several studies, as well as *ex vivo, in vitro*, and *in vivo*, indicate that flavonoid-containing nanoparticles possess higher medicinal potential for the therapy of various ailments. Nanoparticles are mostly represented by polymers, solid lipid nanoparticles, nanostructured lipid carriers, micelles, liposomes, solid dispersions, microsponges, nanosuspensions, and nanoemulsions (Agarawal et al. 2022; Hussain et al. 2020; Joshi et al. 2021; Maity et al. 2022). For example, the flavonoid naringenin exhibits powerful antioxidant, anti-inflammatory, and antidiabetic effects. However, it is poorly soluble in water, so its bioavailability is very low when absorbed orally. In order to increase the bioavailability of naringenin, Joshi and collaborators developed a new method of its delivery in the form of herbosomes and recently introduced lipid-based compositions that improve absorption and provide increased bioavailability. In addition, herbosomes provide better drug delivery and improved storage compared to other carriers. Naringenin herbosomes were obtained by solvent evaporation technique and evaluated for physicochemical characteristics. Herbosomes showed a significant improvement in the naringenin release profile. It has been established that herbosome better reduces the level of glucose in the plasma, and also demonstrates higher release of naringenin, compared to its pure form (Joshi et al. 2021; Lu et al. 2019).

Wickramasinghe et al. analyzed research, conducted during 2009–2020, on the working out of nano-drugs based on medicinal plants for the management of T2DM. The authors report that among the reviewed investigations, 68% were on inorganic herbal, while 17% and 8% concerned polymer- and lipid-based herbal nanomedicines, respectively. The significant biological effects of the nano-drugs comprised improved glycemic control and insulin concentrations, retardation of AGEs generation, and regeneration of pancreatic β-cells. The abovementioned effects were established by screening nano-drugs using cellular and noncellular *in vitro* models as well as *in vivo* animal models of T2DM, which were investigated for the acute or subacute period. Only two clinical studies involving diabetics have been described, which require further investigations on healing herbal-based nano-drugs as a medicinal tool for the treatment of T2DM. It was established that the properties of nano-drugs are superior to the characteristics of the appropriate plant extracts and isolated active principles. Both in the preclinical and clinical phases, there are several understudied areas of investigation in the development and evaluation of the biological action of plant nanomedicines (Wickramasinghe et al. 2022).

Food products containing encapsulated flavonoids may be one way to increase the bioavailability of flavonoids, but the substances should persist stable in the food matrix during the stages of processing, preservation, and use, as well as be bioavailable after consumption and allow the release, absorption, and metabolism of these compounds. Controlling these variables will allow the food industry to develop functional food products selected for diabetics (de Oliveira et al. 2022).

Veeramani et al. developed a rapid and biocompatible method for the synthesis of gold nanoparticles (Au-NPs) by phytofabrication for antidiabetic applications. Au-NPs were synthesized and combined with *Nigella sativa* seed decoction, in which its aqueous extract functioned as a reducing agent and also coated the surface of Au-NPs. *N. sativa* seed extract and Au-NP showed 78% antidiabetic properties, which were verified by the α-amylase method. The phytochemical-based nanoparticles obtained in this research have shown effective antidiabetic activity and might be regarded for medicinal uses. Furthermore, the synthesis of nanoparticles based on phytochemicals is safe, environmentally friendly, and biocompatible (Veeramani et al. 2022).

6.10 Plant-derived Nutraceuticals with Antidiabetic Potentials

6.10.1 Beetroot (*Beta vulgaris* L.)

Beetroot is among the most consumed plants in the world. Previous studies have shown many health benefits of beet, as well as evidence of its powerful hypoglycemic, antioxidant, detoxifying, and anti-inflammatory effects (Al-Harbi et al. 2021; Lysiuk et al. 2020; Oboh et al. 2020). A study by Aliahmadi et al. revealed the impact of consuming raw red beets on metabolic markers and cognitive functions in persons with T2DM. The study involved 44 T2DM patients (57 ± 4.5 years old) who ate raw red beets (100 g/day) for eight weeks. The results of consuming raw red beets led to a significant decrease in fasting plasma glucose concentrations (−13.53 mg/dL), glycated hemoglobin (−0.34%), ApoB (−8.25 mg/dL), aspartate aminotransferase (−1.75 U/L), alanine aminotransferase (−3.7 U/L), homocysteine (−7.88 μmol/l), as well as systolic (−0.73 mmHg) and diastolic blood pressure (−0.34 mmHg), and a considerable raise in total antioxidant capacity (105 μmol/l) and cognitive function tests (Aliahmadi et al. 2021).

Water-soluble polysaccharide complex, isolated from the leaves of fodder beet, exhibited pronounced anabolic activity in rats. The fodder beet leaf extract possesses significant antioxidant and hepatoprotective effects, resulting in the ability to block the processes of lipid peroxygenation and normalize the state of the antioxidant defense mechanisms in rat liver in the model of acute tetrachloromethane hepatitis (Lenchyk et al. 2021).

Abd El-Ghffar et al. evaluated the impact of water–methanol extract of beet leaves on alloxan-induced diabetic rats. It was found that the extract significantly reduces the glucose level in serum and exhibits an immunomodulatory/antioxidant action, which may be useful in slowing the progression of diabetic complications (Abd El-Ghffar et al. 2019). Additionally, water and methanol extracts of *Beta vulgaris* inhibit α-amylase and α-glucosidase and the formation of AGEs, preventing diabetes complications. Methanol extract of *B. vulgaris* also showed high antibacterial activity against *Proteus mirabilis* and *Bacillus subtilis*. Collectively, these results indicate that *B. vulgaris* leaf extracts are valuable food and natural sources of bioactive molecules that can be used in developing new natural drugs against cancer and diabetes (Bouchmaa et al. 2022).

In Daliri et al. study, lactic acid bacteria isolated from Lithuanian traditional fermented products were tested for β-glucosidase activity and used for beet fermentation to develop an antidiabetic formulation. The study results showed that the fermentation process created high levels of dihydroquercetin (an antidiabetic flavonoid), which is absent in non-fermented beets. It indicates that beet fermentation using *Latilactobacillus curvatus* PN39 at 45°C for 72 hours may be an effective method for developing antidiabetic functional food products (Daliri et al. 2023).

The research conducted by Wu et al. aimed to explore the influence and pathways of beetroot impact on metabolic processes of glucolipids. The investigation was carried out on C57BL/KsJ db/db mice, which were separated into four groups: diabetic control (distilled water, db/db-Veh), metformin administration (metformin, db/db-Met, 250 mg/kg), and beetroot consumption (db/db, 1.8 and 3.6 g/kg). Metformin and beetroot were administered intragastrically to the mice. The research outcomes demonstrated that administration of beetroot reduced fasting plasma sugar levels, uric acid, and lipids, as well as improved liver and pancreas anomalies. This research revealed that beet exhibits hypoglycemic and hypolipidemic activities in diabetic mice (Wu et al. 2022a). The consumption of beetroot juice, as suggested by Mirmiran et al., is proposed for managing diabetes, insulin hemostasis, arterial pressure, vascular status, and kidney well-being. Studies have indicated the renoprotective impact of beetroot juice is related to reduction of mortality rates and favorable alterations in renal functional parameters in patients with kidney diseases. Likewise, regular consumption of beetroot juice significantly delays postprandial glycemic response and reduces plasma sugar peaks (Mirmiran et al. 2020).

OS and chronic inflammatory processes are primary factors underlying the inception and advancement of diabetes, as well as the development of diabetes-related complications in patients with T2DM. The findings of the study by Karimzadeh et al. suggest that daily intake of beetroot juice diminishes the levels of certain inflammation markers, including IL-6, TNF-α, and NF-κB, which contribute to the pathogenesis of obstacles in T2DM patients (Karimzadeh et al. 2022).

However, Karimzadeh et al. further reached the opposite conclusion in a randomized, controlled, open-label study with parallel groups, involving 46 patients with T2DM. The results of their recent study indicate that the consumption of 24 ml of concentrated beetroot juice daily for twelve weeks did not reveal remarkable impact on fasting blood sugar, insulin, glycated hemoglobin, total cholesterol, LDLs, triglycerides, and blood pressure. The authors concluded that the daily intake of 24 ml of concentrated beetroot juice did not impact glycemia, blood pressure, and lipid profile, and additional trials are required to prove these outcomes (Karimzadeh et al. 2023).

6.10.2 Cabbage (*Brassica oleracea* var. capitata)

White cabbage (*Brassica oleracea* var. capitata) and red cabbage (*B. oleracea* var. *capitata* f. *alba* and *rubra*, respectively) are among the most marketable and valuable cruciferous products, well known for their unique nutritional properties that categorize them as potential candidates for inclusion in the nutraceutical category (Zayed et al. 2023). Approximately 72 phytochemical compounds have been identified from *B. oleracea* var. *capitata*, encompassing various compound types such as alkaloids, flavonoids, organic acids, glucosinolates, steroids, and carbohydrates. Its total extracts and some individual active principles exhibit diverse therapeutic actions, including antidiabetic, antitumor, hypotensive, anticholesterolemic, antioxidant, anti-inflammatory, antibacterial, anticoagulant, and hepatoprotective properties (Ray et al. 2021; Uuh-Narvaez and Segura-Campos 2021).

OS impacts plasma insulin and generates insulin resistance, leading to a prediabetic state. Adeoye et al. studied the impact of cabbage water extract on the blood insulin level, insulin resistance, and OS in male albino rats induced with prediabetes. The results showed that the fresh and dried cabbage extracts controlled rats' serum insulin levels and insulin resistance. There has been a decrease in MDA and a substantial decline in the activity of superoxide dismutase and catalase in the extract, while glutathione level has been elevated. This implies that cabbage can potentially reduce liver OS in a prediabetic state (Adeoye et al. 2019).

6.10.3 Onion (*Allium cepa* L.)

Edible bulbs of onion are the third most valuable food crop after potatoes and tomatoes and are widely eaten worldwide due to their distinctive flavor (Bjorklund et al. 2022; Teshika et al. 2019). Flavonoids quercetin and kaempferol, phenolic acids, alk(en)yl cysteine sulfoxides, including S-methyl cysteine sulfoxide and S-propyl cysteine sulfoxide, cycloalliin, thiosulfinates, and sulfides, vitamin E, sodium, chromium, potassium, and zinc are among the main compounds found in the plant. Pharmacologically active components of *Allium cepa* exhibit a broad spectrum of actions, including neuroprotective, antioxidant, anti-inflammatory, lipid-modulating, antiplatelet, antihypertensive, and antidiabetic actions (Galavi et al. 2021; Kianian et al. 2021; Mahmood et al. 2023).

Among the key uses of onions in traditional medical practice is their administration for lowering blood glucose levels (El-Soud and Khalil 2010; Kianian et al. 2021). Compounds like S-methyl cysteine sulfoxide and S-allyl cysteine sulfoxide reduce blood glucose levels and possess powerful antioxidant activity, which may explain their hypoglycemic potential. S-methyl cysteine sulfoxide exhibits antidiabetic properties in three different ways: it stimulates insulin production in the body and enhances pancreatic secretion, hinders glucose absorption from food, and allows efficient insulin utilization (Kumari and Augusti 2002; Tran et al. 2020).

Onion extracts also modify the activity of liver hexokinase and glucose-6-phosphatase, and normalize lipid and glucose levels in plasma (Akash et al. 2014). Daily consumption of 100 g of onion by patients with T1DM and T2DM led to a remarkable decline in fasting plasma sugar concentrations in comparison with insulin, which was approximately 89 and 40 mg/dL in patients with T1DM and T2DM, respectively (Kianian et al. 2021). The antidiabetic effect of onions may be associated with the restoration or proliferation of beta cells in the pancreas and improved enzyme activity in the glycolysis and gluconeogenesis pathways (Akash et al. 2014). Eldin et al. note that the oral administration of raw aqueous-alcoholic extract of *A. cepa* in animal model (alloxan-induced diabetic rats) caused significant hypoglycemic activity and favorable health effects, which the regeneration of pancreatic beta cells can explain. *A. cepa* as an antidiabetic agent does not increase insulin levels but directly affects liver tissues, muscles, *etc.* and alters the activity of regulatory enzymes in glycolysis, gluconeogenesis, and other pathways (Eldin et al. 2010). Airaodion et al. reported that in alloxan-induced diabetic rats exposed to *A. cepa* juice at doses of 1, 2, and 3 ml/100 g body weight, a significant decrease ($P < 0.05$) in blood sugar levels was observed, compared to untreated animals. Furthermore, *A. cepa* juice exhibits a hypolipidemic effect, as there was a pronounced reduction in triglyceride levels and total cholesterol in plasma, compared to the untreated group (Airaodion et al. 2020).

Ogbonna et al. evaluated tablets of *A. cepa* L. for their *in vitro* characteristics, hypoglycemic action, and their impact on certain biochemical parameters in diabetic rats. The tablets consisted of extracted lyophilized powder of *A. cepa* and auxiliary substances. The aqueous extract of *A. cepa* L. was obtained by cold maceration method and lyophilization. The prepared tablets exhibited appropriate mechanical integrity and dissolution profiles after being kept for half a year. It has been observed that there is a significant decrease in serum sugar levels with daily intake of tablets in comparison with the negative control ($p = 0.05$). The highest rate of decline in plasma sugar concentration from 20% to 70% was reached after 21 days of daily tablet intake. Blood concentrations of liver enzymes such as alanine transaminase, aspartate transaminase, and alkaline phosphatase were significantly lowered, along with total cholesterol and triglycerides. Catalase, glutathione reductase, and MDA activity were also lowered (Ogbonna et al. 2019). Lyophilized onion powder may protect against hyperglycemia and dyslipidemia, caused by diabetes. However, the thermal processing of onions negatively impacts their protective role (Ülger and Çakiroglu 2020).

The study, conducted by Shakir et al., utilized an ethanol extract of *A. cepa* leaves to determine its antidiabetic activity in rats, induced with diabetes using alloxan monohydrate over two weeks. Results were compared with increasing doses of the standard antidiabetic glibenclamide at doses of 2.5, 3.8, and 5 mg/kg, respectively, as well as with a control, a physiological saline solution. It was observed that with increasing doses of the ethanol extract of *A. cepa* leaves (200, 250, and 300 mg/kg), there was a dose-dependent significant reduction ($p < 0.05$) in the blood glucose level of diabetic rats after a 6-week treatment, compared to control rats, receiving the physiological saline solution. The most effective concentration showing a significant decrease in blood sugar levels was at a dose of 300 mg/kg (Shakir et al. 2023).

It has been established that *A. cepa* L. exhibits antidiabetic action, caused by occurrence of different components. For instance, S-methylcysteine sulfoxide can activate insulin production, control plasma sugar levels, and modulate the actions of liver enzymes such as hexokinase, glucose-6-phosphatase, and 3-hydroxy-3-methylglutaryl (HMG)-CoA reductase. S-allyl cysteine sulfoxide activates insulin production and secretion, reduces glucose absorption from food, and preserves insulin. Allylpropyl disulfide and chromium reduce fasting sugar and insulin levels and enhance glucose tolerance. Quercetin modifies plasma sugar concentrations, ameliorates glucose-induced insulin secretion, and preserves pancreatic islets (Kianian et al. 2021). Further research is necessary to study and confirm the hypoglycemic activity of *A. cepa* and its components or synthetic analogs.

6.10.4 Bilberry (*Vaccinium myrtillus* L.)

Diabetes causes severe complications, such as DR, which is a major cause of visual impairment. Anthocyanins, flavonoid compounds obtained from food plants, demonstrate pharmacological properties, due to their simultaneous impact on numerous metabolic pathways and may act as an optional approach to other treatment methods and arrest further ailment progression. Bilberry anthocyanins can potentially prevent the progression of DR and may be considered candidates for clinical studies aimed at developing medications. The main natural source of anthocyanins is bilberry, also known as blueberries, which have a higher content of anthocyanins than other berries (Cravotto et al. 2010; Kowalczyk et al. 2003).

In accordance with Müller et al., who studied the yield of anthocyanins in commercially available blueberry juices and fresh berries, blueberries comprised up to 1017 mg of anthocyanins per 100 g of fresh weight (7465 mg/100 g dry weight) (Müller et al. 2012). Using ultrahigh-performance liquid chromatography, in fresh blueberries about sixteen anthocyanins were identified and quantitatively determined (Paes et al. 2014). The most common anthocyanins in blueberries are delphinidin-3-O-galactoside, accounting for up to 14.9%; delphinidin-3-O-arabinoside, up to 15.3%; delphinidin-3-O-glucoside, up to 14.0%; cyanidin-3-O-arabinoside, up to 13.6%; and cyanidin-3-O-glucoside (C3G), up to 10.1% of the total anthocyanin content in blueberry fruits (Kähkönen et al. 2003; Petruskevicius et al. 2023; Werner and Merz 2015).

The anthocyanins in blueberry extract inhibit the activation of the protein signal transducer and activator of transcription 3 (STAT3), reducing inflammation-related rhodopsin expression, and lowering the levels of intracellular ROS, arresting damage to photoreceptor cells and preserving visual function in cases of inflammatory processes in retina (Miyake et al. 2012). They support retinal function and structural integrity by enhancing antioxidant defense, impeding lipid peroxidation, pro-inflammatory cytokine suppression, and retinal cell apoptosis (Liu et al. 2021; Wang et al. 2015). One of the main anthocyanins in blueberries, malvidin, and its glycosides are able to preserve human retinal capillary cells from damage caused by high glucose levels (Huang et al. 2018).

Blueberry anthocyanins reduce the suppression of the retinal vascular endothelial growth factor and the degradation of proteins zonula occludens-1, occludin, and claudin-5, which are markers of

DR in streptozotocin-induced diabetic rats (Kim et al. 2015a). Research by Paik et al. showed that one of the anthocyanins in blueberry extract, C3G, reduces retinal degeneration caused by diabetes, indicating that the application of 10 μM C3G performs functions in restoring cellular OS without causing cytotoxicity (Paik et al. 2012).

Thus, the chemical instability and poor bioavailability of anthocyanins limit their application. Environmental impact can markedly affect their stability, as well as temperature, light, oxygen, enzymes, and pH. Enhancing the stability of anthocyanins is an important item that is in need of an immediate solution. Developing effective new systems to enhance the stability and, hence, the bioavailability of anthocyanins in the human body through microencapsulation, nanoencapsulation, or protein complexes will help improve their effectiveness (Chi et al. 2019; Herrera-Balandrano et al. 2021).

Therefore, the analysis of research results suggests that blueberry anthocyanins can potentially prevent the progression of DR and may be considered candidates for further clinical studies aimed at developing medications.

6.10.5 Parsley (*Petroselinum crispum* (Mill.) Nym. ex Hill)

Parsley is a biennial plant that grows several stems from one root with tripartite and egg-shaped leaves. It is typically used in human diets as a source of vitamins, minerals, and phenolic compounds. Additionally, parsley contains flavonoids, carotenoids, myristicin, apiole, bergapten, imperatorin, phthalides, sesquiterpenes, and furanocoumarins. Among the flavonoids present are apigenin, apiin, 6′-acetylapiin, and catechins. The presence of apigenin-glucuronide, luteolin-2, and lithospermic acid, as well as derivatives of p-coumaric acid, has also been demonstrated. This plant is a primary source of apigenin, a flavone with chemoprophylactic properties, and quercetin, a flavonoid that reduces the risk of neurodegenerative disorders (Ertik et al. 2023; Gnintoungbe et al. 2023).

The aqueous extract of parsley leaf powder contains coumarins, flavones, catechins, tannins, saponins, and mucilages. Due to its content, parsley possesses pharmacological potential. Experimental studies have shown that the plant exhibits antiglycemic, antihyperlipidemic, and antioxidant effects. It has been applied medicinally since ancient times for treating DM (Nusair et al. 2022; Upadhyay and Singh 2021, 2023; Vikou et al. 2023). In Spain, *Petroselinum crispum* leaves are used to treat diabetes, amenorrhea, dysmenorrhea, hyperuricemia, kidney stones, prostatitis, hypertension, constipation, odontalgia, and pain (Ouattara-Soro et al. 2022).

Research conducted on humans has shown that consuming parsley for two weeks increases the amount of antioxidants and elevates the concentrations of the enzymes glutathione reductase and superoxide dismutase, due to content of flavonoids acting as antioxidants (Allam et al. 2016). Consuming large quantities of parsley may lower blood sugar levels and increase the body's ability to retain more sodium, leading to elevated blood pressure. Parsley can hurt health during pregnancy, especially in case of excessive consumption, as it contains compounds that can stimulate uterine contractions and spasms, potentially leading to miscarriage and pregnancy termination (Alsaadi et al. 2023).

Gestational diabetes seriously affects the health of both the mother and the child. The study by Abd Rabou evaluated the influence of parsley extract on the kidneys of rats with gestational diabetes. Pregnant rats with diabetes exhibited certain morphological and histological alterations, as well as deep neck abscesses, tumors, and lung bleeding. Conversely, mothers with diabetes who used parsley leaf extract demonstrated normal morphological structure. In accordance with hematological and histopathological research, it was found that parsley leaf extract reduces the damage caused by hyperglycemia in gestational diabetes (Abd Rabou 2018). The study by Bolkent et al. indicates that parsley demonstrates significant hepatoprotective effects in diabetic rats. Rats with diabetes, receiving parsley, showed markedly reduced concentrations of plasma sugar, alanine transaminase, and alkaline phosphatase (Bolkent et al. 2004).

Cheraghi et al. studied the effect of aqueous-alcoholic extract of parsley leaves on spermatogenesis and the hypothalamic-pituitary-gonadal axis in male rats with diabetes, induced by streptozotocin. Comparing the levels of luteinizing hormone, follicle-stimulating hormone, and testosterone in rats, receiving 1 mg/kg of aqueous-alcoholic parsley leaf extract, demonstrated a significant increase, in comparison with the control group with diabetes ($P \leq 0.05$). Additionally, the histopathological evaluation demonstrated a notable elevation in the quantity and motility of spermatozoa, testicular weight, and total spermatozoa count. The extract also improved the condition of germ cells in the experimental group (1 mg/kg), in comparison with the control group of diabetic rats. The study's results suggest that the aqueous-alcoholic extract of parsley may effectively reduce reproductive disorders in patients with diabetes (Cheraghi et al. 2021).

6.10.6 Pumpkin (*Cucurbita pepo* L.)

The common pumpkin is a well-known vegetable within the pumpkin family, appreciated for its importance as a food product. Pumpkin fruits contain various nutritional ingredients, such as polysaccharides, proteins, essential amino acids, carotenoids, and minerals. In recent years, they have garnered significant attention due to these components' nutritional and therapeutic value. Pumpkin fruits possess antidiabetic, antioxidant, anti-inflammatory, antibacterial, hepatoprotective, and chemoprotective properties. Pumpkin seeds yield high quantities of proteins and fats, which range from 14.3% to 38.0% and 21.9% to 54.9%, respectively. Pumpkin, especially its seeds, is also considered a useful source of amino acids. Among the 20 amino acids, incorporated into human proteins, 17 are found in pumpkin seeds (Dotto and Chacha 2020).

Pumpkin seeds contain tocopherols, carotenoids, provitamins, pigments, pyrazine, triterpenoids, squalene, saponins, phytosterols, phenolic compounds and their derivations, flavonoids, coumarins, unsaturated fatty acids, and proteins. Additionally, they are rich in magnesium, potassium, phosphorus, and other minerals, such as zinc, manganese, iron, calcium, sodium, and copper (Abou-Zeid et al. 2018; Aktaş et al. 2018; Amin et al. 2019; Broznić et al. 2016; Datta et al. 2019; Dixit et al. 2021a; Naziri et al. 2016; Seymen et al. 2016). The main fatty acids in its oil are represented by linoleic, oleic, stearic, and palmitic, which comprise above 95% of the total fatty acids, with approximately 75% being unsaturated. Low levels of arachidonic and linolenic acids have been detected (Dotto and Chacha 2020).

Research by Kushawaha et al. showed that pumpkin seed extract exerts hypoglycemic and antidiabetic effects by activating insulin release from pancreatic β-cells (Kushawaha et al. 2017). It was also found that alcohol extracts of pumpkin seeds could inhibit α-amylase, while water extracts demonstrated significant inhibition of amylase and α-glucosidase, both directly associated with the antidiabetic action, recently considered for pumpkin seeds (Rahman et al. 2022).

In another research, Gutierrez and Martha investigated the hypoglycemic and antioxidant activities of a flaxseed and pumpkin seed combination in the kidneys of alloxan-induced diabetic rats. The mixture enhanced the activity of antioxidant enzymes in diabetic rats and reduced MDA levels in plasma and kidneys. It suggests the mixture has significant hypoglycemic and antioxidant effects (Gutierrez and Martha 2016). Pumpkin powder demonstrated a significant elevation of insulin level in plasma and decrease in blood sugar level. Pumpkin seeds are abundant in pectin, a kind of dietary fiber that modulates plasma glycemia and reduces need for insulin if consumed by diabetic individuals (Dotto and Chacha 2020).

In their study, Bharti et al. assessed in a diabetic rat model the antidiabetic activity of tocopherol, present in raw *Cucurbita pepo* L. seed oil, showing the occurrence of tocopherol isomers (α, β, γ, and δ). Its content has been measured as 107.4 ± 2.9 mg/100 g. In comparison with diabetic control animals, rats receiving the extract showed significant improvements in glycemia, insulinemia, and lipid dysmetabolism. Significant reductions in oxidation markers and improvements in the characteristics of the colon and pancreas were also observed (Bharti et al. 2013).

Treatment of diabetic rats with pumpkin seed protein improved the level of glucose, insulin, triglycerides, total cholesterol, alanine aminotransferase, increasing liver and muscle glycogen stores, adiponectin levels and adiponectin/leptin ratio, as well as lowering plasma leptin levels. In addition, rats fed pumpkin protein ameliorated oxidative stress in the liver and kidneys (Abdelkader et al. 2022; Chenni et al. 2022; Tasya et al. 2022).

As evidenced by the abovementioned studies, pumpkin seeds have therapeutic and healing significance. The analysis of the nutritional composition of pumpkin seeds demonstrates their richness in essential nutrients for health. Using pumpkin seeds may mitigate the side effects of diabetes and increase insulin levels. Most studies have been conducted on animals, with very few human trials. Therefore, additional testing on animals and humans, as well as clinical studies, are necessary to precisely determine the molecular pathways of the impact of pumpkin seeds on the impediment and treatment of DM in diabetics.

Data concerning nutraceuticals with antidiabetic potentials that are affordable for daily consumption are summarized in Table 6.1.

Table 6.1 Nutraceuticals with antidiabetic potentials, affordable for daily consumption.

Plant names	Herbal substance	Biological effects	References
Beetroot (*Beta vulgaris* L.)	Raw red beets	Hypoglycemic, decrease in fasting blood sugar and glycated hemoglobin levels; antioxidant, hypotensive	Aliahmadi et al. (2021)
	Beetroot juice	Postprandial glycemic response delay, reduction of blood glucose peaks, insulin hemostasis control, nephroprotective, arterial pressure, and vascular function control, lowering of the concentrations of inflammation markers	Mirmiran et al. (2020) Karimzadeh et al. (2022)
	Water–methanol extract of beet leaves	Reduction of glucose level in serum; immunomodulatory, antioxidant	Abd El-Ghffar et al. (2019)
	Water and methanol extracts	Inhibition of α-amylase, α-glucosidase, and advanced glycation end-products (AGEs) formation; antibacterial	Bouchmaa et al. (2022)
	Sugar beet extract	Hypoglycemic, reduction of fasting blood glucose; hypouricemic, hypolipidemic; improvement of liver and pancreas function	Wu et al. (2022b)
	Water-soluble polysaccharide complex of the leaves	Anabolic	Lenchyk et al. (2021)
	Fodder beet leaves extract	Antioxidant, hepatoprotective	Lenchyk et al. (2021)
Cabbage (*Brassica oleracea var. capitata*)	Total extract	Antidiabetic, anticancer, antihypertensive, anticholesterolemic, antioxidant, anti-inflammatory, antibacterial, anticoagulant, hepatoprotective	Ray et al. (2021); Uuh-Narvaez and Segura-Campos (2021)
	Fresh and dried water extracts	Serum insulin level and insulin resistance control; antioxidant, reduction of liver oxidative stress	Adeoye et al. (2019)

(*Continued*)

Table 6.1 (Continued)

Plant names	Herbal substance	Biological effects	References
Onion (*Allium cepa* L.)	Raw bulbs	Antidiabetic; neuroprotective, antioxidant, anti-inflammatory, lipid-modifying, antiplatelet, antihypertensive	Galavi et al. (2021); Kianian et al. (2021); Mahmood et al. (2023)
	Raw bulbs	Blood glucose level lowering	El-Soud and Khalil (2010); Kianian et al. (2021)
	Raw bulbs	Reduction in fasting blood glucose levels	Kianian et al. (2021)
	Bulb extracts	Normalization of glucose and lipid levels in plasma, restoration and reproduction of beta cells in the pancreas, and improvement of enzyme activity in the glycolysis and gluconeogenesis pathways	Akash et al. (2014); Eldin et al. (2010)
	Raw aqueous-alcoholic extract	Hypoglycemic, the regeneration of pancreatic beta cells	Eldin et al. (2010)
	Bulb juice	Decrease in blood sugar level; hypolipidemic	Airaodion et al. (2020)
	Tablets of extracted lyophilized powder	Hypoglycemic, decrease in blood glucose levels; antioxidant, hypolipidemic, hepatoprotective	Ogbonna et al. (2019)
	Lyophilized onion powder	Hypoglycemic; hypolipidemic	Ülger and Çakiroglu (2020)
	Ethanol extract of leaves	Antidiabetic, reduction in the blood glucose level	Shakir et al. (2023)
Bilberry (*Vaccinium myrtillus* L.)	Extract (malvidin and its glycosides)	Protection of human retinal capillary cells from damage, caused by high glucose levels	Huang et al. (2018)
	Extract (anthocyanins)	Enhancement of antioxidant defense mechanisms, inhibition of lipid peroxidation, pro-inflammatory cytokine expression, and retinal cell apoptosis	Liu et al. (2021); Wang et al. (2015)
	Extract (anthocyanins)	Inhibition of STAT3 activation, reduction of inflammation-related rhodopsin expression, lowering the level of intracellular ROS	Miyake et al. (2012)
	Extract (anthocyanins)	Reduction of the retinal vascular endothelial growth factor (VEGF) expression, degradation of protein zonula occludens-1, occludin and claudin-5	Kim et al. (2015a)
	Extract [cyanidin-3-o-glucoside (C3G)]	Reduction of retinal degeneration, caused by diabetes	Paik et al. (2012)

Table 6.1 (Continued)

Plant names	Herbal substance	Biological effects	References
Parsley (*Petroselinum crispum* (Mill.) Nym. ex Hill)	Juice	Antiglycemic; antihyperlipidemic, antioxidant	Nusair et al. (2022); Vikou et al. (2023)
	Juice	Increase of antioxidant content, elevation of the levels of glutathione reductase and superoxide dismutase	Allam et al. (2016)
	Leaf extract	Reduction of the damage caused by hyperglycemia in gestational diabetes	Abd Rabou (2018)
		Lowering of blood glucose, alanine transaminase, and alkaline phosphatase levels; hepatoprotective	Bolkent et al. (2004)
	Aqueous-alcoholic extract of leaves	Increase of luteinizing hormone, follicle-stimulating hormone, and testosterone levels, in the quantity and motility of spermatozoa, testicular weight, total spermatozoa count, and improvement of condition of germ cells	Cheraghi et al. (2021)
Pumpkin (*Cucurbita pepo* L.)	Fruits	Antidiabetic; antioxidant, anti-inflammatory, antibacterial, hepatoprotective, chemoprotective	Dotto and Chacha (2020)
	Seed extract	Induction of hypoglycemic and antidiabetic activity, stimulating insulin release from pancreatic β-cells	Kushawaha et al. (2017)
	Alcohol extract of seeds Water extract of seeds	Inhibition of α-amylase; inhibition of α-amylase and α-glucosidase	Rahman et al. (2022)
	Powder	Elevation of insulin level in plasma, decrease in blood glucose level	Dotto and Chacha (2020)
	Raw seeds (tocopherol)	Antidiabetic, improvement in glycemia, insulinemia, and lipid dysmetabolism; reduction in oxidation markers, improvement in the characteristics of the colon and pancreas	Bharti et al. (2013)

6.11 Conclusions

Diabetes remains a serious health challenge worldwide currently. New concepts in the treatment of diabetes have emerged, such as nutraceuticals, dietary therapy, phytonutrients, and herbal medicine. Among the promising nutraceuticals with antidiabetic potential affordable for daily consumption by diabetics might be beetroot, cabbage, onion, bilberry, parsley, and pumpkin.

Flavonoids and other natural polyphenolic, which contribute significantly to the total effect of herbal substances, lower blood glucose levels by protecting pancreatic β-cells, stimulating the pancreas to produce insulin, and arresting glycogenolysis, gluconeogenesis, digestive, and carbohydrate-metabolizing enzymes. Administration of the nutraceuticals can reduce the complications of diabetes, such as retinopathy, nephropathy, neuropathy, vascular dysfunction, and cardiomyopathy, by modulating the gut flora, altering the inflammatory response, and inhibiting the oxidative process. The selection of relevant substances and delivery systems is pivotal as they affect retention capacity, particle/matrix interaction, and release profile for the active principles.

Further research concerning substantiation of optimal nutraceuticals, mode, and duration of administration for prophylaxis and therapy of DM or its consequences is required.

References

Abd El-Ghffar, E.A., Hegazi, N.M., Saad, H.H. et al. (2019). HPLC-ESI- MS/MS analysis of beet (*Beta vulgaris*) leaves and its beneficial properties in type 1 diabetic rats. *sss* 120: 109541. https://doi.org/10.1016/j.biopha.2019.109541.

Abd Rabou, M.A. (2018). Therapeutic effects of parsley extract in streptozotocin-induced gestational diabetic rats. *Journal of Biological Sciences* 18: 364–372. https://doi.org/10.3923/jbs.2018.364.372.

Abdelkader, C., Cherif, F.Z.H., Elius, E.A.E. et al. (2022). Pumpkin seed proteins (*Cucurbita pepo* L.) protect against diet-induced metabolic syndrome by improving insulin resistance and markers of oxidative stress and inflammation in rats. *Biologia* 77: 2677–2687. https://doi.org/10.1007/s11756-022-01112-9.

Abou-Zeid, S.M., AbuBakr, H.O., Mohamed, M.A., and El-Bahrawy, A. (2018). Ameliorative effect of pumpkin seed oil against emamectin induced toxicity in mice. *Biomedicine & Pharmacotherapy* 98: 242–251. https://doi.org/10.1016/j.biopha.2017.12.040.

Addepalli, V. and Suryavanshi, S.V. (2018). Catechin attenuates diabetic autonomic neuropathy in streptozotocin induced diabetic rats. *Biomedicine & Pharmacotherapy* 108: 1517–1523. https://doi.org/10.1016/j.biopha.2018.09.179.

Adeoye, B.K., Adeyele, S.O., Ngozi, E.O. et al. (2019). EFfect of white cabbage (*Brassica oleracea*) aqueous extract on oxidative stress in pre-diabetes—induced male albino rats. *Nigerian Journal of Nutritional Sciences* 40 (1): 63–72.

Agarawal, K., Kulkarni, A.Y., and Wairkar, S. (2022). Nanoformulations of flavonoids for diabetes and microvascular diabetic complications. *Drug Delivery and Translational Research.* https://doi.org/10.1007/s13346-022-01174-x.

Airaodion, A., Akaninyene, I.U., Ngwogu, K.O. et al. (2020). Hypolipidaemic and antidiabetic potency of *Allium cepa* (Onions) bulb in alloxan-induced diabetic rats. *Acta Scientific Nutritional Health* 4 (3): 01–08. https://doi.org/10.31080/ASNH.2020.04.0648.

Akash, M.S.H., Rehman, K., and Chen, S. (2014). Spice plant *Allium cepa*: dietary supplement for treatment of type 2 diabetes mellitus. *Nutrition* 30: 1128–1137. https://doi.org/10.1016/j.nut.2014.02.011.

Aktaş, N., Uzlaşır, T., and Tunçil, Y.E. (2018). Pre-roasting treatments significantly impact thermal and kinetic characteristics of pumpkin seed oil thermochim. *Acta* 669: 109–115. https://doi.org/10.1016/j.tca.2018.09.012.

Al Duhaidahawi, D., Hasan, S.A., and Al Zubaidy, H.F.S. (2021). Flavonoids in the treatment of diabetes: clinical outcomes and mechanism to ameliorate blood glucose levels. *Current Diabetes Reviews* 17 (6): e120720188794. https://doi.org/10.2174/1573399817666201207200346.

Al-Dosari, D.I., Ahmed, M.M., Al-Rejaie, S.S. et al. (2017). Flavonoid naringenin attenuates oxidative stress, apoptosis and improves neurotrophic effects in the diabetic rat retina. *Nutrients* 9: 1161. https://doi.org/10.3390/nu9101161.

Al-Harbi, L.N., Alshammari, G.M., Al-Dossari, A.M. et al. (2021). *Beta vulgaris* L. (Beetroot) methanolic extract prevents hepatic steatosis and liver damage in T2DM rats by hypoglycemic, insulin-sensitizing, antioxidant effects, and upregulation of PPARα. *Biology* 10 (12): 1306. https://doi.org/10.3390/biology10121306.

Aliahmadi, M., Amiri, F., Bahrami, L.S. et al. (2021). Effects of raw red beetroot consumption on metabolic markers and cognitive function in type 2 diabetes patients. *Journal of Diabetes and Metabolic Disorders* 20: 673–682. https://doi.org/10.1007/s40200-021-00798-z.

Al-Ishaq, R.K., Abotaleb, M., Kubatka, P. et al. (2019). Flavonoids and their anti-diabetic effects: cellular mechanisms and effects to improve blood sugar levels. *Biomolecules* 9 (9): 430. https://doi.org/10.3390/biom9090430.

Allam, A.A., Maodaa, S.N., Abo-Eleneen, R., and Ajarem, J. (2016). Protective effect of parsley juice (*Petroselinum crispum, Apiaceae*) against cadmium deleterious changes in the developed albino mice newborns (*Mus musculus*) brain. *Oxidative Medicine and Cellular Longevity* 2016: 2646840. https://doi.org/10.1155/2016/2646840.

Al-Lawati, J.A. (2017). Diabetes mellitus: a local and global public health emergency! *Oman Medical Journal* 32 (3): 177–179. https://doi.org/10.5001/omj.2017.34.

Alsaadi, S., Shannon, A., and Abdulazeez, S. (2023). Biophysiological impacts of some herbs extracts as antioxidants, an advanced contemporary view. *Kirkuk University Journal for Agricultural Sciences* 14 (2): 145–161. https://doi.org/10.58928/ku23.14214.

Al-Yaarubi, S., Ullah, I., Sharef, S.W. et al. (2014). Demographic and clinical characteristics of type 1 diabetes mellitus in omani children—single center experience. *Oman Medical Journal* 29 (2): 119–122. https://doi.org/10.5001/omj.2014.29.

Amin, M.Z., Islam, T., Uddin, M.R. et al. (2019). Comparative study on nutrient contents in the different parts of indigenous and hybrid varieties of pumpkin (*Cucurbita maxima* Linn.). *Heliyon* 5 (9): e02462. https://doi.org/10.1016/j.heliyon.2019.e02462.

Arumugam, B., Palanisamy, U.D., Chua, K.H., and Kuppusamy, U.R. (2019). Protective effect of myricetin derivatives from *Syzygium malaccense* against hydrogen peroxide-induced stress in ARPE-19 cells. *Molecular Vision* 25: 47–59.

Balaha, M., Kandeel, S., and Kabel, A. (2018). Phloretin either alone or in combination with duloxetine alleviates the STZ-induced diabetic neuropathy in rats. *Biomedicine & Pharmacotherapy* 101: 821–832. https://doi.org/10.1016/j.biopha.2018.02.135.

Banday, M.Z., Sameer, A.S., and Nissar, S. (2020). Pathophysiology of diabetes: an overview. *Avicenna Journal of Medicine* 10 (4): 174–188. https://doi.org/10.4103/ajm.ajm_53_20.

Barchuk, O.Z., Lysiuk, R.M., Denys, A.I. et al. (2017). Experimental study of goat's rue (*Galega officinalis* L.) herb and its liquid extracts. *The Pharma Innovation Journal* 6 (11): 393–397.

Berrougui, H., Grenier, G., Loued, S. et al. (2009). A new insight into resveratrol as an atheroprotective compound: inhibition of lipid peroxidation and enhancement of cholesterol efflux. *Atherosclerosis* 207: 420–427. https://doi.org/10.1016/j.atherosclerosis.2009.05.017.

Bharat, D., Cavalcanti, R.R.M., Petersen, C. et al. (2018). Blueberry metabolites attenuate lipotoxicity-induced endothelial dysfunction. *Molecular Nutrition & Food Research* 62: doi:10.1002/mnfr.201700601.

Bharti, S.K., Kumar, A., Sharma, N.K. et al. (2013). Tocopherol from seeds of *Cucurbita pepo* against diabetes: validation by *in vivo* experiments supported by computational docking. *Journal of the Formosan Medical Association* 112 (11): 676–690. https://doi.org/10.1016/j.jfma.2013.08.003.

Bjørklund, G., Dadar, M., Chirumbolo, S., and Lysiuk, R. (2017). Flavonoids as detoxifying and pro-survival agents: what's new? *Food and Chemical Toxicology* 110: 240–250. https://doi.org/10.1016/j.fct.2017.10.039.

Bjorklund, G., Lysiuk, R., Butnariu, M. et al. (2022). Low prevalence of Covid-19 in Laos and Cambodia: does diet play a role? *Acta Medica Leopoliensia* 28 (1-2): 161–180. https://doi.org/10.25040/aml2022.1-2.161.

Bolkent, S., Yanardag, R., Ozsoy-Sacan, O., and Karabulut-Bulan, O. (2004). Effects of parsley (*Petroselinum crispum*) on the liver of diabetic rats: a morphological and biochemical study. *Phytotherapy Research* 18 (12): 996–999. https://doi.org/10.1002/ptr.1598.

Bouchmaa, N., Mrid, R.B., Kabach, I. et al. (2022). *Beta vulgaris* subsp. *maritima*: a valuable food with high added health benefits. *Applied Sciences* 12 (4): 1866. https://doi.org/10.3390/app12041866.

Broznić, D., Jurešić, G.Č., and Milin, Č. (2016). Involvement of α-, γ-and δ-tocopherol isomers from pumpkin (*Cucurbita pepo* L.) seed oil or oil mixtures in the biphasic DPPH disappearance kinetics. *Food Technology and Biotechnology* 54 (2): 200–210. https://doi.org/10.17113/ftb.54.02.16.4063.

Caro-Ordieres, T., Marín-Royo, G., Opazo-Ríos, L. et al. (2020). The coming age of flavonoids in the treatment of diabetic complications. *Journal of Clinical Medicine* 9 (2): 346. https://doi.org/10.3390/jcm9020346.

Chen, J., Liu, W., Yi, H. et al. (2018). The natural rotenoid deguelin ameliorates diabetic neuropathy by decreasing oxidative stress and plasma glucose levels in rats via the Nrf2 signaling pathway. *Cellular Physiology and Biochemistry* 48: 1164–1176. https://doi.org/10.1159/000491983.

Chenni, A., Cherif, F.Z.H., Chenni, K. et al. (2022). Effects of pumpkin (*Cucurbita pepo* L.) seed protein on blood pressure, plasma lipids, leptin, adiponectin, and oxidative stress in rats with fructose-induced metabolic syndrome. *Preventive Nutrition and Food Science* 27 (1): 78–88. https://doi.org/10.3746/pnf.2022.27.1.78.

Cheraghi, J., Kridhchi, P., Nasri, S., and Zargooshi, M. (2021). Effects of parsley (*Petroselinum crispum*) hydroalcoholic extract on spermatogenesis and pituitary- gonadal axis in streptozotocin-induced diabetic male rat. *Iranian Journal of Veterinary Medicine* 15 (4): 411–422. https://doi.org/10.22059/IJVM.2021.312948.1005139.

Chi, J., Ge, J., Yue, X. et al. (2019). Preparation of nanoliposomal carriers to improve the stability of anthocyanins. *LWT* 109: 101–107. https://doi.org/10.1016/j.lwt.2019.03.070.

Chirumbolo, S., Bjørklund, G., Lysiuk, R. et al. (2018). Targeting cancer with phytochemicals via their fine tuning of the cell survival signaling pathways. *International Journal of Molecular Sciences* 19 (11): 3568. https://doi.org/10.3390/ijms19113568.

Cravotto, G., Boffa, L., Genzini, L., and Garella, D. (2010). Phytotherapeutics: an evaluation of the potential of 1000 plants. *Journal of Clinical Pharmacy and Therapeutics* 35 (1): 11–48. https://doi.org/10.1111/j.1365-2710.2009.01096.x.

Dai, C., Jiang, S., Chu, C. et al. (2019). Baicalin protects human retinal pigment epithelial cell lines against high glucose-induced cell injury by up-regulation of microRNA-145. *Experimental and Molecular Pathology* 106: 123–130. https://doi.org/10.1016/j.yexmp.2019.01.002.

Daliri, E.B., Balnionytė, T., and Stankevičiūtė J., Lastauskienė E., Meškys R., Burokas A. (2023). High temperature lacto-fermentation improves antioxidant and antidiabetic potentials of Lithuanian red beetroot. *LWT* 185: 115122. https://doi.org/10.1016/j.lwt.2023.115122.

Datta, S., Sinha, B., Bhattacharjee, S., and Seal, T. (2019). Nutritional composition, mineral content, antioxidant activity and quantitative estimation of water soluble vitamins and phenolics by RP-HPLC in some lesser used wild edible plants. *Heliyon* 5 (3): e01431. https://doi.org/10.1016/j.heliyon.2019.e01431.

Deng, Y.H., Alex, D., Huang, H.Q. et al. (2011). Inhibition of TNF-α-mediated endothelial cell-monocyte cell adhesion and adhesion molecules expression by the resveratrol derivative, trans-3,5,4′-trimethoxystilbene. *Phytotherapy Research* 25: 451–457. https://doi.org/10.1002/ptr.3279.

Dixit, S., Shukla, A., Singh, V., and Upadhyay, S.K. (2021a). *Bioprospecting of Natural Compounds for Industrial and Medical Applications*. Wiley https://doi.org/10.1002/9781119718017.ch3.

Dotto, J.M. and Chacha, J.S. (2020). The potential of pumpkin seeds as a functional food ingredient: a review. *Scientific African* 10: e00575. https://doi.org/10.1016/j.sciaf.2020.e00575.

Duncan, B.B., Schmidt, M.I., Pankow, J.S. et al. (2003). Low-grade systemic inflammation and the development of type 2 diabetes: the atherosclerosis risk in communities study. *Diabetes* 52: 1799–1805. https://doi.org/10.2337/diabetes.52.7.1799.

Dyck, G.J.B., Raj, P., Zieroth, S. et al. (2019). The effects of resveratrol in patients with cardiovascular disease and heart failure: a narrative review. *International Journal of Molecular Sciences* 20 (4): 904. https://doi.org/10.3390/ijms20040904.

Eldin, I.M.T., Ahmed, E.M., and Abd, E.H. (2010). Preliminary study of the clinical hypoglycemic effects of *Allium cepa* (red onion) in type 1 and type 2 diabetic patients. *Environmental Health Insights* 4: 71–77. https://doi.org/10.4137/EHI.S5540.

El-Soud, N. and Khalil, M. (2010). Antioxidative effects of *Allium cepa* essential oil in streptozotocin induced diabetic rats. *Macedonian Journal of Medical Sciences* 3: 344–351. https://doi.org/10.3889/MJMS.1857-5773.2010.0118.

Ertik, O., Sacan, O., and Yanardag, R. (2023). Anti-adenosine deaminase, anti-neuraminidase, antixanthine oxidase, anti-acetylcholinesterase and antioxidant activities of parsley extract. *Journal of Herbal Medicine* 42: 100787. https://doi.org/10.1016/j.hermed.2023.100787.

Fan, X., Fan, Z., Yang, Z. et al. (2022). Flavonoids-natural gifts to promote health and longevity. *International Journal of Molecular Sciences* 23 (4): 2176. https://doi.org/10.3390/ijms23042176.

Fowler, M.J. (2011). Microvascular and macrovascular complications of diabetes. *Clinical Diabetes* 29: 116–122. https://doi.org/10.2337/diaclin.29.3.116.

Galavi, A., Hosseinzadeh, H., and Razavi, B.M. (2021). The effects of *Allium cepa* L. (onion) and its active constituents on metabolic syndrome: a review. *Iranian Journal of Basic Medical Sciences* 24 (1): 3–16. https://doi.org/10.22038/ijbms.

Gasmi, A., Mujawdiya, P.K., Lysiuk, R. et al. (2022a). Quercetin in the prevention and treatment of coronavirus infections: a focus on SARS-CoV-2. *Pharmaceuticals (Basel)* 15 (9): 1049. https://doi.org/10.3390/ph15091049.

Gasmi, A., Mujawdiya, P.K., Noor, S. et al. (2022b). Polyphenols in metabolic diseases. *Molecules* 27 (19): 6280. https://doi.org/10.3390/molecules27196280.

Gnintoungbe, G., Medehouenou, T., Adounkpe, F. et al. (2023). Phytochemical screening, antioxidant activity and safety of *Petroselinum crispum* (Mill.) AW Hill *Apiaceae* leaves grown in Benin. *Open Journal of Applied Sciences* 13 (1): 36–50. https://doi.org/10.4236/ojapps.2023.131004.

Guo, F., Amila Abulati, A., Wang, J.W. et al. (2022). Flavonoids of *Coreopsis tinctoria* Nutt alleviate the oxidative stress and inflammation of glomerular mesangial cells in diabetic nephropathy via RhoA/ROCK signaling. *Journal of Functional Foods* 89: 104955. https://doi.org/10.1016/j.jff.2022.104955.

Gutierrez, P. and Martha, R. (2016). Review of *Cucurbita pepo* (pumpkin) its phytochemistry and pharmacology. *Medicinal Chemistry (Los Angeles)* 6 (1): 12–21. https://doi.org/10.4172/2161-0444.1000316.

Hajiaghaalipour, F., Khalilpourfarshbafi, M., and Arya, A. (2015). Modulation of glucose transporter protein by dietary flavonoids in type 2 diabetes mellitus. *International Journal of Biological Sciences* 11: 508–524. https://doi.org/10.7150/ijbs.11241.

Hanhineva, K., Törrönen, R., Bondia-Pons, I. et al. (2010). Impact of dietary polyphenols on carbohydrate metabolism. *International Journal of Molecular Sciences* 11: 1365–1402. https://doi.org/10.3390/ijms11041365.

Herrera-Balandrano, D.D., Chai, Z., Beta, T. et al. (2021). Blueberry anthocyanins: an updated review on approaches to enhancing their bioavailability. *Trends in Food Science & Technology* 118 (B): 808–821. https://doi.org/10.1016/j.tifs.2021.11.006.

Hu, Q., Qu, C., Xiao, X. et al. (2021). Flavonoids on diabetic nephropathy: advances and therapeutic opportunities. *Chinese Medicine* 16: 74. https://doi.org/10.1186/s13020-021-00485-4.

Huang, W.Y., Yan, Z., Li, D.J. et al. (2018). Antioxidant and anti-inflammatory effects of blueberry anthocyanins on high glucose-induced human retinal capillary endothelial cells. *Oxidative Medicine and Cellular Longevity* 1862418. https://doi.org/10.1155/2018/1862462.

Hussain, T., Tan, B., Murtaza, G. et al. (2020). Flavonoids and type 2 diabetes: evidence of efficacy in clinical and animal studies and delivery strategies to enhance their therapeutic efficacy. *Pharmacological Research* 152: 104629. https://doi.org/10.1016/j.phrs.2020.104629.

Intagliata, S., Modica, M.N., Santagati, L.M., and Montenegro, L. (2019). Strategies to improve resveratrol systemic and topical bioavailability: an update. *Antioxidants (Basel)* 8 (8): 244. https://doi.org/10.3390/antiox8080244.

International Diabetes Federation. IDF diabetes atlas 9, 2019. https://www.diabetesatlas.org/upload/resources/2019/IDF_Atlas_9th_Edition_2019.pdf. (accessed 1 December 2023).

Jia, Q., Wang, Y., Liu, X. et al. (2019). Effects of genistein on Nrf2/HO-1 pathway in myocardial tissues of diabetic rats. *Zhong Nan Da Xue Xue Bao. Yi Xue Ban* 44 (8): 850–856. https://doi.org/10.11817/j.issn.1672-7347.2019.08.180429.

Jiang, Y., Liu, J., Zhou, Z. et al. (2018). Diosmetin attenuates Akt signaling pathway by modulating nuclear factor kappa-light-chain-enhancer of activated B cells (NF-kappaB)/inducible nitric oxide synthase (iNOS) in streptozotocin (STZ)-induced diabetic nephropathy mice. *Medical Science Monitor: International Medical Journal of Experimental and Clinical Research* 24: 7007–7014. https://doi.org/10.12659/MSM.910764.

Joshi, R., Laddha, A.P., Kulkarni, Y.A., and Wairkar, S. (2021). Improved performance of naringenin herbosomes over naringenin in streptozotocin-induced diabetic rats: *in vitro* and *in vivo* evaluation. *Asian Pacific Journal of Tropical Biomedicine* 11: 385–393. https://doi.org/10.4103/2221-1691.321131.

Jubaidi, F.F., Zainalabidin, S., Taib, I.S. et al. (2021). The potential role of flavonoids in ameliorating diabetic cardiomyopathy via alleviation of cardiac oxidative stress, inflammation and apoptosis. *International Journal of Molecular Sciences* 22 (10): 5094. https://doi.org/10.3390/ijms22105094.

Kähkönen, M.P., Heinämäki, J., Ollilainen, V., and Heinonen, M. (2003). Berry anthocyanins: isolation, identification and antioxidant activities. *Journal of the Science of Food and Agriculture* 83: 1403--1411. https://doi.org/10.1002/jsfa.1511.

Kang, M.K., Lee, E.J., Kim, Y.H. et al. (2018). Chrysin ameliorates malfunction of retinoid visual cycle through blocking activation of AGE-RAGE-ER stress in glucose-stimulated retinal pigment epithelial cells and diabetic eyes. *Nutrients* 10: https://doi.org/10.3390/nu10081046.

Karimzadeh, L., Behrouz, V., Sohrab, G. et al. (2022). A randomized clinical trial of beetroot juice consumption on inflammatory markers and oxidative stress in patients with type 2 diabetes. *Food Science* 87 (12): 5430–5441. https://doi.org/10.1111/1750-3841.16365.

Karimzadeh, L., Sohrab, G., Hedayati, M. et al. (2023). Effects of concentrated beetroot juice consumption on glycemic control, blood pressure, and lipid profile in type 2 diabetes patients: randomized clinical trial study. *Irish Journal of Medical Science* 192: 1143–1153. https://doi.org/10.1007/s11845-022-03090-y.

Khedher, O., Rigane, G., Riguene, H. et al. (2021). Phenolic profile (HPLC-UV) analysis and biological activities of two organic extracts from *Echinops spinosissimus* Turra roots growing in Tunisia. *Natural Product Research* 35 (24): 5786–5793. https://doi.org/10.1080/14786419.2020.1837812.

Kianian, F., Marefati, N., Boskabady, M. et al. (2021). Pharmacological properties of *Allium cepa*, preclinical and clinical evidences: a review. *Iranian Journal of Pharmaceutical Research* 20 (2): 107–134. https://doi.org/10.22037/ijpr.2020.112781.13946.

Kim, J., Kim, C.S., Lee, Y.M. et al. (2015a). *Vaccinium myrtillus* extract prevents or delays the onset of diabetes-induced blood-retinal barrier breakdown. *International Journal of Food Sciences and Nutrition* 66 (2): 236–242. https://doi.org/10.3109/09637486.2014.979319.

Kim, S.H. (2015b). Maturity-onset diabetes of the young: what do clinicians need to know? *Diabetes and Metabolism Journal* 39 (6): 468–477. https://doi.org/10.4093/dmj.2015.39.6.468.

Kowalczyk, E., Krzesiński, P., Kura, M. et al. (2003). Anthocyanins in medicine. *Polish Journal of Pharmacology* 55 (5): 699–702.

Kumari, K. and Augusti, K.T. (2002). Antidiabetic and antioxidant effects of S-methyl cysteine sulfoxide isolated from onions (*Allium cepa* Linn) as compared to standard drugs in alloxan diabetic rats. *Indian Journal of Experimental Biology* 40 (9): 1005–1009.

Kushawaha, D.K., Yadav, M., Chatterji, S. et al. (2017). Evidence based study of antidiabetic potential of *C. maxima* seeds—*in vivo*. *Journal of Traditional and Complementary Medicine* 7 (4): 466–470. https://doi.org/10.1016/j.jtcme.2016.12.001.

Lemelman, M.B., Letourneau, L., and Greeley, S.A.W. (2018). Neonatal diabetes mellitus: an update on diagnosis and management. *Clinics in Perinatology* 45 (1): 41–59. https://doi.org/10.1016/j.clp.2017.10.006.

Lenchyk, L.V., Senyuk, I.V., and Lysiuk, R.M. (2021). Medicinal herbs for metabolic disorders. In: *An Introduction to Medicinal Herbs* (ed. M. Emerald), 171–204. Nova Science Publishers, Inc.

Letourneau, L.R., Carmody, D., Wroblewski, K. et al. (2017). Diabetes presentation in infancy: high risk of diabetic ketoacidosis. *Diabetes Care* 40 (10): e147–e148. https://doi.org/10.2337/dc17-1145.

Li, J., Kang, S.-W., Kim, J.-L. et al. (2010a). Isoliquiritigenin entails blockade of TGF-β1-SMAD signaling for retarding high glucose-induced mesangial matrix accumulation. *Journal of Agricultural and Food Chemistry* 58 (5): 3205–3212. https://doi.org/10.1021/jf9040723.

Li, W., Li, S., Higai, K. et al. (2013). Evaluation of licorice flavonoids as protein tyrosine phosphatase 1B inhibitors. *Bioorganic and Medicinal Chemistry Letters* 23 (21): 5836–5839. https://doi.org/10.1016/j.bmcl.2013.08.102.

Li, S., Li, W., Wang, Y. et al. (2010b). Prenylflavonoids from *Glycyrrhiza uralensis* and their protein tyrosine phosphatase-1B inhibitory activities. *Bioorganic & Medicinal Chemistry Letters* 20 (18): 5398–5401. https://doi.org/10.1016/j.bmcl.2010.07.110.

Lin, X., Xu, Y., Pan, X. et al. (2020). Global, regional, and national burden and trend of diabetes in 195 countries and territories: an analysis from 1990 to 2025. *Scientific Reports* 10 (1): 14790. https://doi.org/10.1038/s41598-020-71908-9.

Liu, J., Zhou, H., Song, L. et al. (2021). Anthocyanins: promising natural products with diverse pharmacological activities. *Molecules* 26 (13): 3807. https://doi.org/10.3390/molecules26133807.

Lu, X., Liu, M., Dong, H. et al. (2022). Dietary prenylated flavonoid xanthohumol alleviates oxidative damage and accelerates diabetic wound healing via Nrf2 activation. *Food and Chemical Toxicology* 160: 112813. https://doi.org/10.1016/j.fct.2022.112813.

Lu, M., Qiu, Q., Luo, X. et al. (2019). Phyto-phospholipid complexes (phytosomes): a novel strategy to improve the bioavailability of active constituents. *Asian Journal of Pharmaceutical Sciences* 14 (3): 265–274. https://doi.org/10.1016/j.ajps.2018.05.011.

Lv, P., Yu, J., Xu, X. et al. (2019). Eriodictyol inhibits high glucose-induced oxidative stress and inflammation in retinal ganglial cells. *Journal of Cellular Biochemistry* 120: 5644–5651. https://doi.org/10.1002/jcb.27848.

Lysiuk, R. and Hudz, N. (2017). Differential spectrophotometry: application for quantification of flavonoids in herbal drugs and nutraceuticals. *The International Journal of Food Sciences and Nutrition* 1: e102.

Lysiuk, R. and Mansuetusmboya, J. (2020). Herbal drugs for the treatment of diabetic nephropathy: Current status and prospects for the application. *Journal of Kidney Treatment and Diagnosis* 3 (2): 3–4.

Lysiuk, R., Oliynyk, P., Antonyak, H., and Voronenko, D. (2020). Development of phyto-antidotes against adverse chemical agents, Chapter 12. In: *Poisonous Plants and Phytochemicals in Drug Discovery* (ed. A.G. Mtewa, C. Egbuna, and G.N. Rao), 249–268. Wiley. doi: 10.1002/9781119650034.ch12.

Mahmood, N., Muazzam, M.A., Ahmad, M. et al. (2023). Phytochemistry of *Allium cepa* L. (onion): an overview of its nutritional and pharmacological importance. *Scientific Inquiry and Review* 5 (3): 41–49. doi: 10.32350/sir/53.04.

Maity, S., Acharyya, A., and Chakraborti, A.S. (2022). Flavonoid-based polymeric nanoparticles: a promising approach for cancer and diabetes treatment. *European Polymer Journal* 177: 111455. https://doi.org/10.1016/j.eurpolymj.2022.111455.

Mirmiran, P., Houshialsadat, Z., Gaeini, Z. et al. (2020). Functional properties of beetroot (*Beta vulgaris*) in management of cardio-metabolic diseases. *Nutrition & Metabolism (London)* 17 (3): https://doi.org/10.1186/s12986-019-0421-0.

Mishra, A., Oliinyk, P., Lysiuk, R. et al. (2022). Flavonoids and stilbenoids as a promising arsenal for the management of chronic arsenic toxicity. *Environmental Toxicology and Pharmacology* 95: 103970. https://doi.org/10.1016/j.etap.2022.103970.

Miyake, S., Takahashi, N., Sasaki, M. et al. (2012). Vision preservation during retinal inflammation by anthocyanin-rich bilberry extract: cellular and molecular mechanism. *Laboratory Investigation* 92: 102–109. https://doi.org/10.1038/labinvest.2011.132.

Müller, D., Schantz, M., and Richling, E. (2012). High performance liquid chromatography analysis of anthocyanins in bilberries (*Vaccinium myrtillus* L.), blueberries (*Vaccinium corymbosum* L.), and corresponding juices. *Food Science* 77 (4): 340–345. https://doi.org/10.1111/j.1750-3841.2011.02605.x.

Muoio, D.M. and Newgard, C.B. (2008). Mechanisms of disease: molecular and metabolic mechanisms of insulin resistance and beta-cell failure in type 2 diabetes. *Nature Reviews Molecular Cell Biology* 9 (3): 193–205. https://doi.org/10.1038/nrm2327.

Naziri, E., Mitić, M.N., and Tsimidou, M.Z. (2016). Contribution of tocopherols and squalene to the oxidative stability of cold-pressed pumkin seed oil (*Cucurbita pepo* L.). *European Journal of Lipid Science and Technology* 118 (6): 898–905. https://doi.org/10.1002/ejlt.201500261.

Nusair, S.D., Zainalabdeen, E.A., Alshogran, O.Y., and Alkaraki, A. (2022). Evaluation of orellanine-induced toxicity from the mushroom *Cortinarius orellanus* and the antagonistic effect of *Petroselinum crispum*. *Toxicon* 214: 1–7. https://doi.org/10.1016/j.toxicon.2022.04.018.

Oboh, H.A., Obayiuwana, A., Aihie, E.O. et al. (2020). Beetroot (*Beta vulgaris*) juice inhibits key carbohydrate metabolising enzymes associated with type II diabetes. *Nigerian Journal of Basic and Applied Sciences* 28 (1): 1–6. https://doi.org/10.4314/njbas.v28i1.1.

Ogbonna, J.I., Umeh, O.N., Mbah, C.C. et al. (2019). Hypoglycemic activities and biochemical parameters modulation of herbal formulations of *Allium cepa* L. in alloxanized diabetic rats. *Scientific Research and Essays* 14 (10): 74–85. https://doi.org/10.5897/SRE2019.6620.

de Oliveira, Q.W., Neri-Numa, I.A., Arruda, H.S. et al. (2022). Encapsulation of flavonoids in foods for diabetics: the emerging paradigm for an effective therapy. *Trends in Food Science & Technology* 127: 198–206. https://doi.org/10.1016/j.tifs.2022.06.004.

Othman, M.S., Khaled, A.M., Al-Bagawi, A.H. et al. (2022). *Echinops spinosus* effect against diabetes and its hepatorenal complications: total extract and flavonoids fraction. *Environmental Science and Pollution Research International* 29 (25): 38606–38617. https://doi.org/10.1007/s11356-022-18824-9.

Ouattara-Soro, F.S., Acray-Zengbe, P., Coulibaly, Z. et al. (2022). Study of the anti-anemic activity of the aqueous extract of the leaves of *Petroselinum crispum* (*Apiaceae*) in the white mouse (*Mus musculus*) of SWISS strain. *Journal of Blood Disorders and Therapeutics* 2 (1): 101–115.

Oyenihi, O.R., Oyenihi, A.B., Adeyanju, A.A., and Oguntibeju, O.O. (2016). Antidiabetic effects of resveratrol: the way forward in its clinical utility. *Journal Diabetes Research* 2016: 9737483. https://doi.org/10.1155/2016/9737483.

Oza, M.J. and Kulkarni, Y.A. (2019). Formononetin attenuates kidney damage in type 2 diabetic rats. *Life Sciences* 219: 109–121. https://doi.org/10.1016/j.lfs.2019.01.013.

Paes, J., Dotta, R., Barbero, G.F., and Martínez, J. (2014). Extraction of phenolic compounds and anthocyanins from blueberry (*Vaccinium myrtillus* L.) residues using supercritical CO2 and pressurized liquids. *The Journal of Supercritical Fluids* 95: 8–16. https://doi.org/10.1016/j.supflu.2014.07.025.

Paik, S.S., Jeong, E., Jung, S.W. et al. (2012). Anthocyanins from the seed coat of black soybean reduce retinal degeneration induced by N-methyl-N-nitrosourea. *Experimental Eye Research* 97: 55–62. https://doi.org/10.1016/j.exer.2012.02.010.

Pandey, K.B. and Rizvi, S.I. (2009). Plant polyphenols as dietary antioxidants in human health and disease. *Oxidative Medicine and Cellular Longevity* 2: 270–278. https://doi.org/10.4161/oxim.2.5.9498.

Petruskevicius, A., Viskelis, J., Urbonaviciene, D., and Viskelis, P. (2023). Anthocyanin accumulation in berry fruits and their antimicrobial and antiviral properties: an overview. *Horticulturae* 9 (2): 288. https://doi.org/10.3390/horticulturae9020288.

Pickup, J.C. (2004). Inflammation and activated innate immunity in the pathogenesis of type 2 diabetes. *Diabetes Care* 27: 813–823. https://doi.org/10.2337/diacare.27.3.813.

Rahman, M., Dhar, P.S., Sumaia, A.F. et al. (2022). Exploring the plant-derived bioactive substances as antidiabetic agent: an extensive review. *Biomedicine & Pharmacotherapy* 152: 113217. https://doi.org/10.1016/j.biopha.2022.113217.

Ramprasath, V.R. and Jones, P.J. (2010). Anti-atherogenic effects of resveratrol. *European Journal of Clinical Nutrition* 64: 660–668. https://doi.org/10.1038/ejcn.2010.77.

Ray, L.R., Alam, M.S., Junaid, M. et al. (2021). *Brassica oleracea* var. *capitata* f. *alba*: a review on its botany, traditional uses, phytochemistry and pharmacological activities. *Mini Reviews in Medicinal Chemistry* 21 (16): 2399–2417. https://doi.org/10.2174/1389557521666210111150036.

Rice-Evans, C.A., Miller, N.J., and Paganga, G. (1996). Structure-antioxidant activity relationships of flavonoids and phenolic acids. *Free Radical Biology & Medicine* 20: 933–956. https://doi.org/10.1016/0891-5849(95)02227-9.

Sapra, A. and Bhandari, P. (2023). *Diabetes Mellitus*. StatPearls Publishing https://www.ncbi.nlm.nih.gov/books/NBK551501/.

Saravanan, P., Diabetes in Pregnancy Working Group, Maternal Medicine Clinical Study Group, and Royal College of Obstetricians and Gynaecologists, UK (2020). Gestational diabetes: opportunities for improving maternal and child health. *The Lancet Diabetes and Endocrinology* 8: 793–800. https://doi.org/10.1016/S2213-8587(20)30161-3.

Semeraro, F., Morescalchi, F., Cancarini, A. et al. (2019). Diabetic retinopathy, a vascular and inflammatory disease: therapeutic implications. *Diabetes & Metabolism* 45 (6): 517–527. https://doi.org/10.1016/j.diabet.2019.04.002.

Serafini, M., Peluso, I., and Raguzzini, A. (2010). Flavonoids as anti-inflammatory agents. *Proceedings of the Nutrition Society* 69 (3): 273–278. https://doi.org/10.1017/S002966511000162X.

Seymen, M., Uslu, N., Türkmen, Ö. et al. (2016). Chemical compositions and mineral contents of some hull-less pumpkin seed and oils. *Journal of the American Oil Chemists' Society* 93 (8): 1095–1099. https://doi.org/10.1007/s11746-016-2850-5.

Shakir, A., Afroz, S., Masood, A.S.M. et al. (2023). Antidiabetic activity of ethanolic leaves extract of *Allium cepa*. *Journal of Xi'an Shiyou University* 19 (08): 538–549.

Smoliga, J.M. and Blanchard, O. (2014). Enhancing the delivery of resveratrol in humans: if low bioavailability is the problem, what is the solution? *Molecules* 19: 17154–17172. https://doi.org/10.3390/molecules191117154.

Sokolova, L.K., Pushkarev, V.M., and Tronko, M.D. (2020). The effects of resveratrol in norm and various pathologies. [Efekty resveratrolu v normi ta za riznykh patolohii]. *Endokrynologia* 25 (1): 76–88. https://doi.org/10.31793/1680-1466.2020.25-1.76.

Sood, A., Kumar, B., Kumar, S.S. et al. (2020). Flavonoids as potential therapeutic agents for the management of diabetic neuropathy. *Current Pharmaceutical Design* 26 (42): 5468–5487. https://doi.org/10.2174/1381612826666200826164322.

Svintsitskyi, A.S. and Haievski, P. (2018). *Internal Diseases. A Textbook Based on the Principles of Evidence-Based Medicine.[Vnutrishni khvoroby. Pidruchnyk, zasnovanyi na pryntsypakh dokazovoi medytsyny]*, 1632. Kraków: Medycyna Praktyczna https://medkniga.com.ua/25560-vnutrishni-hvorobi-pidruchnik-zasnovanij-na-printsipah-dokazovoyi-meditsini-2018-19/ (Ukrainian).

Tasya, Z., Amiruddin, R., Syam, A., and Thamrin Y. (2022). Pumpkin seed intervention to control diabetes mellitus: a systematic review. *Macedonian Journal of Medical Sciences* 10 (F): 535–540. https://doi.org/10.3889/oamjms.2022.9850.

Teshika, J.D., Zakariyyah, A.M., Zaynab, T. et al. (2019). Traditional and modern uses of onion bulb (*Allium cepa* L): a systematic review. *Critical Reviews in Food Science and Nutrition* 59: S39–S70. https://doi.org/10.1080/10408398.2018.1499074.

Tkachenko, V. (2014). Analysis of the prevalence and incidence of diabetes mellitus among the population of the world and Ukraine, 2003-2013. [Analiz poshyrenosti ta zakhvoriuvanosti na tsukrovyi diabet sered naselennia svitu ta Ukrainy za 2003–2013 rr.]. *Liky Ukrainy* 4 (21): 55–59. (Ukrainian).

Tran, N., Pham, B., and Le, L. (2020). Bioactive compounds in anti-diabetic plants: from herbal medicine to modern drug discovery. *Biology (Basel)* 9 (9): 252. https://doi.org/10.3390/biology9090252.

Ülger, T.G. and Çakiroglu, F.P. (2020). The effects of onion (*Allium cepa* L.) dried by different heat treatments on plasma lipid profile and fasting blood glucose level in diabetic rats. *Avicenna Journal of Phytomedicine* 10 (4): 325–333. https://doi.org/10.22038/AJP.2019.14066.

Ullah, A., Munir, S., Badshah, S.L. et al. (2020). Important flavonoids and their role as a therapeutic agent. *Molecules* 25 (22): 5243. https://doi.org/10.3390/molecules25225243.

Upadhyay, S.K. and Singh, S.P. (2021). *Bioprospecting of Plant Biodiversity for Industrial Molecules.* Wiley https://doi.org/10.1002/9781119718017.

Upadhyay, S.K. and Singh, S.P. (2023). *Plants as Bioreactors for Industrial Molecules.* Wiley https://doi.org/10.1002/9781119875116.

Uuh-Narvaez, J.J. and Segura-Campos, M.R. (2021). Cabbage (*Brassica oleracea* var. *capitata):* a food with functional properties aimed to type 2 diabetes prevention and management. Food. *Science* 86 (11): 4775–4798. https://doi.org/10.1111/1750-3841.15939.

Veeramani, S., Narayanan, A.P., Yuvaraj, K. et al. (2022). *Nigella sativa* flavonoids surface coated gold NPs (Au-NPs) enhancing antioxidant and anti-diabetic activity. *Process Biochemistry* 114: 193–202. https://doi.org/10.1016/j.procbio.2021.01.004.

Vikou, C., Semassa, J., Diaconeasa, Z. et al. (2023). Diversity, chemical compositions and beneficial effects of some spices and aromatic leaves consumed in Benin and in the world: critical review. *American Journal of Plant Sciences* 14: 569–598. https://doi.org/10.4236/ajps.2023.145039.

Vinayagam, R. and Xu, B. (2015). Antidiabetic properties of dietary flavonoids: a cellular mechanism review. *Nutrition & Metabolism (London)* 12: 60. https://doi.org/10.1186/s12986-015-0057-7.

Wang, W., Wu, Q.H., Sui, Y. et al. (2017). Rutin protects endothelial dysfunction by disturbing Nox4 and ROS-sensitive NLRP3 inflammasome. *Biomedicine & Pharmacotherapy* 86: 32–40. https://doi.org/10.1016/j.biopha.2016.11.134.

Wang, W., Zhang, Y., Jin, W. et al. (2018). Catechin weakens diabetic retinopathy by inhibiting the expression of NF-kappaB signaling pathway-mediated inflammatory factors. *Annals of Clinical and Laboratory Science* 48: 594–600.

Wang, Y., Zhao, L., Lu, F. et al. (2015). Retinoprotective effects of bilberry anthocyanins via antioxidant, anti-inflammatory, and anti-apoptotic mechanisms in a visible light-induced retinal degeneration model in pigmented rabbits. *Molecules* 20: 22395–22410. https://doi.org/10.3390/molecules201219785.

Werner, R.D.C. and Merz, A.D.B. (2015). European Medicines Agency (EMA)—Committee on Herbal Medicinal Products (HMPC) Assessment report on *Vaccinium myrtillus* L. fructus recens and *Vaccinium myrtillus* L., fructus siccus. *HMPC Monograph* 555161: 1–83.

Wickramasinghe, A.S.D., Kalansuriya, P., and Attanayake, A.P. (2022). Nanoformulation of plant-based natural products for type 2 diabetes mellitus: from formulation design to therapeutic applications. *Current Therapeutic Research, Clinical and Experimental* 96: 100672. https://doi.org/10.1016/j.curtheres.2022.100672.

Wu, Z., Gong, H., Shen, Y. et al. (2022b). Hypoglycemic and hypolipidemic effects of sugar beet (*Beta vulgaris* L.) extract on C57BL/KsJ db/db mice. *Rendiconti Lincei. Scienze Fisiche e Naturali* 33: 205–212. https://doi.org/10.1007/s12210-022-01054-z.

Wu, Y., Wang, Z., Du, Q. et al. (2022a). Pharmacological effects and underlying mechanisms of licorice-derived flavonoids. *Evidence-based Complementary and Alternative Medicine* 2022: 9523071. https://doi.org/10.1155/2022/9523071.

Xiao, J. (2022). Recent advances in dietary flavonoids for management of type 2 diabetes. *Current Opinion in Food Science* 44: 100806. https://doi.org/10.1016/j.cofs.2022.01.002.

Yahaya, T.O. and Ufuoma, S.B. (2020). Genetics and pathophysiology of maturity-onset diabetes of the young (MODY): a review of current trends. *Oman Medical Journal* 35 (3): e126. https://doi.org/10.5001/omj.2020.44.

Yang, L., Jiang, Y., Zhang, Z. et al. (2020). The anti-diabetic activity of licorice, a widely used Chinese herb. *Journal of Ethnopharmacology* 263. https://doi.org/10.1016/j.jep.2020.113216.113216.

Ye, W., Luo, C., Huang, J. et al. (2022). Gestational diabetes mellitus and adverse pregnancy outcomes: systematic review and meta-analysis. *BMJ* 377: e067946. https://doi.org/10.1136/bmj-2021-067946.

Yen, F.S., Qin, C.S., Xuan, S.T. et al. (2021). Hypoglycemic effects of plant flavonoids: a review. *Evidence-Based Complementary and Alternative Medicine* ID 2057333: 12. https://doi.org/10.1155/2021/2057333.

Yin, M.-S., Zhang, Y.-C., Xu, S.-H. et al. (2019). Puerarin prevents diabetic cardiomyopathy *in vivo* and *in vitro* by inhibition of inflammation. *Journal of Asian Natural Products Research* 21 (5): 476–493. https://doi.org/10.1080/10286020.2017.1405941.

Zayed, A., Sheashea, M., Kassem, I.A., and Farag, M.A. (2023). Red and white cabbages: an updated comparative review of bioactives, extraction methods, processing practices, and health benefits. *Critical Reviews in Food Science and Nutrition* 63 (24): 7025–7042. https://doi.org/10.1080/10408398.2022.2040416.

Zhang, T., Mei, X., Ouyang, H. et al. (2019). Natural flavonoid galangin alleviates microglia-trigged blood-retinal barrier dysfunction during the development of diabetic retinopathy. *The Journal of Nutritional Biochemistry* 65: 1–14. https://doi.org/10.1016/j.jnutbio.2018.11.006.

Zhang, J., Yang, S., Li, H. et al. (2017). Naringin ameliorates diabetic nephropathy by inhibiting NADPH oxidase 4. *European Journal of Pharmacology* 804: 1–6. https://doi.org/10.1016/j.ejphar.2017.04.006.

Zheng, Y., Ley, S.H., and Hu, F.B. (2018). Global aetiology and epidemiology of type 2 diabetes mellitus and its complications. *Nature Reviews. Endocrinology* 14 (2): 88–98. https://doi.org/10.1038/nrendo.2017.151.

7

Herbal Wine and Health Benefits

Sonia Morya[1], Gopika S. Pillai[1], Abishek Dinesh[1], Amenah S. Alotaibi[2], Hanan A. Alatawi[3], and Farid Menaa[4]

[1] *Department of Food Technology and Nutrition, School of Agriculture, Lovely Professional University, Phagwara, Punjab, India*
[2] *Depatment of Biology, Faculty of Science, Genomic & Biotechnlogy, University of Tabuk, Tabuk, Saudi Arabia*
[3] *Department of Biological Sciences, University College of Haqel, Tabuk University, Tabuk, Saudi Arabia*
[4] *Department of Biomedical and Environmental Engineering, California Innovations Corporation, San Diego, CA, USA*

7.1 Introduction

Herbal wine has long been valued for its distinct flavors and fragrances. It is a beautiful blend of traditional winemaking with the infusion of herbs, spices, and botanicals (Karki 2019). Because the plants employed in its manufacture have medicinal qualities, herbal wine has been linked to a host of potential health advantages beyond just its taste (Nazir et al. 2019).

According to Bhise and Morya's (2021) study, herbal wines are a great source of phytochemicals that are known to help the body fight inflammation and oxidative stress, such as flavonoids, polyphenols, and antioxidants. These bioactive substances are extracted from the infused herbs and botanicals, giving the wine unique tastes as well as perhaps health-promoting qualities. In addition, a lot of the herbs that are frequently used to make herbal wine have long been known to have digestive, immune-stimulating, and mood-enhancing properties in traditional medicine. Herbs such as chamomile and ginger are well-known for their relaxing and digestive qualities, while hibiscus and elderflower are highly valued for their antioxidant qualities.

7.2 Defining Concept of Herbal Wines

A base wine is infused with various herbs, spices, flowers, and other botanical components to create herbal wine, a type of alcoholic beverage. It is also known as infused wine or botanical wine (McGovern et al. 2009). The wine has unique flavors and aromas from this infusion process, which may also have health advantages. Herbal wines are made by carefully selecting and blending base wine with various botanicals to produce a beverage that offers a complex and diverse sensory experience.

Sonia Morya and Farid Menaa are co-corresponding authors and contributed equally to this chapter.

Herbal Nutraceuticals: Products and Processes, First Edition. Edited by Santosh Kumar Upadhyay and Sudhir Pratap Singh.

Yang et al. (2009) study mainly focuses on the phytochemical makeup and antioxidant activity of traditional red and white wines; it also highlights the possible advantages of infused herbal wines. The study demonstrates how the antioxidant qualities of wines are enhanced by the presence of specific phytochemicals, such as flavonoids and polyphenols. Numerous herbs and botanicals that are frequently used to make herbal wine have large amounts of these compounds, which implies that adding these ingredients to the finished product could increase its antioxidant content.

Wine can have its taste qualities and nutritional profile enhanced by infusing it with herbs and botanicals. Botanical components can be carefully chosen and blended with base wine to create a beverage with a rich scent and flavor profile. Lavender, rosemary, thyme, chamomile, elderflower, hibiscus, and many more botanicals are frequently used in the manufacturing of herbal wines, and each one adds a unique flavor to the finished product (Maleš et al. 2022a).

Infusing wine with herbs and botanicals can improve both its flavor and nutritional profile (Vilela and Pinto 2019). With careful selection and blending, botanical ingredients can be used with base wine to produce a drink with a complex flavor and aroma. Botanicals such as lavender, rosemary, thyme, chamomile, elderflower, hibiscus, and many more are commonly used to make herbal wines, and each one contributes a distinct flavor to the final product. The utilization of botanical components in herbal wine opens up new research directions for potential health benefits. Numerous herbs and botanicals include high concentrations of phytochemicals, such as flavonoids, polyphenols, and antioxidants, which are well-known for their capacity to promote health and wellness.

7.3 Historical Roots

Wine was regarded as a sacred beverage with ritualistic and symbolic meaning in ancient Egypt (Varriano 2022). According to archaeological findings, the ancient Egyptians flavored and scented wine by infusing it with a range of herbs and botanicals, such as juniper, coriander, myrrh, and thyme. These herbal infusions were prized for their purported therapeutic benefits in addition to their sensual appeal.

Herbal wine remained a common ingredient in European cuisine and medicine throughout the Middle Ages. Wines with rich flavors and supposed health advantages were infused with a wide variety of herbs and botanicals that monks and herbalists grew in their gardens (Balick 2014). The widespread belief in the therapeutic value of botanical elements is reflected in the frequent consumption of herbal wines as tonics or cures for common diseases. Herbal wine continued to be made throughout the renaissance and into the modern period, albeit its cultural significance and level of popularity varied. A growing awareness of artisanal and craft beverages has led to a renaissance of interest in herbal wine in recent decades.

In conclusion, the gastronomic, medical, and cultural traditions of past cultures can be traced back to the historical origins of herbal wine. Herbal wine has developed throughout time from a highly esteemed concoction to a well-liked beverage valued for its tastes and nutritional qualities. Herbal wine is still a vibrant and alluring way for consumers to show their ongoing passion for wine and botanicals, despite the increasing desire for natural and artisanal products.

7.4 Emerging Trends of Herbal Wine

Customers are becoming more aware of the things they eat, especially their beverage selections, in today's health-conscious world. With its blend of botanical elements and apparent health advantages, herbal wine has become a popular choice for consumers who are health-conscious and looking for both enjoyment and possible wellness benefits (Erfurt-Cooper and Cooper 2009).

The fact that herbal wine is linked to natural components and age-old treatments is one of the main factors attracting health-conscious consumers to it. Due to their alleged health benefits, many of the herbs and botanicals used to make herbal wine have a long history of usage in traditional medicine. Because of this, customers view herbal wine as a more natural and holistic substitute for traditional wines, which is in line with their demand for goods that are pure and less processed.

Additionally, herbal wine gives customers a chance to enjoy a tasty libation while maybe benefiting from the herbs and botanicals that are blended with the wine. Ingredients like elderflower, chamomile, and ginger are frequently linked to immune-stimulating, calming, and digestive benefits, which appeal to people who are health-conscious and are looking for functional drinks. Additionally, health-conscious consumers are particularly interested in the antioxidant qualities of herbal wine. Polyphenols, flavonoids, and other antioxidants found in a wide variety of herbs and botanicals have been related to several health advantages, such as decreased inflammation and enhanced cardiovascular health. Customers therefore see herbal wine as a possible source of antioxidant support, especially when ingested in moderation as part of a healthy lifestyle (Lante et al. 2023).

7.5 The Herbal Component

Herbs like lavender, rosemary, thyme, chamomile, elderflower, hibiscus, ginger, lemon balm, sage, and peppermint are frequently used in the making of herbal wine. Bioactive substances, including polyphenols, flavonoids, terpenes, and essential oils, found in these herbs give the wine its distinct flavors and smell, as well as possible health benefits (Table 7.1).

7.6 Selection of Herbs

For herbal wine to have the appropriate flavor profile, aroma, and possibly even health advantages, choosing the right herbs is essential. Every herb contributes distinctive qualities to the mixture, providing a wide range of tastes, scents, and bioactive ingredients. Taste, aroma, potential health advantages, and compatibility with the base wine are all important considerations when selecting herbs for making herbal wine.

7.6.1 Lavender

Lavender, when utilized as a raw ingredient in the making of herbal wine, is a multipurpose herb that has several advantages. The following are some of the main benefits of adding lavender to herbal wine: Lavender gives herbal wines a lovely flowery scent that accentuates the complexity and depth of the flavor profile (Lakićević et al. 2022). The overall drinking experience is enhanced by its floral overtones, which inspire a sensation of calm and tranquility (Koulivand et al. 2013). Lavender is well renowned for its calming and relaxing properties, which are attributed to compounds like linalool and linalyl acetate. Herbal wine with added lavender can help promote calmness and relaxation, making it an excellent choice for unwinding after a demanding day or right before bed (Batiha et al. 2023).

Lavender has long been used to relieve gas, bloating, and other symptoms associated with the digestive system (Radu et al. 2020). In addition to making wine smoother and more enjoyable to drink, lavender infusions may provide some minor gastrointestinal relief. Antioxidant benefits include

Table 7.1 Herbs with bioactive compounds and health benefits.

Herb	Bioactive compounds	Potential health benefits	Citation and reference
Lavender	Linalool, linalyl acetate	Antianxiety, antidepressant analgesic	Bradley et al. (2019)
Rosemary	Carnosic acid, rosmarinic acid	Antioxidant, anti-inflammatory, neuroprotective	Dos Santos et al. (2022)
Thyme	Thymol, carvacrol	Antimicrobial, antioxidant, anti-inflammatory	De Oliveira et al. (2020)
Chamomile	Apigenin, bisabolol	Calming, digestive aid, anti-inflammatory	Sah et al. (2022)
Elderflower	Flavonoids, phenolic acids	Immune-boosting, antioxidant, anti-inflammatory	Ho et al. (2017)
Hibiscus	Anthocyanins, flavonoids	Lower blood pressure, antioxidant, cardiovascular health	Sandhu and Morya (2022)
Ginger	Gingerol, shogaol	Digestive aid, anti-inflammatory, antiemetic	Rahmani et al. (2014)
Lemon Balm	Rosmarinic acid, citronellal	Calming, antioxidant, cognitive enhancement	Ullah and Hassan (2022)
Sage	Carnosic acid, rosmarinic acid	Antioxidant, anti-inflammatory, memory enhancement	Tildesley et al. (2005)
Peppermint	Menthol, menthone	Digestive aid, headache relief, antimicrobial	Nayak et al. (2020)

helping the body combat oxidative stress and reducing inflammation. Lavender contains antioxidants. Flavonoids and polyphenols are examples of these antioxidants, which protect cells from damage caused by free radicals and enhance overall health and wellness. Thus, lavender can boost the antioxidant content of herbal wine and potentially offer further defense against oxidative damage (Faridzadeh et al. 2022). Studies have shown that the fragrant molecules in lavender have a relaxing and uplifting effect on people's emotions. Savoring lavender-infused herbal wine with friends or in a peaceful setting can elevate the spirits and increase the enjoyment of drinking. Lavender is commonly used in aromatherapy and herbal medications because of its capacity to promote sleep. An invigorating and fragrant nightcap option for anyone searching for natural sleep aids is herbal wine infused with lavender. Consuming in moderation before bed may promote a more peaceful slumber (Akgül et al. 2021).

Lavender has a lot of potential benefits when used as a raw component in herbal wines. These include the ability to improve mood, aid with digestion, provide antioxidants, have a calming and relaxing effect, and possibly even help with sleep. By infusing wine with lavender, producers can create a pleasant and potentially beneficial beverage that appeals to consumers seeking both sensory enjoyment and holistic wellness benefits (Maleš et al. 2022b).

7.6.2 Rosemary

The adaptable plant rosemary (*Rosmarinus officinalis*) is prized for its unique flavor, heady perfume, and possible medicinal properties. In addition to adding to the sensory profile of the finished product, rosemary has a number of possible health benefits when employed as a raw material in the manufacture of herbal wines. Here, we examine the advantages of making herbal wine with rosemary. Rosemary is a popular choice among producers looking to create tasty and maybe beneficial beverages since it provides a number of possible benefits as a raw material for herbal wine manufacturing (Dos Santos et al. 2022).

With notes of pine, citrus, and earthiness, rosemary gives herbal wines a distinctive herbal flavor and scent. Its rich flavor profile adds depth and complexity to the finished blend by balancing light-bodied whites and full-bodied reds. Carnosic acid and rosmarinic acid, two fragrant chemicals found in rosemary, give the herb its unique scent and improve the wine's overall bouquet. Rosemary is full of antioxidants, particularly rosmarinic and carnosic acids, which have been shown to be particularly potent antioxidants. These compounds help the body get rid of harmful free radicals, reduce oxidative stress, and protect cells from damage (Aziz et al. 2022). By adding rosemary to their wine, producers of herbal wines can boost the antioxidant content of their product and potentially provide consumers with extra health benefits.

Research has indicated that the presence of anti-inflammatory compounds in rosemary, such as rosmarinic acid and carnosic acid, may help reduce inflammation within the body. Chronic inflammation has been linked to a number of medical conditions, such as neurological problems, cardiovascular disease, and arthritis. Producers can offer consumers a potential medical beverage option by incorporating rosemary, which has anti-inflammatory properties, into herbal wine. Research on rosemary's potential neuroprotective qualities has been prompted by the plant's high content of rosmarinic and carnosic acids (Rahbardar and Hosseinzadeh 2020). These compounds have been shown to possess neuroprotective properties, which help protect brain tissue from oxidative stress and inflammation. According to a study, regular use of rosemary may enhance cognitive function and reduce the risk of age-related cognitive decline. By including rosemary in their blends, makers of herbal wines can give customers a drink that awakens the senses and supports brain health.

Numerous research works have examined the cardiovascular advantages of rosemary, emphasizing the plant's capacity to promote heart health. Compounds like carnosic acid and rosmarinic acid, which are found in rosemary, may help lower blood pressure, enhance circulation, and minimize the risk of heart disease. Furthermore, rosemary's antioxidant qualities may lessen inflammation and shield blood vessels from oxidative damage, improving cardiovascular health in general (Faridzadeh et al. 2022).

In conclusion, using rosemary as a raw material for herbal wine manufacturing has a plethora of potential benefits. Herbal wines get depth and richness from the addition of rosemary, which also has the potential to enhance their health-promoting benefits due to their unique flavor and aroma and antioxidant, anti-inflammatory, neuroprotective, and cardiovascular-supporting capabilities. Herbal wine producers can develop wines that satisfy the palate and potentially provide customers with a number of wellness benefits by adding rosemary to their blends.

7.6.3 Thyme

A well-liked herb for its culinary applications, thyme also has some possible health advantages when utilized as a base for making herbal wines. This aromatic herb has been valued for generations for its medicinal qualities due to its unique earthy and somewhat minty flavor (Martinez-Frances et al. 2021). Herbal wine blends that include thyme benefit from a deeper flavor profile as well as the introduction of several bioactive ingredients that may enhance general health and well-being.

Thyme's antibacterial qualities make it a valuable addition to herbal wines. Compounds found in thyme, such as carvacrol and thymol, have potent antibacterial activity against a variety of bacteria and fungi. In addition to aiding in the wine's preservation, these antibacterial qualities may also be beneficial to health by thwarting pathogenic microbes (Torrens and Castellano 2020).

Thyme is also a great source of antioxidants, such as carotenoids, phenolic acids, and flavonoids, which lessen oxidative stress in the body by scavenging dangerous free radicals. Numerous chronic illnesses, such as cancer, neurological problems, and cardiovascular diseases, are linked to oxidative stress (De Oliveira et al. 2020). Producers of herbal wines can increase the wine's potential health advantages and provide consumers with a tasty beverage that also has extra antioxidant benefits by adding antioxidant-rich thyme to the blend.

Additionally, thyme has anti-inflammatory qualities that can help lessen inflammation and soothe the symptoms of inflammatory diseases such as inflammatory bowel disease, arthritis, and asthma (Dimitrov et al. 2020). Thyme's bioactive components, including luteolin and rosmarinic acid, are thought to have anti-inflammatory qualities because they obstruct the body's inflammatory pathways. For this reason, thyme added to herbal wine blends may help people suffering from inflammatory diseases.

Furthermore, thyme has long been used to support respiratory health and reduce symptoms of cough, cold, and bronchitis (Maleš et al. 2022a). Thyme's volatile oils include expectorant and bronchodilator properties that help to widen the airways and make breathing easier. By adding thyme to their blends, herbal winemakers can create a pleasant beverage that may also promote respiratory health during the cold and flu season.

Beyond its medicinal applications, thyme adds richness and depth of flavor to herbal wine mixtures. Its earthy, slightly minty flavor complements a variety of base wines and heightens the overall sensory experience. Whether added alone or in conjunction with other herbs, thyme lends a distinct character to herbal wines that appeal to consumers seeking savory and unique drinks (Hammoudi Halat et al. 2022).

In conclusion, thyme has a range of potential health advantages as well as sensory attributes when used as a raw material to make herbal wine. Thyme's antibacterial, antioxidant, anti-inflammatory, and respiratory support properties give herbal wines depth of flavor and possible health advantages (Sardari et al. 2021).

7.6.4 Chamomile

Chamomile is a popular raw material for herbal wines due to its adaptability, subtle flowery aroma, and sedative properties. Because of its several potential advantages, it is a useful addition to herbal wine mixes, enhancing the flavor profile and potentially health-promoting aspects of the beverage (Akram et al. 2024).

One of the key benefits of herbal wine is the chamomile's calming and relaxing properties. Chamomile has long been used in traditional medicine as a natural means of promoting relaxation and reducing stress and anxiety. Wine that has been infused with chamomile has a relaxing effect that contributes to a pleasant and tranquil drinking experience. This makes chamomile-infused herbal wines a fantastic choice for unwinding after a busy day or enjoying quiet times of reflection (Sah et al. 2022).

Additionally, chamomile is well-known for its digestive advantages. It has been used for ages to treat indigestion, gas, and bloating, among other digestive discomforts. Compounds with anti-inflammatory and antispasmodic qualities, such as apigenin and bisabolol, found in chamomile, may help calm the digestive tract. Chamomile, when added to herbal wine, can help make the beverage smoother and more balanced while also supporting digestive comfort and wellness (Sah et al. 2022).

Apart from its ability to soothe and aid in digestion, chamomile is highly valued for its possible anti-inflammatory characteristics. Numerous medical disorders, such as diabetes, arthritis, and heart disease, are associated with chronic inflammation. Flavonoids and other bioactive substances found in chamomile have anti-inflammatory properties that aid in lowering inflammation and improving general health. By adding chamomile to their wine, producers may create wines that both stimulate the taste and support the body's natural inflammatory response (Nyamwamu et al. 2020).

Furthermore, chamomile can boost immunity, which is why it's a good addition to herbal wine blends, particularly in the winter or at times when individuals are more susceptible to illness. Quercetin and apigenin, two antioxidants present in chamomile, strengthen the immune system and neutralize free radicals. By incorporating chamomile-infused wine into their daily routine, drinkers may be able to strengthen their immunity and enhance their overall health (Bayliak et al. 2021).

Apart from its health benefits, chamomile is well appreciated for its aroma and taste. Herbal wine blends gain depth and complexity from their subtle, flowery aroma and somewhat sweet flavor, which enhances the overall sensory experience for customers. Chamomile-infused wine is a delightful combination of flowery notes, subtle herbal undertones, and potential health benefits. It can be enjoyed on its own or in combination with other herbs and botanicals (Akram et al. 2024).

In conclusion, chamomile is a very flexible and helpful raw component for herbal wines. Because of its calming, digestive, anti-inflammatory, and immune-boosting properties, it's a perfect choice for flavoring and potentially even functional beverages. Winemakers are able to produce visually beautiful wines that also promote relaxation, well-being, and vitality.

7.6.5 Hibiscus

The flowering plant hibiscus, scientifically known as *Hibiscus sabdariffa*, is native to tropical regions and is prized for both its culinary flexibility and ornamental beauty. Hibiscus has several advantages as a raw material for making herbal wine, from its vivid color and distinct flavor profile to its possible health benefits. We will delve into the many advantages of using hibiscus as a raw material for herbal wine in this extensive research (Figure 7.1) (Izquierdo-Vega et al. 2020).

The vivid red color of hibiscus is one of its most remarkable qualities, and it enhances the aesthetic appeal of herbal wines (Teixeira et al. 2023). The wine's deep red hue is infused with rich pigmentation from hibiscus flowers, giving it an eye-catching and alluring appearance for

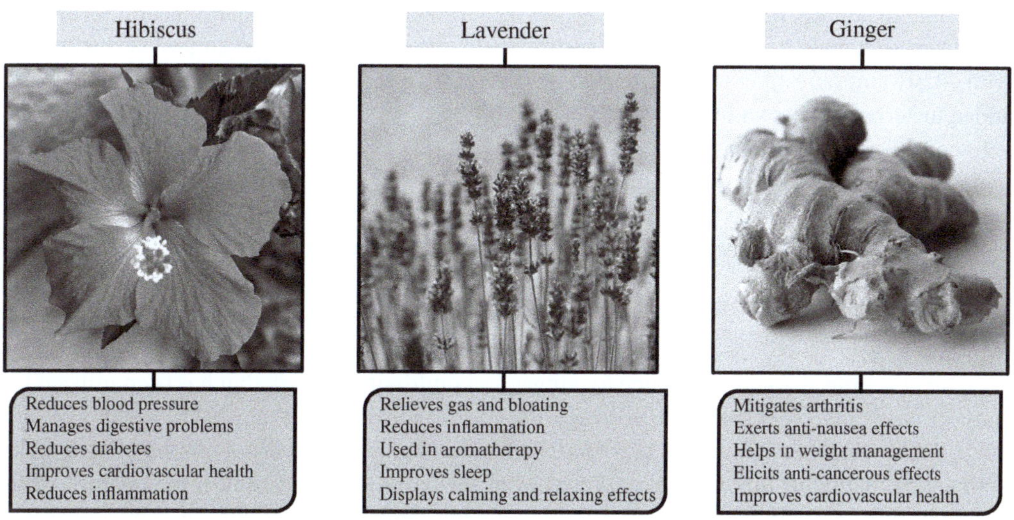

Figure 7.1 Health benefits of hibiscus, lavender, and ginger.

customers. Hibiscus can be added to wine to provide an aesthetically pleasing drink that sticks out on the shelf and improves the entire drinking experience.

The tart, flowery taste profile that hibiscus offers, coupled with undertones of citrus, raspberry, and cranberry, is beneficial to herbal wines. This unique taste profile unites acidity and sweetness to create a well-rounded palate experience, giving the wine more depth and complexity. Since its acidity pairs well with a wide range of base wines, from robust reds to crisp whites, hibiscus is a versatile ingredient for making herbal wines (Suprabha et al. 2024).

Hibiscus contains antioxidants such as anthocyanins, flavonoids, and polyphenols that support the body's defense against inflammation and oxidative stress. These antioxidants play a critical role in promoting cardiovascular health, reducing the risk of chronic disorders, and sustaining overall well-being (Tiwari et al. 2016).

The possible cardiovascular advantages of hibiscus ingestion have been noted by several research studies. According to research, hibiscus may help lower cholesterol and blood pressure, which lowers the risk of heart disease and stroke. Hibiscus is a great ingredient for supporting heart health in herbal wines because of its cardio-protective benefits, which are attributed to its bioactive constituents, such as flavonoids and anthocyanins (Arslan et al. 2021).

Natural anti-inflammatory substances found in hibiscus may aid in lowering inflammation all over the body. Numerous medical disorders, such as diabetes, heart disease, and arthritis, are linked to chronic inflammation. Herbal wine manufacturers may provide consumers with a delightful beverage that also enhances the body's inherent anti-inflammatory processes by adding hibiscus to their blends (Shelke et al. 2021).

Hibiscus has long been used as a treatment for digestive problems, such as bloating, constipation, and indigestion. The natural components in the plant aid in promoting regular bowel movements, relieving gastrointestinal discomfort, and stimulating digestion. Hibiscus can be used in herbal wines to provide customers with a pleasant drink that not only pleases their palates but also promotes digestive health (Jejurkar and Chavan et al. 2023).

Vitamin C, which is abundant in hibiscus and is crucial for immune system support, is one of the important nutrients it contains. Vitamin C enhances immune system performance by fortifying the body's defenses against infections (Ilodibia et al. 2019). Winemakers can make a drink that stimulates the taste buds and strengthens the immune system naturally by adding hibiscus to their wine.

According to several studies, hibiscus may support healthy metabolism and weight control. According to studies, hibiscus extract may aid in improving lipid profiles and insulin sensitivity while lowering body weight, body fat, and waist circumference. Hibiscus can provide customers with a tasty and refreshing way to enhance their overall wellness and weight loss objectives when added to herbal wines (Amin et al. 2020).

In conclusion, hibiscus provides a wealth of advantages as a starting point for the creation of herbal wines, ranging from its eye-catching hue and unique flavor profile to its possible health benefits. A pleasing choice that pleases the tongue and the body, hibiscus-infused wines are a response to consumers' growing desire for natural and functional beverages.

7.6.6 Ginger

Zingiber officinale, the scientific name for ginger, is a multipurpose herb that has been utilized for ages in medicine, cooking, and beverage uses (Mahomoodally et al. 2021; Upadhyay and Singh 2021). Ginger has a number of potential advantages when utilized as a raw material to make herbal wine, including the ability to improve flavor, aid in digestion, have anti-inflammatory qualities, and more. Examining these advantages in greater detail can shed light on why ginger is a preferred ingredient among herbal wine producers.

Ginger is prized for its unique flavor profile, which is characterized by a taste that is strong, spicy, and slightly sweet. Ginger adds greater depth and complexity to herbal wines, enhancing their overall flavor character. Its aromatic qualities lend a warming sensation and a touch of lemony freshness, making it a versatile component that pairs well with a wide range of wine styles (Zadorozhna and Mangieri 2021).

One of ginger's most well-known health benefits is that it can help with digestion. Shogaol and gingerol, two bioactive compounds in ginger, promote the production of digestive enzymes and help food move more easily through the digestive tract. For this reason, ginger is widely used to treat symptoms of indigestion, bloating, and nausea. The anti-inflammatory properties found in ginger can help lower inflammation all over the body. Numerous medical disorders, including arthritis, cardiovascular disease, and several forms of cancer, are linked to chronic inflammation. Customers may profit from the anti-inflammatory qualities of ginger by adding it to herbal wine, but it's vital to remember that the amount of ginger in wine may not be high enough to have a noticeable therapeutic impact (Dissanayake et al. 2020).

Strong antioxidants found in ginger, such as zingerone and gingerol, aid in scavenging dangerous free radicals and shielding the body from oxidative stress. It has been demonstrated that these antioxidants enhance immunological function and may lower the chance of developing chronic illnesses. Ginger can help customers increase their antioxidant intake and strengthen their immune system as a whole when it is added to herbal wines. Ginger may help the cardiovascular system, including lowering blood pressure and promoting heart health, according to some research. Compounds found in ginger aid in blood vessel relaxation, blood circulation improvement, and cholesterol reduction. These outcomes might help reduce the risk of stroke and heart disease. Even though there may not be enough ginger in herbal wine to have a major positive impact on cardiovascular health, it may nevertheless have a little positive impact (Jafarzadeh et al. 2021).

Motion sickness and nausea have long been treated naturally with ginger. Ginger has been demonstrated in studies to help reduce nausea and vomiting brought on by chemotherapy, surgery, and pregnancy. Ginger's capacity to lessen stomach spasms and regulate neurotransmitter activity in the brain is thought to be responsible for its anti-nausea effects. Ginger in herbal wine can have a calming and settling impact on the stomach, which makes it a relaxing option for people who are having digestive issues. Ginger may help with weight management by boosting fat burning, decreasing hunger, and raising metabolism, according to some research. The thermogenic qualities of ginger aid in raising the body's metabolic rate, which raises calorie expenditure. Ginger may also aid in promoting feelings of fullness and suppressing appetite, which may help with portion management and avoid overindulging. Although the effect of adding ginger to herbal wine on weight control is probably insignificant, it may nevertheless help with a holistic plan for maintaining a healthy weight (Crichton et al. 2022).

In conclusion, there are a variety of possible health advantages to utilizing ginger as a raw ingredient in herbal wine. Ginger's taste-enhancing properties, along with its anti-inflammatory, immune-boosting, cardiovascular, anti-nausea, and weight-management benefits, make it a perfect addition to herbal wine blends. As more people turn to natural and holistic remedies, ginger-infused herbal wine presents a delightful and possibly healthful option for those looking to indulge in a health-conscious beverage choice.

7.6.7 Lemon Balm

The fragrant herb lemon balm, scientifically known as *Melissa officinalis*, is a member of the mint family. For millennia, people have valued it for its culinary, medicinal, and fragrant qualities. Lemon balm adds distinct flavors and fragrances to blends and infusions and may have health advantages when used as a raw material in the manufacturing of herbal wines. Lemon balm has a light lemony flavor with undertones of mint and earthiness, as well as a vibrant, citrusy scent. Lemon balm adds richness to wine by infusing it with a refreshing and uplifted taste. Because of its subtle flavor profile, it goes well with a variety of wine varietals, especially light-bodied white wines (Alimoradi et al. 2023).

The aroma of lemon balm is attributed to its aromatic components, which include geraniol, citral, and citronellal. These substances give lemon balm its distinct lemon scent, which has herbal overtones and is reminiscent of lemon zest. Lemon balm contributes a crisp, energizing scent to herbal wines, enhancing their fragrance. Lemon balm has long been used to ease tension and anxiety while encouraging relaxation. Its anxiolytic effects have been linked to chemicals like citronellal and rosmarinic acid, which are responsible for its soothing qualities. Herbal wine mixes that include lemon balm can intensify their calming effects and offer customers a feeling of peace and wellness (Ullah and Hassan 2022).

Due to its digestive qualities, lemon balm is well-known for relieving bloating, indigestion, and gastrointestinal pain. The herb is a good addition to herbal wines that are consumed before or after meals since it stimulates digestive enzymes and encourages a healthy digestive system. Antioxidants found in abundance in lemon balm, such as flavonoids and rosmarinic acid, aid in the fight against oxidative stress and shield cells from harm. Because these antioxidants neutralize free radicals and reduce inflammation, they may improve general health and well-being. According to some research, lemon balm may have benefits for the brain, including increased focus, memory, and cognitive performance. Lemon balm is a promising addition to herbal wines labeled as "brain-boosting" or "nootropic," as it contains compounds like citral and rosmarinic acid that may enhance brain health and cognitive function. Because of its antibacterial qualities, lemon balm prevents the growth of fungi, germs, and viruses. This antimicrobial activity might aid in boosting immunity and warding off diseases. By preventing microbiological development, adding lemon balm to herbal wine compositions may help preserve and extend its shelf life (Doğan et al. 2021).

Lemon balm enhances culinary creations with depth and richness, in addition to its medical virtues. It is a versatile herb in the kitchen because of its fresh, lemony flavor, which goes well with a wide range of dishes. Lemon balm's bright and aromatic profile enhances a variety of foods, including salads, desserts, marinades, sauces, and teas. It is also a great addition to herbal wines (Gabrani et al. 2023).

While moderate consumption of lemon balm is generally regarded as safe, sensitive people may experience negative effects or interactions with specific drugs. Before using lemon balm in your diet, like with any herbal supplement or ingredient, it's crucial to speak with a healthcare provider, particularly if you have any underlying medical concerns or are taking medication.

To sum up, using lemon balm as a raw material for herbal wine production has a lot of advantages. It's an excellent addition to blends and infusions because of its bright flavor, fragrant properties, and possible health advantages. Lemon balm enhances the flavor of herbal wines and improves overall well-being. It can be used for its relaxing effects, digestive assistance, antioxidant protection, cognitive development, or culinary versatility. Lemon balm also adds complexity and depth to herbal wines.

7.6.8 Peppermint

When utilized as a base for herbal wine, peppermint is a versatile herb with a pleasant flavor and energizing aroma that may have several potential health benefits. Peppermint adds a special set of qualities to herbal wine mixes, from its capacity to ease headaches and give antibacterial effects to its ability to aid with digestion (Nayak et al. 2020).

Supporting digestive health is one of peppermint's most well-known advantages. Menthol, a substance found in peppermint, aids in calming the muscles in the digestive system, which lessens gas, bloating, and indigestion symptoms. Therefore, peppermint-infused herbal wines may aid in improving digestion and easing post-meal discomfort. In addition to potentially aiding with digestion, peppermint can make herbal wine taste nice and calming to the tongue. Because peppermint relieves stress and encourages relaxation, it's frequently used as a natural cure for headaches and migraines. Peppermint contains menthol, which has a cooling effect that helps relieve headaches and neck and head muscular tension. Herbal winemakers can make a cool drink that stimulates the senses and may help with headaches and migraines by adding peppermint to their blend. Drinking a glass of herbal wine infused with peppermint may provide a calming experience that lessens headache symptoms and enhances general well-being (Zhao et al. 2022).

Because peppermint has a high concentration of menthol and other bioactive components, it has antibacterial qualities. Because of its antibacterial qualities, peppermint can effectively prevent the growth of pathogenic bacteria and fungi, which makes it a useful component of herbal treatments for digestive problems, respiratory infections, and oral health problems. Producers of herbal wines can potentially improve the microbiological stability of their product and add additional health advantages by adding peppermint to it. Herbal wines infused with peppermint may provide consumers with a revitalizing and wellness-enhancing choice. It has been demonstrated that the scent of peppermint improves mood by encouraging feelings of enthusiasm, alertness, and attention (Zhao et al. 2022). Peppermint essential oil is a popular choice for aromatherapy and natural mood enhancement because it has been shown to increase energy levels and reduce weariness when inhaled. Peppermint may impart its uplifting aroma to herbal wine when utilized as a raw ingredient, making for a revitalizing and refreshing drink. Herbal wine with a peppermint infusion may provide customers with a lovely sensory experience while also uplifting their spirits and attitudes.

Peppermint's expectorant and decongestant qualities make it a popular choice for supporting respiratory health. Peppermint's menthol facilitates better breathing by clearing congestion and widening the airways. Herbal wines with peppermint infusions can assist promote respiratory health by making a calming and cool drink that helps open up nasal passages and reduce cold, allergy, and respiratory infection symptoms. Sipping a glass of herbal wine infused with peppermint oil can help ease congestion and support respiratory comfort, especially in the winter months (Mahendran and Rahman 2020).

In conclusion, using peppermint as a raw material for herbal wine manufacturing has a plethora of potential health benefits. Peppermint adds a pleasant and health-promoting aspect to herbal wine mixes, with qualities that improve respiratory health, aid with digestion, ease headaches, and have antibacterial benefits. Herbal winemakers may make tasty concoctions that satisfy the palate and may even have health advantages by adding peppermint to their formulas. Peppermint-infused herbal wine offers consumers a pleasurable and potentially useful beverage alternative, whether it is used as a headache treatment, digestive help after meals, or refreshing tonic for respiratory comfort.

7.7 Techniques for Incorporating Herbs into the Wine Matrix

Infusion methods from ancient Egypt, the Paleolithic era, and the Neolithic periods from China and the Middle East have been uncovered from archaeological studies, translating their immense cultural and therapeutic significance (McGovern et al. 2009). Incorporating herbs into the profile of wine is a crucial step for enhancing the therapeutic properties of the herbal wine. Hence, the selection of herbs and optimization of the methods are performed to ensure unique combinations of flavors and health benefits (Liang et al. 2021). Herbs, including lavender, chamomile, rosemary, mint, and elderflower, are widely used to tailor the infusion with respect to their various therapeutic potentials (Martinez-Frances et al. 2021). Utilizing high-quality herbs will subsequently impact the flavor and medicinal properties of the final product. More importantly, proper ratios are calculated before infusion to achieve the desired potency. Generally, 11–2 tablespoons of hers are incorporated with 750 ml of wine, although slight alterations can be adopted according to the personal preference and strength of the herbs. The duration of infusion is also another factor impacting the strength and flavor. The typical infusion after 24–48 hours and continuous steeping is preferred, beyond which the taste may end up bitter and unsatisfactory (Chawafambira 2021). Hence, it is essential to assess the quality of the product periodically to maintain consumer acceptance and desire. Once the wine has attained the desired infusion levels, it is important to use a fine-mesh strainer or a cheesecloth to separate the sediments to result in a clear and smooth wine. Finally, the wine is stored in a clean, airtight container to preserve its freshness and flavor (Goulas et al. 2021). Proper storage will also prevent the oxidation of the bioactive compounds, which will maintain the health benefits of wine. Brief details regarding the type of herb and the date of infusion are labeled to keep track of the flavor and expiration date to savor the wine at its peak freshness (Deshmukh and Deshmukh 2021).

Understanding the effects of infusion methods directly correlates to the quality and value of the herbal wine. Presently, several methods with their specific benefits and considerations are used for the same (Liang et al. 2021). Techniques such as cold infusion, where dried or fresh herbs are incorporated directly into the wine and letting it rest for a predetermined period of time at room temperature or in the fridge, are an easy and widely used method (Alti-Palacios et al. 2023). With this method, it is possible to combine one or more herbs depending on the choice of flavor and aromatic profile. Although it is a delicate process that takes longer extraction time, it greatly preserves the flavor and freshness of the wine. Contrary to cold infusion methods, hot infusion involves heating the mixture of herbs and wine in stove or slow-pressure cooker to extract the bioactive compounds (Chawafambira 2021). Application of heat is an efficient way to extract the active compounds quickly but may alter the delicate flavor and richness if not observed vigilantly. To counter this difficulty, hot infusions using a double boiler are used to prevent the direct contact of bioactive compounds with the heat (Deshmukh and Deshmukh 2021). The method involves placing the mixture of herbs and wine in a heat-resistant container over simmering water and slowly allowing them to diffuse into each other. The method is performed slowly and carefully, enabling the infusion of delicate herbs into the wine while maintaining their freshness and therapeutic potential. Similar method for preserving the delicacy of wine is the sun infusion method, which uses the gentle warmth of the sun to delicately transform herbal wine with interesting and soothing aromatic profiles (Saha et al. 2022). For some specific herbs, such as rosemary, the leaves can be crushed or bruised and soaked in wine for a period of time to release their flavors, called the maceration method (Harutyunyan and Malfeito-Ferreira 2022). Although this method is not a delicate one, it is often used for the preparation of large quantities of wine in a short duration of time. Similarly, more advanced techniques, such as percolation method, offer more sophisticated equipment such as a percolator or infusion press, where the wine is passed gently through a chamber filled with the crushed herbs, allowing the liquid to leach out the bioactive compounds (Hidayat and Wulandari 2021). By adjusting the flow rate and the pressure applied to the instrument, careful tailoring of the flavor profile can be achieved with ease. It is recommended to monitor the setup for possible clogging or over-extraction to preserve the quality of the wine. The fastest of the ways adopted, called the instant infusion method, uses a whipped cream dispenser filled with nitrous oxide gas to rapidly infuse the herbs and wine in an airtight container (Naji et al. 2023). Even though the methods result in quick infusion, it is generally difficult to control the flavor and aromatic profile.

7.8 Significance of Herbal Wine in Human Health

Herbal wines have been used for centuries on account of their medicinal properties in various cultures across the planet. Present scientific investigation has cataloged the specific health benefits of the different combinations of herbs and spices (Chawafambira 2021). Notably, the excellent antioxidant properties exhibited help neutralize the harmful free radicals and toxins in the body (Thakur 2020). For instance, infused red wines are rich in resveratrol, a well-known and potent antioxidant from grape skins that has been linked with enhancing the cardiovascular health and reducing inflammation (Liang et al. 2021). The presence of polyphenols such as catechin, epicatechin, flavones, proanthocyanidins, and anthocyanins from red wines when consumed as a part of balanced diet may help lower blood pressure and improve cholesterol levels (Buljeta et al. 2023). Rhamnetin, another antioxidant found in red wines, has been investigated for its ability to protect cardiomyoblasts from apoptosis by eliminating free oxygen radicals (Woo et al. 2021). Consequently, citrus flavonoids are linked to enhancing glucose tolerance and insulin insensitivity, regulating the metabolism of lipids, promoting adipocyte differentiation, and supporting endothelial dysfunction (Zhao et al. 2021). The anti-inflammatory properties have been a valuable property for combating conditions such as arthritis and inflammatory bowel diseases (Al-Khayri et al. 2022). Wine prepared from peppermint and chamomile can reduce indigestion, bloating, and nausea and promote overall digestive health by soothing the irritated lining of the gastrointestinal tract to help control diseases such as gastritis and acid reflux (Jackson 2020). The herbs contain stimulants that favor the production of digestive enzymes and promote gastric motility. Compounds from milk thistle and dandelion are also believed to be helpful in detoxification and reducing inflammation, supporting liver health (Wang et al. 2023). Gentian and wormwood are another group of bitter herbs capable of stimulating the secretion of saliva, gastric acid, and bile translating to improved nutrient absorption, mainly fats (Felix-Cuencas et al. 2022). Certain compounds from cinnamon, turmeric, and thyme have attained scientific interest on account of their possible prebiotic properties (Matera et al. 2023). They prevent the proliferation of harmful bacteria and sustain the growth and balance of the gut microbiota (Shehata et al. 2022). Several studies have shown the immune-boosting properties of herbs like echinacea, elderberry, and astragalus. They are known to reduce the risk of infections contracted during times of seasonal change (Mercola 2021). Some herbs used in herbal winemaking, like thyme, oregano, and cloves, help fight pathogenic bacteria, viruses, and fungi and provide highly desired antimicrobial properties (Stan et al. 2021). Well-known combinations of *Ginkgo biloba* and rosemary have been experimented with for their potential cognitive-enhancing effects. They are recorded to improve memory, concentration, and overall cognitive health from the research conducted in the past years (Su et al. 2022). Fortification of herbs has been an alternative to the conventional therapeutic for their anticarcinogenic, antimutagenic, and chemo-protective properties (Amol Dombe and Jaykumar Shirote 2023). Herbs such as *Carum carvi L, Echium amoenum, Citrus aurantifolia, Zataria multiflora,* and *Falcaria vulgaris* possess organoleptic properties used for the treatment of chronic diseases such as diabetes, kidney disease, cancer, and Alzheimer's disease (Dixit et al. 2021; Upadhyay and Singh 2021, 2023). For women, balanced consumption of wine has been reported to alleviate the symptoms of postpartum depression, obesity, atherosclerosis, myocardial infarction, stroke, and breast cancer (Purcell 2020). Herbs like lavender and chamomile also help sooth the nerves, providing a more calming effect on the

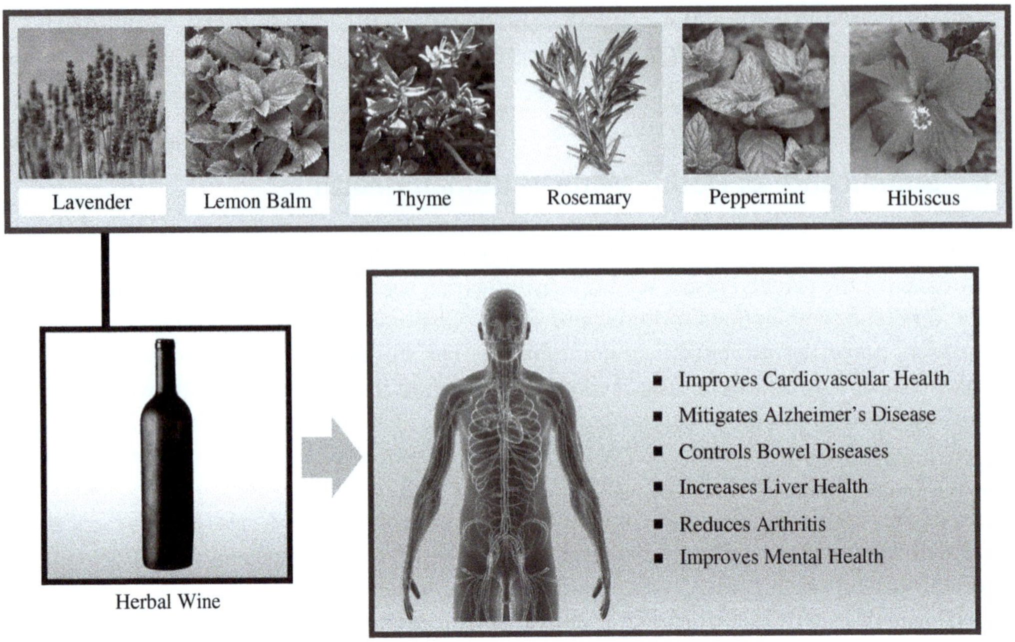

Figure 7.2 Medical significance of herbal wines.

body. They help in reducing stress, promoting relaxation, and more importantly, improving the quality of sleep, which translates to branching mental and physical health benefits (Figure 7.2).

7.9 Future Prospective of Herbal Wine

In recent years, herbal wine, a blend of traditional winemaking and herbalism, has attracted a lot of attention due to its intriguing flavor profiles and possible health advantages. Herbal wine is positioned to make significant advancements in a number of fields as interest in wellness and sustainable living grows globally. This essay explores the possible uses of herbal wine in the gastronomic, health, and environmentally aware sectors, as well as its future prospects.

The increased perception of the health benefits of herbal wine is one of the main factors contributing to its rising popularity. Numerous herbs used in winemaking, like elderflower, chamomile, and lavender, are abundant in bioactive components with anti-inflammatory qualities. With customers being more mindful of their health, herbal wine presents a compelling substitute for conventional alcoholic drinks. Additionally, adding herbs like ashwagandha and ginseng may impart adaptogenic qualities, which relieve stress and improve general well-being. Herbal wine is well-positioned to take a sizable chunk of the market, given the growing demand for functional beverages. It appeals to people who are health-conscious and want to feel both indulgence and well-being.

Herbal wine is becoming more and more of a symbol of environmental conscience in the beverage sector as worries about environmental sustainability grow. Herbal winemakers can promote biodiversity and soil health by reducing their dependency on synthetic additives and pesticides by utilizing the power of herbs and botanicals. Additionally, the production of wine from medicinal plants supports sustainable agricultural practices, boosts local economies, and provides small-scale farmers with economic prospects. Herbal wine production's eco-friendly philosophy is further reinforced by the focus on organic and biodynamic farming practices. Herbal wine stands out as a symbol of environmental care, fostering good change throughout the business as customers prioritize sustainability more and more in their purchase decisions.

7.10 Conclusion

In summary, the various therapeutic properties represented by herbal wine including antioxidant, anti-inflammatory, digestive, cardiovascular, immune-supporting, and mood-enhancing properties grant immense interest on account of their beneficial effects on the human body. As scientific research on the vast potential of herbal wine evolves, novel outcomes will provide valuable insights into their role in promoting overall well-being. It is proven that moderate and careful consumption of such medicinal-rich herbal wines can be a suitable alternative to conventional medicine to counter various diseases and chronic illnesses. The art of herbal wine represents the synergy of traditional herbal medicine and the present culinary arts, offering a powerful brew of flavorful and medicinal value. As the journey toward exploring the full spectrum of the benefits of herbal wine, modern science holds great promise for unraveling the therapeutic potential of this ancient elixir.

References

Akgül, E.A., Karakul, A., Altın, A. et al. (2021). Effectiveness of lavender inhalation aromatherapy on pain level and vital signs in children with burns: a randomized controlled trial. *Complementary Therapies in Medicine* 60: 102758.

Akram, W., Ahmed, S., Rihan, M. et al. (2024). An updated comprehensive review of the therapeutic properties of Chamomile (*Matricaria chamomilla* L.). *International Journal of Food Properties* 27 (1): 133–164.

Alimoradi, Z., Jafari, E., Abdi, F., and Griffiths, M.D. (2023). Therapeutic applications of lemon balm (*Melissa officinalis*) for obstetrics and gynecological health issues: a systematic review. *Journal of Herbal Medicine* 42: 100751.

Al-Khayri, J.M., Sahana, G.R., Nagella, P. et al. (2022). Flavonoids as potential anti-inflammatory molecules: a review. *Molecules* 27 (9): 2901.

Alti-Palacios, L., Martínez, J., Teixeira, J.A. et al. (2023). Influence of cold pre-fermentation maceration on the volatilomic pattern and aroma of white wines. *Food* 12 (6): 1135.

Amin, A.R., Kassab, R.B., Abdel Moneim, A.E., and Amin, H.K. (2020). Comparison among garlic, berberine, resveratrol, *Hibiscus sabdariffa*, genus zizyphus, hesperidin, red beetroot, catha edulis, portulaca oleracea, and mulberry leaves in the treatment of hypertension and type 2 DM: a comprehensive review. *Natural Product Communications* 15 (4): 1934578X20921623.

Amol Dombe, S. and Jaykumar Shirote, P. (2023). Phytonutraceuticals in cancer prevention and therapeutics. *Current Nutrition & Food Science* 19 (3): 209–228.

Arslan, M., Zareef, M., Tahir, H.E. et al. (2021). Medicinal and therapeutic potential of Roselle (*Hibiscus sabdariffa*). In: *Roselle (Hibiscus sabdariffa)*, 155–186. Amsterdam, The Netherlands: Elsevier.

Aziz, E., Batool, R., Akhtar, W. et al. (2022). Rosemary species: a review of phytochemicals, bioactivities and industrial applications. *South African Journal of Botany* 151: 3–18.

Balick, M. (2014). *Rodale's 21st-century Herbal: A Practical Guide for Healthy Living Using Nature's Most Powerful Plants*. Rodale Books.

Batiha, G.E.S., Teibo, J.O., Wasef, L. et al. (2023). A review of the bioactive components and pharmacological properties of Lavandula species. *Naunyn-Schmiedeberg's Archives of Pharmacology* 396 (5): 877–900.

Bayliak, M.M., Dmytriv, T.R., Melnychuk, A.V. et al. (2021). Chamomile as a potential remedy for obesity and metabolic syndrome. *EXCLI Journal* 20: 1261.

Bhise, P. and Morya, S. (2021). The health sustainability of herbal wine bioactives towards different chronic diseases. *The Pharma Innovation Journal* 10 (5): 512–517.

Buljeta, I., Pichler, A., Šimunović, J., and Kopjar, M. (2023). Beneficial effects of red wine polyphenols on human health: comprehensive review. *Current Issues in Molecular Biology* 45 (2): 782–798.

Chawafambira, A. (2021). The effect of incorporating herbal (*Lippia javanica*) infusion on the phenolic, physicochemical, and sensorial properties of fruit wine. *Food Science & Nutrition* 9 (8): 4539–4549.

Crichton, M., Davidson, A.R., Innerarity, C. et al. (2022). Orally consumed ginger and human health: an umbrella review. *The American Journal of Clinical Nutrition* 115 (6): 1511–1527.

De Oliveira, J.R.G. et al. (2020). Biological Activities of Thyme Essential Oil. *Acta Scientiarum Biological Sciences* 42: e48400.

Deshmukh, A.W. and Deshmukh, S.A. (2021). Herbal wine production from fruits and vegetable wastes and peels. *International Journal of Engineering Applied Sciences and Technology* 5 (9): 2455–2143.

Dimitrov, D., Parzhanova, A., Ivanova, S., & Haygarov, V. (2020). Volatile Composition of Wine Distillates with Added Thyme Extracts (Thymus callieri).

Dissanayake, K.G.C., Waliwita, W.A.L.C., and Liyanage, R.P. (2020). A review on medicinal uses of *Zingiber officinale* (ginger). *International Journal of Health Sciences and Research* 10 (6): 142–148.

Dixit, S., Shukla, A., Singh, V., and Upadhyay, S.K. (2021). *Bioprospecting of Natural Compounds for Industrial and Medical Applications*. Wiley https://doi.org/10.1002/9781119718017.ch3.

Doğan, H., Uskutoğlu, T., Hatice, B.A.Ş. et al. (2021). Phytochemical composition of wild lemon balm (*Melissa officinalis* L.) from the flora of Bulgaria. *Anatolian. Journal of Botany* 5 (2): 112–119.

Dos Santos, M.D. et al. (2022). Rosemary as a potential therapeutic agent: promising neuromodulator. *Frontiers in Pharmacology* 13: 633775.

Erfurt-Cooper, P. and Cooper, M. (2009). The health and wellness concept: a global overview. In: *Health and Wellness Tourism: Spas and Hot Springs*, 25–48. Channel View Publications.

Faridzadeh, A., Salimi, Y., Ghasemirad, H. et al. (2022). Neuroprotective potential of aromatic herbs: rosemary, sage, and lavender. *Frontiers in Neuroscience* 16: 909833.

Felix-Cuencas, L., Delis-Hechavarria, E., Jarro, A. et al. (2022). Bioactivity characterization of herbal molecules. In: *Herbal Biomolecules in Healthcare Applications*, 145–183. Academic Press.

Gabrani, R., Ishaque, M., and Jain, K. (2023). Phytochemical and pharmacological activities of *Melissa Officinalis* (lemon balm). In: *Bioactives and Pharmacology of Lamiaceae*, 331–342. Apple Academic Press.

Goulas, V., Stavrou, K., Michael, C. et al. (2021). The potential of sun-dried grape pomace as a multi-functional ingredient for herbal infusion: effects of brewing parameters on composition and bioactivity. *Antioxidants* 10 (4): 586.

Hammoudi Halat, D., Krayem, M., Khaled, S., and Younes, S. (2022). A focused insight into thyme: biological, chemical, and therapeutic properties of an indigenous Mediterranean herb. *Nutrients* 14 (10): 2104.

Harutyunyan, M. and Malfeito-Ferreira, M. (2022). Historical and heritage sustainability for the revival of ancient wine-making techniques and wine styles. *Beverages* 8 (1): 10.

Hidayat, R. and Wulandari, P. (2021). Methods of extraction: maceration, percolation and decoction. *Eureka Herba Indonesia* 2 (1): 68–74.

Ho, G.T.T., Wangensteen, H., and Barsett, H. (2017). Elderberry and elderflower extracts, phenolic compounds, and metabolites and their effect on complement, RAW 264.7 macrophages and dendritic cells. *International Journal of Molecular Sciences* 18 (3): 584.

Ilodibia, C.V., Achebe, U.A., Arubalueze, C.U., and Ibeh, E. (2019). A comparative study on morphology, mineral and vitamin compositions of two variants of *Hibiscus sabdariffa* (L.). *Archives of Agriculture and Environmental Science* 4 (1): 50–56.

Izquierdo-Vega, J.A., Arteaga-Badillo, D.A., Sánchez-Gutiérrez, M. et al. (2020). Organic acids from Roselle (*Hibiscus sabdariffa* L.)—a brief review of its pharmacological effects. *Biomedicine* 8 (5): 100.

Jackson, R.S. (2020). Wine, food, and health. *Wine Science* 947.

Jafarzadeh, A., Jafarzadeh, S., and Nemati, M. (2021). Therapeutic potential of ginger against COVID-19: is there enough evidence. *Journal of Traditional Chinese Medical Sciences* 8 (4): 267–279.

Jejurkar, G. and Chavan, M. (2023). Therapeutic benefits of gossypin as an emerging phytoconstituents of Hibiscus spp.: a critical review. *Future Journal of Pharmaceutical Sciences* 9 (1): 95.

Karki, S. (2019). Preparation and quality analysis of yacon (smallanthus sonchifolius) based herbal wine. Doctoral dissertation. Department of Food Technology Central Campus of Technology Institute of Science and Technology Tribhuvan University, Nepal 2019.

Koulivand, P.H., Khaleghi Ghadiri, M., Gorji, A. (2013). Lavender and the nervous system. *Evidence Based Complementary and Alternative Medicine, 2013*, 681304. doi: https://doi.org/10.1155/2013/681304.

Lakićević, S.H., Karabegović, I.T., Cvetković, D.J. et al. (2022). Insight into the aroma profile and sensory characteristics of 'Prokupac'red wine aromatised with medicinal herbs. *Horticulturae* 8 (4): 277.

Lante, A., Marangon, M., Vincenzi, S. et al. (2023). Wine and health: from the perspective of *Alvise cornaro* to the latest scientific opinions. *Agriculture* 13 (2): 415.

Liang, Z., Zhang, P., Zeng, X.A., and Fang, Z. (2021). The art of flavored wine: tradition and future. *Trends in Food Science & Technology* 116: 130–145.

Mahendran, G. and Rahman, L.U. (2020). Ethnomedicinal, phytochemical and pharmacological updates on peppermint (Mentha × piperita L.)—a review. *Phytotherapy Research* 34 (9): 2088–2139.

Mahomoodally, M.F., Aumeeruddy, M.Z., Rengasamy, K.R. et al. (2021). Ginger and its active compounds in cancer therapy: from folk uses to nano-therapeutic applications. In: *Seminars in Cancer Biology*, vol. 69, 140–149. Academic Press.

Maleš, I., Dragović-Uzelac, V., Jerković, I. et al. (2022a). Non-volatile and volatile bioactives of *Salvia officinalis* L., *Thymus serpyllum* L. and *Laurus nobilis* L. extracts with potential use in the development of functional beverages. *Antioxidants* 11 (6): 1140.

Maleš, I., Pedisić, S., Zorić, Z. et al. (2022b). The medicinal and aromatic plants as ingredients in functional beverage production. *Journal of Functional Foods* 96: 105210.

Martinez-Frances, V., Rivera, D., Obon, C. et al. (2021). Medicinal plants in traditional herbal wines and liquors in the east of Spain and the Balearic Islands. *Frontiers in Pharmacology* 12: 713414.

Matera, R., Lucchi, E., and Valgimigli, L. (2023). Plant essential oils as healthy functional ingredients of nutraceuticals and diet supplements: a review. *Molecules* 28 (2): 901.

McGovern, P.E., Mirzoian, A., and Hall, G.R. (2009). Ancient Egyptian herbal wines. *Proceedings of the National Academy of Sciences* 106 (18): 7361–7366.

Mercola, J. (2021). *Upgrade Your Immunity with Herbs: Herbal Tonics, Broths, Brews, and Elixirs to Supercharge Your Immune System*. Hay House, Inc.

Naji, A.M., Başyiğit, B., Alaşalvar, H. et al. (2023). Instant soluble roselle (*Hibiscus sabdariffa* L.) powder rich in bioactive compounds: effect of the production process on volatile compounds. *Journal of Food Measurement and Characterization* 17 (1): 108–120.

Nayak, P., Kumar, T., Gupta, A.K., and Joshi, N.U. (2020). Peppermint a medicinal herb and treasure of health: a review. *Journal of Pharmacognosy and Phytochemistry* 9 (3): 1519–1528.

Nazir, M., Arif, S., Khan, R.S. et al. (2019). Opportunities and challenges for functional and medicinal beverages: current and future trends. *Trends in Food Science & Technology* 88: 513–526.

Nyamwamu, N.C., Okari, O.J., and Gisesa, W.N.O. (2020). A survey of medicinal plants used by the gusii community in the treatment of digestive disorders and other inflammatory conditions. *Journal of Medicinal Plants* 8: 21–33.

Purcell, N. (2020). Women and wine in ancient Rome. In: *Gender, Drink and Drugs*, 191–208. Routledge.

Radu, D., Alexe, P., and Stănciuc, N. (2020). Overview on the potential role of phytochemicals from lavender as functional ingredients. *The Annals of the University Dunarea de Jos of Galati. Fascicle VI-Food Technology* 44 (2): 173–188.

Rahbardar, M.G. and Hosseinzadeh, H. (2020). Therapeutic effects of rosemary (*Rosmarinus officinalis* L.) and its active constituents on nervous system disorders. *Iranian Journal of Basic Medical Sciences* 23 (9): –1100.

Rahmani, A.H., Shabrmi, F.M., and Aly, S.M. (2014). Active ingredients of ginger as potential candidates in the prevention and treatment of diseases via modulation of biological activities. *International Journal of Physiology Pathophysiology and Pharmacology* 6 (2): 125–136.

Sah, A., Naseef, P.P., Kuruniyan, M.S. et al. (2022). A comprehensive study of therapeutic applications of chamomile. *Pharmaceuticals* 15 (10): 1284.

Saha, S., Ghosh, A., Acharyya, S., and Sarkar, S. (2022). Fusion of tea infusion and Bakhar (starter of ethnic liquor Haria) to develop "tea Haria": a novel approach to ferment tea with insight into in vitro biochemical attributes and metabolomics. *Journal of Food Technology* 9 (1): 1–17.

Sandhu, D. and Morya, S. (2022). A review on the comparative study of nutraceutically activated fruits and herbs based wines: fruit vs. herbal wine. *Journal of Applied and Natural Science* 14 (2): 500–511.

Sardari, S., Mobaiend, A., Ghassemifard, L. et al. (2021). Therapeutic effect of thyme (*Thymus vulgaris*) essential oil on patients with covid19: a randomized clinical trial. *Journal of Advances in Medical and Biomedical Research* 29 (133): 83–91.

Shehata, A.A., Yalçın, S., Latorre, J.D. et al. (2022). Probiotics, prebiotics, and phytogenic substances for optimizing gut health in poultry. *Microorganisms* 10 (2): 395.

Shelke, M., Parjane, S., Mankar, S.D., and Siddheshwar, S.S. (2021). Therapeutic potential of *Hibiscusrosa sinensis*-a review. *Research Journal of Science and Technology* 13 (2): 151–156.

Stan, D., Enciu, A.M., Mateescu, A.L. et al. (2021). Natural compounds with antimicrobial and antiviral effect and nanocarriers used for their transportation. *Frontiers in Pharmacology* 12: 723233.

Su, X., Shi, R., Hu, H. et al. (2022). Medicinal values and potential risks evaluation of ginkgo biloba leaf extract (GBE) drinks made from the leaves in autumn as dietary supplements. *Molecules* 27 (21): 7479.

Suprabha, D., Avilash, R., Paul, A. et al. (2024). Food and medicinal properties of hibiscus (*Hibiscus sabdariffa* & hibiscus rosa-sinensis). *Sustainability, Agri, Food and Environmental Research* 12: https://doi.org/10.7770/safer-V12N1-art2528.

Teixeira, M., Tao, W., Fernandes, A. et al. (2023). Anthocyanin-rich edible flowers, current understanding of a potential new trend in dietary patterns. *Trends in Food Science & Technology* 138: 708–725.

Thakur, R. (2020). Herbal wine preparation from agricultural wastes. *Journal of Current Research in Food Science* 1 (1): 52–55.

Tildesley, N.T.J. et al. (2005). Sage (*Salvia officinalis*) enhances memory in healthy young volunteers. *Pharmacology, Biochemistry, and Behavior* 75 (3): 669–674.

Tiwari, S., Shukla, S., Morya, S., and Kishor, K. (2016). Production of herbal wine from *Hibiscus rosasinenesis* by using different strains of *Saccharomyces cerevisae*. *Advances in Life Sciences* 5: 20.

Torrens, F. and Castellano, G. (2020). Herbs brandies, wines, liquors, origanum onites, clove and cinnamon oils. In: *Research Methodologies in Applied Chemistry with Multidisciplinary Perspectives*, 37. Nova Science Publishers.

Ullah, M.A. and Hassan, A. (2022). Medicinal benefits of lemon balm (*Melissa officinalis*) for human health. *World* 1 (01): 028–033.

Upadhyay, S.K. and Singh, S.P. (2021). *Bioprospecting of Plant Biodiversity for Industrial Molecules*. Wiley doi:10.1002/9781119718017.

Upadhyay, S.K. and Singh, S.P. (2023). *Plants as Bioreactors for Industrial Molecules*. Wiley https://doi.org/10.1002/9781119875116.

Varriano, J. (2022). *Wine: A Cultural History*, 202–204. Reaktion Books.

Vilela, A. and Pinto, T. (2019). Grape infusions: the flavor of grapes and health-promoting compounds in your tea cup. *Beverages* 5 (3): 48.

Wang, S., Cai, R., Liu, X. et al. (2023). The detoxification of ochratoxin A in wine and grape juice by different enzymes and evaluation of their effects on the quality. *Food* 4 (1): e61.

Woo, S.H., Kim, J.C., Eslenur, N. et al. (2021). Modulations of cardiac functions and pathogenesis by reactive oxygen species and natural antioxidants. *Antioxidants* 10 (5): 760.

Yang, J., Martinson, T.E., and Liu, R.H. (2009). Phytochemical profiles and antioxidant activities of wine grapes. *Food Chemistry* 116 (1): 332–339.

Zadorozhna, M. and Mangieri, D. (2021). Mechanisms of chemopreventive and therapeutic proprieties of ginger extracts in cancer. *International Journal of Molecular Sciences* 22 (12): 6599.

Zhao, L., Ouyang, H., Zhang, N. et al. (2021). Effects of huangjiu, baijiu and red wine combined with high-fat diet on glucose and lipid metabolism: aggravate or alleviate? *Alcohol and Alcoholism* 56 (3): 334–347.

Zhao, H., Ren, S., Yang, H. et al. (2022). Peppermint essential oil: its phytochemistry, biological activity, pharmacological effect and application. *Biomedicine & Pharmacotherapy* 154: 113559.

8

Plant-based Nutraceuticals with Anticarcinogenic Potential

Omar S. Al-Odat[1,2], *Gabriella Yao*[1], *Nicole K. Schmalbach*[1], *Daniel A. Guirguis*[1], *Osama Aloudat*[1], *Patrick A. Newport-Ratiu*[1,2], *Subash C. Jonnalagadda*[2], *and Manoj K. Pandey*[1]

[1] *Department of Biomedical Sciences, Cooper Medical School of Rowan University, Camden, NJ, USA*
[2] *Department of Chemistry and Biochemistry, Rowan University, Glassboro, NJ, USA*

8.1 Journey of Nutraceuticals: From Folklores to Linchpins

Cancer remains a pressing issue on a global scale, representing a grave danger to life and a substantial obstacle to public health. It continues to be the leading cause of death worldwide. Addressing cancer treatment is a significant challenge, especially in the United States. The American Cancer Society predicts that approximately two million new cancer cases will be diagnosed, resulting in an estimated 609,000 cancer-related deaths in the US alone (Siegel et al. 2022).

During the early 1900s, Paul Ehrlich, a renowned German scientist, dedicated his efforts to developing drugs that could effectively treat infectious illnesses. He coined the term "chemotherapy," describing it as the application of chemicals to combat illness, marking a significant milestone in medical history. He was the first to show the effectiveness of animal models in testing different chemical compounds for potential anti-disease properties. Since then, there has been a significant impact on cancer drug development and remarkable progress in chemotherapy for cancer treatment. Chemotherapeutic regimens offer much-needed relief from cancer. Chemotherapeutic drug resistance is a significant concern in treatment of cancer that requires immediate attention. Regrettably, as cancer advances, the effectiveness of all chemotherapy treatments, including other combination strategies, diminishes.

Nutraceuticals are a type of food, dietary supplement, or medicinal food that harnesses the power of natural substances to promote health benefits. First introduced by Stephen DeFelice in 1989, these compounds cover a wide range of active substances that are often linked to their ability to fight against cancer (Calvani et al. 2020). Nutraceuticals have garnered considerable attention for their low toxicity and potential efficacy in addressing cancer and cancer stem cells. Many small molecules approved in the field of cancer between the 1940s and the end of 2014 were either natural compounds or derived from them (Newman and Cragg 2016). There has been significant research conducted on the therapeutic potential of nutraceuticals in cancer prevention and therapy (Gold et al. 2019; Padhye et al. 2012; Yusufi et al. 2013). Furthermore, recent investigations have highlighted the significant impact of these findings on improving the sensitivity of cancer

Herbal Nutraceuticals: Products and Processes, First Edition. Edited by Santosh Kumar Upadhyay and Sudhir Pratap Singh.

cells in overcoming multidrug resistance (MDR) challenges (Angulo et al. 2017; Dandawate et al. 2013; Dandawate et al. 2016). Scientists have shown significant interest in investigating the pharmacological uses of nutraceuticals, particularly their potential to inhibit cell growth, reduce inflammation, and combat cancer. The precise biological targets responsible for these various effects are still not completely understood. As a result, cancer research has emerged as an incredibly promising area of study. This is due to the presence of previously unknown active compounds and the fact that natural compounds tend to have lower toxicity and fewer side effects. Consequently, an increasing number of patients are opting for alternative therapy over conventional chemotherapy and radiation therapy. There is a wide range of promising nutraceuticals due to their potential in fighting cancer and their ability to target different types of cancer. Table 8.1 presents an overview of the primary natural compounds that are currently being utilized in the treatment of different forms of cancer.

8.2 Therapeutic Nutraceuticals for the Treatment of Chronic Inflammation and Cancer

The body's immune system reacts to various threats, such as pathogens, damaged cells, toxins, or radiation, by triggering inflammation (Chen et al. 2018). The progression from uncontrolled acute inflammation to chronic inflammation can have serious consequences. The aforementioned phenomenon has the potential to contribute to the onset of chronic inflammatory disorders or establish a milieu that facilitates cellular mutation and proliferation, hence potentially resulting in the development of cancer (Pandey et al. 2022). Research suggests that a significant portion of human cancers may be linked to chronic infection and inflammation (Antoni et al. 2017; Balkwill and Mantovani 2001; Busuttil et al. 2014; Ernst and Gold 2000; Tsukuma et al. 1993; zur Hausen 2002). In addition, chronic inflammation and the development of cancer have been associated with noninfectious inflammatory factors such as tobacco smoking, alcohol consumption, and obesity (Howe et al. 2013; Meadows and Zhang 2015; Tsukuma et al. 1993). The extensive research conducted globally has resulted in the discovery of a significant "connection between inflammation and cancer" (Aggarwal et al. 2006; Coussens and Werb 2002; Lussana and Rambaldi 2017). Figure 8.1 illustrates the principal pathways implicated in the pathogenesis of cancer and inflammation.

Dysregulation of certain inflammatory signaling pathways can lead to the initiation of cancer. These pathways include NF-κβ, MAPK, Wnt/β-catenin, mTOR, TLR4, and NRF2. The downstream effects of these pathways consequence in the production of pro-inflammatory cytokines and the upregulation of pro-inflammatory genes. These processes ultimately support cell survival, proliferation, and growth. The MAPK pathway responds to various triggers such as mitogens, heat shock, osmotic stress, and inflammatory cytokines like TNF-α, IL-1, and -6. This pathway plays an essential role in controlling cell survival, cell proliferation, cell differentiation, and apoptosis. The MAPK pathway consists of three subfamilies: ERK1/2, which responds to growth factors, and p38 MAPK and JNK, which react to stress. The MAPK pathway is composed of a kinase core that involves the activation of MAP3K, MAP2K, and MAPK (ERK, p38, and JNK), leading to the stimulation of NF-κβ. The MAPK signaling pathway is activated by norepinephrine, a stress hormone, in different types of tumors (Chang and Karin 2001; Huang et al. 2012; Yaniv et al. 2010).

NF-κβ, a key factor in inflammation and cell functions, including growth, apoptosis, and stress, plays a significant role (Figure 8.2). There are five genes in this family of transcription factors: NF-κβ1, NF-κβ2, c-Rel, RelA, and RelB. The dimers containing RelA or c-Rel bind with NF-κβ1 in the classical pathway and remain inactive in the cytoplasm due to their association with the

Table 8.1 The key naturally occurring compounds and their anticarcinogenic properties against specific cancers.

Nutraceutical	Spice	Treatment	Molecular targets
[6]-Gingerol	Ginger	Breast cancer, colorectal cancer	↓IL-1β, ↓IL-2, ↓TNF-α
1,8-Cineole	Sage, cardamom, basil	Colon cancer	↓NF-κB, ↓IL-1β, ↓IL-6, ↓TNF-α, ↓COX-2
Ajoene	Garlic	Melanoma	↓TNF-α
Allicin	Garlic	Gastric cancer	↓IL-1β, ↓IL-8, ↓TNF-α
Allyl isothiocyanate	Horseradish, mustard	Bladder cancer	↓IL-1β, ↓IL-8, ↑IL-2, ↓TNF-α
Anethole	Anise	Lung cancer	↓NF-κB
Apigenin	Celery seed, parsley	Gastrointestinal cancer	↓TNF-α, ↓IL-1β
Capsaicin	Red pepper, paprika	Bladder cancer, multiple myeloma (MM)	↓Stat3, ↓IL-6
Carnosol	Rosemary	Breast cancer	↓NF-κB, ↓iNOS
Caryophyllene	Bay leaf, savory	Colorectal cancer	↓NF-κB
Cinnamaldehyde	Cinnamon, clove	Breast cancer, colorectal cancer	↓NF-κB, ↓COX-2, ↓iNOS
Curcumin	Turmeric	Melanoma, hepatocellular carcinoma (HCC), MM (Tables 8.2 and 8.3).	↓NF-κB, ↓COX-2, ↓TNF-α, ↓P-Stat3
Diallyl sulfide (DAS)	Garlic, chive	Melanoma, colon cancer, prostate cancer	↓TNF-α, ↓IL-1β, ↓IL-6
Diosgenin	Fenugreek	Brest cancer, chronic myeloid leukemia	↓MMP-3, ↓COX2
Eugenol	Allspice, basil, clove, cinnamon, bay leaf	Colon cancer	↓IL-1β, ↓TNF-α, ↓COX-2
Garcinol	Garcinia	Oral cancer, lung cancer, breast cancer	↓NF-κB, ↓Wnt, ↓Stat3
Perillyl alcohol	Caraway	Breast cancer	↓NF-κB
Phytic acid	Sesame seed	Colon cancer	↓IL-1β, ↓IL-8, ↓IL-6
Piperine	White pepper, black pepper	Melanoma	↓NF-κB, ↓Stat3, ↓IL-1β, ↓IL-6, ↓TNF-α, ↓COX-2
Quercetin	Onion, allspice, horseradish	Breast cancer, ovarian cancer, prostate cancer	↓NF-κB
Sulforaphane	Mustard	Prostate cancer	↓NF-κB
Thymoquinone	Back cumin	Bladder cancer, lung cancer, gastric cancer, ovary cancer	↓NF-κB, ↓Stat-3, ↓IL-1β
α-Pinene	Rosemary, eucalyptus	HCC, prostate cancer	↓NF-κB, ↓IL-1β, ↓IL-6, ↓TNF-α

inhibitor IkappaB kinase (IκB or IKK). After activation, the pathway triggers the phosphorylation of IκB kinase-beta (IKKβ), resulting in NF-κβ translocation to the nucleus and degradation of IκB (Liu et al. 2017). In cancer, NF-κβ plays a crucial role in connecting inflammation and cancer by overexpressing tumor-promoting cytokines such as IL-6, TNFα, and pro-survival genes like Bcl-xL (Xia et al. 2018). In addition, NF-κβ has been found to regulate epithelial-to-mesenchymal transition, metastasis, and vascularization by increasing the expression of vascular endothelial growth

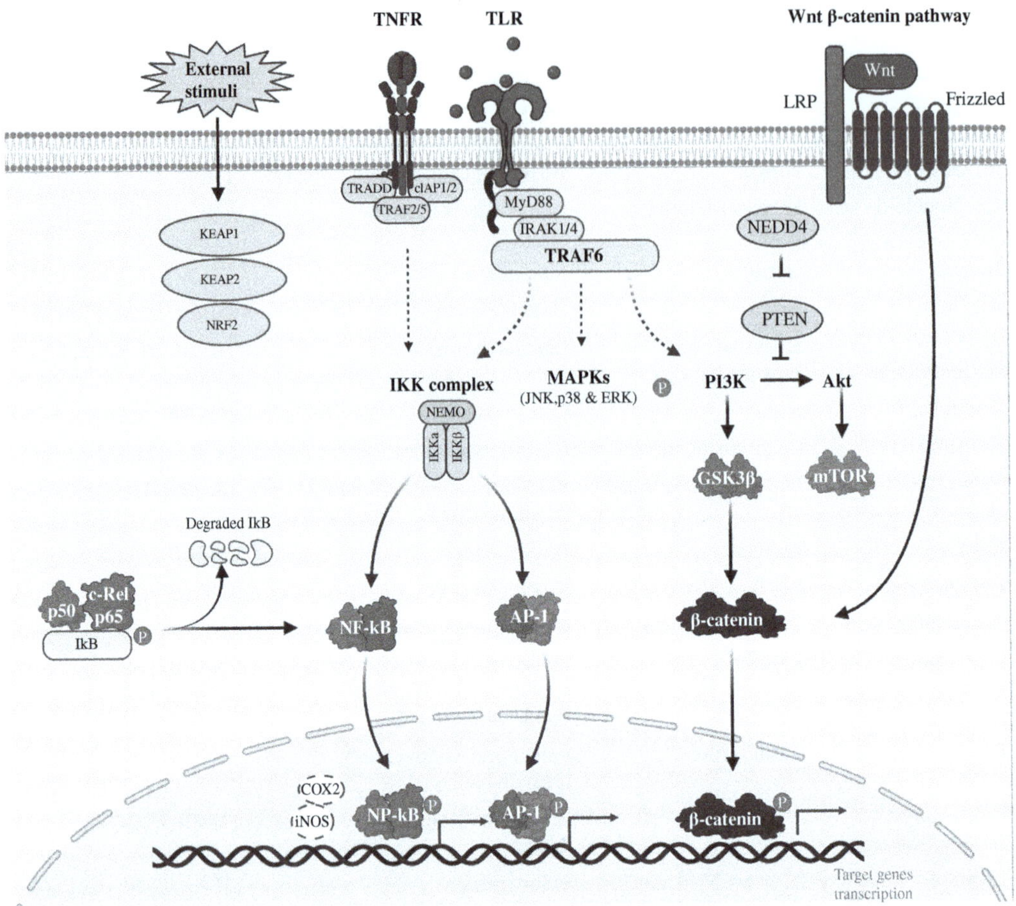

Figure 8.1 Inflammatory pathway and interconnection in cancer. NF-κβ plays major role in the inflammatory pathway of cancer. The increase in NF-κβ is a significant contributor to the development of multidrug-resistant (MDR) malignancies and resistance to chemotherapy. This is achieved by the modulation of apoptotic factors, including Bcl-2 family members, FLIP, and XIAP. The activation of cyclooxygenases (COX), inducible nitric oxide (iNOS), interleukin-1 (IL-1), and activator protein 1 (AP-1) by this transcription factor has been linked to the processes of tumor growth and proliferation. Additionally, NF-κβ is involved in the process of angiogenesis by promoting the overexpression of vascular endothelial growth factors (VEGFs).

factor (VEGF) (Huber et al. 2004; Xie et al. 2010). NF-κβ can activate various inflammatory cytokines and enzymes, such as TNF-α, COX, and iNOS, downstream. When dysregulated and secreted in the circulation, TNF-α can contribute to the development of various disease states, including cancer. This molecule forms a significant function in inflammation pathways by triggering the manufacture of other inflammatory mediators, such as MAPK and NF-κβ activation. In addition, TNF-α can trigger apoptosis by recruiting caspase-8, which in turn activates caspase-3 and leads to apoptosis (Zelová and Hošek 2013).

It is worth noting that NF-κβ is not only upregulated in certain cancers, as previously mentioned, but can also be activated by chemotherapy drugs such as doxorubicin, cisplatin, docetaxel, and paclitaxel. It was shown that there is constitutive activity of NF-κβ in chemotherapeutic-resistant multiple myeloma (MM) cell lines (Ma et al. 2003). Chemoresistance is facilitated by the overexpression of NF-κβ, which functions as a transcription factor for multiple genes including COX-2, Bcl-2, Bcl-xL, and XIAP (Catz and Johnson 2001; Stehlik et al. 1998; Tamatani et al. 1999; Yamamoto

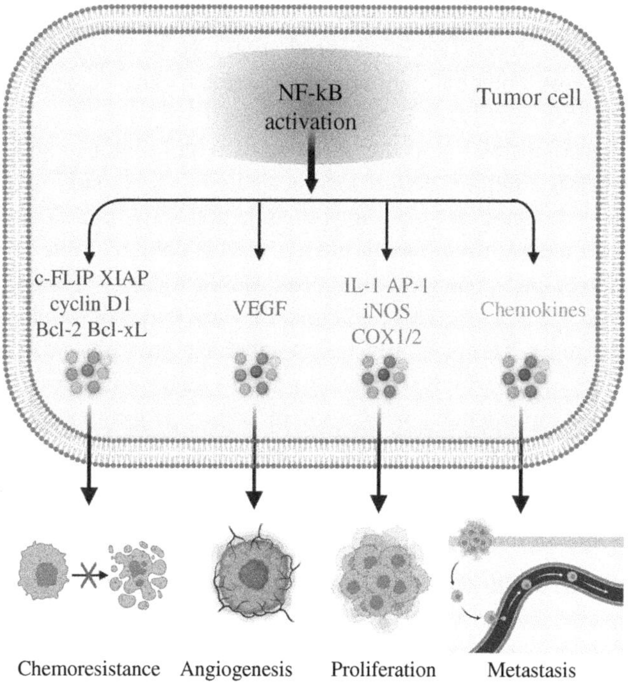

Figure 8.2 Inflammatory and cancer resistance pathways. IKK complex activates NF-κβ, which is able to phosphorylate IκB and allow translocation of NF-κβ. In this same NF-κβ pathway, c-Rel/p65 expression also allows activation of NF-κβ. Levels of AKT also affect IκB and downstream NF-κβ through an alternative pathway. Moreover, the overexpression of NF-κβ helps trigger the IL-6/STAT3 pathway. Furthermore, COX2 and iNOS levels are indirectly affected by NF-κβ. In the MAPK pathway, increasing phosphorylation levels of p38 MAPK, JNK, and ERK1/2 play an essential role in inflammatory and treatment resistance. Upstream of these pathways, activation of TLR4 receptor/MyD88 protein upregulates the inflammatory response and NF-κβ signaling. PI3K/AKT/mTOR pathway through activation of mTOR, AKT, and GSK-3β. This pathway can also be upregulated by increasing the expression of NEDD4, which promotes the degradation of PTEN, causing stimulation of the PI3K/AKT/mTOR pathway.

et al. 1995). The upregulation of Bcl2 family members and FLIP, a caspase-8 inhibitor, by NF-κβ also leads to the suppression of intrinsic and extrinsic apoptosis pathways (Micheau et al. 2001). In tumor development and resistance mechanisms, protein kinase B (AKT) is often upregulated to provide additional protection against apoptosis. This upregulation leads to the inactivation of caspase-8 and phosphorylated IκB proteins, which are accountable for allowing NF-κβ translocation (Madrid et al. 2000).

The PI3K/AKT/mTOR pathway is an essential inflammatory mechanism that plays a crucial role in the advancement of tumors and the development of cancer. When activated, PI3K transforms phosphatidylinositol into PIP2 and PIP3. In addition, AKT is recruited to the cellular membrane and undergoes phosphorylation by phosphatidylinositol-dependent kinase 1 (PDK1). This phosphorylation enables AKT to phosphorylate downstream targets such as mTOR and GSK-3 (Porta et al. 2014).

All inflammatory pathways ultimately lead to the production of cytokines. The cytokines play a crucial role in various processes such as tumorigenesis, tumor development, metastasis, inflammation, cell migration, apoptosis, and chemoresistance. They can play a crucial role in drug-resistant mechanisms when directly secreted by cancer cells. As an illustration, IL-6 is a crucial cytokine that enhances the expression of MM (Dankbar et al. 2000; Ravoet et al. 1994). This statement highlights the impact of IL-6 on tumor cells, indicating its association with disease progression and its relationship to treatment response and survival (Nachbaur et al. 1991). IL-10 is frequently observed to be excessively produced in tumors, potentially enhancing the defense of cancer against immune-mediated destruction (Steinbrink et al. 1999).

In our exploration of the molecular mechanisms connecting inflammation and tumor progression, we have discovered that simply targeting a single inflammatory molecule will not be sufficient. Perhaps, rather than fixating on a solitary targeted drug, we should contemplate the notion of a mystical injection to efficiently avert and manage tumors (Aneja et al. 2014). In addition, it is worth noting that FDA-approved drugs that target a single condition can be quite costly and may come with significant side effects (Mencher and Wang 2005). This highlights the importance of developing new, affordable, and multifunctional agents. It is noteworthy to mention that organically derived agents, commonly referred to as nutraceuticals, exhibit significant promise owing to their cost-effectiveness, safety, and appropriateness for extended periods of utilization (Kunnumakkara et al. 2018; Pandey et al. 2018). A wide range of sources, including vegetables, fruits, species, nuts, grains, and other resources, can be utilized to get nutraceuticals (Dixit et al. 2021; Kunnumakkara et al. 2018; Pandey et al. 2017; Upadhyay and Singh 2021, 2023). There is a substantial body of research indicating that nutraceuticals are effective in targeting inflammatory pathways (Pandey et al. 2017). Clinical trials have utilized nutraceuticals both as standalone agents and in combination with other agents (Kunnumakkara et al. 2017a; Pandey et al. 2017). Here are some significant nutraceuticals that have demonstrated potential as anti-inflammatory and anticancer agents.

8.2.1 Capsaicin

Red peppers contain capsaicin as their main ingredient (O'Neill et al. 2012). Capsaicin has demonstrated efficacy in the prevention of inflammation and cancer through its targeting of many variables, including NF-κB, IL-6, PGE2, STAT-3, TNF-α, and other associated molecules (Fernandes et al. 2016; Georgescu et al. 2017). The aforementioned treatment has demonstrated notable effectiveness in addressing a range of cancer types, encompassing stomach, lung, pancreatic, cervical, and breast cancer (Clark and Lee 2016; Friedman et al. 2018). Numerous studies have demonstrated the diverse mechanisms through which capsaicin demonstrates anti-inflammatory and anticancer properties (Fernandes et al. 2016). Research has demonstrated that capsaicin has various effects on the body. Research has revealed that it enhances the production of liver X receptor α (LXRα), which aids in the inhibition of inflammatory cytokines such as TNF-α, IL-1β, and IL-6 (Tang et al. 2015). Moreover, capsaicin has the ability to inhibit the NF-κB pathway, potentially mitigating inflammation within the salivary glands (Shin et al. 2013). Moreover, capsaicin has the ability to bring about cell cycle arrest in bladder cancer by inhibiting the pathway that is controlled by FOXO3a (Qian et al. 2016).

8.2.2 Diallyl Sulfide

The primary constituent present in garlic is diallyl sulfide (DAS) (Sato et al. 2020). From ancient times, garlic has been employed for the treatment of coughs, colds, and inflammatory diseases (Arreola et al. 2015; Bayan et al. 2014). DAS is effective in treating inflammatory conditions such as asthma (Arreola et al. 2015). Previous studies have shown that the suppression of pro-inflammatorypathways,suchasreactiveoxygenspecies(ROS),NF-κB,8-hydroxy-2′-deoxyguanosine, and 8-iso-prostaglandin F2α, along with the promotion of NRF2 activation, can effectively mitigate allergic asthma produced by ovalbumin (OVA) in animal models (Ho et al. 2016). Similarly, it has been observed that DAS exhibits inhibitory effects on osteoarthritis by reducing the expression of COX-2 through the NF-κB pathway (Lee et al. 2009). The study showed that DAS exhibited anticancer properties against various cancer types, including prostate cancer, colon cancer, and skin cancer. The regulation of inflammatory pathways is responsible for achieving this outcome (Lai et al. 2015; Saini et al. 2017; Wang et al. 2012; Zheng et al. 2019).

8.2.3 Cinnamaldehyde

The primary constituent present in cinnamon (*Cinnamomum zeylanicum*) is cinnamonaldehyde (CM) (Rao and Gan 2014). This substance is widely recognized for its remarkable array of advantages, such as cholesterol reduction, tumor development inhibition, inflammation suppression, infection prevention, and immune system enhancement (Kim et al. 2018; Mateen et al. 2019; Muhammad et al. 2015). The administration of CM resulted in a significant reduction in gastric inflammation through the inhibition of NF-κB activation (Muhammad et al. 2015). Furthermore, it has been observed in vivo that cinnamon exhibits a reduction in allergic encephalomyelitis through the regulation of T cells (Mondal and Pahan 2015). The efficacy of cinnamon bark as a therapeutic intervention for rheumatic conditions has been substantiated through empirical investigations, wherein it has been observed to significantly mitigate inflammation in arthritic models through the inhibition of specific cytokines, namely interferon γ (IFNγ), IL-2, and IL-4 (Liu et al. 2020). In the treatment of neurodegenerative illnesses like AD (Khasnavis and Pahan 2014), cinnamon is effective.

8.2.4 6-Gingerol

The ginger spice contains a naturally occurring chemical known as 6-Gingerol, which exhibits several biological properties, including antiproliferative, anti-inflammatory, and antioxidant activities (de Lima et al. 2018; Wang et al. 2014). This particular molecule possesses the capacity to decrease the expression of multiple inflammatory modulators, including TNF-α, NF-κB, IL-6, and monocyte chemoattractant protein-1 (MCP-1) (Li et al. 2017). Moreover, 6-gingerol has been found to have proapoptotic and anti-tumorigenic properties (Chen et al. 2018; Rastogi et al. 2015). It has been observed that 6-gingerol has the ability to augment the secretion of nonsteroidal anti-inflammatory drug (NSAID)-activated gene-1, leading to an increased rate of apoptosis in colorectal cancer cells (Lee et al. 2008).

8.2.5 Eugenol

Clove extract contains eugenol, a compound known for its well-documented anti-inflammatory properties (Barboza et al. 2018; Cortes-Rojas et al. 2014). There have been numerous studies demonstrating the anti-inflammatory properties of eugenol. The influence of this phenomenon has been observed on multiple variables, such as NF-κB, IL-1, IL-6, COX-2, TNF-α, PGE2, and others (Magalhaes et al. 2019). The chemical demonstrated a notable inhibitory effect on the NF-κB pathway, resulting in a substantial reduction in the development of stomach cancer cells in an *in vivo* setting (Manikandan et al. 2011). Research findings have demonstrated that eugenol possesses anti-inflammatory characteristics in cervical cancer cells and augments the effectiveness of gemcitabine, an anticancer medication (Hussain et al. 2011). Furthermore, it has been demonstrated that eugenol exhibits a protective effect against skin cancer through the inhibition of certain proteins implicated in cellular apoptosis. The present work demonstrated a significant reduction in breast cancer growth through the regulation of E2F1/surviving (Ng et al. 2018).

8.2.6 Diosgenin

Fenugreek spice contains diosgenin as its main ingredient (Jesus et al. 2016). This spice has been used for centuries as a powerful agent with numerous health benefits. Its antioxidant, anti-inflammatory, antidiabetic, anticancer, and hypocholesterolemic properties have been acknowledged (Selim and Al Jaouni 2016; Sethi et al. 2018). The anti-inflammatory effects of diosgenin have demonstrated promise as a therapeutic intervention for various types of cancer, such as breast cancer and chronic myeloid leukemia (Guo et al. 2019; Tian et al. 2020). Research has demonstrated that diosgenin possesses the capacity to diminish inflammatory markers such as iNOS, COX-2, MMP-3, and MMP-13. This characteristic renders it a highly promising contender for a chemopreventive medication (Wang et al. 2015). Furthermore, previous studies have demonstrated that diosgenin has the capacity to mitigate the tumor growth of breast cancer in controlled laboratory environments as well as trigger apoptosis in hepatocellular carcinoma and prostate cancer (Yu et al. 2018). The inhibitory effect of diosgenin on the mTOR signaling pathway was observed to augment cytotoxicity and ROS-dependent autophagy in CML cells (Jiang et al. 2016).

8.2.7 Garcinol

Garcinol, also known as kokum, is a potent compound found in the Garcinia indica plant (Padhye et al. 2009). Garcinol is a polyisoprenylated benzophenone (Padhye et al. 2009). Numerous studies have provided evidence supporting the antioxidative, anti-inflammatory, and anticarcinogenic properties of Ganciclovir (Behera et al. 2016; Saadat and Gupta 2012). An *in vitro* investigation revealed that garcinol has the capacity to inhibit the growth of breast cancer cells (Aggarwal et al. 2020). This medicine was observed to decrease the activity of caspase-3/iPLA(2) and NF-κB/Twist1 signaling pathways in a mouse 4T1 breast carcinoma model. As a result, breast cancer became more responsive to the taxol chemotherapy medication (Tu et al. 2017). It is noteworthy that garcinol has demonstrated a significant ability to mitigate the onset of colon cancer linked to inflammation (Tsai et al. 2014). Furthermore, it has exhibited anticancer properties by inhibiting the activation of NF-κB and STAT3 in head and neck cancer (Li et al. 2013). Moreover, research has demonstrated that garcinol possesses the capacity to impede the p38/MAPK signaling pathway in lung cancer, resulting in its anticancer characteristics (Yu et al. 2014). Additionally, it has been noted that it has the ability to suppress NF-κB in oral cancer (Aggarwal and Das 2016), as well as impact Wnt signaling and epithelial-mesenchymal transition (EMT) in breast cancer (Ahmad et al. 2012).

8.2.8 Thymoquinone

The chemopreventive, antioxidant, and anti-inflammatory properties of thymoquinone, a prominent constituent of black cumin, have been demonstrated in previous studies (Kundu et al. 2013; Shaterzadeh-Yazdi et al. 2018). Thymoquinone has demonstrated efficacy in the treatment of several malignancies, such as bladder, lung, ovarian, gastric, and other forms of tumors (Bashmail et al. 2018; Majdalawieh et al. 2017). The antitumor effect of thymoquinone has been demonstrated by its capacity to inhibit the NF-κB, ERK, PI3K/AKT, and STAT3 pathways (Al-Malki and Sayed 2014; Iskender et al. 2016). A recent study revealed that the bioactive component showed significant efficacy in reducing MAPKs and NF-κB activation in chondrocytes affected by osteoarthritis. As a result, it led to a decrease in IL-1β-induced inflammation (Wang et al. 2015). Furthermore, it effectively suppressed vascular remodeling, neoangiogenic factors, and inflammation in vivo in asthma (Su et al. 2016).

8.2.9 Quercetin

Onions contain the flavonoid quercetin. Several research studies have demonstrated the anti-inflammatory, anticancer, and antioxidant characteristics of flavonoids (Li et al. 2016). This flavonoid is effective against inflammatory conditions and has been found to have anticancer properties (Darband et al. 2018). The anti-inflammatory and anticancer characteristics of quercetin are thought to be attributed to its capacity to suppress the MAPK and NF-κB pathways while simultaneously increasing the NRF2 and PI3K/AKT pathways (Kim et al. 2019; Liu et al. 2017; Meng et al. 2018; Wang et al. 2017).

8.2.10 Sulforaphane

Cruciferous vegetables contain a wide range of nutrients, including sulforaphane. This chemical contains sulfur (isothiocyanate) (Miekus et al. 2020). Multiple studies have highlighted the powerful impact of sulforaphane in reducing inflammation and fighting cancer (Chen et al. 2018; Soni and Kohli 2019; Ullah 2015). It has proven to be effective in combating various inflammatory conditions as well (Chen et al. 2018). Furthermore, various studies suggest that sulforaphane has the potential to reduce pro-inflammatory signals by inhibiting the NF-κB pathway (Clarke et al. 2008; Lee et al. 2013).

8.2.11 α-Pinene

The oils of fragrant plants like rosemary and eucalyptus contain a significant amount of monoterpene α-pinene. This compound has been found to have antibacterial, antimetastatic, apoptotic, and antimicrobial characteristics (Cote et al. 2017). Studies have shown that α-pinene has promise in the treatment of inflammatory diseases due to its ability to inhibit the NF-κB pathway and MAPKs (Kim et al. 2015). In vivo, therapy with α-pinene greatly reduces the inflammation associated with acute pancreatitis. The administration of this medication results in the downregulation of IL-1β, IL-6, and TNF-κ, which are pivotal elements in the inflammatory response (de Cassia da Silveria e Sa et al. 2013).

8.2.12 Piperine

Piperine (*Piper longum*) is the primary plant alkaloid that has been isolated from long pepper and black pepper (*Piper nigrum*). Piperine has a range of biological qualities, such as anticonvulsant, anticancer, analgesic, and anti-inflammatory effects (Derosa et al. 2016; Jaisin et al. 2020; Smilkov et al. 2019). There is evidence suggesting that piperine may help reduce inflammation in various chronic conditions, such as Alzheimer's, chronic gastritis, endometritis, asthma, arthritis, Parkinson's, and more (Zadorozhna et al. 2019). The anti-inflammatory activities of piperine have been seen in several chronic illnesses through the downregulation of inflammatory pathways such as NF-κB, NOS-2, AP-1, PGE2, STAT3, IL-1β, MAPK, TNF-α, COX-2, and other related pathways (Doucette et al. 2015; Guo et al. 2020; Han et al. 2017; Xia et al. 2015).

8.2.13 1,8-Cineole (Cin)

A monoterpene oxide known as 1,8-cineole (Cin) is present in a variety of spices, such as sage, cardamom, and basil. Cinnamon has been employed in the treatment of many inflammatory ailments, such as asthma, chronic rhinitis, sinusitis, and bronchitis (Juergens 2014; Yadav and Chandra 2017).

It has been shown that Cin has the potential to reduce the activity of NF-κB, NOS-2, and COX-2, suggesting that it may possess anti-inflammatory properties (Greiner et al. 2013). In addition, Cin demonstrated positive effects in reducing acute pulmonary inflammation, alleviating colonic damage in rats with colitis induced by trinitrobenzene sulfonic acid (TNBS), and improving acute pancreatitis by suppressing NF-κB, cytokines, and oxidative stress (Lima et al. 2013; Santos et al. 2004; Zhao et al. 2014). In amyloid-β-toxicated PC12 cells, 1,8-cineole was found to drastically reduce the release of pro-inflammatory cytokines IL-1β, IL-6, and TNF-α (Khan et al. 2014). Spices contain a variety of chemicals, such as ursolic acid found in basil, carvone in mint, myristicin in nutmeg, crocin in saffron, linalool in coriander, sesamin in sesame seed, anethole in fennel, and carnosol in rosemary, among others. In addition, they can be found in spices. This research has demonstrated the effectiveness of various spices in preventing and treating chronic diseases. The aforementioned spices possess potent constituents capable of impeding signaling pathways such as ERK/MAPK, NF-κB, and STAT3 (Kunnumakkara et al. 2018; Pandey et al. 2017; Pandey et al. 2018; Pandey et al. 2022).

8.3 A Cancer and Inflammation Savior: Curcumin

Curcumin, derived from the spice turmeric, is widely acknowledged as a renowned nutraceutical with significant recognition. Turmeric possesses dual functionality as both a conventional medicine and a culinary component. This plant is a part of the Zingiberaceae family and is extremely valued in Ayurveda and Unani systems of medicine in South Asia (Trinh et al. 2009). Curcumin has been the subject of extensive research in both preclinical and clinical trials, as it shows promise in treating various chronic diseases. Within the realm of biological studies, there has been extensive research on compounds with anti-inflammatory and anticancer properties. Curcumin, in particular, has garnered significant attention in the field of cancer research, with over 7000 results found when searching for it on PubMed. Curcumin has been found to effectively target chronic inflammation and multiple types of cancer. Additionally, it has been reported to enhance the effectiveness of cancer drugs in cases of multidrug-resistant cancers. Curcumin has a diverse array of biological targets. Studies conducted worldwide, including our own previous research, have shown that curcumin plays a role in regulating various factors associated with chronic inflammation, cancer development, and resistance. These factors include IL-6, STAT3, TNF-α, NF-κB, PI3K/AKT/mTOR, MAPK, MMP-9, NRF2, TLR4, and more. Studies have demonstrated that curcumin can effectively reduce inflammation and reduce the growth and spread of cancer in multiple ways, including through the regulation of different signaling pathways (Gupta et al. 2013; Kunnumakkara et al. 2017b; Kunnumakkara et al. 2017b).

8.3.1 Curcumin Against Solid Cancers

Conventional chemotherapy and radiotherapies frequently fall short of completely eliminating cancer in patients, primarily due to MDR. The cytotoxic and chemopreventive effects of curcumin have been documented in many cancer cells and animal models, suggesting its capacity to selectively target multiple pathways implicated in the development of cancer. Curcumin has the ability to enhance the sensitivity of cells that were previously resistant to chemotherapeutics. After curcumin treatment, the doxorubicin-resistant cell lines showed a notable increase in the intracellular doxorubicin's accumulation (Wen et al. 2019). In addition, the sensitivity of resistant cancer cell lines was significantly enhanced when curcumin was combined with paclitaxel, cisplatin, or doxorubicin, resulting in a 5–15-fold increase (Zhou et al. 2015). The releasing amounts of drug-resistant associated proteins and cancer stem cells were extensively inhibited in resistant treated cell lines with curcumin (Chen et al. 2019; Su et al. 2018). In resistant cells, the evaluation of MDR-associated protein 5 (MRP5) suggests that it may contribute to resistance by actively removing drugs like 5-fluorouracil and gemcitabine (Li et al. 2011). Curcumin was observed to boost the effectiveness of 5-fluorouracil in pancreatic cells, as demonstrated by proliferation assays (Li et al. 2011). The combination of curcumin and cisplatin effectively reduced cell proliferation and reversed cisplatin resistance in A549 cell lines. This was achieved by degrading hypoxia-inducible factor 1-alpha (HIF-1α) (Ye et al. 2012). Table 8.2 provides a summary of the preclinical studies on curcumin, both as a standalone treatment and in combination with other therapies. These studies highlight the significant impact of curcumin on enhancing the efficacy of chemotherapy drugs for solid cancers.

8.3.2 Curcumin Against Hematological Malignancies

It is worth noting that curcumin has shown promise in treating various hematologic malignancies such as CLL, ALL, CML, AML, and MM. The pathogenesis of CML is primarily controlled by the BCR–ABL fusion protein. This protein triggers cancer proliferation via activating MAPK pathway as well as preventing cell death (Wu et al. 2003). Nevertheless, studies have demonstrated that curcumin combination therapy can effectively reduce the expression of NF-κβ, surviving, heat shock protein 90 (Hsp90), and BCR-ABL (Zhang et al. 2007). The JAK2 V617F mutations can lead to constant activation of the JAK/STAT pathway in BCR-ABL-negative patients. This pathway also interacts with the mTOR pathway. Curcumin was observed to trigger apoptosis in mutated cells by inhibiting the mTORC1 and JAK2/STAT pathways (Petiti et al. 2019). Curcumin was found to inhibit MAPK signaling in AML by upregulating the JNK/ERK pathway and MMP signaling (Yang et al. 2012; Zhu et al. 2016). In addition, curcumin has been found to induce apoptosis by inhibiting the AKT/mTOR and MEK/ERK pathways (Guo et al. 2014). Curcumin was found to inhibit NF-κβ, STAT3, and AKT in CLL (Ghosh et al. 2009). Curcumin was found to inhibit NF-κβ and activate caspase-3 (Pimentel-Gutiérrez et al. 2016; William et al. 2008). Additionally, it can induce autophagic cell death by upregulating the MAPK pathway RAF/MEK/ERK (Guo et al. 2018). Curcumin has been shown to effectively inhibit IL-6, a significant factor in the development of MM, as well as the phosphorylation of STAT3 induced by IL-6. In addition, the study revealed a decrease in the level of NF-κβ-related proteins, including IκB, cyclin D1, Bcl-2, and Bcl-xL (Bharti et al. 2003). Table 8.3 presents a comprehensive overview of the powerful preclinical applications of curcumin and curcumin derivatives against hematologic malignancies.

8.3.3 Development of Curcumin Clinical Trials

Both laboratory and animal investigations have shown that curcumin exhibits unique effects. Moreover, curcumin has demonstrated potential in suppressing cancer in multiple clinical trials, whether employed as a standalone treatment or in conjunction with existing chemotherapeutic medications (Table 8.4). It is worth noting that nearly 120 clinical trials have demonstrated the potential of this treatment in effectively managing various chronic diseases, all while avoiding any negative side effects (Tan and Norhaizan 2019). On the other hand, curcumin derivatives have been shown to be more potent than curcumin in its natural state, thanks to their enhanced water solubility and quicker metabolism. These derivatives have already been extensively studied (Dandawate et al. 2012; Vyas et al. 2013).

Curcumin was found to show activity against cachexia, a challenging condition associated with weight loss caused by cancer (Tisdale 2002). In 2011, a randomized clinical trial was conducted on 126 patients diagnosed with colorectal cancer. During the trial, the patients were given curcumin

Table 8.2 Curcumin treatment alone and in combination with chemotherapy drugs in solid cancer therapy preclinical studies.

Types of cancer	Cancer cell line	Treatment	Outcome	Ref.
Breast cancer	MCF-7	Albumin/paclitaxel and curcumin co-delivery (nanoparticle) by liposome	Combination of both drugs was found to be more cytotoxic than individual drugs.	Ruttala and Ko (2015)
	• In vitro: MCF7 • In vivo: 6–8 weeks BALB/c nude mice bearing MCF-7	PH-sensitive prodrug of curcumin and doxorubicin co-delivery by NP (transferrin-targeted nanoparticles)	Co-encapsulation of doxorubicin and curcumin in NPs can promote the cytotoxicity of both drugs in preclinical. Co-encapsulation approach had superior tumor-targeted drug delivery, stronger antitumor effects, and less cytotoxic effects.	Cui et al. (2017)
	MDR MCF-7	Doxorubicin and curcumin and co-delivery by albumin nanoparticles	Simultaneous co-administration of doxorubicin and curcumin-loaded albumin nanoparticles resulted in a higher intracellular concentration of doxorubicin and increased cell killing.	Motevalli et al. (2019)
	BT-20, T-47D, SK, BR3, and MCF-7 MCF-7 ADR (adriamycin-R) BT-20TNF (tumor necrosis factor 24-R)	Highly purified curcumin	Each of the cell lines exhibited high sensitivity. The observed repressive effect of curcumin on cellular development can be attributed to its ability to decrease the activity of ornithine decarboxylase. The cell death induced by curcumin was not attributed to apoptosis.	Mehta et al. (1997)
	Human mammary epithelial cell line (MCF-10A), MDR MCF-7/TH	Curcumin	MCF-7/TH cells are 3.5-fold more susceptible to curcumin than the MCF-10A cells. Curcumin treatment triggered a decrease in the level of PCNA, Ki67, and p53 mRNAs in breast cancer.	Ramachandran and You (1999)
	MCF-7/Dox (doxorubicin-R)	Curcumin and curcumin analogs (PGV-0 and PGV-1) alone and with doxorubicin	Curcumin and curcumin analogs improve the cytotoxicity of doxorubicin to MCF-7 cells via reduction of NF-kB activation and HER2 activity.	Meiyanto et al. (2014)

	MCF-7/LCC2 and MCF-7/LCC 9 (tamoxifen and tamoxifen/ fulvestrant-R)	Curcumin alone and with tamoxifen	Curcumin may kill the release of progrowth and antiapoptosis proteins, inactivate NF-κB, AKT/mTOR, and Src signaling pathways, and inhibit the crucial epigenetic modifier EZH2. Combination therapy increases tamoxifen's activity and overcomes endocrine resistance.	Jiang et al. (2013)
	Resistant MCF-7/ADM	Curcumin and photodynamic therapy (PDT)	Combination-induced apoptosis mainly via intrinsic apoptosis pathway. The opposite of MDR was recognized as the reduction of P-gp level of expression.	Li et al. (2020)
Breast cancer and liver cancer	• In vitro: MCF-7 (doxorubicin sensitive with extreme CD44 receptor expression), HepG2 (minimal expression of CD44 receptor), MCF-7/ADR (doxorubicin-R) • In vivo: mice with 4T1 tumor	Doxorubicin and curcumin co-encapsulated by hyaluronic acid-vitamin E succinate graft copolymer.	• Doxorubicin and curcumin co-encapsulating in micelles could improve cellular uptake, induce cell apoptosis, enhance cytotoxicity, and reverse MDR. • In vivo, it could minimize side effects and improve antitumor effects.	Ma et al. (2017)
Liver cancer	HepG2	Curcumin and doxorubicin delivered in a lipid nanoparticle	Synergistic effects of co-delivery of doxorubicin and curcumin by nanoparticles compared with doxorubicin nanoparticles alone. Increased apoptosis in HepG2 cells.	Zhao et al. (2015)
	• In vitro: HepG2 and HL-7702 • In vivo: Female nude BALB/c bearing HepG2 cells.	Curcumin and 5-fluorouracil (5-FU) delivered by BLPP (biotin-/lactobionic acid)-modified poly (ethylene glycol)-poly (lactic-co-glycolic acid)-poly (ethylene glycol)	• The dual-targeted nanoparticles have higher cytotoxicity and uptake efficiency. • In vivo, dual-targeted nanoparticles effectively inhibited tumor growth and decreased side effects of the chemotherapeutic agents.	Ni et al. (2019)
	• In vitro: BEL 7402 MDR BEL 7402/5-FU • In vivo: 24 mice received diethylnitrosamine solution in sesame oil (diethylnitrosamine-induced HCC mice). 8 mice received sesame oil only (normal mice)	Curcumin-doxorubicin-loaded lipid nanoparticles and doxorubicin/curcumin-loaded lipid nanoparticles	• Curcumin and doxorubicin nanoparticles showed synergistic effects on HCC, induced apoptosis, and reduced C-myc, PCNA, and VEGF • In vivo, enhanced antitumor effects in mice with DNE-induced HCC.	Zhao et al. (2015)

(*Continued*)

Table 8.2 (Continued)

Types of cancer	Cancer cell line	Treatment	Outcome	Ref.
Lung cancer	• In vitro: LL/2 and MS1 • In vivo: C57 mice received LL/2 cell suspension injections	Curcumin and doxorubicin by MPEG-PCL micelles (methoxy poly (ethylene glycol)-poly(caprolactone)	Curcumin and doxorubicin inhibited the growth of LL/2 lung cancer both in vitro and in vivo and had fewer side effects.	Wang et al. (2013)
	MDR A549/ADR and A549/ADR (doxorubicin-R)	Curcumin and doxorubicin by polyethylene glycol (PEG)-linked (PEGylated) amphiphilic copolymer and d-tocopheryl PEG1000 succinate	Doxorubicin and curcumin loaded in polymeric micelle reduced doxorubicin side effects and showed extreme efficiency inhibition in tumor growth. Curcumin and TPGS1000 may synergistically reverse A549/Adr cells' doxorubicin resistance.	Gu et al. (2016)
Colorectal cancer	(HCT116; MMR-deficient) and HCT116+ch3 (MMR-proficient), HCT116R, and HCT116+ch3R (5-FU-R)	Curcumin or/and 5-fluorouracil (5-FU)	Curcumin alone and in combination with 5-FU may prevent chemotherapy resistance in colon cancer via reduction of cancer stem cells (CSC).	Shakibaei et al. (2014)
	HT29, LoVo, and DLD1 HTOXAR3, LoVOXAR3, and DLDOXAR3 (oxaliplatin-R)	Curcumin and oxaliplatin	Combination of curcumin with oxaliplatin can be effective in reverting oxaliplatin resistance in colorectal cancer through the reduction of the NF-κB signaling cascade.	Ruiz de Porras et al. (2016)
	HCT116 wild-type and HCT116 cells complemented with chromosome 3 (HCT116+ch3)	Curcumin and 5-fluorouracil (5-FU)	Curcumin-induced apoptosis and increased 5-FU-induced expression. Combination with 5-FU could be more beneficial in treating resistant colon cancer. This may be regulated by NF-κB, PI-3K, and Src signaling pathways and NF-κB-mediated gene.	Shakibaei et al. (2013)
Colon cancer	C26	Curcumin and camptothecin delivered in cationic polymeric nanoparticles	Combination therapy of curcumin and camptothecin in a nanoparticle enhances both drugs' synergistic effects.	Xiao et al. (2015)
	C26	Curcumin and doxorubicin delivered in LCL-PEGylated (long-circulating liposomes)	Curcumin with doxorubicin in PEGylated LCL increases their cytotoxicity on C26 murine colon cancer cells by downregulating tumor-associated angiogenesis and inflammation factors.	Sesarman et al. (2018)

Pancreatic cancer	• In vitro: HPAF-II and Panc-1 • In vivo: orthotopic mouse model	Curcumin and gemcitabine loaded into nanoparticles of superparamagnetic iron oxide	Nanoparticles of superparamagnetic iron oxide loaded with curcumin (SP-CUR) facilitate gemcitabine uptake. Gemcitabine and SP-CUR have high combination effects in inhibiting the pancreatic cancer cells as well as CSC.	Khan et al. (2019)
Prostate cancer	C4-2B (castration-R)	Curcumin and nelfinavir	Curcumin and nelfinavir with an adjuvant therapy sensitize the castration-resistant prostate cancer to docetaxel therapy.	Mathur et al. (2014)
	PC3DTX-R and DU145DTX-R (docetaxel-R)	Curcumin loaded in Miglyol 812 and D-alpha-tocopheryl poly succinate 1000 nanoparticles (CUR MT NPs)	Curcumin may help castration-resistant prostate cancer overcome docetaxel resistance.	Tanaudommongkon et al. (2020)
Cervical cancer	MDR KBV20C	Unsymmetrical curcumin mimics with various amide moieties	Inhibition of P-gp function and almost recovered the cytotoxicity of vincristine and paclitaxel.	Um et al. (2008)

Table 8.3 Curcumin treatment alone and in combination with chemotherapy drugs in hematologic malignancies therapy preclinical studies.

Types of cancer	Cancer cell line	Treatment	Outcome	Ref.
ALL	MDR CEM/ ADR5000	Chloro and asymmetrical series of synthetic curcumin derivatives	Inhibition of P-gp function	Ooko et al. (2016)
CML	MDR K562/ADR	Diketone and cyclohexanone curcumin analogs	Inhibition of P-gp function and expression	Mapoung et al. (2016)
AML	Daunorubicin-insensitive CD34+ AML cells (KG1a and Kasumi-1)	Curcumin and daunorubicin	Induction of apoptosis and downregulation of Bcl-2 (antiapoptotic protein)	Rao et al. (2011)
	HL60	Curcumin	Induction of mitochondrial-mediated apoptosis and autophagy	Li et al. (2020)
CLL	CLL B cells	Curcumin	Induction of apoptosis and inhibition of NF-κB, STAT3, and AKT. In addition, curcumin inhibited the level of antiapoptotic proteins Mcl-1and XIAP while upregulating BIM (proapoptotic protein)	Ghosh et al. (2009)
MM	MDR MOLP-2/R	Curcumin with melphalan	Induction of apoptosis and inhibition of FA/ BRCA pathway	Xiao et al. (2010)
	U266	Curcumin with carfilzomib	Downregulation of NF-κβ and induction of p53/p21	Allegra et al. (2018)
	H929	Curcumin with bortezomib	Inactivation of NF-κB while activating JNK signaling	Bai and Zhang (2012)

three times per day for 10–30 days. Patients experienced positive outcomes, including an increase in cancer cell apoptosis and an improvement in body weight (He et al. 2011). In addition, dose-escalation phase I study demonstrated that curcumin can be safely and well tolerated when used alongside chemotherapy FOLFOX in 12 patients with colorectal liver metastases. The study found that curcumin was administered at doses up to 2 g/day (James et al. 2015). The present investigation, a phase I clinical trial, investigated the potential and tolerability of a combination therapy involving curcumin and docetaxel in a cohort of 14 patients diagnosed with advanced or metastatic breast cancer (Bayet-Robert et al. 2010).

Curcumin showed promising results with advanced pancreatic cancer in a phase II clinical trial involving 21 patients. The study found that curcumin was able to downregulate NF-κB, p-STAT3, and COX-2 (Dhillon et al. 2008). A prospective phase II trial enrolled 52 consecutive patients. A total of 44 patients, consisting of 13 with locally advanced cancer and 31 with metastatic, were included in the study to assess the safety and effectiveness of gemcitabine (10 mg/m^2) and curcumin

Table 8.4 Clinical trials performed in treatment of curcumin monotherapy and in combination with chemotherapy.

Types of cancer	Number of patients	Dosage, duration	Outcome	Ref.
Colorectal cancer	126	360 mg curcumin, 3× a day, 10–30 days	Body weight increase (good prognostic factor) TNF-α decrease increase in cancer cell apoptosis increase in expression of p53 increase in Bax, decrease in Bcl-2	He et al. (2011)
Colorectal liver metastases	12	Curcumin doses of up to 2 g daily plus oxaliplatin or plus 5-FU	Curcumin plus oxaliplatin/5-FU increased antiproliferative and apoptosis effects in a proportion of patient-derived explants, while downregulating the level of stem cell-related markers aldehyde dehydrogenases (ALDH) and CD133	James et al. (2015)
Advanced pancreatic cancer	25	8 g curcumin daily, restaging every 2 months	21 evaluable for response, no toxicities were observed. Curcumin reduced levels of NF-κB, p-STAT3, and COX-2 in peripheral blood mononuclear cells	Dhillon et al. (2008)
Advanced pancreatic cancer	17	Curcumin 8 g daily plus gemcitabine 1 g/m^2 weekly	Tumor progression was shown after 1–12 months, and overall survival was 1–24 months	Epelbaum et al. (2010)
Pancreatic cancer gemcitabine-resistant	21	Curcumin 1 g daily plus gemcitabine 1 mg/m^2 (on day 1 and 8), and 60 mg/m^2 of S-1 orally for 14 consecutive days every 3 weeks	Median survival time after starting curcumin treatment was 161 days and a year survival rate was 19%	Kanai et al. (2011)
Advanced and metastatic pancreatic cancer	44 patients (13 advanced and 31 metastatic)	Curcumin 2 g/die continuously (4 of 500 mg capsules every day) plus gemcitabine 10 mg/m^2	Overall survival and median progression-free survival were 10.2 and 8.4 months, respectively	Pastorelli et al. (2018)
Breast cancer	150	Paclitaxel (80 mg/m^2) plus curcumin (CUC-1, 300 mg solution, once per week) IV (intravenously) for 12 weeks, 3 month follow up	ORR of curcumin notably greater than placebo (51 versus 33%)	Saghatelyan et al. (2020)
Advanced or metastatic breast cancer	14	Curcumin 500 mg daily, escalated until dose-limiting toxicity, plus docetaxel 100 mg/m^2 for 7 days every 3 weeks	Improves biological and clinical responses	Bayet-Robert et al. (2010)
Prostate cancer	85	Isoflavones (40 mg) and/or curcumin (100 mg), 6 months	Decrease in PSA in group receiving supplement	Ide et al. (2010)

(Continued)

Table 8.4 (Continued)

Types of cancer	Number of patients	Dosage, duration	Outcome	Ref.
Lung cancer	62	Curcumin 1.5 g/day for 30 days	Decrease in urinary mutagens in smokers is an effective anti-mutagen, and it may be useful in chemoprevention	Polasa et al. (1992)
CML	50	Imatinib (400 mg twice a day) for 6 week or curcumin (5 g three times a day) plus imatinib (400 mg twice a day) for 6 weeks	NO levels decreased in the group taking the combination curcumin treatment	Ghalaut et al. (2012)
MM	Fifteen were being treated with either an immunomodulatory drug or proteasome inhibitor plus dexamethason	Replace the dex with curcumin 3-4g daily, 1–3 months	Decrease in paraprotein load, plasmacytosis, and without any other adverse effects of steroid-based	Ramakrishna et al. (2020)
Orbital pseudotumor	8	Curcumin 375 mg three times a day orally for 6–22 months	Complete recovery was observed with no adverse reactions noted and no recurrence	Lal et al. (2000)

(4 capsules, each containing 0.5 g/day). Furthermore, a comprehensive examination of inflammatory biomarkers was undertaken. The overall survival and progression-free median durations were 10.2 and 8.4 months, respectively (Pastorelli et al. 2018). In the context of treating advanced pancreatic cancer, the combination of gemcitabine with a phytosome complex containing curcumin has been found to be both safe and effective, leading to a commendable response rate (Pastorelli et al. 2018).

8.4 Conclusion and Future Direction

The development of cancer can be ascribed to the disruption of inflammatory pathways. Cytokine production is a result of inflammatory pathways. The cytokines play a crucial role in various processes such as tumorigenesis, tumor development, metastasis, cell migration, apoptosis, and chemoresistance. Although there have been significant advancements in cancer chemotherapy, chemotherapeutic agents' resistance remains a significant problem in achieving successful treatment. Another issue to consider is the potential negative side effects associated with the use of these chemicals.

Today, researchers are increasingly intrigued by natural compounds that possess the ability to target multiple pathways while minimizing any potentially harmful adverse reactions. Furthermore, research has shown that combining natural anticancer compounds with chemotherapy treatment can enhance its cytotoxic effects and overcome drug resistance. The impact of nutraceuticals on various signaling pathways and molecular mechanisms is significant. They have the ability to regulate the levels of numerous mediators involved in inflammation, cell cycles, the immune system, apoptosis, metastasis, angiogenesis, and carcinogenesis. These proteins include oncogenes, proapoptotic and antiapoptotic proteins, tumor suppressor genes, NF-κB, STAT3, AKT, VEGF, EGFR, MAPKs, HER-2, and COX-2.

Curcumin has gained immense popularity as a nutraceutical and shows promising potential as a standalone drug rather than just an adjuvant. Targeting cancer initiation and resistant pathways shows great potential and could pave the way for a new treatment option for various types of cancer. Furthermore, it is crucial to conduct a thorough and extensive clinical investigation to explore the combined effects of nutraceuticals and chemotherapeutic drugs. Nutraceuticals have the potential to become a promising therapeutic approach for several forms of cancer in the near future.

Author's Contributions

Pandey MK and Jonnalagadda SC conceived the idea, provided layout, instructed students for literature search, and edited the article. Yao G, Schmalbach NK, Guirguis DA, Aloudat O, and Newport-Ratiu PA did literature search, wrote sections, and made tables. Al-Odat OS did literature search, assembled all sections, made figures, and wrote and edited sections. All figures were created with BioRender.com.

Financial Support and Sponsorship

This work was supported by the Camden Research Initiative fund (M.K.P. and S.J.), the New Jersey Health Foundation (S.J., M.K.P.), NIH R16 (1R16GM153547-01) (M.K.P, S.J.), and inter-department fund from Cooper Medical School of Rowan University, Camden, NJ (M.K.P.).

Abbreviations

ABC	The ATP-binding cassette
AKT	Protein kinase B, PKB
ALDH	Aldehyde dehydrogenases
AML	Acute myeloid leukemia
AP-1	Activator protein 1
Bcl-2	B-cell lymphoma 2
Bcl-xL	B-cell lymphoma-extra large
CLL	Chronic lymphocytic leukemia
CLRs	C-type lectin receptors
CM	Cinnamonaldehyde
CML	Chronic myeloid leukemia
COX	Cyclooxygenase
DAMPs	Danger-associated molecular patterns
DAS	Diallyl sulfide
EGFR	Epidermal growth factor receptor
ERK	Extracellular signal-regulated kinases
FANCD2	Fanconi anemia complementation group D2
FDA	Food and Drug Administration
FOLFOX	combination of chemotherapy drugs (folinic acid, fluorouracil, and oxaliplatin)
FOXO3a	Forehead box O3a
FZD	Frizzled
GSK-3	Glycogen synthase kinase-3
GSTP1-1	Glutathione-S-transferase P1-1
HBV, HCV	Hepatitis B and C viruses
HER2	Human epidermal growth factor receptor 2
HIF-1α	Hypoxia-inducible factor 1-alpha HIF-1α
HO-1	Heme oxygenase-1
HPVs	Papillomaviruses
IGF-1R	Insulin-like growth factor-1 receptor
IKKβ	IκB kinase-beta
IL-1	Interleukin-1
IL-6	Interleukin-6
iNOS	Inducible nitric oxide synthase
IRAKs	Interleukin-1 receptor-associated kinase
IκB or IKK	IkappaB kinase
JNK	c-Jun N-terminal kinases
Keap1	Kelch-like ECH-associated protein 1
LPS	Lipopolysaccharide coat of bacteria
LRP	Low-density lipoprotein receptor-related protein
MAL	MyD88-adapter-like
MAP 2	Microtubule-associated protein 2
MAPK	Mitogen-associated protein kinase
MCP-1	Monocyte chemoattractant protein-1
MDR	Multidrug resistance
MDS	Myelodysplastic syndromes
MM	Multiple myeloma

MMP-9	Matrix metalloproteinase-9
MRD1	Multidrug resistance mutation 1
MRP5	Multidrug resistance-associated protein 5
MyD88	Myeloid differentiation primary response 88
NF-κβ	Nuclear factor-κB
NLRs	NOD-like receptors
NRF2	The nuclear factor erythroid 2-related factor 2
OVA	Ovalbumin
PAMPs	Pathogen-associated molecular patterns
PDK-1	Phosphatidylinositol-dependent kinase 1
P-gp	P-glycoprotein
PI3K/AKT/mTOR	Phosphatidylinositol-3-kinase (PI3K)/AKT/mammalian target of rapamycin (mTOR)
PIP2	Phosphatidylinositol 4,5-bisphosphate or PtdInsP
PRRs	Pattern recognition receptors
PSA	Prostate-specific antigen
PTEN	Phosphatase and tensin homolog
RelA	REL-associated protein
RLRs	Retinoic acid-inducible gene (RIG)-I-like receptors
STAT3	Signal transducer and activator of transcription 3
TAK1	Transforming growth factor beta-activated kinase 1
TLRs	Toll-like receptors
TNF-α	Tumor necrosis factor α
TRAF6	TNF receptor-associated factor 6
TRAIL	TNF-related apoptosis-inducing ligand
VEGF	Vascular endothelial growth factors
XIAP	X-linked inhibitor of apoptosis protein

References

Aggarwal, S. and Das, S.N. (2016). Garcinol inhibits tumour cell proliferation, angiogenesis, cell cycle progression and induces apoptosis via NF-kappaB inhibition in oral cancer. *Tumour Biology* 37 (6): 7175–7184.

Aggarwal, B.B., Shishodia, S., Sandur, S.K. et al. (2006). Inflammation and cancer: how hot is the link? *Biochemical Pharmacology* 72 (11): 1605–1621.

Aggarwal, V., Tuli, H.S., Kaur, J. et al. (2020). Garcinol exhibits anti-neoplastic effects by targeting diverse oncogenic factors in tumor cells. *Biomedicine* 8 (5): 103.

Ahmad, A., Sarkar, S.H., Bitar, B. et al. (2012). Garcinol regulates EMT and Wnt signaling pathways in vitro and in vivo, leading to anticancer activity against breast cancer cells. *Molecular Cancer Therapeutics* 11 (10): 2193–2201.

Allegra, A., Speciale, A., Molonia, M.S. et al. (2018). Curcumin ameliorates the in vitro efficacy of carfilzomib in human multiple myeloma U266 cells targeting p53 and NF-κB pathways. *Toxicology In Vitro* 47: 186–194.

Al-Malki, A.L. and Sayed, A.A. (2014). Thymoquinone attenuates cisplatin-induced hepatotoxicity via nuclear factor kappa-beta. *BMC Complementary and Alternative Medicine* 14: 282.

Aneja, P., Rahman, M., Beg, S. et al. (2014). Cancer targeted magic bullets for effective treatment of cancer. *Recent Patents on Anti-Infective Drug Discovery* 9 (2): 121–135.

Angulo, P., Kaushik, G., Subramaniam, D. et al. (2017). Natural compounds targeting major cell signaling pathways: a novel paradigm for osteosarcoma therapy. *Journal of Hematology & Oncology* 10 (1): 1–13.

Antoni, S., Ferlay, J., Soerjomataram, I. et al. (2017). Bladder cancer incidence and mortality: a global overview and recent trends. *European Urology* 71 (1): 96–108.

Arreola, R., Quintero-Fabián, S., López-Roa, R.I. et al. (2015). Immunomodulation and anti-inflammatory effects of garlic compounds. *Journal of Immunology Research* 2015: 401630.

Bai, Q.X. and Zhang, X.Y. (2012). Curcumin enhances cytotoxic effects of bortezomib in human multiple myeloma H929 cells: potential roles of NF-κB/JNK. *International Journal of Molecular Sciences* 13 (4): 4831–4838.

Balkwill, F. and Mantovani, A. (2001). Inflammation and cancer: back to Virchow? *The Lancet* 357 (9255): 539–545.

Barboza, J.N., da Silva Maia Bezerra Filho, C., Silva, R.O. et al. (2018). An overview on the anti-inflammatory potential and antioxidant profile of Eugenol. *Oxidative Medicine and Cellular Longevity* 2018: 3957262.

Bashmail, H.A., Alamoudi, A.A., Noorwali, A. et al. (2018). Thymoquinone synergizes gemcitabine anti-breast cancer activity via modulating its apoptotic and autophagic activities. *Scientific Reports* 8 (1): 11674.

Bayan, L., Koulivand, P.H., and Gorji, A. (2014). Garlic: a review of potential therapeutic effects. *Avicenna Journal of Phytomedicine* 4 (1): 1–14.

Bayet-Robert, M., Kwiatkowski, F., Leheurteur, M. et al. (2010). Phase I dose escalation trial of docetaxel plus curcumin in patients with advanced and metastatic breast cancer. *Cancer Biology & Therapy* 9 (1): 8–14.

Behera, A.K., Swamy, M.M., Natesh, N. et al. (2016). Garcinol and its role in chronic diseases. *Advances in Experimental Medicine and Biology* 928: 435–452.

Bharti, A.C., Donato, N., Singh, S. et al. (2003). Curcumin (diferuloylmethane) down-regulates the constitutive activation of nuclear factor-kappa B and IkappaBalpha kinase in human multiple myeloma cells, leading to suppression of proliferation and induction of apoptosis. *Blood* 101 (3): 1053–1062.

Busuttil, R.A., Zapparoli, G.V., Haupt, S. et al. (2014). Role of p53 in the progression of gastric cancer. *Oncotarget* 5 (23): 12016–12026.

Calvani, M., Pasha, A., and Favre, C. (2020). Nutraceutical boom in cancer: inside the labyrinth of reactive oxygen species. *International Journal of Molecular Sciences* 21 (6): 1936.

de Cassia da Silveria e Sa, R., Andrade, L.N., and de Sousa, D.P. (2013). A review on anti-inflammatory activity of monoterpenes. *Molecules* 18 (1): 1227–1254.

Catz, S.D. and Johnson, J.L. (2001). Transcriptional regulation of bcl-2 by nuclear factor kappa B and its significance in prostate cancer. *Oncogene* 20 (50): 7342–7351.

Chang, L. and Karin, M. (2001). Mammalian MAP kinase signalling cascades. *Nature* 410 (6824): 37–40.

Chen, L., Deng, H., Cui, H. et al. (2018). Inflammatory responses and inflammation-associated diseases in organs. *Oncotarget* 9 (6): 7204–7218.

Chen, P., Huang, H.P., Wang, Y. et al. (2019). Curcumin overcome primary gefitinib resistance in non-small-cell lung cancer cells through inducing autophagy-related cell death. *Journal of Experimental & Clinical Cancer Research* 38 (1): 1–17.

Chen, C.Y., Kao, C.L., and Liu, C.M. (2018). The cancer prevention, anti-inflammatory and anti-oxidation of bioactive phytochemicals targeting the TLR4 signaling pathway. *International Journal of Molecular Sciences* 19 (9): 2729.

Clark, R. and Lee, S.H. (2016). Anticancer properties of capsaicin against human cancer. *Anticancer Research* 36 (3): 837–843.

Clarke, J.D., Dashwood, R.H., and Ho, E. (2008). Multi-targeted prevention of cancer by sulforaphane. *Cancer Letters* 269 (2): 291–304.

Cortes-Rojas, D.F., de Souza, C.R., and Oliveira, W.P. (2014). Clove (*Syzygium aromaticum*): a precious spice. *Asian Pacific Journal of Tropical Biomedicine* 4 (2): 90–96.

Cote, H. et al. (2017). Anti-inflammatory, antioxidant, antibiotic, and cytotoxic activities of *Tanacetum vulgare* L essential oil and its constituents. *Medicines (Basel)* 4 (2).

Coussens, L.M. and Werb, Z. (2002). Inflammation and cancer. *Nature* 420 (6917): 860–867.

Cui, T., Zhang, S., and Sun, H. (2017). Co-delivery of doxorubicin and pH-sensitive curcumin prodrug by transferrin-targeted nanoparticles for breast cancer treatment. *Oncology Reports* 37 (2): 1253–1260.

Dandawate, P., Padhye, S., Ahmad, A. et al. (2013). Novel strategies targeting cancer stem cells through phytochemicals and their analogs. *Drug Delivery and Translational Research* 3 (2): 165–182.

Dandawate, P.R., Subramaniam D, Jensen RA, Anant S. Targeting cancer stem cells and signaling pathways by phytochemicals: novel approach for breast cancer therapy. *Seminars in Cancer Biology*. 2016. 40–41:192–208. doi: https://doi.org/10.1016/j.semcancer.2016.09.001.

Dandawate, P.R., Vyas, A., Ahmad, A. et al. (2012). Inclusion complex of novel curcumin analogue CDF and β-cyclodextrin (1, 2) and its enhanced in vivo anticancer activity against pancreatic cancer. *Pharmaceutical Research* 29 (7): 1775–1786.

Dankbar, B., Padró, T., Leo, R. et al. (2000). Vascular endothelial growth factor and interleukin-6 in paracrine tumor-stromal cell interactions in multiple myeloma. *Blood* 95 (8): 2630–2636.

Darband, S.G., Kaviani, M., Yousefi, B. et al. (2018). Quercetin: a functional dietary flavonoid with potential chemo-preventive properties in colorectal cancer. *Journal of Cellular Physiology* 233 (9): 6544–6560.

Derosa, G., Maffioli, P., and Sahebkar, A. (2016). Piperine and its role in chronic diseases. *Advances in Experimental Medicine and Biology* 928: 173–184.

Dhillon, N., Aggarwal, B.B., Newman, R.A. et al. (2008). Phase II trial of curcumin in patients with advanced pancreatic cancer. *Clinical Cancer Research* 14 (14): 4491–4499.

Dixit, S., Shukla, A., Singh, V., and Upadhyay, S.K. (2021). Bioprospecting of natural compounds for industrial and medical applications: current scenario and bottleneck. In: *Bioprospecting of Plant Biodiversity for Industrial Molecules*. Wiley; Portico. https://doi.org/10.1002/9781119718017.ch3.

Doucette, C.D., Greenshields, A.L., Liwski, R.S. et al. (2015). Piperine blocks interleukin-2-driven cell cycle progression in CTLL-2 T lymphocytes by inhibiting multiple signal transduction pathways. *Toxicology Letters* 234 (1): 1–12.

Epelbaum, R., Schaffer, M., Vizel, B. et al. (2010). Curcumin and gemcitabine in patients with advanced pancreatic cancer. *Nutrition and Cancer* 62 (8): 1137–1141.

Ernst, P.B. and Gold, B.D. (2000). The disease spectrum of helicobacter pylori: the immunopathogenesis of gastroduodenal ulcer and gastric cancer. *Annual Review of Microbiology* 54 (1): 615–640.

Fernandes, E.S., Cerqueira, A.R., Soares, A.G. et al. (2016). Capsaicin and its role in chronic diseases. *Advances in Experimental Medicine and Biology* 929: 91–125.

Friedman, J.R., Nolan, N.A., Brown, K.C. et al. (2018). Anticancer activity of natural and synthetic capsaicin analogs. *The Journal of Pharmacology and Experimental Therapeutics* 364 (3): 462–473.

Georgescu, S.R., Sârbu, M.I., Matei, C. et al. (2017). Capsaicin: friend or foe in skin cancer and other related malignancies? *Nutrients* 9 (12).

Ghalaut, V.S., Sangwan, L., Dahiya, K. et al. (2012). Effect of imatinib therapy with and without turmeric powder on nitric oxide levels in chronic myeloid leukemia. *Journal of Oncology Pharmacy Practice* 18 (2): 186–190.

Ghosh, A.K., Kay, N.E., Secreto, C.R. et al. (2009). Curcumin inhibits prosurvival pathways in chronic lymphocytic leukemia B cells and may overcome their stromal protection in combination with EGCG. *Clinical Cancer Research* 15 (4): 1250–1258.

Gold, M., Mujahid, Y., Ahmed, K. et al. (2019). A new 4-(pyridinyl)-4H-benzo [g] chromene-5, 10-dione ruthenium (II) complex inducing senescence in 518A2 melanoma cells. *JBIC Journal of Biological Inorganic Chemistry* 24 (5): 647–657.

Greiner, J.F., Müller, J., Zeuner, M.T. et al. (2013). 1,8-Cineol inhibits nuclear translocation of NF-kappaB p65 and NF-kappaB-dependent transcriptional activity. *Biochimica et Biophysica Acta* 1833 (12): 2866–2878.

Gu, Y., Li, J., Li, Y. et al. (2016). Nanomicelles loaded with doxorubicin and curcumin for alleviating multidrug resistance in lung cancer. *International Journal of Nanomedicine* 11: 5757.

Guo, W., Chen, Y., Gao, J. et al. (2019). Diosgenin exhibits tumor suppressive function via down-regulation of EZH2 in pancreatic cancer cells. *Cell Cycle* 18 (15): 1745–1758.

Guo, Y., Shan, Q., Gong, Y. et al. (2014). Curcumin induces apoptosis via simultaneously targeting AKT/mTOR and RAF/MEK/ERK survival signaling pathways in human leukemia THP-1 cells. *Pharmazie* 69 (3): 229–233.

Guo, Y., Shan, Q.Q., Gong, P.Y. et al. (2018). The autophagy induced by curcumin via MEK/ERK pathway plays an early anti-leukemia role in human Philadelphia chromosome-positive acute lymphoblastic leukemia SUP-B15 cells. *Journal of Cancer Research and Therapeutics* 14 (8): 125–131.

Guo, G., Shi, F., Zhu, J. et al. (2020). Piperine, a functional food alkaloid, exhibits inhibitory potential against TNBS-induced colitis via the inhibition of IkappaB-alpha/NF-kappaB and induces tight junction protein (claudin-1, occludin, and ZO-1) signaling pathway in experimental mice. *Human & Experimental Toxicology* 39 (4): 477–491.

Gupta, S.C., Patchva, S., and Aggarwal, B.B. (2013). Therapeutic roles of curcumin: lessons learned from clinical trials. *The AAPS Journal* 15 (1): 195–218.

Han, S.Z., Liu, H.X., Yang, L.Q. et al. (2017). Piperine (PP) enhanced mitomycin-C (MMC) therapy of human cervical cancer through suppressing Bcl-2 signaling pathway via inactivating STAT3/NF-kappaB. *Biomedicine & Pharmacotherapy* 96: 1403–1410.

He, Z.-Y., Shi, C.B., Wen, H. et al. (2011). Upregulation of p53 expression in patients with colorectal cancer by administration of curcumin. *Cancer Investigation* 29 (3): 208–213.

Ho, C.Y., Lu, C.C., Weng, C.J. et al. (2016). Protective effects of diallyl sulfide on ovalbumin-induced pulmonary inflammation of allergic asthma mice by MicroRNA-144, -34a, and -34b/c-modulated Nrf2 activation. *Journal of Agricultural and Food Chemistry* 64 (1): 151–160.

Howe, L.R., Subbaramaiah, K., Hudis, C.A. et al. (2013). Molecular pathways: adipose inflammation as a mediator of obesity-associated cancer. *Clinical Cancer Research* 19 (22): 6074–6083.

Huang, X.Y., Wang, H.C., Yuan, Z. et al. (2012). Norepinephrine stimulates pancreatic cancer cell proliferation, migration and invasion via β-adrenergic receptor-dependent activation of P38/MAPK pathway. *Hepato-Gastroenterology* 59 (115): 889–893.

Huber, M.A., Azoitei, N., Baumann, B. et al. (2004). NF-kappaB is essential for epithelial-mesenchymal transition and metastasis in a model of breast cancer progression. *The Journal of Clinical Investigation* 114 (4): 569–581.

Hussain, A., Brahmbhatt, K., Priyani, A. et al. (2011). Eugenol enhances the chemotherapeutic potential of gemcitabine and induces anticarcinogenic and anti-inflammatory activity in human cervical cancer cells. *Cancer Biotherapy & Radiopharmaceuticals* 26 (5): 519–527.

Ide, H., Tokiwa, S., Sakamaki, K. et al. (2010). Combined inhibitory effects of soy isoflavones and curcumin on the production of prostate-specific antigen. *The Prostate* 70 (10): 1127–1133.

Iskender, B., Izgi, K., and Canatan, H. (2016). Novel anti-cancer agent myrtucommulone-A and thymoquinone abrogate epithelial-mesenchymal transition in cancer cells mainly through the inhibition of PI3K/AKT signalling axis. *Molecular and Cellular Biochemistry* 416 (1-2): 71–84.

Jaisin, Y., Ratanachamnong, P., Wongsawatkul, O. et al. (2020). Antioxidant and anti-inflammatory effects of piperine on UV-B-irradiated human HaCaT keratinocyte cells. *Life Sciences* 263: 118607.

James, M.I., Iwuji, C., Irving, G. et al. (2015). Curcumin inhibits cancer stem cell phenotypes in ex vivo models of colorectal liver metastases, and is clinically safe and tolerable in combination with FOLFOX chemotherapy. *Cancer Letters* 364 (2): 135–141.

Jesus, M., Martins, A.P., Gallardo, E. et al. (2016). Diosgenin: recent highlights on pharmacology and analytical methodology. *Journal of Analytical Methods in Chemistry* 2016: 4156293.

Jiang, S., Fan, J., Wang, Q. et al. (2016). Diosgenin induces ROS-dependent autophagy and cytotoxicity via mTOR signaling pathway in chronic myeloid leukemia cells. *Phytomedicine* 23 (3): 243–252.

Jiang, M., Huang, O., Zhang, X. et al. (2013). Curcumin induces cell death and restores tamoxifen sensitivity in the antiestrogen-resistant breast cancer cell lines MCF-7/LCC2 and MCF-7/LCC9. *Molecules* 18 (1): 701–720.

Juergens, U.R. (2014). Anti-inflammatory properties of the monoterpene 1.8-cineole: current evidence for co-medication in inflammatory airway diseases. *Drug Research (Stuttg)* 64 (12): 638–646.

Kanai, M., Yoshimura, K., Asada, M. et al. (2011). A phase I/II study of gemcitabine-based chemotherapy plus curcumin for patients with gemcitabine-resistant pancreatic cancer. *Cancer Chemotherapy and Pharmacology* 68 (1): 157–164.

Khan, S., Setua, S., Kumari, S. et al. (2019). Superparamagnetic iron oxide nanoparticles of curcumin enhance gemcitabine therapeutic response in pancreatic cancer. *Biomaterials* 208: 83–97.

Khan, A., Vaibhav, K., Javed, H. et al. (2014). 1,8-cineole (eucalyptol) mitigates inflammation in amyloid Beta toxicated PC12 cells: relevance to Alzheimer's disease. *Neurochemical Research* 39 (2): 344–352.

Khasnavis, S. and Pahan, K. (2014). Cinnamon treatment upregulates neuroprotective proteins Parkin and DJ-1 and protects dopaminergic neurons in a mouse model of Parkinson's disease. *Journal of Neuroimmune Pharmacology* 9 (4): 569–581.

Kim, D.S., Lee, H.J., Jeon, Y.D. et al. (2015). Alpha-pinene exhibits anti-inflammatory activity through the suppression of MAPKs and the NF-kappaB pathway in mouse peritoneal macrophages. *The American Journal of Chinese Medicine* 43 (4): 731–742.

Kim, M.E., Na, J.Y., and Lee, J.S. (2018). Anti-inflammatory effects of trans-cinnamaldehyde on lipopolysaccharide-stimulated macrophage activation via MAPKs pathway regulation. *Immunopharmacology and Immunotoxicology* 40 (3): 219–224.

Kim, S.H., Yoo, E.S., Woo, J.S. et al. (2019). Antitumor and apoptotic effects of quercetin on human melanoma cells involving JNK/P38 MAPK signaling activation. *European Journal of Pharmacology* 860: 172568.

Kundu, J.K., Liu, L., Shin, J.W. et al. (2013). Thymoquinone inhibits phorbol ester-induced activation of NF-kappaB and expression of COX-2, and induces expression of cytoprotective enzymes in mouse skin in vivo. *Biochemical and Biophysical Research Communications* 438 (4): 721–727.

Kunnumakkara, A.B., Bordoloi, D., Harsha, C. et al. (2017a). Curcumin mediates anticancer effects by modulating multiple cell signaling pathways. *Clinical Science* 131 (15): 1781–1799.

Kunnumakkara, A.B., Bordoloi, D., Padmavathi, G. et al. (2017b). Curcumin, the golden nutraceutical: multitargeting for multiple chronic diseases. *British Journal of Pharmacology* 174 (11): 1325–1348.

Kunnumakkara, A.B., Sailo, B.L., Banik, K. et al. (2018). Chronic diseases, inflammation, and spices: how are they linked? *Journal of Translational Medicine* 16 (1): 14.

Lai, K.C., Hsu, S.C., Yang, J.S. et al. (2015). Diallyl trisulfide inhibits migration, invasion and angiogenesis of human colon cancer HT-29 cells and umbilical vein endothelial cells, and suppresses murine xenograft tumour growth. *Journal of Cellular and Molecular Medicine* 19 (2): 474–484.

Lal, B., Kapoor, A.K., Agrawal, P.K. et al. (2000). Role of curcumin in idiopathic inflammatory orbital pseudotumours. *Phytotherapy Research* 14 (6): 443–447.

Lee, S.H., Cekanova, M., and Baek, S.J. (2008). Multiple mechanisms are involved in 6-gingerol-induced cell growth arrest and apoptosis in human colorectal cancer cells. *Molecular Carcinogenesis* 47 (3): 197–208.

Lee, H.S., Lee, C.H., Tsai, H.C. et al. (2009). Inhibition of cyclooxygenase 2 expression by diallyl sulfide on joint inflammation induced by urate crystal and IL-1beta. *Osteoarthritis and Cartilage* 17 (1): 91–99.

Lee, Y.R., Noh, E.M., Han, J.H. et al. (2013). Sulforaphane controls TPA-induced MMP-9 expression through the NF-kappaB signaling pathway, but not AP-1, in MCF-7 breast cancer cells. *BMB Reports* 46 (4): 201–206.

Li, Y., Revalde, J.L., Reid, G. et al. (2011). Modulatory effects of curcumin on multi-drug resistance-associated protein 5 in pancreatic cancer cells. *Cancer Chemotherapy and Pharmacology* 68 (3): 603–610.

Li, F., Shanmugam, M.K., Chen, L. et al. (2013). Garcinol, a polyisoprenylated benzophenone modulates multiple proinflammatory signaling cascades leading to the suppression of growth and survival of head and neck carcinoma. *Cancer Prevention Research (Philadelphia, Pa.)* 6 (8): 843–854.

Li, N., Wen, S., Chen, G. et al. (2020). Antiproliferative potential of piperine and curcumin in drug-resistant human leukemia cancer cells are mediated via autophagy and apoptosis induction, S-phase cell cycle arrest and inhibition of cell invasion and migration. *Journal of BUON* 25: 401–406.

Li, Y., Xiao, P., Huang, Z. et al. (2020). Evaluation of curcumin-mediated photodynamic therapy on the reverse of multidrug resistance in tumor cells. *RSC Advances* 10 (1): 298–306.

Li, Y., Xu, B., Xu, M. et al. (2017). 6-Gingerol protects intestinal barrier from ischemia/reperfusion-induced damage via inhibition of p38 MAPK to NF-kappaB signalling. *Pharmacological Research* 119: 137–148.

Li, Y., Yao, J., Han, C. et al. (2016). Quercetin, inflammation and immunity. *Nutrients* 8 (3): 167.

Lima, P.R., de Melo, T.S., Carvalho, K.M. et al. (2013). 1,8-cineole (eucalyptol) ameliorates cerulein-induced acute pancreatitis via modulation of cytokines, oxidative stress and NF-kappaB activity in mice. *Life Sciences* 92 (24-26): 1195–1201.

de Lima, R.M.T., Dos Reis, A.C., de Menezes, A.P.M. et al. (2018). Protective and therapeutic potential of ginger (*Zingiber officinale*) extract and [6]-gingerol in cancer: a comprehensive review. *Phytotherapy Research* 32 (10): 1885–1907.

Liu, Y., Gong, W., Yang, Z.Y. et al. (2017). Quercetin induces protective autophagy and apoptosis through ER stress via the p-STAT3/Bcl-2 axis in ovarian cancer. *Apoptosis* 22 (4): 544–557.

Liu, P., Wang, J., Wen, W. et al. (2020). Cinnamaldehyde suppresses NLRP3 derived IL-1beta via activating succinate/HIF-1 in rheumatoid arthritis rats. *International Immunopharmacology* 84: 106570.

Liu, T., Zhang, L., Joo, D. et al. (2017). NF-κB signaling in inflammation. *Signal Transduction and Targeted Therapy* 2 (1): 17023.

Lussana, F. and Rambaldi, A. (2017). Inflammation and myeloproliferative neoplasms. *Journal of Autoimmunity* 85: 58–63.

Ma, W., Guo, Q., Li, Y. et al. (2017). Co-assembly of doxorubicin and curcumin targeted micelles for synergistic delivery and improving anti-tumor efficacy. *European Journal of Pharmaceutics and Biopharmaceutics* 112: 209–223.

Ma, M.H., Yang, H.H., Parker, K. et al. (2003). The proteasome inhibitor PS-341 markedly enhances sensitivity of multiple myeloma tumor cells to chemotherapeutic agents. *Clinical Cancer Research* 9 (3): 1136–1144.

Madrid, L.V., Wang, C.Y., Guttridge, D.C. et al. (2000). Akt suppresses apoptosis by stimulating the transactivation potential of the RelA/p65 subunit of NF-κB. *Molecular and Cellular Biology* 20 (5): 1626–1638.

Magalhaes, C.B., Casquilho, N.V., Machado, M.N. et al. (2019). The anti-inflammatory and anti-oxidative actions of eugenol improve lipopolysaccharide-induced lung injury. *Respiratory Physiology & Neurobiology* 259: 30–36.

Majdalawieh, A.F., Fayyad, M.W., and Nasrallah, G.K. (2017). Anti-cancer properties and mechanisms of action of thymoquinone, the major active ingredient of *Nigella sativa*. *Critical Reviews in Food Science and Nutrition* 57 (18): 3911–3928.

Manikandan, P., Vinothini, G., Vidya Priyadarsini, R. et al. (2011). Eugenol inhibits cell proliferation via NF-kappaB suppression in a rat model of gastric carcinogenesis induced by MNNG. *Investigational New Drugs* 29 (1): 110–117.

Mapoung, S., Pitchakarn, P., Yodkeeree, S. et al. (2016). Chemosensitizing effects of synthetic curcumin analogs on human multi-drug resistance leukemic cells. *Chemico-Biological Interactions* 244: 140–148.

Mateen, S., Rehman, M.T., Shahzad, S. et al. (2019). Anti-oxidant and anti-inflammatory effects of cinnamaldehyde and eugenol on mononuclear cells of rheumatoid arthritis patients. *European Journal of Pharmacology* 852: 14–24.

Mathur, A., Abd Elmageed, Z.Y., Liu, X. et al. (2014). Subverting ER-stress towards apoptosis by nelfinavir and curcumin coexposure augments docetaxel efficacy in castration resistant prostate cancer cells. *PLoS One* 9 (8): e103109.

Meadows, G.G. and Zhang, H. (2015). Effects of alcohol on tumor growth, metastasis, immune response, and host survival. *Alcohol Research* 37 (2): 311–322.

Mehta, K., Pantazis, P., McQueen, T. et al. (1997). Antiproliferative effect of curcumin (diferuloylmethane) against human breast tumor cell lines. *Anti-Cancer Drugs* 8 (5): 470–481.

Meiyanto, E., Putri, D.D., Susidarti, R.A. et al. (2014). Curcumin and its analogues (PGV-0 and PGV-1) enhance sensitivity of resistant MCF-7 cells to doxorubicin through inhibition of HER2 and NF-kB activation. *Asian Pacific Journal of Cancer Prevention* 15 (1): 179–184.

Mencher, S.K. and Wang, L.G. (2005). Promiscuous drugs compared to selective drugs (promiscuity can be a virtue). *BMC Clinical Pharmacology* 5: 3.

Meng, L.Q., Yang, F.Y., Wang, M.S. et al. (2018). Quercetin protects against chronic prostatitis in rat model through NF-kappaB and MAPK signaling pathways. *Prostate* 78 (11): 790–800.

Micheau, O., Lens, S., Gaide, O. et al. (2001). NF-κB signals induce the expression of c-FLIP. *Molecular and Cellular Biology* 21 (16): 5299–5305.

Miekus, N., Marszałek, K., Podlacha, M. et al. (2020). Health benefits of plant-derived sulfur compounds, glucosinolates, and organosulfur compounds. *Molecules* 25 (17): 3804.

Mondal, S. and Pahan, K. (2015). Cinnamon ameliorates experimental allergic encephalomyelitis in mice via regulatory T cells: implications for multiple sclerosis therapy. *PLoS One* 10 (1): e0116566.

Motevalli, S.M., Eltahan, A.S., Liu, L. et al. (2019). Co-encapsulation of curcumin and doxorubicin in albumin nanoparticles blocks the adaptive treatment tolerance of cancer cells. *Biophysics Reports* 5 (1): 19–30.

Muhammad, J.S., Zaidi, S.F., Shaharyar, S. et al. (2015). Anti-inflammatory effect of cinnamaldehyde in Helicobacter pylori induced gastric inflammation. *Biological & Pharmaceutical Bulletin* 38 (1): 109–115.

Nachbaur, D.M., Herold, M., Maneschg, A. et al. (1991). Serum levels of interleukin-6 in multiple myeloma and other hematological disorders: correlation with disease activity and other prognostic parameters. *Annals of Hematology* 62 (2-3): 54–58.

Newman, D.J. and Cragg, G.M. (2016). Natural products as sources of new drugs from 1981 to 2014. *Journal of Natural Products* 79 (3): 629–661.

Ng, C.Y., Yen, H., Hsiao, H.Y. et al. (2018). Phytochemicals in skin cancer prevention and treatment: an updated review. *International Journal of Molecular Sciences* 19 (4): 941.

Ni, W., Li, Z., Liu, Z. et al. (2019). Dual-targeting nanoparticles: codelivery of curcumin and 5-fluorouracil for synergistic treatment of hepatocarcinoma. *Journal of Pharmaceutical Sciences* 108 (3): 1284–1295.

O'Neill, J., Brock, C., Olesen, A.E. et al. (2012). Unravelling the mystery of capsaicin: a tool to understand and treat pain. *Pharmacological Reviews* 64 (4): 939–971.

Ooko, E., Alsalim, T., Saeed, B. et al. (2016). Modulation of P-glycoprotein activity by novel synthetic curcumin derivatives in sensitive and multidrug-resistant T-cell acute lymphoblastic leukemia cell lines. *Toxicology and Applied Pharmacology* 305: 216–233.

Padhye, S., Ahmad, A., Oswal, N. et al. (2009). Emerging role of Garcinol, the antioxidant chalcone from Garcinia indica Choisy and its synthetic analogs. *Journal of Hematology & Oncology* 2: 38.

Padhye, S., Dandawate, P., Yusufi, M. et al. (2012). Perspectives on medicinal properties of plumbagin and its analogs. *Medicinal Research Reviews* 32 (6): 1131–1158.

Pandey, M.K., Gupta, S.C., Karelia, D. et al. (2018). Dietary nutraceuticals as backbone for bone health. *Biotechnology Advances* 36 (6): 1633–1648.

Pandey, M.K., Gupta, S.C., Nabavizadeh, A. et al. (2017). Regulation of cell signaling pathways by dietary agents for cancer prevention and treatment. *Seminars in Cancer Biology* 46: 158–181.

Pandey, M.K., Von Suskil, M., Chitren, R. et al. (2022). Cancer on fire: role of inflammation in prevention and treatment. In: *Current Advances for Development of Functional Foods Modulating Inflammation and Oxidative Stress* (ed. B. Hernández-Ledesma and C. Martínez-Villaluenga), 605–626. Elsevier.

Pastorelli, D., ASC, F., Giovanis, P. et al. (2018). Phytosome complex of curcumin as complementary therapy of advanced pancreatic cancer improves safety and efficacy of gemcitabine: results of a prospective phase II trial. *Pharmacological Research* 132: 72–79.

Petiti, J., Rosso, V., Lo Iacono, M. et al. (2019). Curcumin induces apoptosis in JAK2-mutated cells by the inhibition of JAK2/STAT and mTORC1 pathways. *Journal of Cellular and Molecular Medicine* 23 (6): 4349–4357.

Pimentel-Gutiérrez, H.J., Bobadilla-Morales, L., Barba-Barba, C.C. et al. (2016). Curcumin potentiates the effect of chemotherapy against acute lymphoblastic leukemia cells via downregulation of NF-κB. *Oncology Letters* 12 (5): 4117–4124.

Polasa, K., Raghuram, T.C., Krishna, T.P. et al. (1992). Effect of turmeric on urinary mutagens in smokers. *Mutagenesis* 7 (2): 107–109.

Porta, C., Paglino, C., and Mosca, A. (2014). Targeting PI3K/Akt/mTOR signaling in cancer. *Frontiers in Oncology* 4: 64.

Qian, K., Wang, G., Cao, R. et al. (2016). Capsaicin suppresses cell proliferation, induces cell cycle arrest and ROS production in bladder cancer cells through FOXO3a-mediated pathways. *Molecules* 21 (10).

Ramachandran, C. and You, W. (1999). Differential sensitivity of human mammary epithelial and breast carcinoma cell lines to curcumin. *Breast Cancer Research and Treatment* 54 (3): 269–278.

Ramakrishna, R., Diamond, T.H., Alexander, W. et al. (2020). Use of curcumin in multiple myeloma patients intolerant of steroid therapy. *Clinical Case Reports* 8 (4): 739–744.

Rao, P.V. and Gan, S.H. (2014). Cinnamon: a multifaceted medicinal plant. *Evidence-based Complementary and Alternative Medicine* 2014: 642942.

Rao, J., Xu, D.R., Zheng, F.M. et al. (2011). Curcumin reduces expression of Bcl-2, leading to apoptosis in daunorubicin-insensitive CD34+ acute myeloid leukemia cell lines and primary sorted CD34+ acute myeloid leukemia cells. *Journal of Translational Medicine* 9 (1): 1–15.

Rastogi, N., Duggal, S., Singh, S.K. et al. (2015). Proteasome inhibition mediates p53 reactivation and anti-cancer activity of 6-gingerol in cervical cancer cells. *Oncotarget* 6 (41): 43310–43325.

Ravoet, C., DeGreve, J., Vandewoude, K. et al. (1994). Tumour stimulating effects of recombinant human interleukin-6. *The Lancet* 344 (8936): 1576–1577.

Ruiz de Porras, V., Bystrup, S., Martínez-Cardús, A. et al. (2016). Curcumin mediates oxaliplatin-acquired resistance reversion in colorectal cancer cell lines through modulation of CXC-Chemokine/NF-κB signalling pathway. *Scientific Reports* 6 (1): 1–17.

Ruttala, H.B. and Ko, Y.T. (2015). Liposomal co-delivery of curcumin and albumin/paclitaxel nanoparticle for enhanced synergistic antitumor efficacy. *Colloids and Surfaces B: Biointerfaces* 128: 419–426.

Saadat, N. and Gupta, S.V. (2012). Potential role of garcinol as an anticancer agent. *Journal of Oncology* 2012: 647206.

Saghatelyan, T., Tananyan, A., Janoyan, N. et al. (2020). Efficacy and safety of curcumin in combination with paclitaxel in patients with advanced, metastatic breast cancer: a comparative, randomized, double-blind, placebo-controlled clinical trial. *Phytomedicine* 70: 153218.

Saini, V., Manral, A., Arora, R. et al. (2017). Novel synthetic analogs of diallyl disulfide triggers cell cycle arrest and apoptosis via ROS generation in MIA PaCa-2 cells. *Pharmacological Reports* 69 (4): 813–821.

Santos, F.A., Silva, R.M., Campos, A.R. et al. (2004). 1,8-cineole (eucalyptol), a monoterpene oxide attenuates the colonic damage in rats on acute TNBS-colitis. *Food and Chemical Toxicology* 42 (4): 579–584.

Sato, S., Sekine, Y., Kakumu, Y. et al. (2020). Measurement of diallyl disulfide and allyl methyl sulfide emanating from human skin surface and influence of ingestion of grilled garlic. *Scientific Reports* 10 (1): 465.

Selim, S. and Al Jaouni, S. (2016). Anti-inflammatory, antioxidant and antiangiogenic activities of diosgenin isolated from traditional medicinal plant, *Costus speciosus* (Koen ex.Retz.) Sm. *Natural Product Research* 30 (16): 1830–1833.

Sesarman, A., Tefas, L., Sylvester, B. et al. (2018). Anti-angiogenic and anti-inflammatory effects of long-circulating liposomes co-encapsulating curcumin and doxorubicin on C26 murine colon cancer cells. *Pharmacological Reports* 70 (2): 331–339.

Sethi, G., Shanmugam, M.K., Warrier, S. et al. (2018). Pro-apoptotic and anti-cancer properties of diosgenin: a comprehensive and critical review. *Nutrients* 10 (5).

Shakibaei, M., Buhrmann, C., Kraehe, P. et al. (2014). Curcumin chemosensitizes 5-fluorouracil resistant MMR-deficient human colon cancer cells in high density cultures. *PLoS One* 9 (1): e85397.

Shakibaei, M., Mobasheri, A., Lueders, C. et al. (2013). Curcumin enhances the effect of chemotherapy against colorectal cancer cells by inhibition of NF-κB and Src protein kinase signaling pathways. *PLoS One* 8 (2): e57218.

Shaterzadeh-Yazdi, H., Noorbakhsh, M.F., Hayati, F. et al. (2018). Immunomodulatory and anti-inflammatory effects of thymoquinone. *Cardiovascular & Hematological Disorders Drug Targets* 18 (1): 52–60.

Shin, Y.H., Namkoong, E., Choi, S. et al. (2013). Capsaicin regulates the NF-kappaB pathway in salivary gland inflammation. *Journal of Dental Research* 92 (6): 547–552.

Siegel, R.L., Miller, K.D., Fuchs, H.E. et al. (2022). Cancer statistics, 2022. *CA: A Cancer Journal for Clinicians* 72 (1): 7–33.

Smilkov, K., Ackova, D.G., Cvetkovski, A. et al. (2019). Piperine: old spice and new nutraceutical? *Current Pharmaceutical Design* 25 (15): 1729–1739.

Soni, K. and Kohli, K. (2019). Sulforaphane-decorated gold nanoparticle for anti-cancer activity: in vitro and in vivo studies. *Pharmaceutical Development and Technology* 24 (4): 427–438.

Stehlik, C., de Martin, R., Kumabashiri, I. et al. (1998). Nuclear factor (NF)-kappaB-regulated X-chromosome-linked iap gene expression protects endothelial cells from tumor necrosis factor alpha-induced apoptosis. *The Journal of Experimental Medicine* 188 (1): 211–216.

Steinbrink, K., Jonuleit, H., Müller, G. et al. (1999). Interleukin-10–treated human dendritic cells induce a melanoma-antigen–specific anergy in CD8+ T cells resulting in a failure to lyse tumor cells. *Blood* 93 (5): 1634–1642.

Su, X., Ren, Y., Yu, N. et al. (2016). Thymoquinone inhibits inflammation, neoangiogenesis and vascular remodeling in asthma mice. *International Immunopharmacology* 38: 70–80.

Su, P., Yang, Y., Wang, G. et al. (2018). Curcumin attenuates resistance to irinotecan via induction of apoptosis of cancer stem cells in chemoresistant colon cancer cells. *International Journal of Oncology* 53 (3): 1343–1353.

Tamatani, M., Che, Y.H., Matsuzaki, H. et al. (1999). Tumor necrosis factor induces Bcl-2 and Bcl-x expression through NFkappaB activation in primary hippocampal neurons. *The Journal of Biological Chemistry* 274 (13): 8531–8538.

Tan, B.L. and Norhaizan, M.E. (2019). Curcumin combination chemotherapy: the implication and efficacy in cancer. *Molecules* 24 (14): 2527.

Tanaudommongkon, I., Tanaudommongkon, A., Prathipati, P. et al. (2020). Curcumin nanoparticles and their cytotoxicity in docetaxel-resistant castration-resistant prostate cancer cells. *Biomedicine* 8 (8): 253.

Tang, J., Luo, K., Li, Y. et al. (2015). Capsaicin attenuates LPS-induced inflammatory cytokine production by upregulation of LXRalpha. *International Immunopharmacology* 28 (1): 264–269.

Tian, Y., Gong, G.Y., Ma, L.L. et al. (2020). Anti-cancer effects of polyphyllin I: an update in 5 years. *Chemico-Biological Interactions* 316: 108936.

Tisdale, M.J. (2002). Cachexia in cancer patients. *Nature Reviews Cancer* 2 (11): 862–871.

Trinh, H.-T., Bae, E.A., Lee, J.J. et al. (2009). Inhibitory effects of curcuminoids on passive cutaneous anaphylaxis reaction and scratching behavior in mice. *Archives of Pharmacal Research* 32 (12): 1783–1787.

Tsai, M.L., Chiou, Y.S., Chiou, L.Y. et al. (2014). Garcinol suppresses inflammation-associated colon carcinogenesis in mice. *Molecular Nutrition & Food Research* 58 (9): 1820–1829.

Tsukuma, H., Hiyama, T., Tanaka, S. et al. (1993). Risk factors for hepatocellular carcinoma among patients with chronic liver disease. *New England Journal of Medicine* 328 (25): 1797–1801.

Tu, S.H., Chiou, Y.S., Kalyanam, N. et al. (2017). Garcinol sensitizes breast cancer cells to Taxol through the suppression of caspase-3/iPLA2 and NF-kappaB/Twist1 signaling pathways in a mouse 4T1 breast tumor model. *Food & Function* 8 (3): 1067–1079.

Ullah, M.F. (2015). Sulforaphane (SFN): an isothiocyanate in a cancer chemoprevention paradigm. *Medicines (Basel)* 2 (3): 141–156.

Um, Y., Cho, S., Woo, H.B. et al. (2008). Synthesis of curcumin mimics with multidrug resistance reversal activities. *Bioorganic & Medicinal Chemistry* 16 (7): 3608–3615.

Upadhyay, S.K. and Singh, S.P. (Éds.) (2021). *Bioprospecting of Plant Biodiversity for Industrial Molecules*. Wiley. https://doi.org/10.1002/9781119718017.

Upadhyay, S.K. and Singh, S.P. (2023). *Plants as Bioreactors for Industrial Molecules*. Wiley. https://doi.org/10.1002/9781119875116.

Vyas, A., Dandawate, P., Padhye, S. et al. (2013). Perspectives on new synthetic curcumin analogs and their potential anticancer properties. *Current Pharmaceutical Design* 19 (11): 2047–2069.

Wang, L., Ma, T., Zheng, Y. et al. (2015). Diosgenin inhibits IL-1beta-induced expression of inflammatory mediators in human osteoarthritis chondrocytes. *International Journal of Clinical and Experimental Pathology* 8 (5): 4830–4836.

Wang, H.C., Pao, J., Lin, S.Y. et al. (2012). Molecular mechanisms of garlic-derived allyl sulfides in the inhibition of skin cancer progression. *Annals of the New York Academy of Sciences* 1271: 44–52.

Wang, D., Qiao, J., Zhao, X. et al. (2015). Thymoquinone inhibits IL-1beta-induced inflammation in human osteoarthritis chondrocytes by suppressing NF-kappaB and MAPKs signaling pathway. *Inflammation* 38 (6): 2235–2241.

Wang, B.-L., Shen, Y.M., Zhang, Q.W. et al. (2013). Codelivery of curcumin and doxorubicin by MPEG-PCL results in improved efficacy of systemically administered chemotherapy in mice with lung cancer. *International Journal of Nanomedicine* 8 (1): 3521–3531.

Wang, R., Zhang, H., Wang, Y. et al. (2017). Inhibitory effects of quercetin on the progression of liver fibrosis through the regulation of NF-small ka, CyrillicB/Ismall ka, CyrillicBalpha, p38 MAPK, and Bcl-2/Bax signaling. *International Immunopharmacology* 47: 126–133.

Wang, S., Zhang, C., Yang, G. et al. (2014). Biological properties of 6-gingerol: a brief review. *Natural Product Communications* 9 (7): 1027–1030.

Wen, C., Fu, L., Huang, J. et al. (2019). Curcumin reverses doxorubicin resistance via inhibition the efflux function of ABCB4 in doxorubicin-resistant breast cancer cells. *Molecular Medicine Reports* 19 (6): 5162–5168.

William, B.M., Goodrich, A., Peng, C. et al. (2008). Curcumin inhibits proliferation and induces apoptosis of leukemic cells expressing wild-type or T315I-BCR-ABL and prolongs survival of mice with acute lymphoblastic leukemia. *Hematology* 13 (6): 333–343.

Wu, L.X., Xu, J.H., Wu, G.H. et al. (2003). Inhibitory effect of curcumin on proliferation of K562 cells involves down-regulation of p210(bcr/abl) initiated Ras signal transduction pathway. *Acta Pharmacologica Sinica* 24 (11): 1155–1160.

Xia, Y., Khoi, P.N., Yoon, H.J. et al. (2015). Piperine inhibits IL-1beta-induced IL-6 expression by suppressing p38 MAPK and STAT3 activation in gastric cancer cells. *Molecular and Cellular Biochemistry* 398 (1-2): 147–156.

Xia, L., Tan, S., Zhou, Y. et al. (2018). Role of the NFκB-signaling pathway in cancer. *OncoTargets and Therapy* 11: 2063–2073.

Xiao, B., Si, X., Han, M.K. et al. (2015). Co-delivery of camptothecin and curcumin by cationic polymeric nanoparticles for synergistic colon cancer combination chemotherapy. *Journal of Materials Chemistry B* 3 (39): 7724–7733.

Xiao, H., Xiao, Q., Zhang, K. et al. (2010). Reversal of multidrug resistance by curcumin through FA/BRCA pathway in multiple myeloma cell line MOLP-2/R. *Annals of Hematology* 89 (4): 399–404.

Xie, T.-X., Xia, Z., Zhang, N. et al. (2010). Constitutive NF-κB activity regulates the expression of VEGF and IL-8 and tumor angiogenesis of human glioblastoma. *Oncology Reports* 23 (3): 725–732.

Yadav, N. and Chandra, H. (2017). Suppression of inflammatory and infection responses in lung macrophages by eucalyptus oil and its constituent 1,8-cineole: role of pattern recognition receptors TREM-1 and NLRP3, the MAP kinase regulator MKP-1, and NFkappaB. *PLoS One* 12 (11): e0188232.

Yamamoto, K., Arakawa, T., Ueda, N. et al. (1995). Transcriptional roles of nuclear factor kappa B and nuclear factor-interleukin-6 in the tumor necrosis factor alpha-dependent induction of cyclooxygenase-2 in MC3T3-E1 cells. *The Journal of Biological Chemistry* 270 (52): 31315–31320.

Yang, C.W., Chang, C.L., Lee, H.C. et al. (2012). Curcumin induces the apoptosis of human monocytic leukemia THP-1 cells via the activation of JNK/ERK pathways. *BMC Complementary and Alternative Medicine* 12: 22.

Yaniv, S.P., Lucki, A., Klein, E. et al. (2010). Dexamethasone enhances the norepinephrine-induced ERK/MAPK intracellular pathway possibly via dysregulation of the alpha2-adrenergic receptor: implications for antidepressant drug mechanism of action. *European Journal of Cell Biology* 89 (9): 712–722.

Ye, M.-X., Zhao, Y.L., Li, Y. et al. (2012). Curcumin reverses cis-platin resistance and promotes human lung adenocarcinoma A549/DDP cell apoptosis through HIF-1α and caspase-3 mechanisms. *Phytomedicine* 19 (8-9): 779–787.

Yu, S.Y., Liao, C.H., Chien, M.H. et al. (2014). Induction of p21(Waf1/Cip1) by garcinol via downregulation of p38-MAPK signaling in p53-independent H1299 lung cancer. *Journal of Agricultural and Food Chemistry* 62 (9): 2085–2095.

Yu, H., Liu, Y., Niu, C. et al. (2018). Diosgenin increased DDX3 expression in hepatocellular carcinoma. *American Journal of Translational Research* 10 (11): 3590–3599.

Yusufi, M., Banerjee, S., Mohammad, M. et al. (2013). Synthesis, characterization and anti-tumor activity of novel thymoquinone analogs against pancreatic cancer. *Bioorganic & Medicinal Chemistry Letters* 23 (10): 3101–3104.

Zadorozhna, M., Tataranni, T., and Mangieri, D. (2019). Piperine: role in prevention and progression of cancer. *Molecular Biology Reports* 46 (5): 5617–5629.

Zelová, H. and Hošek, J. (2013). TNF-α signalling and inflammation: interactions between old acquaintances. *Inflammation Research* 62 (7): 641–651.

Zhang, K.-Z., Xu, J.H., Huang, X.W. et al. (2007). Curcumin synergistically augments bcr/abl phosphorothioate antisense oligonucleotides to inhibit growth of chronic myelogenous leukemia cells. *Acta Pharmacologica Sinica* 28 (1): 105–110.

Zhao, X., Chen, Q., Li, Y. et al. (2015). Doxorubicin and curcumin co-delivery by lipid nanoparticles for enhanced treatment of diethylnitrosamine-induced hepatocellular carcinoma in mice. *European Journal of Pharmaceutics and Biopharmaceutics* 93: 27–36.

Zhao, X., Chen, Q., Liu, W. et al. (2015). Codelivery of doxorubicin and curcumin with lipid nanoparticles results in improved efficacy of chemotherapy in liver cancer. *International Journal of Nanomedicine* 10: 257.

Zhao, C., Sun, J., Fang, C. et al. (2014). 1,8-cineol attenuates LPS-induced acute pulmonary inflammation in mice. *Inflammation* 37 (2): 566–572.

Zheng, J., Cheng, X., Xu, S. et al. (2019). Diallyl trisulfide induces G2/M cell-cycle arrest and apoptosis in anaplastic thyroid carcinoma 8505C cells. *Food & Function* 10 (11): 7253–7261.

Zhou, Q., Ye, M., Lu, Y. et al. (2015). Curcumin improves the tumoricidal effect of mitomycin C by suppressing ABCG2 expression in stem cell-like breast cancer cells. *PLoS One* 10 (8): e0136694.

Zhu, G.H., Dai, H.P., Shen, Q. et al. (2016). Curcumin induces apoptosis and suppresses invasion through MAPK and MMP signaling in human monocytic leukemia SHI-1 cells. *Pharmaceutical Biology* 54 (8): 1303–1311.

zur Hausen, H. (2002). Papillomaviruses and cancer: from basic studies to clinical application. *Nature Reviews Cancer* 2 (5): 342–350.

9

Nano-based Herbal Nutraceuticals Against Different Diseases

Priyanku Pradip Das[1], Mehak Thakur[2], Dheeraj Bisht[3], Rajeshwar Kamal Kant Arya[4], Shavkatjon Azizov[5], H. Lalhlenmawia[6], and Deepak Kumar[1]

[1] *Department of Pharmaceutical Chemistry, School of Pharmaceutical Sciences, Shoolini University, Solan, Himachal Pradesh, India*
[2] *School of Biological and Environmental Sciences, Shoolini University, Solan, Himachal Pradesh, India*
[3] *Devsthali Vidyapeeth College of Pharmacy, Rudrapur, Uttarakhand, India*
[4] *Department of Pharmaceutical Sciences, Sir J. C. Bose Technical Campus, Bhimtal, Kumaun University, Nainital, Uttarakhand, India*
[5] *Faculty of Life Sciences, Pharmaceutical Technical University, Tashkent, Uzbekistan*
[6] *Department of Pharmacy, Regional Institute of Paramedical and Nursing Sciences, Zemabawk, Aizawl, Mizoram, India*

9.1 Introduction

Nutraceuticals are dietary foods or components that offer health and medical advantages, such as preventing and treating illnesses. In 1989, Stephen DeFelice, MD, the founder and head of the Foundation for Innovation in Medicine, coined the phrase "nutraceutical" by merging "nutrition" with "pharmaceutical" (Kalra 2003; Puri et al. 2022). Nutraceuticals include various kinds of items, from single nutrients and supplements to special diets and genetically modified foods, as well as herbal products and processed foods like cereals and drinks (Puri et al. 2022; Singh et al. 2023). The World Health Organization (WHO) maintains a Global Health Expenditure database that shows that the United States and India have spent 17% and 4% of their GDP on healthcare, respectively (WHO Indicators). The significant healthcare expenditures have affected the research and development sectors of multinational companies and consumers, who now choose food products with health benefits. Nutraceuticals are categorized as antioxidants, dietary fibers, prebiotics, inorganic mineral supplements, phytochemicals, probiotics, and herbs as useful foods based on their chemical composition (Dureja et al. 2003; Hussain et al. 2015; Sachdeva et al. 2020).

Herbal nutraceuticals are plant-based foods or food parts that contain berries, oils, roots, lowers, or seeds that help maintain health and fight against short-term and long-term diseases caused by poor nutrition. Many old civilizations had high trust in the use of herbs for both nourishment and healing (Bommakanti et al. 2023; Priya and Satheesh Kumar 2020; Das et al. 2024). Some examples of herbal nutraceuticals are aloe vera, ginseng, garlic, ginger, turmeric, and green tea (Dixit et al. 2021; Mishra et al. 2022; Upadhyay and Singh 2021, 2023). Herbal nutraceuticals are globally developing in the sphere of health care services, focused on health promotion and illness reduction. They are recognized for their nutritional benefits, medicinal effects, and safety profiles. Herbal nutraceuticals have been employed not solely for nourishment but also as a supplementary form of therapy to prevent and treat diverse ailments, such as diminishing the adverse effects of

Herbal Nutraceuticals: Products and Processes, First Edition. Edited by Santosh Kumar Upadhyay and Sudhir Pratap Singh.

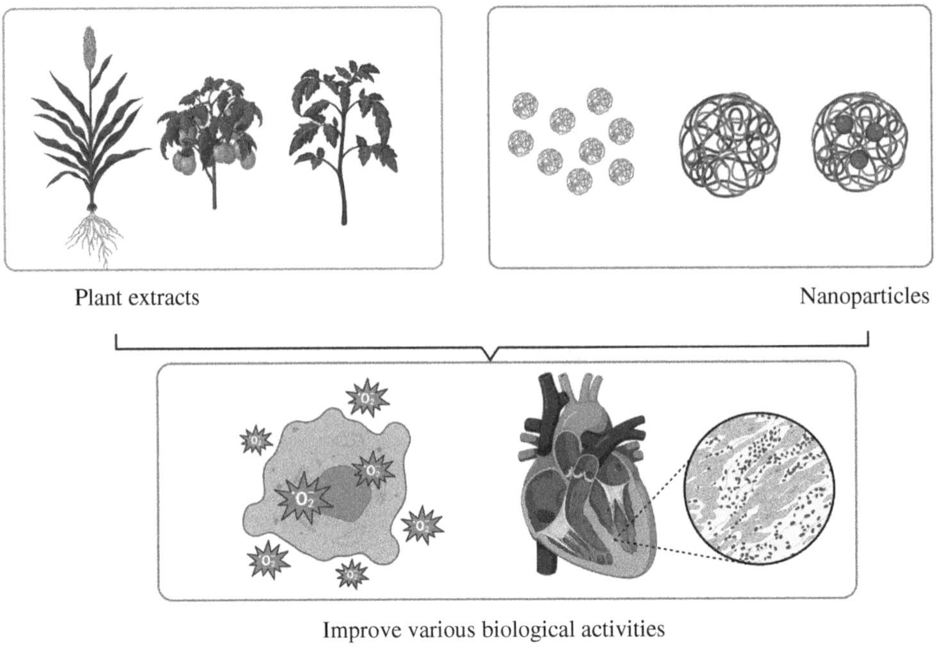

Figure 9.1 Effect of nanophytomedicine on various biological activities.

cancer chemotherapy and radiotherapy (Puri et al. 2022; Shukla et al. 2024). Herbal nutraceuticals encompass a considerable concentration of biologically active substances, sourced from nature, which yield physiological advantages and facilitate the prevention and management of illnesses. Common physiologically bioactive substances found in herbal nutraceuticals include angelicin, aloesin, cholesteric-7-en-3β-ol, and trans-anethole. These substances have displayed promising outcomes in various complications, such as hypercholesterolemia, diabetes, hypertension, obesity, and dementia, as well as serving as immunomodulators and antimicrobial agents (Bommakanti et al. 2023).

Nanotechnology is a term that describes the creation, manipulation, and use of materials or devices that are as small as a few nanometers (Singh et al. 2021). Nanotechnology can enhance the effectiveness and efficiency of nutritional medicine by improving the absorption and biokinetics of the nutrients and medicinal properties, as shown in Figure 9.1. This technology is one of the most innovative and fast-growing areas of R&D, with great potential to boost the health and well-being of people. A novel use of nanotechnology in nutritional medicine is to optimize the delivery of nutraceuticals, which are natural substances with health benefits. The challenges of absorption and solubility of the nutrients – the main issue of nutraceuticals that are excreted by the body without providing significant medical efficacy – can be overcome by the use of nanotechnology (Bertero et al. 2021). Hence, in this review, we aim to highlight the importance of incorporating nanotechnology with nutraceuticals to enhance their efficacy, medicinal benefits, ease of delivery, treatment strategies, and quality of life.

9.2 Plant-mediated Nanoparticles as Antioxidant Agents

The antioxidant properties of plant polyphenolic chemicals are mainly due to their unique structural characteristics (Parcheta et al. 2021). Polyphenols can avoid numerous diseases by neutralizing free radicals produced, thereby preventing DNA from oxidative injury (Zhang et al. 2022).

Chemical compounds like carboxylic acids, alkaloids, alcohol, flavonoids, phenols, esters, terpenoids, aldehydes, and proteins modify the nanoparticles through synthesis, which produces a variety of morphological features and variations in pharmacological activity that can be anticipated (Kandar 2021; Szczyglewska et al. 2023). Natural extracts include abundant biomolecules that have several applications, including their usage as antioxidant medications, dietary supplements, and intricate components for the biogenic production of metallic nanoparticles (Soni et al. 2021). The natural parts of the product may work as strong reducing and capping agents (Sidhu et al. 2022). Nanomaterials provide distinctive characteristics that offer novel opportunities in the domains of electronics, catalysis, materials chemistry, energy, sensors, agriculture, medicine, and biology. The dimensions, morphology, and arrangement of the particles determine the properties of nanomaterials, which depend on the methods used in their synthesis (Yadav et al. 2021).

Yen et al. (2009) utilized a nanosuspension approach to produce nanoparticles of *Cuscuta chinensis* Lam. (Convolvulaceae) for addressing oxidative stress. A comparison was made between the antioxidant activities of an ethanolic extract from *C. Chinensis* seeds and acetaminophen-induced acute liver injury in rats due to oxidative stress. Brewer et al. (2011) conducted research to assess the antioxidant properties of catechins, ascorbic acid, and glutathione. Catechins were shown to be the most efficient in scavenging radicals and showed effectiveness in protecting against oxidative stress. The ability of silver nanoparticles (AgNPs) and gold nanoparticles (AuNPs) to neutralize free radicals using 1,1-diphenyl-2-picrylhydrazyl (DPPH) was used to test their antioxidant and antibacterial properties. The results indicated that the AgNPs synthesized by biological means exhibited greater antioxidant activity compared to AuNPs and the extract from *Pimpinella anisum* (Zayed et al. 2020). According to Mushtaq et al. (2023), green zinc oxide nanoparticles (ZnO-NPs) in aqueous and methanolic extracts at 100 g/ml were able to remove 94% and 98% of DPPH free radicals, respectively. Like this, Singh et al. (2013) showed that the use of *Couroupita guianensis* entire flower bud extract facilitated the environmentally friendly production of silver-based nanoparticles while demonstrating the distinct antioxidant activity of metal-based nanoparticles (MNPs). Magnetic nanoparticles are known to be very good at acting as antioxidants in medicine. This is because organic antioxidants and MNPs work well together in the nanostructures (Adeyemi et al. 2022). A summary of further studies on the antioxidant characteristics of various metals, metal oxides, and bimetallic nanoparticles is included in Table 9.1. Detailed information about these plants is given in their botanical names, which are authenticated from the plant list's website (worldfloraonline.org).

9.3 Plant-mediated Nanoparticles as Anti-inflammatory Agents

Phytoconstituents have strong effects that stop and kill certain inflammatory cells. They do this by changing the way secretions work or by changing cytokines and receptors. Anti-inflammatory medications are widely used across the globe and are integral components of treatment plans for numerous diseases (Saleh et al. 2021). Several NSAIDs and SAIDs have been produced as conventional pharmaceuticals. The most prevalent NSAIDs are classified based on their chemical structures and sensitivity (Bindu et al. 2020). The compounds encompass acetylated salicylates (such as aspirin), propionic acids, non-acetylated salicylates (including indomethacin, ibuprofen, and diclofenac), enolic acids (like piroxicam and meloxicam), anthranilic acids (such as mefenamic acid and meclofenamate), and selective cyclo-oxygenase 2 inhibitors (such as celecoxib and etoricoxib). However, these medications have failed to attain optimal levels of effectiveness and safety, mostly owing to their severe adverse effects and compounded health complications (Anand et al. 2022).

Nanomedicine has emerged as an efficient and secure treatment for inflammation. It provides enhanced transport, targeted accumulation, and regulated release of medication inside the affected

Table 9.1 Some herbal-based nanoparticles and their biological activities.

Biological activity	Plant name/ compounds	Family	Types of NPs	Plant parts	References
Antimicrobial	Catechin		Copper oxide nanoparticles (CuO-NPs)	—	Li and Schluesener (2017)
	Lantana camara L.	Verbenaceae	Silver nanoparticles (AgNPs)	Leaves	Ajitha et al. (2015)
	Taraxacum officinale F.H.Wigg.	Asteraceae	AgNPs	Leaves	Sathishkumar et al. (2016)
	Cissus quadrangularis L.	Vitaceae	AgNPs	—	Ishwarya et al. (2017)
	Melia azedarach L.	Meliaceae	AgNPs	Leaves	Chinnasamy et al. (2019)
	Arisaema flavum (Forssk.) Schott	Araceae	AgNPs	Tuber	Rahman et al. (2019)
	Citrus clementina hort.	Rutaceae	AgNPs	Peel	Saratale et al. (2018a)
	Amomum villosum Lour.	Zingiberaceae	AuNPs	Fruit	Soshnikova et al. (2018)
Antioxidant	*Couroupita guianensis* Aubl.	Lecythidaceae	Gold nanoparticles (AuNPs)	Fruit	Senthilkumar et al. (2016)
	Curcuma longa L.	Zingiberaceae	AuNPs	Rhizomes	Al-Ani et al. (2019)
	M. azedarach L.	Meliaceae	AgNPs	Leaves	Chinnasamy et al. (2019)
	Perilla frutescens (L.) Britton	Lamiaceae	AgNPs	Leaves	Reddy et al. (2021)
	Embelia ribes **Burm.f.**	Primulaceae	AuNPs, AgNPs	Seeds	Dhayalan et al. (2016)
	Backhousia citriodora F.Muell.	Myrtaceae	AuNPs	Leaves	Khandanlou et al. (2018)
	Olea europaea L.	Oleaceae	zinc oxide nanoparticles (ZnO-NPs)	Leaves	Sulaiman et al. (2018)
Antidiabetic	*Punica granatum* L.	Lythraceae	AgNPs	Leaves	Saratale et al. (2018b)
	M. azedarach L.	Meliaceae	AgNPs	Leaves	Chinnasamy et al. (2019)
	Pisum sativum L.	Fabaceae	AgNPs	Leaves	Patra et al. (2019)
Anti-obesity	*Vitis vinifera* L.	Vitaceae	Nanocellulose	Seeds	Abdelbaky et al. (2016)
	Salacia chinensis L.	Celastraceae	AuNPs	—	Gao et al. (2020)
	Smilax glabra **Roxb.**	Smilacaceae	AuNPs	—	Ansari et al. (2019)
	Panax ginseng **C.A.Mey.**	Araliaceae	AuNPs	—	Simu et al. (2018)
	Attalea humilis Mart. ex Spreng.	Arecaceae	AuNPs	Leaves	Li et al. (2020)

Table 9.1 (Continued)

Biological activity	Plant name/ compounds	Family	Types of NPs	Plant parts	References
Anti-hypersensitive	*Dracocephalum moldavica* L.	Lamiaceae	TFDM-SLNs	—	Tan et al. (2017)
	Paeonia emodi Royle	Paeoniaceae	AuNPs	—	Ibrar et al. (2018)
	Curcumin		Nanoemulsion	—	Rachmawati et al. (2016)
	Curcumin		Polymer-based nanoparticle	—	Boarescu et al. (2019)
	Panax tridactylon Decne. & Planch.	Araliaceae	Nanoemulsion	—	Zhang et al. (2012)
	Syzygium cumini **(L.) Skeels**	Yrtaceae	Ag NPs	Seeds	Abdulrahman and Hama (2023)

cells and tissues, therefore reducing toxicity and adverse reactions (Xu et al. 2023). Despite this background, several nanomedicines have been created to address inflammatory-related conditions, including chronic wounds, microbial infections, cancer, cardiovascular and neurological illnesses, and metabolic syndrome (Brusini et al. 2020). The impact of the functional groups connected to medicinal plant MNPs and their extremely tiny size at the nanoscale suggest that these MNPs are physiologically active. An example of the beneficial impact of selenium nanoparticles (SeNPs) is that they can reduce rheumatism-related inflammation. SeNPs have the potential to reduce oxidative stress-related inflammation, which makes this effect significant. They achieve this by decreasing the levels of harmful radicals and nonradical substances that contribute to inflammation while also increasing the activity of the body's natural antioxidant defenses (Rehman et al. 2021). Other examples of its anti-inflammatory properties have been documented in Table 9.1.

Singhai et al. (2013) found that rutin-containing hydrogels have similar anti-inflammatory properties. The anti-inflammatory effects of hesperidin were assessed by liver damage, which was caused by CCl_4 and RAW2647 cells. The findings indicated that hesperidin had potent anti-inflammatory action by significantly decreasing the levels of interleukin (IL-6), tumor necrosis factor (TNF-α), and nitric oxide (NO) in both *in vivo* and *in vitro* environments. Similarly, Suresh and Abraham (2020) enhanced the solubility and absorption of triptolide (TP) by incorporating it into solid lipid nanoparticles (SLNs) composed of tristearin, glyceride, and stearic acid. TP is a refined substance derived from a specific Chinese herbal remedy obtained from the shrub-like vine, *Tripterygium wilfordii* Hook. F (Celastraceae). The findings demonstrated that TP-SLN enhanced the immediate anti-inflammatory efficacy due to the deeper penetration of TP into the skin.

9.4 Plant-mediated Nanoparticles as Antimicrobial Agents

Multidrug-resistant (MDR) pathogenic bacteria are becoming more prevalent and an increasing hazard to public health. The rising prevalence of microbial infections worldwide and the challenges in maintaining effective doses of tissue antibiotics while minimizing systemic medication exposure (Morris and Cerceo 2020; Kumar et al. 2015). Methicillin-resistant *Staphylococcus aureus* (MRSA) is a dominant MDR bacteria that has shifted from mostly being found in hospitals to causing illness outbreaks (Giulieri et al. 2020). Hence, there is a need for novel, efficient, less harmful, and cost-effective antimicrobial drugs (antibiotics). Nanobiotics (NBs), which include using nanotechnology to create antibiotics, are playing a crucial role in antimicrobial medication development (Adeyemi et al. 2022; Sharma et al. 2024).

Polyphenols have potent antibacterial properties against an extensive range of microbes. Polyphenolic compounds have been shown in several studies to enhance the effectiveness of drugs and possess outstanding antibacterial effects (Shamsudin et al. 2022). Liu et al. (2012) found that chitosan-film-loaded curcumin has potent antibiotic effects against *Rhizopus solani* and *S. aureus*. Rajendran et al. (2013) discovered that nanoparticles isolated from a methanol extract *of Ocimum sanctum* L. (Lamiaceae) exhibited enhanced antibacterial activity when encapsulated compared to the extract in its free-form condition. This was observed when testing against *Escherichia coli, Bacillussubtilis, Staphylococcus aureus, and Pseudomonas aeruginosa*. Table 9.1 summarizes other antimicrobial analyses of metals, metal oxide, and bimetallic nanoparticles.

9.5 Plant-mediated Nanoparticles as Antidiabetic Agents

Polyphenol-rich diets may lower diabetes risk. Some research suggests that polyphenols may modify the insulin cascade and boost peripheral tissue insulin sensitivity. (Shahwan et al. 2022). A variety of polyphenolic compounds inhibit the actions of alpha-amylase and alpha-glucosidase, two enzymes that control the absorption of glucose in the intestines and regulate blood sugar levels (Zhao et al. 2020). Ahangarpour et al. (2018) conducted a study to determine the effectiveness of SLNs containing myricitrin in the treatment of type II diabetes caused by streptozotocin-nicotinamide in mice and hyperglycemic myotubes. Diabetic circumstances and the effects of hyperglycemia were seen to improve in both laboratory research (*in vitro*) and investigations involving real beings (*in vivo*). The nanoencapsulation of myricitrin exhibited augmented antioxidant, antidiabetic, and antiapoptotic characteristics in both mouse and myotube cells. Berberine is a phytochemical that comes from the *Coptis chinensis* plant. It was found that SLNs that were enhanced with berberine had better oral bioavailability, stability, and antidiabetic activity (Wang et al. 2011). Additionally, in both *in vitro* and *in vivo* experiments, SLNs with berberine resulted in a liver drug concentration that was roughly 20 times higher than in the plasma and successfully reduced type II diabetes-induced hepatosteatosis (Xue et al. 2015).

Popli et al. (2018) documented the antidiabetic properties of AgNPs derived from a fungus called *Cladosporium*. Some enzymes, like α-amylase and α-glucosidase, worked less well after being treated with these nanoparticles. Additionally, there was an increase in insulin sensitivity and mimetic activity. Additional research has shown the ability of AgNPs derived from two different weed species, *Enhalus acoroids* and *Lonicera japonica*, to potentially treat diabetes. It was discovered that these AgNPs reduced the enzymatic activity of α-glucosidase and α-amylase. Furthermore, the AgNPs derived from *L. japonica* were seen to inhibit both enzymes in a noncompetitive manner. *Solanum nigrum* AgNPs have been shown to be very good at treating diabetes, lowering blood sugar levels to normal levels ($P < 0.001$) in rats that had been given alloxan (Sengottaiyan et al. 2016). Additional antidiabetic studies of herbal-based nanoparticles are summarized in Table 9.1.

9.6 Plant-mediated Nanoparticles as Antihypertensive Agents

Hypertension is responsible for a large proportion of deaths caused by cardiovascular diseases (CVDs) globally. Conventional medication and synthetic pharmaceuticals are less often used due to their high prices and accompanying cardiovascular problems (Campbell and Fleenor 2018, Jindal et al. 2021). Natural products have garnered a lot of interest since they delay the growth of CVDs and are cost-effective. Wang et al. (2017) discuss the detailed favorable benefits of tannins, flavonoids, anthocyanins, resveratrol, and other polyphenolics and their acidic substances on vasodilation and other hypertension-regulating processes. In the same way, Joshi et al. (2022) studied the impact of curcumin, amlodipine, and their combination on isolated rat aortic rings. They discovered that people with hypertension who were prescribed amlodipine can use curcumin or turmeric for medical or nutritional reasons without affecting the blood pressure-lowering effects of amlodipine.

Drug delivery devices are used to transfer medications that have low solubility or high toxicity to target locations (Ovidiu et al. 2008). Rapid advancements in nanotechnology have addressed all the issues related to CVD treatment approaches (Wong et al. 2013). Nanotechnology helps detect and treat cardiovascular illnesses early, reducing the strain on the healthcare system. In conclusion, nanoparticles play essential roles in CVDs through enhanced medical imaging, tailored medicine administration, and nanoparticle delivery to eliminate diseased cells (Mauricio et al. 2018). Nanoparticles can be effectively cleared from blood and tissue because they transport medications via them. Rapid breakthroughs in nanotechnology improve CVD imaging by designing competent cardiovascular molecular imaging probes based on nanotechnology in personalized cardiomedicine. Natural products in CVD treatment provide more focused benefits and fewer adverse effects than synthetic medications (Pala et al. 2020).

Fancher et al. (2019) have shown that numerous Food and Drug Administration (FDA)-approved medications from the various medicinal herbs loaded onto nanoparticles, such as felodipine, lercanidipine, and superoxide dismutase improvement, are new agents for treating resistant hypertension. Similarly, Tian et al. (2008) studied the hypersensitive activity of glycyrrhizic acid-loaded nanoparticles employing the rotary-evaporated film ultrasonication process. Atanasov et al. (2015) found that the nanoformation of polyurethane samples with ginger extract leads to hypersensitive activity. Atale et al. (2017) discovered that the synthesis of *Syzygium cumini* Ag NPs at the nanoscale improved hypertension. Other plant-mediated nanoparticles as agents for lowering blood pressure are mentioned in Table 9.1.

9.7 Plant-mediated Nanoparticles as Anti-obesity Agents

Obesity is a significant health issue worldwide. Weight gain can lead to several health problems, including CVD, type II diabetes, respiratory troubles, cancer, and musculoskeletal illnesses, including osteoarthritis disease. Novel anti-obesity treatments with long-term effectiveness and fewer side effects are required, as current allopathic medicines and drugs only provide partial treatment for obesity (Dixon 2010). Nanoencapsulation enhances the efficacy of secondary metabolites in combating obesity by improving their delivery to the target site and enhancing their functionality (Bhardwaj et al. 2021).

Nanotechnology and nanoencapsulation of secondary metabolites are promising approaches for treating obesity with increased effectiveness and fewer adverse effects. Numerous phytochemicals, including glycosides, polyphenols, terpenoids, flavonoids, and tannins, have been suggested as potentially effective treatments for obesity. Nevertheless, their lack of target selectivity, poor water solubility, stability issues, and severe toxicity at elevated dosages restrict their use in clinical settings (Murthy et al. 2018). These biomolecules mainly aid in the transformation of metal ions from their metallic state and the environmentally friendly creation of metallic nanoparticles. These problems can be fixed by using nanoencapsulated phytochemicals and metallic nanoparticles, which improve target selectivity and bioavailability and stop the breakdown process (El-Seedi et al. 2019).

Polyphenols can help decrease obesity by reducing adipocyte development, increasing adipocyte death, enhancing lipolysis, and promoting oxidation of fat. Green tea offers several health benefits due to its high levels of catechin, gallocatechin gallate, and other polyphenolic compounds such as catechins, which might aid in weight control (Almatroodi et al. 2020). Soybeans can potentially aid in combating obesity by reducing the activity of pancreatic lipase and protein lipase. This stops the growth of adipocytes and turns on hormone-sensitive lipase to break down fat (Bajes et al. 2020). Similarly, Ting et al. (2018) discovered that quercetin supplementation had a substantial inhibitory impact on the growth of fat mass in obese rats, suggesting possible anti-obesity properties.

Trandafir et al. (2022) found that *Dendropanax morbifera* AuNPs adversely affected the adipogenesis process by reducing the expression of PPARγ, STAT3, ap2, CEBPα, and Jak2 in 3T3-L1 cells, HepG2, and ACC levels in cells, as well as lowering FAS. *D. morbifera* AuNPs demonstrate antiadipogenic effects. Similarly, Kim et al. (2016) found that nanoencapsulating curcumin with a phospholipase and polyethylene glycol (PLA-PEG) molecule improved its anti-obesity effect by modifying serum lipid profile ($p \leq 0.01$) and modulating transcription factors ($p \leq 0.01$) compared to native curcumin molecules. Resveratrol and oxyresveratrol are chemical components of *Ramulus mori* twigs (Park et al. 2020), *Polygonum cuspidatum* root extract (Choi et al. 2020), and a nanocellulose molecule produced by *Vitis vinifera*.Trans-resveratrol-encapsulated NPs (Fu et al. 2015) controlled the signaling pathways, lipid levels, and transformation of white adipose tissue (WAT) to brown adipose tissue (BAT) related to obesity management. The substances, namely NPs and crude/extract, were used either singly or in combination for the purpose of managing obesity. Wan et al. (2018) used poly lactic-co-glycolic acid (PLGA) to encapsulate resveratrol and assessed the anti-obesity effectiveness of these nanoencapsulated structures using HepG2 cells. A substantial reduction was discovered. Table 9.1 provides a summary of further studies that have examined the anti-obesity capabilities of various metals, metal oxides, and bimetallic nanoparticles.

9.8 Conclusion and Future Direction

There is a growing interest in enhancing medicine delivery for treating disorders. Candidates like biogenic metal-based nanoparticles provide a unique possibility because of their enhanced biodistribution and pharmacokinetics (Devi et al. 2023). Metallic nanoparticles with nature-inspired characteristics are a novel class of advanced nanomedicines engineered to imitate the behavior of natural circulatory cells (Ji et al. 2023). These materials have demonstrated the ability to increase blood circulation duration and boost the delivery of medicine to cells and tissues (Souri et al. 2022). Nanotechnology's application in precision medical treatment, with less severe side effects, might lead to a positive change in clinical practice favoring life-saving therapies. Additionally, the gap between the laboratory where innovative materials are developed and the industrial facility where standardized preparation and manufacturing processes are executed must be bridged for these materials to serve as effective biological tools (Germaini et al. 2022). Nonetheless, despite the limitations, it is exciting because numerous new discoveries in nanotechnology are rapidly developing. Among them is the construction of several distinctive nanoparticles that communicate with specific illnesses.

References

Abdelbaky, M.S.M.E., Ibrahim, H.S., Hassan, M.L., and Sayed, Z.E. (2016). Nanoparticles effects of red grape (*Vitis vinifera*) seeds and grape seeds powder on obese hyperlipidemic rats. *ARC Journal of Nutrition and Growth* 2: 1–15.

Abdulrahman, M.D. and Hama, H.A. (2023). Anticancer of genus Syzygium: a systematic review. *Exploration of Targeted Anti-tumor Therapy* 4 (2): 273.

Adeyemi, J.O., Oriola, A.O., Onwudiwe, D.C., and Oyedeji, A.O. (2022 Apr). Plant extracts mediated metal-based nanoparticles: synthesis and biological applications. *Biomolecules.* 12 (5): 627.

Ahangarpour, A., Oroojan, A., Khorsandi, L. et al. (2018). Solid lipid nanoparticles of myricitrin have antioxidant and antidiabetic effects on streptozotocin-nicotinamide-induced diabetic model and myotube cell of male mouse. *Oxidative Medicine and Cellular Longevity* 2018: 1–18.

Ajitha, B., Ashok Kumar Reddy, Y., Shameer, S. et al. (2015). Lantana camara leaf extract mediated silver nanoparticles: antibacterial, green catalyst. *Journal of Photochemistry and Photobiology B Biology* 149: 84–92.

Al-Ani, L.A., Yehye, W.A., Kadir, F.A. et al. (2019). Hybrid nanocomposite curcumin-capped gold nanoparticle-reduced graphene oxide: anti-oxidant potency and selective cancer cytotoxicity. *PLoS ONE* 14: e0216725.

Almatroodi, S.A., Almatroudi, A., Khan, A.A. et al. (2020). Potential therapeutic targets of epigallocatechin gallate (EGCG), the most abundant catechin in green tea, and its role in the therapy of various types of cancer. *Molecules* 25 (14): 3146.

Anand U, Dey A, Chandel AK, Sanyal R, Mishra A, Pandey DK, De Falco V, Upadhyay A, Kandimalla R, Chaudhary A, Dhanjal JK. Cancer chemotherapy and beyond: current status, drug candidates, associated risks and progress in targeted therapeutics. *Genes & Diseases.* 2022; 10(4): 1367–1401Mar.

Ansari, S., Bari, A., Ullah, R. et al. (2019). Gold nanoparticles synthesized with *Smilax glabra* rhizome modulates the anti-obesity parameters in high-fat diet and streptozotocin induced obese diabetes rat model. *Journal of Photochemistry and Photobiology B: Biology* 201: 111643.

Atale, N., Saxena, S., Nirmala, J.G. et al. (2017). Synthesis and characterization of *Sygyzium cumini* nanoparticles for its protective potential in high glucose-induced cardiac stress: a green approach. *Applied Biochemistry and Biotechnology* 181 (3): 1140–1154.

Atanasov, A.G., Waltenberger, B., Pferschy-Wenzig, E.-M. et al. (2015). Discovery and resupply of pharmacologically active plant-derived natural products: a review. *Biotechnology Advances* 33 (8): 1582–1614.

Bajes, H.R., Almasri, I., and Bustanji, Y. (2020). Plant products and their inhibitory activity against pancreatic lipase. *Revista Brasileira de Farmacognosia.* 30: 321–330.

Bertero, A., Fossati, P., Coccini, T. et al. (2021). Application of "nano" nutraceuticals in medicine. In: *Nutraceuticals*, 263–270. Academic Press.

Bhardwaj, M., Yadav, P., Vashishth, D. et al. (2021). A review on obesity management through natural compounds and a green nanomedicine-based approach. *Molecules.* 26 (11): 3278.

Bindu, S., Mazumder, S., and Bandyopadhyay, U. (2020). Non-steroidal anti-inflammatory drugs (NSAIDs) and organ damage: a current perspective. *Biochemical Pharmacology.* 1 (180): 114147.

Boarescu, P.-M., Boarescu, I., Bocșan, I.C. et al. (2019). Curcumin nanoparticles protect against isoproterenol induced myocardial infarction by alleviating myocardial tissue oxidative stress, electrocardiogram, and biological changes. *Molecules.* 24 (15): 2802.

Bommakanti, V., Puthenparambil Ajikumar, A., Sivi, C.M. et al. (2023). An overview of herbal nutraceuticals, their extraction, formulation, therapeutic effects and potential toxicity. *Separations.* 10 (3): 177.

Brewer, E., Coleman, J., and Lowman, A. (2011). Emerging technologies of polymeric nanoparticles in cancer drug delivery. *Journal of Nanomaterials* 2011 (1): 1–10.

Brusini, R., Varna, M., and Couvreur, P. (2020). Advanced nanomedicines for the treatment of inflammatory diseases. *Advanced Drug Delivery Reviews.* 1 (157): 161–178.

Campbell, M.S. and Fleenor, B.S. (2018). The emerging role of curcumin for improving vascular dysfunction: a review. *Critical Reviews in Food Science and Nutrition* 58 (16): 2790–2799. https://doi.org/10.1080/10408398.2017.1341865.

Chinnasamy, G., Chandrasekharan, S., and Bhatnagar, S. (2019). Biosynthesis of silver nanoparticles from *Melia azedarach*: enhancement of antibacterial, wound healing, antidiabetic and antioxidant activities. *International Journal of Nanomedicine* 14: 9823–9836.

Choi, D.H., Han, J.H., Yu, K.H. et al. (2020). Antioxidant and anti-obesity activities of *Polygonum cuspidatum* extract through alleviation of lipid accumulation on 3T3-L1 adipocytes. *Journal of Microbiology and Biotechnology* 30: 21–30.

Das, A.K., Gandhi, H., Panwar, V. et al. (2024). Natural product-loaded nanoparticles for wound healing. In: *Nanotechnological Aspects for Next-Generation Wound Management*, 319–331. Academic Press.

Devi, L., Kushwaha, P., Ansari, T.M. et al. (2023). Recent trends in biologically synthesized metal nanoparticles and their biomedical applications: a review. *Biological Trace Element Research.* 25: 1–7.

Dhayalan, M., Denison, M.I.J., Anitha Jegadeeshwari, L. et al. (2016). In vitro antioxidant, antimicrobial, cytotoxic potential of gold and silver nanoparticles prepared using *Embelia Ribes. Natural Product Research* 31: 465–468.

Dixit, S., Shukla, A., Singh, V., and Upadhyay, S.K. (2021a). *Bioprospecting of Natural Compounds for Industrial and Medical Applications.* Wiley. 10.1002/9781119718017.ch3.

Dixon, J.B. (2010). The effect of obesity on health outcomes. *Molecular and Ccellular Eendocrinology.* 316 (2): 104–108.

Dureja, H., Kaushik, D., and Kumar, V. (2003). Developments in nutraceuticals. *Indian Journal of Pharmacology.* 35 (6): 363–372.

El-Seedi, H.R., El-Shabasy, R.M., Khalifa, S.A. et al. (2019). Metal nanoparticles fabricated by green chemistry using natural extracts: biosynthesis, mechanisms, and applications. *RSC Aadvances.* 9 (42): 24539–24559.

Fancher, M.E., Fournier, S., Townsend, J., and Lallier, T.E. (2019 Jun). Cytotoxic effects of silver diamine fluoride. *American Journal of Dentistry.* 32 (3): 152–156.

Fu, J., Li, Z., Zhang, H. et al. (2015). Molecular pathways regulating the formation of brown-like adipocytes in white adipose tissue. *Diabetes/Metabolism Research and Reviews* 31 (5): 433–452.

Gao, L., Hu, Y., Hu, D. et al. (2020). Anti-obesity activity of gold nanoparticles synthesized from *Salacia chinensis* modulates the biochemical alterations in high-fat diet-induced obese rat model via AMPK signaling pathway. *Arabian Journal of Chemistry* 13: 6589–6597.

Germaini, M.M., Belhabib, S., Guessasma, S. et al. (2022). Additive manufacturing of biomaterials for bone tissue engineering–a critical review of the state of the art and new concepts. *Progress in Materials Science.* 1 (130): 100963.

Giulieri, S.G., Tong, S.Y., and Williamson, D.A. (2020). Using genomics to understand meticillin-and vancomycin-resistant *Staphylococcus aureus* infections. *Microbial Genomics.* 6 (1).

Hussain, S.A., Panjagari, N.R., Singh, R.R., and Patil, G.R. (2015 Jan). Potential herbs and herbal nutraceuticals: food applications and their interactions with food components. *Critical Reviews in Food Science and Nutrition.* 55 (1): 94–122.

Ibrar, M., Khan, M.A., and Imran, M. (2018). Evaluation of *Paeonia emodi* and its gold nanoparticles for cardioprotective and antihyperlipidemic potentials. *Journal of Photochemistry and Photobiology B: Biology* 189: 5–13.

Ishwarya, R., Vaseeharan, B., Shanthi, S. et al. (2017). Green synthesized silver nanoparticles: toxicity against *Poecilia Reticulata* fishes and ceriodaphnia cornuta crustaceans. *Journal of Cluster Science* 28: 519–527.

Ji, W., Zhang, Y., Deng, Y. et al. (2023). Nature-inspired nanocarriers for improving drug therapy of atherosclerosis. *Regenerative Biomaterials.* 1 (10): rbad069.

Jindal, D.K., Sah, P., Bisht, D. et al. (2021). *Role of Medicinal Plants in Pulmonary Hypertension. Medicinal Plants for Lung Diseases: A Pharmacological and Immunological Perspective*, 303–316.

Joshi, P., Joshi, S., Semwal, D.K. et al. (2022 Oct). Role of curcumin in ameliorating hypertension and associated conditions: a mechanistic insight. *Molecular and Cellular Biochemistry.* 477 (10): 2359–2385.

Kalra, E.K. (2003). Nutraceutical-definition and introduction. *AAPS PharmSciTech* 5 (3): 25.

Kandar, C.C. (2021). Secondary metabolites from plant sources. *Bioactive Natural Products for Pharmaceutical Applications.* 329–377.

Khandanlou, R., Murthy, V., Saranath, D., and Damani, H. (2018). Synthesis and characterization of gold-conjugated *Backhousia citriodora* nanoparticles and their anticancer activity against MCF-7 breast and HepG2 liver cancer cell lines. *Journal of Materials Science* 53: 3106–3118.

Kim, J.H., Kim, O.K., Yoon, H.G. et al. (2016). Anti-obesity effect of extract from fermented *Curcuma longa* L. through regulation of adipogenesis and lipolysis pathway in high-fat diet-induced obese rats. *Food & Nutrition Research* 60: 30428.

Kumar, D., Poornima, M., Kushwaha, R.N. et al. (2015). Antimicrobial and docking studies of (−)-catechin derivatives. *Journal of the Korean Society for Applied Biological Chemistry* 58: 581–585.

Li, C. and Schluesener, H. (2017). Health-promoting effects of the citrus flavanone hesperidin. *Critical Reviews in Food Science and Nutrition* 57: 613–631.

Li, H., Chen, Q., Zhao, J., and Urmila, K. (2015). Enhancing the antimicrobial activity of natural extraction using the synthetic ultrasmall metal nanoparticles. *Scientific Reports* 5: 11033.

Li, W., Wan, H., Yan, S. et al. (2020). Gold nanoparticles synthesized with *Poria cocos* modulates the anti-obesity parameters in high-fat diet and streptozotocin induced obese diabetes rat model. *Arabian Journal of Chemistry* 13: 5966–5977.

Liu, Q., Liu, H., Yuan, Z. et al. (2012). Evaluation of antioxidant activity of chrysanthemum extracts and tea beverages by gold nanoparticles-based assay. *Colloids and Surfaces B: Biointerfaces* 92: 348–352.

Mauricio, M.D., Guerra-Ojeda, S., Marchio, P. et al. (2018). Nanoparticles in medicine: a focus on vascular oxidative stress. *Oxidative Medicine and Cellular Longevity* 2018: 6231482.

Mishra, A.P., Srivastav, N., Singh, A. et al. (2022). The role of nutraceuticals as food and medicine, types and sources. In: *Food and Agricultural Byproducts as Important Source of Valuable Nutraceuticals*, vol. 12, 1–18. Cham: Springer International Publishing.

Morris, S. and Cerceo, E. (2020). Trends, epidemiology, and management of multi-drug resistant gram-negative bacterial infections in the hospitalized setting. *Antibiotics.* 9 (4): 196.

Murthy, K.C., Monika, P., Jayaprakasha, G.K., and Patil, B.S. (2018). Nanoencapsulation: an advanced nanotechnological approach to enhance the biological efficacy of curcumin. In: *Advances in Plant Phenolics: From Chemistry to Human Health*, 383–405. American Chemical Society.

Mushtaq, W., Ishtiaq, M., Maqbool, M. et al. (2023 May 27). Green synthesis of zinc oxide nanoparticles using *Viscum album* extracts: unveiling bioactive compounds, antibacterial potential, and antioxidant activities. *Plants.* 12 (11): 2130.

Ovidiu B, Felicia T, Burta O, et al. Phytotherapy in cardiovasculardiseases: from ethnomedicine to evidence based medicine. *Journal of Biological Sciences*J BiolSci.2008;8 (2): 242–247.

Pala, R., Anju, V.T., Dyavaiah, M. et al. (2020 May). Nanoparticle-mediated drug delivery for the treatment of cardiovascular diseases. *International Journal of Nanomedicine.* 27: 3741–3769.

Parcheta, M., Świsłocka, R., Orzechowska, S. et al. (2021). Recent developments in effective antioxidants: the structure and antioxidant properties. *Materials.* 14 (8): 1984.

Park, Y.H., An, M., Kim, J.K., and Lim, Y.H. (2020). Antiobesity effect of ethanolic extract of Ramulus mori in differentiated 3T3-L1 adipocytes and high-fat diet-induced obese mice. *Journal of Ethnopharmacology* 251: 112542.

Patra, J.K., Das, G., and Shin, H.S. (2019). Facile green biosynthesis of silver nanoparticles using *Pisum sativum* L. outer peel aqueous extract and its antidiabetic, cytotoxicity, antioxidant, and antibacterial activity. *International Journal of Nanomedicine* 14: 6679–6690.

Popli, D., Anil, V., Subramanyam, A.B. et al. (2018). Endophyte fungi, *Cladosporium* species-mediated synthesis of silver nanoparticles possessing in vitro antioxidant, anti-diabetic and anti-Alzheimer activity. *Artificial Cells, Nanomedicine, and Biotechnology.* 46 (suppl 1): 676–683.

Priya, S. and Satheeshkumar, P.K. (2020 Jan). Natural products from plants: recent developments in phytochemicals, phytopharmaceuticals, and plant-based neutraceuticals as anticancer agents. *Functional and Preservative Properties of Phytochemicals.* 1: 145–163.

Puri, V., Nagpal, M., Singh, I. et al. (2022). A comprehensive review on nutraceuticals: therapy support and formulation challenges. *Nutrients.* 14 (21): 4637.

Rachmawati, H., Soraya, I.S., Kurniati, N.F., and Rahma, A. (2016). In vitro study on antihypertensive and antihypercholesterolemic effects of a curcumin nanoemulsion. *Scientia Pharmaceutica* 84 (1): 131–140.

Rahman, A.U., Khan, A.U., Yuan, Q. et al. (2019). Tuber extract of *Arisaema flavum* eco-benignly and effectively synthesize silver nanoparticles: photocatalytic and antibacterial response against multidrug resistant engineered *E. Coli* QH4. *Journal of Photochemistry and Photobiology B: Biology* 193: 31–38.

Rajendran, R., Radhai, R., Kotresh, T.M., and Csiszar, E. (2013). Development of antimicrobial cotton fabrics using herb loaded nanoparticles. *Carbohydrate Polymers* 91 (2): 613–617.

Reddy, N.V., Li, H., Hou, T. et al. (2021). Phytosynthesis of silver nanoparticles using *Perilla frutescens* leaf extract: characterization and evaluation of antibacterial, antioxidant, and anticancer activities. *International Journal of Nanomedicine* 16: 15–29.

Rehman, A., John, P., and Bhatti, A. (2021). Biogenic selenium nanoparticles: potential solution to oxidative stress mediated inflammation in rheumatoid arthritis and associated complications. *Nanomaterials.* 11 (8): 2005.

Sachdeva, V., Roy, A., and Bharadvaja, N. (2020). Current prospects of nutraceuticals: a review. *Current Pharmaceutical Biotechnology.* 21 (10): 884–896.

Saleh, H.A., Yousef, M.H., and Abdelnaser, A. (2021 Mar). The anti-inflammatory properties of phytochemicals and their effects on epigenetic mechanisms involved in TLR4/NF-κB-mediated inflammation. *Frontiers in Immunology.* 26 (12): 606069.

Saratale, R.G., Benelli, G., Kumar, G. et al. (2018a). Bio-fabrication of silver nanoparticles using the leaf extract of an ancient herbal medicine, dandelion (*Taraxacum oOfficinale*), evaluation of their antioxidant, anticancer potential, and antimicrobial activity against phytopathogens. *Environmental Science and Pollution Research* 25: 10392–10406.

Saratale, R.G., Shin, H.S., Kumar, G. et al. (2018b). Exploiting fruit byproducts for eco-friendly nanosynthesis: citrus × clementina peel extract mediated fabrication of silver nanoparticles with high efficacy against microbial pathogens and rat glial tumor C6 cells. *Environmental Science and Pollution Research* 25: 10250–10263.

Sathish Kumar, G., Jha, P.K., Vignesh, V. et al. (2016). Cannonball Fruit (*Couroupita guianensis*, Aubl.) Extract mediated synthesis of gold nanoparticles and evaluation of its antioxidant activity. *Journal of Molecular Liquids* 215: 229–236.

Sengottaiyan, A., Aravinthan, A., Sudhakar, C. et al. (2016). Synthesis and characterization of *Solanum nigrum*-mediated silver nanoparticles and its protective effect on alloxan-induced diabetic rats. *Journal of Nanostructure in Chemistry.* 6: 41–48.

Senthilkumar, P., Santhosh Kumar, D.R., Sudhagar, B. et al. (2016). Seagrass-mediated silver nanoparticles synthesis by Enhalus acoroides and its α-glucosidase inhibitory activity from the Gulf of Mannar. *Journal of Nanostructure in Chemistry.* 6: 275–280.

Shahwan, M., Alhumaydhi, F., Ashraf, G.M. et al. (2022 May). Role of polyphenols in combating Type 2 diabetes and insulin resistance. *International Journal of Biological Macromolecules.* 1 (206): 567–579.

Shamsudin NF, Ahmed QU, Mahmood S, Ali Shah SA, Khatib A, Mukhtar S, Alsharif MA, Parveen H, Zakaria ZA. Antibacterial effects of flavonoids and their structure-activity relationship study: a comparative interpretation. *Molecules.* 2022 9;27 (4): 1149.

Sharma, A., Akash, K., Kumari, S. et al. (2024 Apr 10). Biogenic zinc oxide nanoparticles: an insight into the advancements in antimicrobial resistance. *ECS Journal of Solid State Science and Technology.* 13 (4): 047002.

Shukla, M.K., Sharma, A., Kumar, R. et al. (2024). Anticancer potential of Myrica esculenta plant: a comprehensive review. *Handbook of Oncobiology: From Basic to Clinical Sciences* 15: 1289–1301.

Sidhu, A.K., Verma, N., and Kaushal, P. (2022 Jan). Role of biogenic capping agents in the synthesis of metallic nanoparticles and evaluation of their therapeutic potential. *Frontiers in Nanotechnology.* 31 (3): 801620.

Simu, S.Y., Ahn, S., Castro-Aceituno, V. et al. (2018). Gold nanoparticles synthesized with fresh panax ginseng leaf extract suppress adipogenesis by downregulating PPAR/CEBP signaling in 3T3-L1 mature adipocytes. *Journal of Nanoscience and Nanotechnology* 18: 1–8.

Singh, R., Tiwari, S., and Tawaniya, J. (2013). Review on nanotechnology with several aspects. *International Journal of Computer Engineering Electron* 2 (3): 1–8.

Singh, A.R., Desu, P.K., Nakkala, R.K. et al. (2021 Mar). Nanotechnology-based approaches applied to nutraceuticals. *Drug Delivery and Translational Research.* 18: 1–5.

Singh, M., Kumar, R., Sharma, S. et al. (2023). Hedychium spicatum: a comprehensive insight into its ethnobotany, phytochemistry, pharmacological and therapeutic attributes. *South African Journal of Botany* 161: 638–647.

Singhai, A.K., Malik, J., and Sont, H. (2013). Antimicrobial and antiinflammatory activity of the hydrogels containing rutin delivery. *Asian Journal of Chemistry* 25: 8371–8373.

Soni, V., Raizada, P., Singh, P. et al. (2021 Nov). Sustainable and green trends in using plant extracts for the synthesis of biogenic metal nanoparticles toward environmental and pharmaceutical advances: a review. *Environmental Research.* 1 (202): 111622.

Soshnikova, V., Kim, Y.J., Singh, P. et al. (2018). Cardamom fruits as a green resource for facile synthesis of gold and silver nanoparticles and their biological applications. *Artificial Cells, Nanomedicine, and Biotechnology* 46: 108–117.

Souri, M., Soltani, M., Kashkooli, F.M., and Shahvandi, M.K. (2022 Jan). Engineered strategies to enhance tumor penetration of drug-loaded nanoparticles. *Journal of Controlled Release.* 1 (341): 227–246.

Sulaiman, G.M., Tawfeeq, A.T., and Jaaffer, M.D. (2018). Biogenic synthesis of copper oxide nanoparticles using *Olea europaea* leaf extract and evaluation of their toxicity activities: an in vivo and in vitro study. *Biotechnology Progress* 34: 218–230.

Suresh, A. and Abraham, J. (2020). Phytochemicals and their role in pharmaceuticals. *Advances in Pharmaceutical Biotechnology: Recent Progress and Future Applications.* 193–218.

Szczyglewska, P., Feliczak-Guzik, A., and Nowak, I. (2023). Nanotechnology–general aspects: a chemical reduction approach to the synthesis of nanoparticles. *Molecules.* 28 (13): 4932.

Tan, M., He, C., Jiang, W. et al. (2017). Development of solid lipid nanoparticles containing total flavonoid extract from *Dracocephalum moldavica* L. and their therapeutic effect against myocardial ischemia–reperfusion injury in rats. *International Journal of Nanomedicine* 12: 3253.

Tian, M., Yan, H., and Row, K.H. (2008). Extraction of glycyrrhizic acid and glabridin from licorice. *International Journal of Molecular Sciences* 9: 571–577. https://doi.org/10.3390/ ijms9040571.

Ting, Y., Chang, W.T., Shiau, D.K. et al. (2018). Antiobesity efficacy of quercetin-rich supplement on diet-induced obese rats: effects on body composition, serum lipid profile, and gene expression. *Journal of Agricultural and Food Chemistry* 66 (1): 70–80.

Trandafir, L.M., Dodi, G., Frasinariu, O. et al. (2022). Tackling dyslipidemia in obesity from a nanotechnology perspective. *Nutrients* 14: 3774.

Upadhyay, S.K. and Singh, S.P. (2021). *Bioprospecting of Plant Biodiversity for Industrial Molecules.* Wiley. https://doi.org/10.1002/9781119718017.

Upadhyay, S.K. and Singh, S.P. (2023). *Plants as Bioreactors for Industrial Molecules.* Wiley. https://doi.org/10.1002/9781119875116.

Wan, S., Zhang, L., Quan, Y., and Wei, K. (2018). Resveratrol-loaded PLGA nanoparticles: enhanced stability, solubility and bioactivity of resveratrol for non-alcoholic fatty liver disease therapy. *Royal Society Open Science* 5: 181457.

Wang, T., Wang, N., Song, H. et al. (2011). Preparation of an anhydrous reverse micelle delivery system to enhance oral bioavailability and anti-diabetic-diabetic efficacy of berberine. *European Journal of Pharmaceutical Sciences* 44: 127–135.

Wang, J., Tan, J., Luo, J. et al. (2017). Enhancement of scutellarin oral delivery efficacy by vitamin B12-modified amphiphilic chitosan derivatives to treat type II diabetes induced-retinopathy. *Journal of Nanobiotechnology* 15: 18.

Wong, I.Y., Bhatia, S.N., and Toner, M. (2013). Nanotechnology: emerging tools for biology and medicine. *Genes & Development* 27 (22): 2397–2408.

Xu, M., Han, X., Xiong, H. et al. (2023 Jun 30). Cancer nanomedicine: emerging strategies and therapeutic potentials. *Molecules.* 28 (13): 5145.

Xue, M., Zhang, L., Yang, M. et al. (2015). Berberine-loaded solid lipid nanoparticles are concentrated in the liver and ameliorate hepatosteatosis in db/db mice. *International Journal of Nanomedicine* 10: 5049–5057.

Yadav, V.K., Malik, P., Khan, A.H. et al. (2021). Recent advances on properties and utility of nanomaterials generated from industrial and biological activities. *Crystals.* 11 (6): 634.

Yen, F.Y., Wu, T.H., Lin, L.T. et al. (2009). Naringenin-loaded nanoparticles improve the physicochemical properties and the hepatoprotective effects of naringenin in orally-administered rats with CCl4-induced acute liver failure. *Pharmaceutical Research* 26 (4): 893–902.

Zayed, M.F., Mahfoze, R.A., El-Kousy, S.M., and Al-Ashkar, E.A. (2020 Jan). In-vitro antioxidant and antimicrobial activities of metal nanoparticles biosynthesized using optimized *Pimpinella anisum* extract. *Colloids and Surfaces A: Physicochemical and Engineering Aspects.* 20 (585): 124167.

Zhang. J., Han, X., Li, X. et al. (2012). Core-shell hybrid liposomal vesicles loaded with panax notoginsenoside: preparation, characterization and protective effects on global cerebral ischemia/reperfusion injury and acute myocardial ischemia in rats. *International Journal of Nanomedicine* 7: 4299. https://doi.org/10.2147/IJN.S32385.

Zhang, Z., Li, X., Sang, S. et al. (2022). Polyphenols as plant-based nutraceuticals: health effects, encapsulation, nano-delivery, and application. *Foods* 11 (15): 2189.

Zhao, C., Wan, X., Zhou, S., and Cao, H. (2020). Natural polyphenols: a potential therapeutic approach to hypoglycemia. *EFood.* 1 (2): 107–118.

10

The Flavorful World: Exploring the Applications of Spices in Nutraceuticals

Anuradha and Navneeta Bharadvaja

Plant Biotechnology Laboratory, Department of Biotechnology, Delhi Technological University, New Delhi, India

10.1 Introduction

For centuries, individuals have utilized spices to enhance the taste of their food and elevate the culinary experience. These ingredients provide health benefits when consumed, in addition to giving the food its true color and flavor. The history of the world has been greatly influenced by spices. Traveling to Arabia to purchase valuable incense, the Egyptians were pioneers in the maritime trade in antiquity. Navigators from Greece and Rome headed for India in search of pepper and ginger, and Columbus, drawn by the lure of spices, headed west. Vasco de Gama set out in parallel to the east, navigating treacherous waters in search of the same fragrant riches. Driven by a common desire to find spices, Magellan also set out on a Pacific Ocean expedition. These examples demonstrate the ways in which the search for spices has impacted not only global exploration but also the mingling of cultures across great distances, leaving a lasting impression on the chronicles of our times (Kannappan et al. 2011). Indigenous medical systems have employed their pharmacological properties, which include tonic, carminative, stomachic, diuretic, and antispasmodic effects, which are largely empirical but effective. Due to their diverse physiological effects, different bioactive constituents like *curcumin*, *capsaicin*, *allicin, gingerol, saponin,* and fiber from fenugreek have been shown to have significant health benefits (Shylaja and Peter 2007). This chapter emphasizes the role of bioactive compounds derived from spices in managing and preventing major health conditions, as well as some potential applications for them. The goal is to investigate these compounds' potential for use in the nutraceutical sector.

Nutraceuticals have gained popularity because of their alleged safety, potential nutritional benefits, and medicinal properties. A new category that blends the distinction between food and medicine is called nutraceuticals. By definition, a nutraceutical is defined as a food or food ingredient that provides both health benefits and contribution to nutritional content (Santini et al. 2017). Nutraceuticals are mainly made up of nutrients, herbal extracts, and dietary supplements. They are essential for maintaining health and treating a variety of diseases associated with an unhealthy lifestyle, which in turn improves the quality of life in general. Nutraceutical use as a preventive measure is a proactive and preemptive approach to preclinical health problems. In

Herbal Nutraceuticals: Products and Processes, First Edition. Edited by Santosh Kumar Upadhyay and Sudhir Pratap Singh.

general, the idea of "nutraceuticals" has brought about a new era in healthcare and medicine and turned the food industry into a field that is heavily focused on research (Das et al. 2012).

Plants are essential providers of both food and medicine for humans. A revolutionary shift in food, health, and agriculture has been sparked by the growing understanding of nutrition, medicine, and plant biotechnology. Different plant parts, including flowers, bark, roots, fruits, and buds, are used to extract spices. A wide range of medical fields, such as cosmetics, pharmaceuticals, and fragrance production, find their use in them (Dixit et al. 2021; Kumar 1997; Upadhyay and Singh 2021, 2023; Verma et al. 2009). Due in part to the presence of secondary metabolites, or phytochemicals, which are recognized for a wide range of biological activities, some plant-based diets have been shown to reduce the risk of chronic diseases. Bioactive compounds found in nutraceutical or functional foods, especially those made from herbs and spices, have been linked to the prevention and treatment of a number of illnesses, such as cancer, heart disease, diabetes, obesity, osteoporosis, and allergies (Bansal et al. 2023).

Some long-term physiological effects may be observed from the regular and substantial ingestion of these metabolites as part of the diet, even though they generally show lower potency as bioactive compounds when compared to pharmaceutical drugs.

10.2 Chemistry of Compounds Specific to Spices

A vast variety of plant parts are classified as spices, such as buds, leaves, bark, roots, berries, seeds, and even stigma, as shown in Table 10.1. Spices consist of proteins, fiber, carbohydrates, minerals, essential oils, pigments, and bioactive compounds such as terpenes, flavonoids, phenolic compounds, and sterols, as shown in Table 10.2.

Table 10.1 Summary of spices derived from different parts of plants.

Part of plant	Spices	Botanical name	Bioactive constituent
Rhizome	*Chilli*	*Capsicum frutescens*	Phenols, vitamin C
	Greater galanga	*Alpinia galanga*	1,8-cineol, α-fenchyl acetate, β-pinene
	Sweet flag	*Acorus calamus*	Asarones
Seed	*Fenugreek*	*Trigonella foenum-graecum*	Graecunins, fenugreek B
	Pomegranate seed	*Punica granatum*	Ellagic acid, punicalagin
	Poppyseed	*Papaver somniferum*	Codeine, paperverine
Leaf	*Basil*	*Ocimum basilicum*	Linalool, eugenol
	Sage	*Salvia officinalis*	1,8-cineole, β-pinene
	Rosemary	*Rosmarinus officinalis*	Rosmanol, ursolic acid
Fruit	*Tamarind*	*Tamarindus indica*	Safrole, geraniol
	Star anise	*Illicium verum*	Flavonoid anetheole
	Cumin	*Cuminum cyminum*	Cuminaldehyde
Bark	*Cassia*	*Cinnamomum aromaticum*	Cinnamaldehyde
	Cinnamon	*Cinnamom umzeylanicum*	Eugenol, cinnamaldehyde
Bud	*Clove*	*Syzgium aromaticum*	Eugenol
Resin	*Asafoetida*	*Ferula asofoetida*	Umbelliferone, ferulic acid

Table 10.2 Summary of different bioactive constituents of spices.

Bioactive constituents	Sources	Botanical name	References
Gambogic acid	Gamboge	*Garcinia* sp.	Hatami et al. (2020)
Curcumin	Turmeric	*Curcuma longa*	Soleimani et al. (2018)
Thymoquinone	Blackseed	*Nigella sativa*	Gholamnezhad et al. (2016)
Myristicin, apiole, α-pinene	Parsley	*Petroselinum crispum*	Craft and Setzer (2017)
Cineole, as an abortifacient, laxative, sengupta, and (small) amomum linn fruit carminative in asthma and bronchitis, Sengupta and Bhattacharjee (2009) (large) Amomum limonene, hemorrhoids, strangury, scabies, subulatum roxba-terpinyl pruritis, diseases of the bladder, acetate	Cardamom	*Elettaria cardamomum*	Morsy (2015)
Cineole, as an abortifacient, laxative, sengupta, and (small) amomum linn Fruit is carminative in asthma, bronchitis, Bhattacharjee (2009) (large) Amomum limonene, hemorrhoids, strangury, scabies, subulatum roxba-terpinyl pruritis; diseases of the bladder, acetate, cineole, Limonene, α-terpinyl acetate			
Carvacrol, used for digestive problems including Keshavarz et al. carvone, heartburn, bloating, gas, loss of (2014) a-pinene, appetite, and mild spasms of the limonene, stomach, and intestines a-t erpinene, linaloo, linalool, carvacrol, carvone, limonene	Caraway	*Carum carvi*	Sachan et al. (2016)
Zegxanthin, safranal	Saffron	*Crocus sativus*	Negbi (1999)
Menthol, menthofuran	Mint	*Mentha piperita*	Shah and Mello (2004)
1,8-cineol, α-fenchyl acetate, β-pinene	Greater galanga	*A. galanga*	Chouni and Paul (2018)

Monoterpenes and phenolic compounds are important ingredients in many spices. Interestingly, the active ingredients in many plant-based medicinal extracts are flavonoids.

10.2.1 Phenolic Compounds

Primary antioxidants include phenolic compounds, which mainly function as free radical scavengers. Their function entails hindering the initiation or propagation stages of lipid oxidation, thereby slowing down or stopping its onset. As a result, less volatile breakdown products, such as ketones and aldehydes, are produced, which prevents rancidity from developing (Shahidi and Ambigaipalan 2015).

Additionally, phenolic acids including *caffeic, chlorogenic, and ferulic* are present in dried herbs like *oregano, thyme, fennel, and lovage*. However, dried spices such as *star anise* and *cinnamon* act as storage for chlorogenic, protocatechuic, and *p*-coumaric acids. While flavonoid content in dried herbs and spices varies, it is important to note that certain dried spices, such as cinnamon and ginger, have relatively low flavonoid content.

10.2.2 Terpenes

Terpenoids are classified into groups of natural products according to their biosynthesis and structure. The 5-carbon precursor compound dimethylallyl diphosphate and its isomer isopentenyl diphosphate are the sources of all terpenoids. These vary from single-unit hemiterpenes (5C) to polyterpenes exceeding C40, which are made up of 5×103 to 4 units of C5. They also include mono- (C10), sesqui- (C15), di- (C20), sester- (C25), tri- (C30), and tetra- (C40) (Tetali 2019).

Terpenes and their compounds, like the oils from clove leaves and eucalyptus, have therapeutic uses, especially in dentistry. Eucalyptus oil functions as an expectorant, encouraging the secretion of mucus. Similar to how eucalyptus oil works, terpenes also have diuretic effects and help reduce spasms in the gastrointestinal tract (Kandi et al. 2015). Numerous aromas, such as fruity, floral, minty, citrusy, fresh, herbaceous, camphoraceous, woody, piney, and spicy notes, are linked to monoterpenes such as limonene, myrcene, and pinene. The flavor attributes and general acceptability of food products are greatly influenced by these aromatic profiles.

10.2.3 Alkaloids

Alkaloids are mainly produced by biosynthesis from amino acids, and as a result, they have a wide variety of chemical structures that are mainly found in plants. Alkaloids are nitrogen-containing molecules with small molecular weights that usually have alkaline characteristics because they have a nitrogen-containing heterocyclic ring. Their wide range of pharmacological effects on vertebrates is well known. Alkaloids comprise approximately 20% of the secondary metabolites found in plants.

The strong alkaloid *capsaicin,* which is present in *red pepper* (*Capsicum annuum*), has garnered significant attention in scientific studies due to its pharmacological importance. Research has examined a wide range of its biological effects, such as its potential cardioprotective qualities, capacity to stop urinary tract stones from forming, analgesic and anti-inflammatory effects, thermogenic qualities, and effects on the gastrointestinal tract (Srinivasan 2013).

10.3 Role of Spices as Nutraceuticals

The term "spice" is used by the American Spice Trade Association to refer to a broad range of products—herbs, spices, and aromatic seeds. Numerous phytochemicals, including antioxidant and anti-inflammatory properties, can be found in abundance in both herbs and spices. These botanicals are important for supporting general well-being because they may also contain vitamins, minerals, antioxidants, and essential oils (Dini 2018). Owing to the synergistic and cumulative effects of these compounds, it is thought that the wide range of phytochemicals present in fruits, vegetables, herbs, and spices contribute significantly to their health effects, as illustrated in Figure 10.1.

Diabetes, heart disease, arthritis, and cancer are currently the most important health issues facing humanity. Because of their ability to treat these health issues, including diabetes, hypercholesterolemia, cancer, digestive stimulation, anti-lithogenic, anti-inflammatory, antimutagenic, and anti-inflammatory qualities, spices are of utmost importance (Rabade et al. 2022), as shown in Table 10.3. Essential oils derived from *fennel (Foeniculum vulgare Mill.), rosemary (Rosmarinus officinalis L.), ginger (Zingiber officinale Ros.), garlic (Allium sativum L.), peppermint (Mentha × piperita L.), bitter orange (Citrus aurantium L.), cardamom (Elettaria cardamomum (L.) Maton), and dill (Anethum*

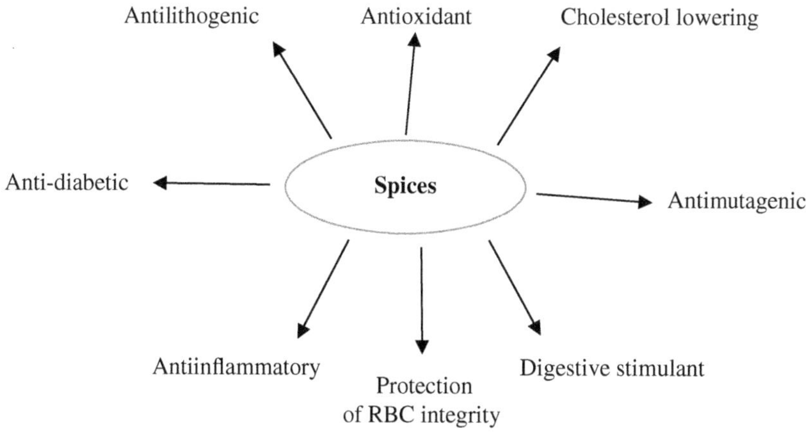

Figure 10.1 Summary of different biological roles of spices.

Table 10.3 Nutraceutical potential of spices.

Spices	Bioactive compounds	Biological role	References
Laurus nobilis	Linalool, α-terpineol acetate, sabinene	Promotes digestive health Manages anxiety and stress	Dönmez (n.d.)
Piper nigrum	Piperine	Acts as analgesic Antihypertensive	Ghosh et al. (2020)
Elletaria cardamomum	Limonene, β-selinene, linayl acetate	Possess anti-alzheimer activity Anticancer	Kumar and Kumari (2021)
Cinnamomum	E-nerolidol, cinnaldehyde	Help lower cancer risk Fights infections and viruses	Błaszczyk et al. (2021)
Coriandrum sativum	Citronelol	Calms intestinal spasms Treats smallpox	Mandal and Mandal (2015)
Syzygium aromaticum	Citronelol	Boosts digestive functioning Useful in gastric irritability	Singletary (2014)
C. cyminum	Cuminaldehyde pyrazines	Prevents anemia Decreases cognitive disorders	Mnif and Aifa (2015)
Murrayakoenigii	Glycosides, carbazole alkaloids, koenigin, girinimbin	Lowers cholesterol levels Good for hair growth	Bhandari (2012)
Foeniculum vulgare	Alpha pinene, beta myrcene, beta pinene, anethole	Boosts immunity Regulates menstruation	Rafieian et al. (2023)
T. foenum-graecum	Saponins, 4-hydroxyisoleucine, trigonelline	Helps prevent colon cancer Helps soothe skin inflammation	Nathiya et al. (2014)
Allium sativum	Allicin	Aids digestion Treats acne	Santhosha et al. (2013)
Zingiber officinale	Gingerol	Treats diarrhea	Butt and Sultan (2011)

(*Continued*)

Table 10.3 (Continued)

Spices	Bioactive compounds	Biological role	References
Mentha	Menthol, Menthone	Counteracts painful periods Cures morning sickness and iron deficiency anemia	Brahmi et al. (2017)
Brassica juncea	Sinigrin, Myrosin, Sinalbin	Controls diabetes Cholesterol lowering ability	Tian and Deng (2020)
Myristica fragrans	Eugenol, Myristicin	Improves memory Insomnia	Naeem et al. (2016)
Allium cepa	Quercetin	Relieves earache Helps in glowing skin	Sagar et al. (2022)
C. longa	Curcumin	Helps to treat arthritis Prevents Alzheimer's disease	Chattopadhyay et al. (2004)

graveolens L.) have proven to be effective in stopping the development of gastric ulcers, lowering the production of gastric acid, and promoting healing within the upper gastrointestinal tract (Tanveer et al. 2020).

10.3.1 Digestion

Spices have long been recognized for their ability to stimulate digestion. Indian and traditional medical systems also use a variety of spices in their medicinal formulations to treat digestive disorders (Platel and Srinivasan 2004). Spices such as *Z. officinale, Piper nigrum, and F. vulgare* are commonly used in herbal remedies or added to prescription medications to treat digestive issues. Many widely used spices, such as *C. annuum, Trachyspermum ammi, Brassica juncea,* and *Cuminum cyminum,* have been demonstrated to aid the liver's bile acid synthesis. In order to aid in digestion and break down fats, bile acids are essential (Adefegha and Oboh 2013).

10.3.2 Antidiabetic

Continuing research for novel plant-derived medications with antidiabetic properties has resulted from the proven significance of bioactive constituents of spices. These have been shown to be beneficial in improving glucose tolerance and glycemic status in both type 2 diabetic people and animals (Srinivasan 2005). Additionally, by inhibiting the activity of digestive enzymes like α-amylase and disaccharidases within the gastrointestinal tract, spices have an additional antidiabetic effect (Cazzola and Cestaro 2014). Scientific research has demonstrated the antidiabetic potential of spices such as *A. sativum, Curcuma longa, Allium cepa*, and *Trigonella foenum-graecum*. In addition, a small number of studies suggest that coriander *Coriandrum sativum, Brassica nigra, Murraya koenigii, Cuminum cyminum, Z. officinale*, and *B nigra* have also been reported to have hypoglycemic effects.

10.3.3 Anti-inflammatory

Numerous diseases that are common throughout the world, such as asthma, atherosclerosis, and rheumatoid arthritis, are significantly impacted by inflammation. Spices have been long employed as a traditional medicine to treat rheumatic disorders such as *C. longa* exhibiting dual effects as an anti-inflammatory and anti-atherosclerotic agent. The main active ingredient in Zingiber *zerumbet* has been proven to be an effective anti-inflammatory agent (Krishnaswamy 2008). The compounds extracted from fenugreek seeds have been shown to exhibit anti-inflammatory and antioxidant qualities, supporting the traditional health claims made about them.

10.3.4 Antioxidant Property

Oxidative processes are slowed down or mitigated by antioxidants, which are substances that block them. Plants are a natural source of antioxidants, an increasingly fascinating subject (Srinivasan 2014). The antioxidant potential of spices and vegetables can be applied to protect lipids and reduce lipid peroxidation in biological systems. Numerous spices possess antioxidant qualities, such as *Salvia rosmarinus, Syzygium aromaticum, A. sativum,* and *Myristica fragrans*. The ability of phenolic compounds to quench singlet oxygen, chelate transition metals, scavenge free radicals, and take advantage of their redox properties is connected to their antioxidant activity (Widowati 2015).

10.3.5 Anti-lithogenic Effect

Bile components that either encourage or impede the crystallization process control the development of cholesterol gallstones in the gallbladder (Vidyashankar et al. 2010). *C. longa* has long been recognized for its beneficial effects on cholesterol levels. Research has demonstrated that bioactive components like *capsaicin and curcumin* have hypotriglyceridemic effects. Moreover, *fenugreek, onions, ginger, capsaicin, and curcumin* encourage the secretion of bile acids into bile. By activating hepatic cholesterol-7α-hydroxylase, these spices or their active components with hypocholesterolemic properties lower blood and liver cholesterol levels by stimulating the conversion of cholesterol into bile acids (Srinivasan 2013).

10.3.6 Anticancer Effect

Free radical damage can be avoided by utilizing the antioxidant properties of the bioactive components found in spices (Bhagat and Chaturvedi 2016). Bioactive compounds found in *A. sativum* have the capacity to specifically inhibit tumor growth by means of chromosomal stability, cell cycle progression regulation, and regulation of DNA repair processes (Shukla and Kalra 2007). Similarly, numerous studies have confirmed the anticancer properties of common spices like *A. sativum, Zingeber officinale, C. annuum, Crocus sativus,* etc., which are used to flavor food all over the world (Gao et al. 2021).

10.4 Conclusion

Sustaining optimal health is essential for feeling well overall and protects against prolonged conditions such as cancer, high blood pressure, and inflammation. The phytochemicals present in a variety of spices have been shown to have a favorable relationship with the prevention of chronic illnesses and to be highly effective in managing conditions like cancer, heart disease, diabetes, inflammation, microbial infections, and cholesterol-related problems. An excellent source of antioxidants are spices and herbs that have been enhanced with phenolic compounds, like clove, rosemary, sage, oregano, and cinnamon. Further stimulating the digestive system and increasing appetite are the effects of spices. One can enhance flavor and vibrancy in life by including a small proportion of spices in their daily diet. It is assumed that a large portion of the health benefits linked to Indian and Mediterranean diets stem from the cumulative and synergistic effects resulting from the complex combination of phytochemicals found in fruits, vegetables, herbs, and spices. Certain geriatric, prolonged disorders, like cardiovascular problems and some types of cancer, are frequently associated with lower incidence rates when following these diets. It is well established that a liberal dose of spices is both safe and beneficial. Including a lot of spices in your diet can improve your overall health in addition to adding flavor, as each spice has several health-promoting qualities and can work synergistically when consumed together.

References

Adefegha, S.A. and Oboh, G. (2013). Phytochemistry and mode of action of some tropical spice in the management of type-2 diabetes and hypertension. *African Journal of Pharmacy and Pharmacology* 7 (7): 332–346.

Bansal, P., Maithani, M., Gupta, V. et al. (2023). Future prospective of nutraceutical and functional food with herbs and spices. In: *Herbs, Spices and their Roles in Nutraceuticals and Functional Foods* (ed. A. Amalraj, S. Kuttappan, A.C. Karthak Verma, and A. Matharu), 361–381. Academic Press.

Bhagat, N. and Chaturvedi, A. (2016). Spices as an alternative therapy for cancer treatment. *Systematic Reviews in Pharmacy* 7 (1): 46–56.

Bhandari, P.R. (2012). Curry leaf (*Murraya koenigii*) or cure leaf: review of its curative properties. *Journal of Medical Nutrition and Nutraceuticals* 1 (2): 92.

Błaszczyk, N., Rosiak, A., and Kałużna-Czaplińska, J. (2021). The potential role of cinnamon in human health. *Forests* 12 (5): 648.

Brahmi, F., Khodir, M., Mohamed, C., and Pierre, D. (2017). Chemical composition and biological activities of Mentha species. *Aromatic and Medicinal Plants-Back to Nature* 15: 47–79.

Butt, M.S. and Sultan, M.T. (2011). Ginger and its health claims: molecular aspects. *Critical Reviews in Food Science and Nutrition* 51 (5): 383–393.

Cazzola, R. and Cestaro, B. (2014). Antioxidant spices and herbs used in diabetes. In: *Diabetes: Oxidative Stress and Dietary Antioxidants* (ed. V.R. Preedy), 89–97. Academic Press.

Chattopadhyay, I., Biswas, K., Bandyopadhyay, U., and Banerjee, R.K. (2004). Turmeric and curcumin: biological actions and medicinal applications. *Current Science* 10: 44–53.

Chouni, A. and Paul, S. (2018). A review on phytochemical and pharmacological potential of *Alpinia galanga*. *Pharmacognosy Journal* 10 (1).

Craft, J.D. and Setzer, W.N. (2017). The volatile components of parsley, *Petroselinum crispum* (Mill.) Fuss. *American Journal of Essential Oils and Natural Products* 5: 27–32.

Das, L., Bhaumik, E., Raychaudhuri, U., and Chakraborty, R. (2012). Role of nutraceuticals in human health. *Journal of Food Science and Technology* 49: 173–183.

Dini, I. (2018). Spices and Herbs as Therapeutic Foods. In: *Food Quality: Balancing Health and Disease* (ed. A.M. Grumezescu and A.M. Holban), 433–469. Academic Press.

Dixit, S., Shukla, A., Singh, V., and Upadhyay, S.K. (2021). Bioprospecting of natural compounds for industrial and medical applications: current scenario and bottleneck. In: *Bioprospecting of Plant Biodiversity for Industrial Molecules*. Wiley; Portico. https://doi.org/10.1002/9781119718017.ch3.

Dönmez, C. (2023). *Laurus nobilis* L. In: *Medicinal Plants of Turkey* (ed. U.K. Caliskan), 128–138. CRC Press.

Gao, J., Mamouni, K., Zhang, L., and Lokeshwar, B.L. (2021). Spice up your food for cancer prevention: cancer chemo-prevention by natural compounds from common dietary spices. In: *Evolutionary Diversity as a Source for Anticancer Molecules* (ed. A.K. Srivastava, V.K. Kannaujiya, and D. Singh), 275–308. Academic Press.

Gholamnezhad, Z., Havakhah, S., and Boskabady, M.H. (2016). Preclinical and clinical effects of Nigella sativa and its constituent, thymoquinone: a review. *Journal of Ethnopharmacology* 190: 372–386.

Ghosh, S., Kumar, A., Sachan, N., and Chandra, P. (2020). Re-exploring an epicentre spice with immense therapeutic potentials: black pepper (*Piper nigrum*). *Current Nutrition & Food Science* 16 (9): 1326–1337.

Hatami, E., Jaggi, M., Chauhan, S.C., and Yallapu, M.M. (2020). Gambogic acid: a shining natural compound to nanomedicine for cancer therapeutics. *Biochimica et Biophysica Acta (BBA)-Reviews on Cancer* 1874 (1): 188381.

Kandi, S., Godishala, V., Rao, P., and Ramana, K.V. (2015). Biomedical significance of terpenes: an insight. *Biomedicine* 3 (1): 8–10.

Kannappan, R., Gupta, S.C., Kim, J.H. et al. (2011). Neuroprotection by spice-derived nutraceuticals: you are what you eat! *Molecular Neurobiology* 44: 142–159. https://doi.org/10.1007/s12035-011-8168-2.

Krishnaswamy, K. (2008). Traditional indian spices and their health significance. *Asia Pacific Journal of Clinical Nutrition* 17 (S1): 265–268.

Kumar, N. (1997). *Introduction to Spices, Plantation Crops, Medicinal and Aromatic Plants*. Oxford and IBH Publishing.

Kumar, S. and Kumari, R. (2021). Traditional, phytochemical and biological activities of *Elettaria cardamomum* (L.) maton–a review. *International Journal of Pharmaceutical Sciences and Research* 12: 2320–5148.

Mandal, S. and Mandal, M. (2015). Coriander (*Coriandrum sativum* L.) essential oil: chemistry and biological activity. *Asian Pacific Journal of Tropical Biomedicine* 5 (6): 421–428.

Mnif, S. and Aifa, S. (2015). Cumin (*Cuminum cyminum* L.) from traditional uses to potential biomedical applications. *Chemistry & Biodiversity* 12 (5): 733–742.

Morsy, N.F. (2015). A short extraction time of high quality hydrodistilled cardamom (*Elettaria cardamomum* L. Maton) essential oil using ultrasound as a pretreatment. *Industrial Crops and Products* 65: 287–292.

Naeem, N., Rehman, R., Mushtaq, A., and Ghania, J.B. (2016). Nutmeg: a review on uses and biological properties. *International Journal of Pharmaceutical Sciences and Research* 9: 107–110.

Nathiya, S., Durga, M., and Devasena, T. (2014). Therapeutic role of *Trigonella foenum-graecum* [fenugreek]–a review. *International Journal of Pharmaceutical Sciences and Research* 27 (2): 74–80.

Negbi, M. (ed.) (1999). *Saffron: Crocus sativus L*. CRC Press.

Platel, K. and Srinivasan, K. (2004). Digestive stimulant action of spices: a myth or reality. *Indian Journal of Medical Research* 119 (5): 167.

Rabade, V., Ingole, A., Gangane, P. et al. (2022). Tasty spices as nutraceuticals. In: *Herbs and Spices* (ed. E. Ivanišová). Intechopen.

Rafieian, F., Amani, R., Rezaei, A. et al. (2023). Exploring fennel (*Foeniculum vulgare*): composition, functional properties, potential health benefits, and safety. *Critical Reviews in Food Science and Nutrition* 2: 1–8.

Sachan, A.K., Das, D.R., and Kumar, M. (2016). *Carum carvi*-an important medicinal plant. *Journal of Chemical and Pharmaceutical Research* 8 (3): 529–533.

Sagar, N.A., Pareek, S., Benkeblia, N., and Xiao, J. (2022). Onion (*Allium cepa* L.) bioactives: chemistry, pharmacotherapeutic functions, and industrial applications. *Food Frontiers* 3 (3): 380–412.

Santhosha, S.G., Jamuna, P., and Prabhavathi, S.N. (2013). Bioactive components of garlic and their physiological role in health maintenance: a review. *Food Bioscience* 1 (3): 59–74.

Santini, A., Tenore, G.C., and Novellino, E. (2017). Nutraceuticals: a paradigm of proactive medicine. *European Journal of Pharmaceutical Sciences* 96: 53–61.

Sengupta, A. and Bhattacharjee, S. (2009). Cardamom (Elettaria cardamomum) and its active constituent, I, 8-cineole. In: *Molecular Targets and Therapeutic Uses of Spices: Modern Uses for Ancient Medicine*, 65–85.

Shah, P.P. and Mello, P.M. (2004). A review of medicinal uses and pharmacological effects of *Mentha piperita*. *Medicine* 4: 214–221.

Shahidi, F. and Ambigaipalan, P. (2015). Phenolics and polyphenolics in foods, beverages and spices: antioxidant activity and health effects–a review. *Journal of Functional Foods* 18: 820–897.

Shukla, Y. and Kalra, N. (2007). Cancer chemoprevention with garlic and its constituents. *Cancer Letters* 247 (2): 167–181.

Shylaja, M.R. and Peter, K.V. (2007). Spices in the nutraceutical and health food industry. *International Symposium on Medicinal and Nutraceutical Plants* 756: 369–378.

Singletary, K. (2014). Clove: overview of potential health benefits. *Nutrition Today* 49 (4): 207–224.

Soleimani, V., Sahebkar, A., and Hosseinzadeh, H. (2018). Turmeric (*Curcuma longa*) and its major constituent (curcumin) as nontoxic and safe substances. *Phytotherapy Research* 32 (6): 985–995.

Srinivasan, K. (2005). Plant foods in the management of diabetes mellitus: spices as beneficial antidiabetic food adjuncts. *International journal of Food Sciences and Nutrition* 56 (6): 399–414.

Srinivasan, K. (2013). Biological activities of pepper alkaloids 45. In: *Natural Products* (ed. K.G. Ramawat and J.M. Mer'illon). Springer-Verlag Berlin Heidelberg. https://doi.org/10.1007/978-3-642-22144-6_184.

Srinivasan, K. (2013). Dietary spices as beneficial modulators of lipid profile in conditions of metabolic disorders and diseases. *Food & Function* 4 (4): 503–521.

Srinivasan, K. (2014). Antioxidant potential of spices and their active constituents. *Critical Reviews in Food Science and Nutrition* 54 (3): 352–372.

Tanveer, M., Wagner, C., Ul Haq, M.I. et al. (2020). Spicing up gastrointestinal health with dietary essential oils. *Phytochemistry Reviews* 19: 243–263.

Tetali, S.D. (2019). Terpenes and isoprenoids: a wealth of compounds for global use. *Planta* 249: 1–8.

Tian, Y. and Deng, F. (2020). Phytochemistry and biological activity of mustard (*Brassica juncea*): a review. *CyTA-Journal of Food* 18 (1): 704–718.

Upadhyay, S.K. and Singh, S.P. (Éds.) (2021). *Bioprospecting of Plant Biodiversity for Industrial Molecules*. Wiley. https://doi.org/10.1002/9781119718017.

Upadhyay, S.K. and Singh, S.P. (Éds.) (2023). *Plants as Bioreactors for Industrial Molecules*. Wiley. https://doi.org/10.1002/9781119875116.

Verma, P., Trivedi, I., Singh, H. et al. (2009). Efficient production of gossypol from hairy root cultures of cotton (Gossypium hirsutum L.). *Current Pharmaceutical Biotechnology* 10 (7): 691–700. https://doi.org/10.2174/138920109789542048.

Vidyashankar, S., Sambaiah, K., and Srinivasan, K. (2010). Effect of dietary garlic and onion on biliary proteins and lipid peroxidation which influence cholesterol nucleation in bile. *Steroids* 75 (3): 272–281.

Widowati, W. (2015). Antioxidant properties of spice extracts. *Biomedical Engineering* 1 (1): 6.

11

Carotenoids: Their Sources, Bioactivity, and Application in Industry

Irwandi Jaswir[1,2], Dedi Noviendri[3], Mohd Aznan Md Aris[4], Soraya Ismail[4], Widya Lestari[5], Noraihan Mat Harun[4], and Ridar Hendri[6]

[1] *Department of Chemistry, Padang State University, Padang, Indonesia*
[2] *International Institute for Halal Research and Training, International Islamic University Malaysia, Kuala Lumpur, Malaysia*
[3] *Bioprocess and Molecular Engineering Research Unit, Kulliyyah of Engineering, International Islamic University Malaysia, Kuala Lumpur, Malaysia*
[4] *Faculty of Medicine, International Islamic University Malaysia, Kuantan Campus, Kuantan, Pahang, Malaysia*
[5] *Faculty of Dentistry, International Islamic University Malaysia, Kuantan Campus, Kuantan, Pahang, Malaysia*
[6] *Faculty of Fisheries, Riau University, Pekanbaru, Indonesia*

11.1 Introduction

Carotenoids are ubiquitous natural pigments found in confer red, orange, and yellow colors to various organisms in the plant and animal kingdoms (Basu et al. 2001; Sugawara et al. 2009). Over 700 naturally occurring pigments found in higher plants and animals are classified as carotenoid compounds (Godinho and Bhosle 2008; Konishi et al. 2008). Plants, algae, fungi, and bacteria biosynthesize carotenoid pigments de novo, while animals are unable to biosynthesize carotenoids, instead obtain them from their diet (Okada et al. 2008; Rock 1997). The carotenoids give color to different parts of the plants, such as flowers, leaves, and fruits (Mattea et al. 2009), and in animals, such as the muscles and feathers of certain animals (Mortensen 2006). In plants, microorganisms, and animals, carotenoids accumulate in the cells where they not only impart color, but also exert important bioactivity (Fujisawa et al. 2008).

Carotenoids are made up of approximately 40 carbon atoms of $8C_5$ isoprene units connected from head to tail (Bonnie and Choo 1999; Mattea et al. 2009). Biosynthesis of carotenoid begins with the addition of C_5 isoprenoid units to form a C_{20} long chain, which undergoes dimerization to form a C_{40} carotenoid. This basic structure then undergoes various modifications, such as cyclization, dehydrogenation, rearrangement, oxidation, and epoxidation, giving rise to various other carotenoid chemical structures (Figure 11.1) (Yano et al. 2005). The two major groups of carotenoids are carotenes and xanthophylls, divided according to their chemical structure (Perez-Rodrıguez 2009). The carotenes such as α-carotene, β-carotene, and lycopene are devoid of any oxygen molecules and consist solely of hydrocarbons, while xanthophylls such as lutein, fucoxanthin, astaxanthin, and zeaxanthin contain oxygenated molecules such as hydroxyl, methoxy, carboxyl, keto, and epoxy groups (Aizawa and Inakuma 2007; Britton 1995;

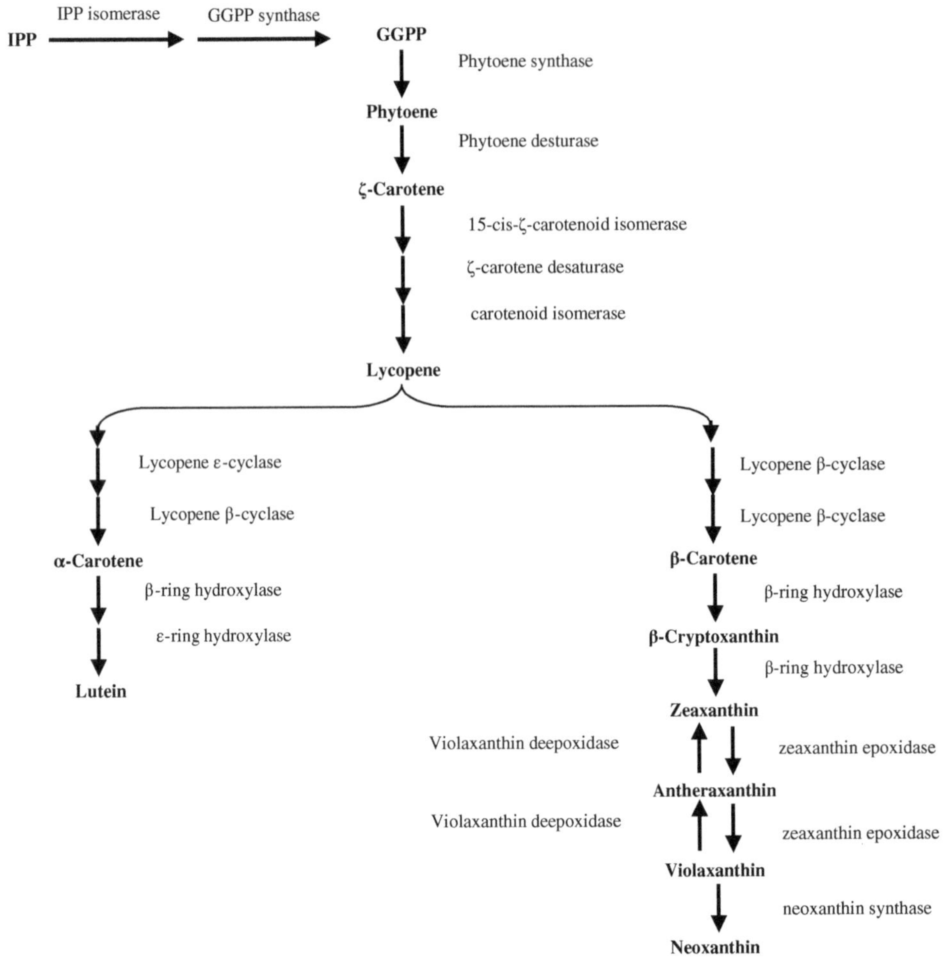

Figure 11.1 Carotenoids biosynthesis pathway in plants. Note: IPP, Isopentenyl pyrophosphate; GGPP, geranylgeranyl pyrophosphate. *Source:* Apel and Bock (2009), Diretto et al. (2006), Jaswir and Noviendri (2010), Kato et al. (2004), Kato et al. (2007), Liu et al. (2009), Yano et al. (2005), and Yamamizo et al. (2008).

Perez-Rodrıguez 2009). All carotenoids contain conjugated double bonds arranged in a long chain with acyclic or cyclic substituents in their structure (Stahl and Sies 2007). Figure 11.2 shows the chemical structures of carotenes and xanthophylls.

Carotenoids not only give plants and animals their vibrant colors (Aoki et al. 2002), but also have important physiological functions in the organism (Umeno et al. 2005). Both photosynthetic and non-photosynthetic organisms have been reported to synthesize carotenoids in their cells, resulting in their various shades of color ranging from yellow to red (Misawa 2009). In plants, carotenoids facilitate photosynthesis, prevent light oxidation, protect the plant from high amounts of light and oxidative damage during light harvest (Demmig-Adams and Adams III 2002), and regulate the fluidity of the cell lipid bilayer (Umeno et al. 2005). Their ability to absorb light is attributed to the multiple conjugated double bonds in their chemical structure, which carry cyclic or acyclic groups (Stahl and Sies 2007).

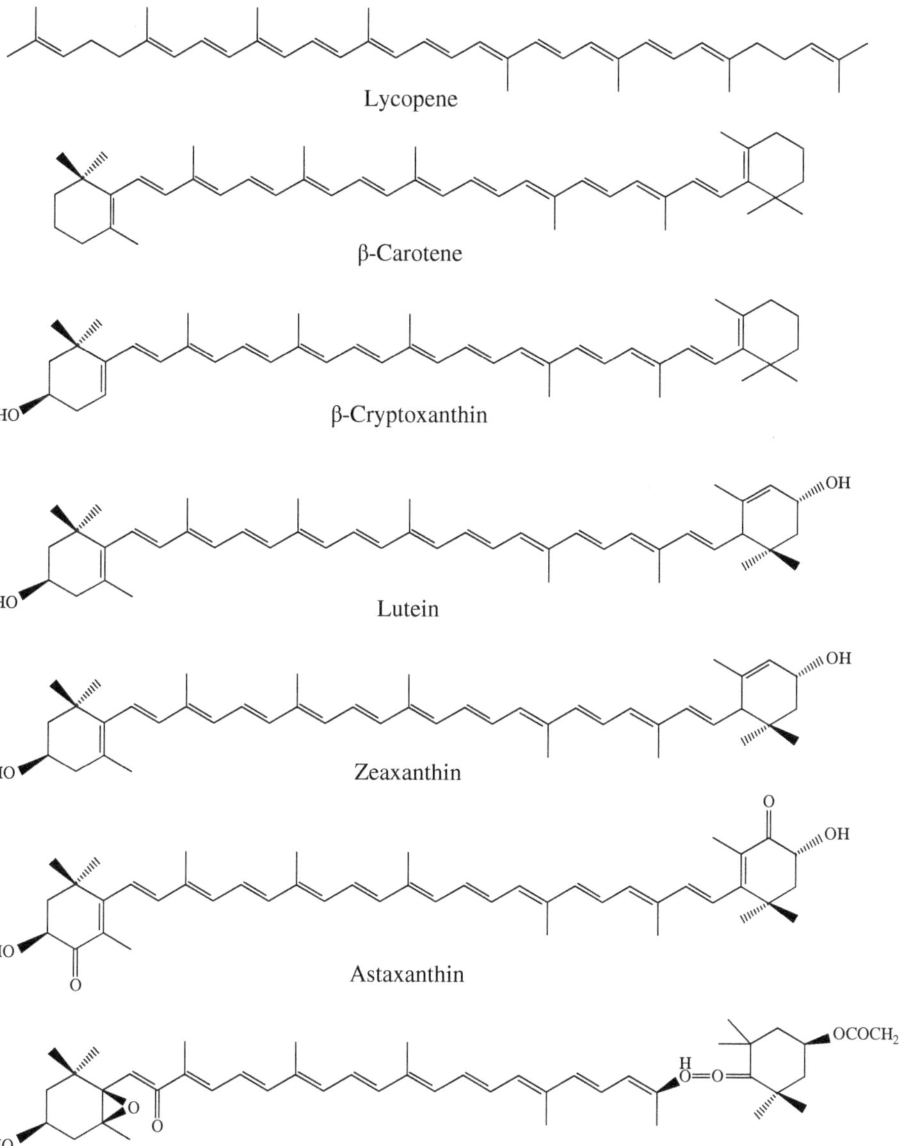

Figure 11.2 Chemical structures of several selected carotenoids. *Source:* Jaswir and Noviendri (2010) and Noviendri et al. (2011).

11.2 Sources of Carotenoid

In nature, bacteria, algae, fungi, and plants biosynthesize carotenoids intracellularly, while animals obtain carotenoids by ingesting them from their diet (Armstrong and Hearst 1996; Singh and Upadhyay 2021, 2023; Upadhyay and Singh 2021, 2023). The carotenoids found in vegetables, fruits, and grains include β-carotene, lycopene, lutein, zeaxanthin, astaxanthin, canthaxanthin, and β-cryptoxanthin. Meanwhile, the human plasma contains α-carotene, β-carotene, lutein, lycopene, β-cryptoxanthin, and zeaxanthin (Aizawa and Inakuma 2007) (Table 11.1).

Table 11.1 Common dietary sources of carotenoids in regular vegetable foods (μg/100 fresh weight).

Food	β-Carotene	Lutein	β-Cryptoxanthin	Lycopene	α-Carotene	Zeaxanthin
Carrot	7975	271	—	—	2186	—
Spinach	4489	6265	—	—	—	—
Broccoli	1580	2560	—	—	—	—
Lettuce	890	1250	—	—	—	—
Green peas	548	1840	—	—	—	—
Watercress	5919	10713	—	—	—	—
Tomato	608	77	—	4375	—	—
Orange*	250	120	700	—	200	—
Orange juice*	375	1180	1980	—	—	—
Mandarin*	275	50	1775	—	—	140
Sweet corn	45	520	—	—	60	440
Red pepper*	1700	270	250	—	30	600
Apricot*	3500	70	120	Trace	Trace	—
Mango*	3100	—	800	—	—	—
Papaya*	640	—	770	3400	30	—
Watermelon	180	20	300	4750	Trace	—

* Carotenoid esters present.
Source: Adapted from Southon and Faulks (2003).

11.2.1 α-Carotene and β-Carotene

β-Carotene is found in the seed fractions, pulp, and peel of *Canarium odontophyllum* Miq. fruit (Prasad et al. 2011). The microalgae *Dunaliella salina* has also been reported to synthesize β-carotene naturally (Pisal and Lele 2005). Sweet potato, pumpkin, carrot, cantaloupe, pink grapefruit, watermelon, tomatoes, and apricot are all sources of the α- and β-carotenes.

11.2.2 β-Cryptoxanthin

β-Cryptoxanthin is the result of enzymatic conversion of β-carotene in plants (Uchiyama et al. 2004a; Yamaguchi et al. 2005). It is abundant in Satsuma mandarin (*Citrus unshiu* Marc), and is also found in peach, papaya, tangerines, and oranges (Sugiura et al. 2002). Strong potential of this carotenoid compound in promoting health via reduction of risk of obesity, hepatic dysfunction, osteoporosis in aging women, and arteriosclerosis has been described.

11.2.3 Lycopene

Lycopene is the carotenoid conferring red color in various fruits and vegetables, such as tomatoes, watermelon, red grapes, papayas, and apricots. Tomatoes and tomato products are an especially important source of lycopene in our diet (Nishino et al. 2002).

11.2.4 Zeaxanthin

Zeaxanthin is the dihydroxy form of β-carotene; it is found abundantly in yellow fruits, vegetables, and grains such as corn, alfalfa, and marigold flowers (Nelis and DeLeenheer 1991; Nishino et al. 2002). Zeaxanthins belong to the xanthophyll group of carotenoids, as they contain 40 oxygenated carbon atoms. They are present in high concentrations in wolfberry (Chang et al. 2010) and have applications in cosmetics and animal feed (Masetto et al. 2001). Masetto et al. (2001) reported biosynthesis of zeaxanthin in Flavobacterium sp.

11.2.5 Astaxanthin

Astaxanthin is the carotenoid conferring red color to various sea animals such as lobster, shrimp, crab, and red fishes (Misawa 2009). The concentration of astaxanthin present in these organisms determines the shade of red they turn to when cooked; higher concentrations of astaxanthin give deeper red shades (Okada et al. 1994). While abundant in nature, most astaxanthin available in the market is currently synthetically derived. However, recently, natural biosynthesis of astaxanthin has been explored, for example, in the green alga *Haematococcus pluvialis* and in *Xanthophyllomyces dendrorhous* (formerly *Phaffia rhodozyma*) that would provide alternative sources of commercial astaxanthin (Misawa 2009).

11.2.6 Fucoxanthin

The brown pigment carotenoid, fucoxanthin is a ubiquitous carotenoid found in marine organisms such as seaweeds (Nishino et al. 2002). Its function as an anti-obesity agent *in vitro* and *in vivo* has been widely reported (Firdaus et al. 2015; Hosokawa et al. 2010). Fucoxanthin has also been explored as a skin supplement due to its ability to suppress melanin formation (Shimoda et al. 2010), and as an anticancer agent (Kotake-Nara et al. 2005). Due to these functions and more, fucoxanthin is considered an important marine nutraceutical with many important benefits to health.

Generally, colored fruits and vegetables can contain various pigments at varying concentrations, e.g. spinach contains a mix of α- and β-carotenes, lutein, and zeaxanthin, while pink grapefruit, tomatoes, watermelon, prunes, squash, orange, mango, papaya, and peaches contain phytofluene and phytoene. Britton and Khachik (2009) have reported the carotenoid content of some plants and classified these concentrations as low (0–0.1 mg/100 g), medium (0.1–0.5 mg/100 g), high (0.5–2.0 mg/100 g), and very high (>2.0 mg/100 g), as shown in Table 11.2.

Table 11.2 Food sources of the nutritionally important carotenoids.

Sources	Latin name	Carotenoids	Content
Apricot	*Prunus armeniaca*	β-Carotene	H–VH
		Lycopene	L
Broccoli	*Brassica oleracea (Italiaca)*	β-Carotene	VH
		Lutein	VH
Carrot	*Daucus carota*	β-Carotene	VH
		Lycopene	H
Grapefruit	*Citrus paradisi*	β-Carotene	L–M
		Lycopene	M–H
Green leafy vegetables		β-Carotene	M–VH
		Lutein	VH
Guava	*Psidium guajava*	β-Carotene	M
Kale	*B. oleracea (Acephala)*	β-Carotene	V–H
"Karat" banana	*Musa troglodytarum*	β-Carotene	H

(*Continued*)

Table 11.2 (Continued)

Sources	Latin name	Carotenoids	Content
Lettuce	*Lactuca sativa*	β-Carotene	M–H
Loquat	*Eriobotrya japonica*	β-Carotene	M
		β-Cryptoxanthin	L–M
Mango	*Mangifera indica*	β-Carotene	H–VH
Orange and juice	*Citrus* spp. and hybrids	β-Carotene	L–M
Papaya	*Carica papaya*	β-Carotene	M
		β-Cryptoxanthin	M–H
		Lycopene	M–H
Peach	*Prunus persica*	β-Carotene	H
Pepper (red, orange, green)	*Capsicum annuum*	β-Carotene	H
Pepper (red, orange)	*C. annuum*	β-Cryptoxanthin	M
Pepper (yellow, green)	*C. annuum*	Lutein	VH
Pepper (orange, red)	*C. annuum*	Zeaxanthin	VH
Red palm oil	*Elaeis guineensis*	β-Carotene	VH
Spinach	*Spinacia oleracea*	β-Carotene	VH
Squash/pumpkin.	*Cucurbita* spp	β-Carotene	L–M
		β-Cryptoxanthin	M–H
		Lutein	M–VH
		Zeaxanthin	M
Sweet potato	*Ipomoea batatas*	β-Carotene	VH
Tomato	*Lycopersicon esculentum*	β-Carotene	M
		Lycopene	VH
Persimmon	*Diospyros kaki*	β-Cryptoxanthin	H
		Lycopene	L–H
		Zeaxanthin	M
Pitanga	*Eugenia uniflora*	Lycopene	H
Egg yolk		Lutein	M–H
Watermelon	*Citrullus lanatus*	Lycopene	H–VH
Sweetcorn	*Zea mays*	Zeaxanthin	M

Note: L = low: 0–0.1 mg/100 g; M = medium: 0.1–0.5 mg/100 g; H = high: 0.5–2.0 mg/100 g; VH = very high: >2.0 mg/100 g.
Source: Adapted from Britton and Khachik (2009).

Carotenoid pigments are highly hydrophobic and have an affinity for lipids or hydrophobic structures in the cell, such as the hydrophobic faction of human tissue and cells, as well as the lipid core of the cellular bilayer membrane (Britton et al. 2008). They have been reported in the mitochondrial membrane, endoplasmic reticulum, or chloroplasts of the cell (Margalith 1999). Due to their fat-soluble nature, carotenoids ingested from various sources end up in the lipid fractions and lipid portions of the cell (El-Qudah 2008). The structural characteristic of the carotenoid, i.e. size, shape, functional groups, decides its orientation and effect in the membrane (Britton et al. 2008).

11.3 Carotenoids Bioactivity

11.3.1 Provitamin A Activity

Vitamin A is a fat-soluble vitamin, essential for maintaining proper eye function; its deficiency results in serious eye conditions and diseases, including nyctalopia (Takahashi et al. 2006). Vitamin A in the diet is obtained from foods containing the precursor provitamin A compounds, which are transformed in the human body into vitamin A or retinol. Carotenoids have been long documented as compounds with provitamin A function (Rodriguez-Amaya 1997); 50 out of 700 carotenoids in nature are reported to contain provitamin A activity (Okada et al. 2008; Rodriguez-Amaya 1997). The most important provitamin carotenoids in humans are α-carotene, β-carotene, and β-cryptoxanthin (Carrillo-Lopez et al. 2010; Park et al. 2009; Thane and Reddy 1997) (Table 11.3). β-Carotene has been reported widely as provitamin A compound with various studies reporting its activity (Imamura et al. 2006). Due to their hydrophobic nature, complete digestion of the carotenoids is necessary to allow carotenoids to be released into cells to carry out vitamin A function (Carrillo-Lopez et al. 2010).

11.3.2 Antioxidant/Prooxidant Activity

Carotenoids are potent antioxidant compounds that confer resistance to oxidative damage (Tian et al. 2007). The role of carotenoids is important in humans as part of the human antioxidant defense machinery and in plants to protect against photooxidative stress (Stahl and Sies 2003; Tian et al. 2007). Carotenoids quench single oxygens and scavenge reactive oxygen species (Stahl et al. 2000) and directly react with superoxides and other free radicals to prevent oxidation. The scavenging activity of carotenoids occurs via three reactions consisting of addition, electron transfer, addition, and hydrogen atom transfer (equations 11.1–11.3) (El-Agamey and McGarvey 2008).

Table 11.3 The sources of carotenoids in fruit and vegetables and their medicinal properties.

Carotenoid type	Example	Abundant sources	Medicinal properties
Provitamin A	α-Carotene	Green leafy vegetables	Antioxidants, anticarcinogenic
	β-Carotene	Yellow–orange fruit and vegetables, e.g. carrot, broccoli, spinach, parsley, celery, tomatoes	Antioxidant boosts activity of Natural Killer immune cell, gives cornea protection against UV light
	β-Cryptoxanthin	Peaches, oranges, tangerines, mangoes, papayas	Antioxidants, anticancer
Non-provitamin A	Lycopene	Tomatoes, pink-red grapefruit, red-fleshed papayas, watermelon	Powerful antioxidant, protects against the formation of cancers (prostate, bladder, cervical, leukemia), reduces cholesterol levels
	Lutein	Green and dark green, leafy vegetables: spinach, parsley, kale, broccoli, corn, avocado, brussels sprouts, beans, peans	Protects the eyes against the development of Age-related macular degeneration (AMD), cataracts, anticancer (colon)
	Zeaxanthin	As for lutein, plus some fruits, e.g. mandarins, peaches, oranges	Protects the eye from macular degeneration and cataracts

Source: Dharti et al. (2010); Keservani et al. (2010); Thane and Reddy (1997).

The reaction of carotenoids (CAR) with lipid peroxyl radicals (ROO*) results in the formation of resonance-stabilized carbon-centered radicals (Smirnoff 2005).

$$CAR + ROO^{*} \rightarrow ROO - CAR^{*} \quad \textit{Addition} \tag{11.1}$$

The electrons are transferred to form a radical cation (CAR^{*+}):

$$CAR + ROO^{*} \rightarrow CAR^{*+} + ROO^{-} \quad \textit{Electron transfer} \tag{11.2}$$

Abstraction of allylic hydrogen abstraction may also occur:

$$CAR + ROO^{*} \rightarrow CAR^{*} + ROOH \quad \textit{Hydrogen atom transfer} \tag{11.3}$$

Alternatively, carotenoids accept electrons to form a radical anion. In the case of lycopene:

$$\text{Lycopene} + O_2^{-*} \rightarrow \text{Lycopene}^{-*} + O_2$$

Among the dietary carotenoids, lycopene shows the most potent antioxidant activity (Rao and Agarwal 1999). A single lycopene molecule can quench more than one free radical at a time due to low energy levels in lycopene in the excited state (^{3}lycopene), thus being unable to excite other molecules to generate more reactive species. This characteristic makes lycopene a superior antioxidant for quenching single oxygen molecules (Rao and Rao 2007). The antioxidant action of lycopene and β-carotene is evident during photosensitized oxidation where the carotenoids effectively quench single oxygen molecules (^{1}O2) (Bonnie and Choo 1999). An example of lycopene antioxidant action for quenching single oxygen molecules is as follows (Krinsky 1992):

$$\text{Lycopene} + {}^1O_2 \rightarrow {}^3\text{Lycopene} + {}^3O_2$$

Lycopene can be found in the diet in apricots, papaya, guava, watermelon, and pink grapefruit, but its biggest source in the United States is from tomatoes, from which over 80% of lycopene is derived (Table 11.4). Tomato ripeness and variety are two factors significantly affecting lycopene content in the fruit (Stahl and Sies 2007).

Besides lycopene, other carotenoids have also been studied for their antioxidant potential *in vitro* and *in vivo*. However, antioxidant studies in animal model are limited due to significant differences in the metabolism of carotenoid between humans and animals. α-Carotene was reported to have similar antioxidant activities to lycopene, while β-carotene antioxidant activity was slightly lower (Nishino et al. 2002). Synergistic activities have also been reported between carotenoids and other antioxidants such as tocopherols, which improve the carotenoid's antioxidant activity. Cooking or processing tomatoes also improves the bioavailability of dietary lycopene, as it allows the lycopene molecule to more readily access dietary lipids and bile acids to form lipid micelles (Stahl and Sies 2007). Thus, addition of a cooking/processing step in tomatoes improves lycopene uptake levels in the body when compared to unprocessed tomatoes (Gartner et al. 1997).

11.3.3 Carotenoids as Anticancer

Ten million cancer cases are diagnosed each year, with numbers on the rise worldwide. Food plays an important role in the prevention of cancer, where plant-based diets and limited sugar consumption have been suggested to decrease cancer rates up to 20% (Basu et al. 2001). For health

Table 11.4 Examples of the major contributor of carotenoids in United States diets.

Carotenoids	Food sources	Amount (μg/100 g)
β-Carotene	Apricot, dried	17,600
	Carrots, cooked	9771
	Green collard	5400
	Spinach, cooked	5300
	Canteloupe	3000
	Beet green	2560
	Broccoli, cooked	1300
	Tomato, raw	520
α-Carotene	Carrots, cooked	3723
Lycopene	Tomatoes paste	36,500
	Tomatoes sauce	13,060
	Tomatoes ketchup	12,390
	Tomatoes juice	10,000
	Tomatoes, raw	3100
Lutein	Green collard	16,300
	Spinach, cooked	12,475
	Beet green	7700
	Broccoli, cooked	1839
	Green peas, cooked	1690
β-Cryptoxanthin	Tangerine	1060
	Papaya	470

Source: Mangels et al. (1993); Rao and Rao (2007).

policymakers, encouraging a carotenoid-rich diet filled with various fruits and vegetables is a low-risk strategy to manage cancer in the population (Rock 2009).

The antioxidant activity of carotenoids supports immune system function and lowers the risk of cancer (Das et al. 2007) (Table 11.5). These benefits of carotenoid consumption in the diet have increased interest in its use as a supplement for improving health (Das et al. 2007). α-Carotene, lycopene, lutein, zeaxanthin, fucoxanthin, β-cryptoxanthin, and astaxanthin have all reported anticancer activities (Nishino et al. 2002).

Table 11.5 Incidence of medical conditions treated with carotenoids (nutraceutical).

Medical conditions	Incidence of condition (%)	Nutraceuticals used
Enlarge prostate	16.9	Lycopene
Degenerative eye conditions	9.9	Lutein
Lactose intolerance	7.2	Lycopene
Diabetes	5.8	Lycopene
Prostate cancer	5.3	Lycopene

Source: Adapted from Gunther et al. (2004).

The xanthophylls, lycopene and fucoxanthin, have been reported for their anticancer activity (Bhosale and Bernstein 2005). Lycopene exerts anticancer activity via various means, such as through its antioxidant action, simulation of anticancer enzyme production, initiation of apoptosis, preventing cell proliferation and progression, and modulating gap junctional communication. Lycopene enhances connexin43 gene expression that encodes the major gap junction protein; upregulation of the gap junction communication inhibits cell growth and cancer formation (Zhang et al. 1991, 1992). Studies of human blood show large amounts of lycopene in the serum (Rao and Agarwal 1999) and lutein, cryptoxanthtin, and zeaxanthin in the plasma.

Fucoxanthin, the brown pigment found abundantly in brown seaweed *Undaria pinnatifida* (Liu et al. 2009a) has been reported to exert anticancer activity. The mechanisms of action of fucoxanthin in human leukemia HL-60 cells include inhibition of proliferation and induction of apoptosis via DNA fragmentation (Hosokawa et al. 1999). Similarly, prevention of tumor proliferation was observed in hepatoma SK-Hep-1 and murine embryonic liver BNL CL.2 cells treated with fucoxanthin (Liu et al. 2009a). Fucoxanthin also exerted anticancer activity in human colon adenocarcinoma WiDr cells and human hepatocarcinoma HepG2 cells via arrest of the cell G0/G1 cycle phase (Das et al. 2005), and reduced viability in human colon cancer cell lines, including Caco-2, HT-29, and DLD-1 (Hosokawa et al. 2004).

11.3.4 Carotenoids for Obesity

The abnormal accumulation of lipids in adipose tissue is known as obesity (Ikeuchi et al. 2007). In chronic cases, obese patients present with high fat accumulation around internal organs, which poses a great risk to health, and is associated with other diseases, such as diabetes and hyperlipidemia (Miyashita 2006). Therapies against obesity have targeted the uncoupling protein (UCP) families UCP1, UCP2, and UCP3 responsible for adaptive thermogenesis (Jezek 2002). Mitochondrial UCP1 found in brown adipose tissue (BAT) has especially been the target of various studies, as its dysregulation has been directly linked to obesity (Lowell et al. 1993; Maeda et al. 2007a; Miyashita 2006).

The administration of carotenoid fucoxanthin in rats and mice increased the expression of UCP1 protein and mRNA in white adipose tissue (WAT) compared to no fucoxanthin in the diet (Maeda et al. 2007a). Fucoxanthin administered with fish oil improved the effectiveness of fucoxanthin at reducing WAT in rats and mice (Maeda et al. 2007b). Upon daily administration, fucoxanthin significantly reduced mice body weight.

Another carotenoid astaxanthin was also reported for its anti-obesity effect. Astaxanthin decreased body weight and adipose tissue in rats fed a high-fat diet (Ikeuchi et al. 2007) and had beneficial effects on liver weight and triglyceride, plasma triglyceride, and total cholesterol. The bioactivity of astaxanthin may address the symptoms of obesity and prevent its advancement.

11.3.5 Anabolic Activity on Bone Components

The carotenoid β-cryptoxanthin has unique activities for the maintenance of bone health and prevention of osteoporosis (Yamaguchi 2008). β-Cryptoxanthin induced anabolic effect in bone components, which stimulates bone formation. Bone mass loss is associated with osteoporosis, causing bone fracture (Uchiyama et al. 2004a). Application of β-cryptoxanthin in 50-week-old female rats at 10^{-7} or 10^{-6} M concentration increased calcium content and alkaline phosphatase activity in the femoral-diaphysial and femoral-metaphysial tissues and presented an anabolic effect in femoral tissue bone components when administered orally in senior female rats (Uchiyama et al. 2004b;

Yamaguchi 2008). β-Cryptoxanthin was also shown to have an anabolic effect *in vitro* and *in vivo* on bone calcification in rat cortical and trabecular bone tissues (Yamaguchi and Uchiyama 2003). Studies have also shown that consumption of juice containing higher concentrations of β-cryptoxanthin prevented bone loss associated with aging (Yamaguchi et al. 2006), stimulated bone formation, and inhibited bone reabsorption when consumed over long durations, which may be beneficial in menopausal women.

11.4 Carotenoid Applications in Industry

While historically used in the feed and food industries, the application of carotenoids in nutraceutical and pharmaceutical industries is relatively newer (Bhosale and Bernstein 2005; Bonnie and Choo 1999; Dixit et al. 2021). Carotenoids such as β-carotene and lycopene specifically are prized commercially and scientifically (Liu et al. 2009b) as natural food coloring agents, health supplements (Bramley 2003), and nutraceuticals for use in the pharmaceutical and cosmetic industries (Lee and Schmidt-Dannert 2002; Schmidt-Dannert 2000).

Their widespread application in the nutraceutical and pharmaceutical industries due to their health benefits has spurred the chemical synthesis of carotenoids with various structural elements (Lee and Schmidt-Dannert 2002). However, currently, few carotenoids are produced synthetically. Lycopene, β-carotene, lutein, astaxanthin, capsanthin, canthaxanthin, β-apo-8-carotenal, and β-apo-8 carotenal-ester are some examples of chemically synthesized carotenoids (Johnson and Schroeder 1995). There is currently increased interest in the manufacture of carotenoids using microorganisms to par levels produced via chemical synthesis. To do this, processes optimizing biosynthesis are explored in order to improve the carotenoid yield from microorganism cells (Bhosale 2004).

Table 11.6 shows the carotenoids and their applications in industry.

Humans primarily obtain carotenoids from ingesting fruits and vegetables, where almost 40 types of carotenoids can be obtained from dietary foods (Rao and Rao 2007). Natural produce with deeper colors reflects a higher concentration of carotenoid compounds present in the cells. Yelloworange foods typically contain α- and β-carotenes, orange foods contain β-cryptoxanthin, green leafy vegetables contain lutein, and tomato and tomato products are huge sources of lycopene. The amount of carotenoids in common food sources is shown in Table 11.4 (Rao and Rao 2007).

The use of food coloring additives is strictly regulated by law in many countries around the world. The law governs the source, purity, application, and concentration of carotenoid used in the food (Mortensen 2006). In Europe and the United States, laws dictate what natural and nature-identical

Table 11.6 Global market for carotenoids in 2004 and 2009 (US$ million).

Types of Product	β-Carotene		Lycopene		Lutein		Canthaxanthin		Astaxanthin		Zeaxanthin	
	2004	2009	2004	2009	2004	2009	2004	2009	2004	2009	2004	2009
Supplements	125.0	128.0	50.5	68.7	54.0	85.0	3.0	3.0	3.5	4.8	22.0	35.0
Food	98.0	103.0	3.5	9.3	18.0	20.0	7.0	8.0	0	0	0	0
Cosmetics	6.0	7.0	0	3.0	0	0	0	0	0	0	0	0
Feed	13.0	15.0	0	0	67.0	82.0	138.0	145.0	230.5	252.2	0	0
Total	242.0	253.0	54.0	81.0	139.0	187.0	148.0	156.0	234.0	257.0	22.0	35.0

Source: Adapted from Mortensen (2009).

pigments are allowed to be used in animal feed. There are strict laws against use of natural pigments for adding color to eggs, but rules for pigmenting chicken and fish flesh are more relaxed.

Lycopene is an example of a carotenoid permitted for use as a food colorant. Lycopene was first approved in the EU and later in the USA in 2005 for use as natural food colorant. However, lycopene intended for this application needs to be strictly sourced from tomatoes (*Lycopersicon esculentum*). Besides lycopene, tomatoes are also a rich source of other carotenoids, including β-carotene, phytoene, and phytofluene found in the tomato oleoresin (Mortensen 2006). However, the use of lycopene as a food colorant is limited by its high cost and susceptibility to oxidation and degradation compared to other carotenoids such as β-carotene (Mortensen 2006). The demand for natural food colorants is driven by customer demand with increased awareness of the health dangers of synthetic food colorants in foods and foodstuffs (Mattea et al. 2009) (Table 11.7).

Among the carotenoids, astaxanthin, and canthaxanthin are of significance industrially, where they are used to confer color to fish and crustaceands in aquaculture industries, as supplements in the pharmaceutical industry, and colorants and supplements in the food industry (Bhosale and Bernstein 2005). Astaxanthin extracted from the green alga *H. pluvialis*, which produces astaxanthin in ester form (Misawa 2009), is gaining popularity for its benefits as a food supplement or nutraceutical. Despite a small percentage of carotenoids being extracted from natural sources, the majority of industrially important carotenoids, such as astaxanthin, canthaxanthin, and β-carotene are chemically synthesized (Johnson and Schroeder 1995) (Table 11.8).

Table 11.7 Approved carotenoids for food and feed industries by different organizations.

Color	EU	FDA	WHO
Mixed carotenes (carrot oil)	Yes (non-legislated)	Yes	Not specified
β-Carotene	Yes (E 160a)	Yes	Yes
Annatto extract, bixin, norbixin	Yes (E 160b)	Yes	Yes
Canthaxanthin	No (E 161g)	Yes	Not allocated
β-Apo-8′-carotenal	No (E 160e)	Yes	Yes
Citranaxanthin	No	Yes	Yes
Xanthophylls, flavoxanthin, rubixanthin, zeoxanthin, and other natural products with some of these carotenoids	Not all	Not all	Not specified

Source: EU (European union); FDA (food and drug administration) and WHO (world health organization). Mortensen (2006); Mattea et al. (2009).

Table 11.8 Examples of plant extracts (contain carotenoids) used commercially.

Plant	Carotenoids	Use
Bixa orellana seeds	Bixin, norbixin (annatto)	Food coloring
Carrot root	Carotenes, mainly β-carotene	Dietary supplement
Capsicum annum fruit	Capsanthin, capsorubin (paprika)	Food coloring
Crocus sativus petals	Crocin, crocetin (saffron)	Food coloring
Marigold petals	Lutein, zeaxanthin	Dietary supplement
Tomato fruit	Lycopene	Dietary supplement
Palm oil	Carotenes	Dietary supplement, coloring

Source: Bramley (2003)/with permission of ELSEVIER.

11.5 Conclusions

The carotenoid pigments are widely found in nature, conferring yellow, orange, and red colors to various organisms ranging from microorganisms to higher plants and animals. They do not only provide visual appeal to foods and feeds but also confer interesting benefits to human health, explaining their widespread use in nutraceutical and therapeutic applications. The majority of carotenoids available commercially are produced via chemical synthesis; however, biosynthesis in microorganism is being developed as an alternative to chemical synthesis. While attractive for their application as food additives and food colorants, the use of carotenoids in the EU and USA is still strictly regulated.

Acknowledgments

This work was financially supported by a research grant (IIUM RMGS 09-05) from the Research Management Center (RMC), International Islamic University Malaysia, and a travel grant from Padang State University, Indonesia.

References

Aizawa, K. and Inakuma, T. (2007). Quantitation of carotenoids in commonly consumed vegetables in Japan. *Food Science and Technology Research* 13 (3): 247–252.

Aoki, H., Kien, N.T.M., Kuze, N. et al. (2002). Carotenoid pigments in GAC fruit (*Momordica cochinchinensis* SPRENG). *Bioscience, Biotechnology, and Biochemistry* 66 (11): 2479–2482.

Apel, W. and Bock, R. (2009). Enhancement of carotenoid biosynthesis in transplastomic tomatoes by induced lycopene-to-provitamin A conversion[1[OA]]. *Plant Physiology* 151: 59–66.

Armstrong, G.A. and Hearst, J.E. (1996). Genetics and molecular biology of carotenoid biosynthesis. *FASEB Journal* 10: 228–237.

Basu, H.N., Vecchio, A.J.D., Flider, F., and Orthoefer, F.T. (2001). Nutritional and potential disease prevention properties of carotenoids. *Journal of the American Oil Chemists' Society* 78 (7): 665–675.

Bhosale, P. (2004). Environmental and cultural stimulants in the production of carotenoids from microorganisms. *Applied Microbiology and Biotechnology* 63: 351–361.

Bhosale, P. and Bernstein, P.S. (2005). Microbial xanthophylls. *Applied Microbiology and Biotechnology* 68: 445–455.

Bonnie, T.Y.P. and Choo, Y.M. (1999). Oxidation and thermal degradation of carotenoids. *Journal of Oil Palm Research* II (1): 62–78.

Bramley, P. (2003). The genetic enhancement of phytochemicals: the case of carotenoids. In: *Phytochemical Functional Foods* (ed. I. Johnson and G. Williamson) Ch. 13, 253–274. Woodhead Publishing Limited. CRC Press. ISBN 0-8493-1754-1.

Britton, G. (1995). Structure and properties of carotenoids in relation to function. *FASEB Journal* 9: 1551–1558.

Britton, G. and Khachik, F. (2009). Carotenoid in food. In: *Carotenoids: Nutrition and Health*, vol. 5 Ch. 3 (ed. G. Britton, F. Liaaen-Jensen, and H. Pfander), 45–66. Birkhauser Verlag Basel, ISBN 978-3-7643-7500-3.

Britton, G., Liaaen-Jensen, S., and Pfender, H. (2008). Special molecules, special properties. In: *Carotenoids; Natural Functions*, vol. 4. Ch. 1 (ed. G. Britton, F. Liaaen-Jensen, and H. Pfander), 1–6. Birkhauser Verlag Basel.

Carrillo-Lopez, A., Yahia, E.M., and Ramirez-Padilla, G.K. (2010). Bioconversion of carotenoids in five fruits and vegetables to vitamin A measured by retinol accumulation in rat livers. *American Journal of Agricultural and Biological Sciences* 5 (2): 215–221.

Chang, R.C.C., Ho, Y.S., Yu, M.S., and So, K.F. (2010). Medicinal and nutraceutical uses of wolfberry in preventing neurodegeneration in Alzheimer's disease. In: *Recent Advances on Nutrition and the Prevention of Alzheimer's Disease*, vol. 2010 (ed. C. Ramassamy and S. Bastianetto), 169–185. Transworld Research Network.

Das, S.K., Hashimoto, T., Shimizu, K. et al. (2005). Fucoxanthin induces cell cycle arrest at G_0/G_1 phase in human colon carcinoma cells through up-regulation of $p21^{WAF1/Cip1}$. *Biochimica et Biophysica Acta* 1726: 328–335.

Das, A., Yoon, S.H., Lee, S.H. et al. (2007). An update on microbial carotenoid production: application of recent metabolic engineering tools. *Applied Microbiology and Biotechnology* 77: 505–512.

Demmig-Adams, B. and Adams, W.W. III (2002). Antioxidants in photosynthesis and human nutrition. *Science* 298: 2149–2153.

Dharti, S.T., Gandhi, S., and Shah, M. (2010). Nutraceuticals-portmanteau of science and nature. *International Journal of Pharmaceutical Sciences Review and Research* 5 (3): 33–38.

Diretto, G., Tavazza, R., Welsch, R. et al. (2006). Metabolic engineering of potato tuber carotenoids through tuber-specific silencing of lycopene epsilon cyclase. *BMC Plant Biology* 6 (13): 1–11.

Dixit, S., Shukla, A., Singh, V., and Upadhyay, S.K. (2021). Bioprospecting of natural compounds for industrial and medical applications: current scenario and bottleneck. In: *Bioprospecting of Plant Biodiversity for Industrial Molecules*. Wiley; Portico. https://doi.org/10.1002/9781119718017.ch3.

El-Agamey, A. and Mcgarvey, D.J. (2008). Carotenoid radicals and radical ions. In: *Carotenoids: Natural Function*, vol. 4. Ch. 7 (ed. G. Britton, F. Liaaen-Jensen, and H. Pfander), 119–154. Birkhauser Verlag Basel, ISBN 978-3-7643-7500-3.

El-Qudah, J.M. (2008). Dietary intake of selected common vegetable foods and their total carotenoids determination. *American Journal of Agricultural and Biological Sciences* 3 (4): 729–733.

Firdaus, M., Agustin, F.P., Kholifatul, A.J. et al. (2015). *Antiobesity of Fucoxanthin from Sargassum echinocarpum by Increasing β-oxidation in Adipocyte*, 97–102. KnE Life Sciences.

Fujisawa, M., Watanabe, M., Choi, S.K. et al. (2008). Enrichment of carotenoids in flaxseed (*Linum usitatissimum*) by metabolic engineering with introduction of bacterial phytoene synthase gene *crtB*. *Journal of Bioscience and Bioengineering* 105 (6): 636–641.

Gartner, C., Stahl, W., and Sies, H. (1997). Lycopene is more bioavailable from tomato paste than from fresh tomatoes. *The American Journal of Clinical Nutrition* 66: 116–122.

Godinho, A. and Bhosle, S. (2008). Carotenes produced by alkaliphilic orange-pigmented strain of *Microbacterium arborescens*-AGSB isolated from coastal sand dunes. *Indian Journal of Geo-Marine Sciences* 37 (3): 307–312.

Gunther, S., Patterson, R.E., Kristal, A.R. et al. (2004). Demographic and health-related correlates of herbal and specialty supplement use. *Journal of the American Dietetic Association* 104 (1): 27–34.

Hosokawa, M., Kudo, M., Maeda, H. et al. (2004). Fucoxanthin induces apoptosis and enhances the antiproliverative effect of the PPARγ ligand, troglitazone, on colon cancer cells. *Biochimica et Biophysica Acta* 1675: 113–119.

Hosokawa, M., Miyashita, T., Nishikawa, S. et al. (2010). Fucoxanthin regulates adipocytokine mRNA expression in white adipose tissue of diabetic/obese KK-Ay mice. *Archives of Biochemistry and Biophysics* 504 (1): 17–25.

Hosokawa, M., Wanazaki, S., Miyauchi, K. et al. (1999). Apoptosis-inducing effect of fucoxanthin on human leukemia cell line HL-60. *Food Science and Technology Research* 5 (3): 243–246.

Ikeuchi, M., Koyama, T., Takahashi, J., and Yazawa, K. (2007). Effects of astaxanthin in obese mice fed a high-fat diets. *Bioscience, Biotechnology, and Biochemistry* 71 (4): 839–899.

Imamura, T., Bando, N., and Yamanishi, R. (2006). β-carotene modulates the immunological function of RAW264, a murine macrophage cell line, by enhancing the level of intracellular glutathione. *Bioscience, Biotechnology, and Biochemistry* 70 (9): 2112–2120.

Jaswir I, Noviendri D (2010). Carotenoids for food and nutraceutical application. *Presented in International Conference on Food Safety and Food Security, Faculty of Agricultural, Universitas Gadjah Mada*, Yogyakarta, Indonesia (1–2 December 2010). 12 p.

Jezek, P. (2002). Possible physiological roles of mitochondrial uncoupling proteins-UCPn. *The International Journal of Biochemistry & Cell Biology* 34: 1190–1206.

Johnson, E. and Schroeder, W. (1995). Microbial carotenoids. *Advances in Biochemical Engineering/ Biotechnology* 53: 119–178.

Kato, M., Ikoma, Y., Matsumoto, H. et al. (2004). Accumulation of carotenoids and expression of carotenoid biosynthetic genes during maturation in citrus fruit1. *Plant Physiology* 134: 824–837.

Kato, M., Matsumoto, H., Ikoma, Y. et al. (2007). Accumulation of carotenoids and expression of carotenoid biosynthetic genes and carotenoid cleavage dioxygenase genes during fruit maturation in the juice sacs of 'Tamami,' 'Kiyomi' Tangor, and 'Wilking' mandarin. *Japanese Society for Horticultural Science* 76 (2): 103–111.

Keservani, R.K., Kesharwani, R.K., Vyas, N. et al. (2010). Nutraceutical and functional food as future food: a review. *Der Pharmacia Lettre* 2 (1): 106–116.

Konishi, I., Hosokawa, M., Sashima, T. et al. (2008). Suppressive effects of alloxanthin and diatoxanthin from *Halocynthia roretzi* on LPS-induced expression of pro-inflammatory genes in RAW264.7 cells. *Journal of Oleo Science* 57 (3): 181–189.

Kotake-Nara, E., Terasaki, M., and Nagao, A. (2005). Characterization of apoptosis induceed by fucoxanthin in human promyelocytic leukemia cells. *Bioscience, Biotechnology, and Biochemistry* 69 (1): 224–227.

Krinsky, N.I. (1992). Mechanism of action of biological antioxidants. *Proceedings of the Society for Experimental Biology and Medicine* 200: 248–254.

Lee, P.C. and Schmidt-Dannert, C. (2002). Metabolic engineering towards biotechnological production of carotenoids in microorganisms. *Applied Microbiology and Biotechnology* 60: 1–11.

Liu, C.L., Huang, Y.S., Hosokawa, M. et al. (2009a). Inhibition of proliferation of hepatoma cell line by fucoxanthin in relation to cell cycle arrest and enhanced gap junctional intercellular communication. *Chemico-Biological Interactions* 182: 165–172.

Liu, G.N., Zhu, Y.H., and Jiang, J.G. (2009b). The metabolomic of carotenoids in engineered cell factory. *Applied Microbiology and Biotechnology* 83: 989–999.

Lowell, B.B., S-Susullc, V., Hamann, A. et al. (1993). Development of obesity in trabsgenic mice after genetic ablation of brown adipose tissue. *Nature* 366: 740–742.

Maeda, H., Hosokawa, M., Sashima, T. et al. (2007a). Effect of medium-chain triacylglycerols on anti-obesity effect of fucoxanthin. *Journal of Oleo Science* 56 (12): 615–621.

Maeda, H., Hosokawa, M., Sashima, T., and Miyashita, K. (2007b). Dietary combination of fucoxanthin and fish oil attenuates the weight gain of white adopose tissue and decrease blood glucose in obese/ diabetic KK-A^y mice. *Journal of Agricultural and Food Chemistry* 55: 7701–7706.

Mangels, A.R., Holden, J.M., Beecher, G.R. et al. (1993). Carotenoid contents of fruits and vegetables: an evaluation of analytical data. *Journal of the American Dietetic Association* 93: 284–296.

Margalith, P.Z. (1999). Production of ketocarotenoids by microalgae. *Applied Microbiology and Biotechnology* 51: 431–438.

Masetto, A., Flores-Cotera, L.B., Diaz, C. et al. (2001). Application of a complete factorial design for the production of zeaxanthin by *Flavobacterium* sp. *Journal of Bioscience and Bioengineering* 92 (1): 55–58.

Mattea, F., Martin, A., and Cocero, M.J. (2009). Carotenoid processing with supercritical fluids. *Journal of Food Engineering* 93: 255–265.

Misawa, N. (2009). Pathway engineering of plants toward astaxanthin production. *Plant Biotechnology* 26: 93–99.

Miyashita, K. (2006). Seaweed carotenoid, fucoxanthin, with highly bioactive and nutritional activities. *Journal of Marine Bioscience and Biotechnology* 1 (1): 48–58.

Mortensen, A. (2006). Carotenoids and other pigments as natural colorants. *Pure and Applied Chemistry* 78 (8): 1477–1491.

Mortensen, A. (2009). *Supplements*. In: *Carotenoids: Nutrition and Helath*, vol. 5. Ch.4 (ed. G. Britton, F. Liaaen-Jensen, and H. Pfander), 67–82. Birkhauser Verlag Basel, ISBN 978-3-7643-7500-3.

Nelis, J.H. and DeLeenheer, P.A. (1991). Microbial sources of carotenoid pigments used in foods and feeds. *The Journal of Applied Bacteriology* 70: 181–191.

Nishino, H., Murakoshi, M., Ii, T. et al. (2002). Carotenoids in cancer chemoprevention. *Cancer Metastasis Reviews* 21: 257–264.

Noviendri D, Jaswir I, Salleh HM, Taher M, Miyashita K (2011). Techniques of extraction and purification of carotenoid (fucoxanthin) from brown seaweed. *Workshop on seaweed processing for pharmaceutical applications. Organized By Bioprocess and Molecular Engineering Research Unit (BPMERU), Department of Biotechnology Engineering, Faculty of Engineering, International Islamic University Malaysia*, Kuala Lumpur, Malaysia (16 May 2011). 26p.

Okada, T., Nakai, M., Maeda, H. et al. (2008). Suppressive effect of neoxanthin on the differentiation of 3T3-L1 adipose cells. *Journal of Oleo Science* 57 (6): 345–351.

Okada, S., Nur-E-Borhan, S.A., and Yamaguchi, K. (1994). Carotenoid composition in the exoskeleton of commercial black tiger prawns. *Fisheries Science* 60 (2): 213–215.

Park, S.Y., Nomura, S.M.Y., Murphy, S.P. et al. (2009). Carotenoid intake and colorectal cancer risk: the multiethnic cohort study. *Journal of Epidemiology* 19 (2): 63–71.

Perez-Rodrıguez, L. (2009). Carotenoids in evolutionary ecology: re-evaluating the antioxidant role. *BioEssays* 31: 1116–1126.

Pisal, D.S. and Lele, S.S. (2005). Carotenoid production from microalga, *Dunaliella salina*. *Indian Journal of Biotechnology* 4: 476–483.

Prasad, K.N., Chew, L.Y., Kho, H.E. et al. (2011). Carotenoids and antioxidant capacities from *Canarium odontophyllum* Miq. fruit. *Food Chemistry* 124: 1549–1555.

Rao, A.V. and Agarwal, S. (1999). Role of lycopene as antioxidant carotenoid in the prevention of chronic diseases: a reviev. *Nutrition Research* 19 (2): 305–323.

Rao, A.V. and Rao, L.G. (2007). Carotenoids and human health. *Pharmacological Research* 55: 207–216.

Rock, C.L. (1997). Carotenoids: biology and treatment. *Pharmacology & Therapeutics* 75 (3): 185–197.

Rock, C.L. (2009). *Carotenoids and cancer*. In: *Carotenoids: Nutrition and Helath*, vol. 5. Ch.13 (ed. G. Britton, F. Liaaen-Jensen, and H. Pfander), 269–286. Birkhauser Verlag Basel, ISBN 978-3-7643-7500-3.

Rodriguez-Amaya DB (1997). Carotenoids and food preparation: the retention of provitamin A carotenoids in prepared, processed, and stored foods. *Departamento de Ciências de Alimentos Faculdade de Engenharia de Alimentos Universidade Estadual de Campinas*, Brazil. John Snow, Inc/ OMNI Project (1997). 93 p.

Schmidt-Dannert, C. (2000). Engineering novel carotenoids in microorganisms. *Current Opinion in Biotechnology* 11: 255–261.

Shimoda, H., Tanaka, J., Shan, S.J., and Maoka, T. (2010). Anti-pigmentary activity of fucoxanthin and its influence on skin mRNA expression of melanogenic molecules. *The Journal of Pharmacy and Pharmacology* 62 (9): 1137–1145.

Singh, S.P. and Upadhyay, S.K. (Éds.) (2021). *Bioprospecting of Microorganism-Based Industrial Molecules*. Wiley. https://doi.org/10.1002/9781119717317.

Singh, S.P. and Upadhyay, S.K. (Éds.) (2023). *Microbial Bioreactors for Industrial Molecules*. Wiley. https://doi.org/10.1002/9781119874096.

Smirnoff, N. (2005). *Ascorbate, tocopherol and carotenoids: metabolism, pathway engineering and functions*. In: *Antioxidants and Reactive Oxygen Species in Plants* (ed. N. Smirnoff) Ch.3, 53–87. Blackwell Publishing. ISBN-13:978-1-4051-2529-1.

Southon, S. and Faulks, R. (2003). *Carotenoids in food: bioavailability and functional benefits*. In: *Phytochemical Functional Foods* (ed. I. Johnson and G. Williamson) Ch. 7, 107–127. Woodhead Publishing Limited. CRC Press. ISBN 0-8493-1754-1.

Stahl, W., Heinrich, U., Jungmann, H. et al. (2000). Carotenoids and carotenoids plus vitamin E protect against ultraviolet light-induced erythema in humans. *The American Journal of Clinical Nutrition* 71: 795–798.

Stahl, W. and Sies, H. (2003). Antioxidant activity of carotenoids. *Molecular Aspects of Medicine* 24: 345–351.

Stahl, W. and Sies, H. (2007). Carotenoids and flavonoids contribute to nutritional protection against skin damage from sunlight. *Molecular Biotechnology* 37: 26–30.

Sugawara, T., Yamashita, K., Asai, A. et al. (2009). Esterification of xanthophylls by human intestinal Caco-2 cells. *Archives of Biochemistry and Biophysics* 483: 205–212.

Sugiura, M., Kato, M., Matsumoto, H. et al. (2002). Serum concentration of β-cryptoxanthin in Japan reflects the frequency of satsuma mandarin (*Citrus unshiu* March.) consumption. *Journal of Health Science* 48 (4): 350–353.

Takahashi, M., Watanabe, H., Kikkawa, J. et al. (2006). Carotenoids extraction from Japanese persimmon (*Hachiyakaki*) peels by supercritical CO_2 with ethanol. *Analytical Sciences* 22: 1441–1447.

Thane, C. and Reddy, S. (1997). Processing of fruit and vegetables: effect on carotenoids. *Nutrition and Food Science* 97 (2): 58–65.

Tian, B., Xu, Z., Sun, Z. et al. (2007). Evaluation of the antioxidant effects of carotenoids from *Deinococcus radiodurans* through targeted mutagenesis, chemiluminescence, and DNA damage analyses. *Biochimica et Biophysica Acta* 1770: 902–911.

Uchiyama, S., Hashizuma, M., Hokari, Y. et al. (2004b). Characterization of active component in marine alga *Sargassum horneri* extract in stimulating bone calcification *in vitro*. *Journal of Health Science* 50 (6): 634–639.

Uchiyama, S., Sumida, T., and Yamaguchi, M. (2004a). Anabolic effect of β-cryptoxanthin on bone components in femoral tissues of aged rats *in vivo* and *in vitro*. *Journal of Health Science* 50 (5): 491–496.

Umeno, D., Tobias, A.V., and Arnold, F.H. (2005). Diversifying carotenoid biosynthetic pathways by directed evolution. *Microbiology and Molecular Biology Reviews* 69 (1): 51–78.

Upadhyay, S.K. and Singh, S.P. (Éds.) (2021). *Bioprospecting of Plant Biodiversity for Industrial Molecules*. Wiley. https://doi.org/10.1002/9781119718017.

Upadhyay, S.K. and Singh, S.P. (Éds.) (2023). *Plants as Bioreactors for Industrial Molecules*. Wiley. https://doi.org/10.1002/9781119875116.

Yamaguchi, M. (2008). β-cryptoxanthin and bone metabolism: the preventive role in osteoporosis. *Journal of Health Science* 54 (4): 356–369.

Yamaguchi, M., Igarashi, A., Morita, S. et al. (2005). Relationship between serum β-cryptoxanthin and circulating bone metabolic markers in healthy individuals with the intake of juice (*Citrus unshiu*) containing β-cryptoxanthin. *Journal of Health Science* 51 (6): 738–743.

Yamaguchi, M., Igarashi, A., Uchiyama, S. et al. (2006). Effect of β-cryptoxanthin on circulating bone markers: intake of juice (*Citrus Unshiu*) supplemented with β-cryptoxanthin has effect in menopausal woman. *Journal of Health Science* 52 (6): 758–768.

Yamaguchi, M. and Uchiyama, S. (2003). Effect of carotenoid on calcium content and alkaline phosphatase activity in rat femoral tissues *in vitro*: the unique anabolic effect of β-cryptoxanthin. *Biological & Pharmaceutical Bulletin* 26 (8): 1188–1191.

Yamamizo, C., Kishimoto, S., and Ohmiya, A. (2008). Carotenoid composition and carotenogenic gene expression during ipomoea petal development. *Journal of Experimental Botany* 61 (3): 709–719.

Yano, M., Kato, M., Ikoma, Y. et al. (2005). Quantitation of carotenoids in raw and processed fruits in Japan. *Food Science and Technology Research* 11 (1): 13–18.

Zhang, L.X., Cooney, R.V., and Bertram, J.S. (1991). Carotenoid enhance gap junction communication and inhibit lipid peroxidation in CH3/10T1/2 cells: relationship to their cancer chemopreventive action. *Carcinogenesis* 12: 2109–2114.

Zhang, L.X., Cooney, R.V., and Bertram, J.S. (1992). Carotenoids up-regulate connexin43 gene expression independent of their provitamin A or antioxidant properties. *Cancer Research* 52: 5707–5712.

12

Nutraceuticals in Legumes

Rajan Katoch[1], Ankur Tripathi[2], Neelam Thakur[1], and Kiran[3]

[1] Biochemistry Laboratory, Department of Chemistry & Biochemistry, CSKHPKV, Palampur, Himachal Pradesh, India
[2] College of Dairy & Food Technology, Agriculture University, Jodhpur, Rajasthan, India
[3] Department of Genetics & Plant Breeding, CSKHPKV, Palampur, Himachal Pradesh, India

12.1 Introduction

Health benefits of foods are becoming more widely accepted and consumers are very keen about quality and quality components in food, which may provide prevention against diseases or improve health. This understanding, together with the knowledge of how nutrition affects health, has prompted a quest for inclusion of natural plant-based health products in the diet. Legumes contain approximately 20–40% proteins, low GI (Glycemic Index), less fat and cholesterol content (~1–3%), and a significant level of carbohydrates (~50–65%). Oligomeric globulins, which belong to the 7S and 11S protein fractions, are major storage proteins present in legumes. Legumes also have a substantial amount of essential amino acids in addition to providing high protein content. Furthermore, a considerable amount of vitamins and minerals are also present in legumes (Kutoš et al. 2003; Osorio-Díaz et al. 2003). The positive health effects of legumes consumption have led to an increase interest in the development of legumes-based functional food products and their utilization in the diet. Because of their excellent nutritional composition, legumes are considered a future superfood that can alleviate hunger and combat protein-calorie malnutrition (Martín-Cabrejas 2019).

The increasing occurrence of chronic diseases, such as diabetes and obesity, cancer, cardiovascular diseases like stroke, and inflammatory diseases in individuals is directly linked to their changing dietary habits and Western-style lifestyles, poses a significant challenge to global health (Priego-Poyato et al. 2021). This has attracted the research interest of nutritionists for plant-based health products. Since legumes contain a variety of bioactive components, such as polyphenols, alkaloids, fiber, saponins, lectins, and terpenoids, their consumption has been linked to lower the risk of diabetes, cancer, obesity, and coronary heart disease. Additionally, legumes contain nutraceutically significant enzyme inhibitors such as α-amylase, α-glucosidase, and γ-aminobutyric acid (GABA). Usually, legumes are consumed after being processed. Following processing, there is a decrease in levels of antinutrients and an increase in the bioavailability of essential nutrients (Priego-Poyato et al. 2021). If consumed in limited amounts, antinutrients in legumes that are considered as potentially harmful or toxic may also provide significant health benefits, thus play crucial role as nutraceuticals.

Herbal Nutraceuticals: Products and Processes, First Edition. Edited by Santosh Kumar Upadhyay and Sudhir Pratap Singh.

12.2 Legumes: A Superfood for Combating Malnutrition and Hunger

Legumes provide high protein content, low fat content, complex carbohydrates such as resistant starch and dietary fiber, and significant amount of vitamins and minerals content. They also provide bioactive compounds, which help in the prevention of chronic diseases. Bioactive principles present in legumes may help in reducing the risk of diabetes, heart diseases and various cancers, including colorectal cancer (Maphosa and Jideani 2017).

Due to numerous health and nutritional benefits, along with other economic and environmental factors essential for a sustainable future, legumes are an integral component of human nutrition and global food production system. For millions of people worldwide, legumes are reasonably affordable source of proteins. Therefore, cropping systems must incorporate them in order to maximize their value and efficient utilization in global food chain system. The Food and Agriculture Organization (FAO) announced 2016 as the International Year of Pulses in order to highlight the significance of legumes in improving global food and nutritional security. The FAO has promoted the legumes in the following areas (Martín-Cabrejas 2019):

i) Economic accessibility
ii) Excellent nutritive and nutraceutical values
iii) Contribution to global food security
iv) Climatic resilience
v) Biodiversity conservation

Millions of people around the world rely on legumes as their main source of nutrition, as they are among the healthiest foods available worldwide. This is particularly true in the regions where access to animal food products is limited. Protein content in legumes is twice the amount of proteins present in cereals. Nutritional quality of proteins is significantly enhanced when consumed with cereals. Legumes are an inexpensive ally against malnutrition since they offer a multitude of nutritional advantages. They have been widely recognized as a future superfood to achieve the goal of global nutrition security and zero hunger. Lower-income households can more easily obtain legumes and can improve the overall health and well-being as they prevent risk of the many chronic diseases. Due to changes in consumer dietary habits and increasing awareness of health benefits of legumes and their significance in reducing the risk of diabetes, cardiovascular disease, and certain cancers, most notably colorectal cancer, there has been an increase in demand for legumes globally (Pham and Luan 2021). Though promising improvements have been done in the past for improving the production of legumes worldwide, it is still less than cereals.

It is essential to emphasize the nutritional profiles of different legumes, use of underutilized legumes, generate value-added products from legumes, educate consumers about the nutritional value of legumes, and develop new strategies to promote the use of already-existing legumes in order to meet the ever-increasing demand for legumes. Due to their nutrient density, resilience to changing climates, economic viability, availability locally and adaptability, underutilized legumes have the potential to significantly reduce the overreliance on staple food crops.

Global food and nutrition security would be benefited from extensive research, especially on underutilized legumes, to uncover their beneficial qualities including high yield, good nutritional and sensory quality, and resilience to climate change. The use of underutilized beans as a functional food provides diet-related health advantages. Underutilized legumes have received little scientific attention; however, they may receive more attention if global partnerships are formed to integrate old knowledge with new developments in plant breeding, transformation, genome editing, and metabolomics. An integrated strategy, comprising thorough research efforts and effective extension activities, is needed to mainstream underutilized legumes into the human diet (Katoch et al. 2023a).

12.3 Nutritional Profile of Legumes

For vegetarians, legumes are primary source of protein due to their high protein content. They also provide a range of essential macro- and micronutrients. Carbohydrate and proteins from legumes have functional properties. These qualities maybe utilized for food and nonfood applications. The nutritional composition of legumes varies in different legumes. The nutritional profile of legumes is presented in the following sections.

12.3.1 Proteins

Legumes can fix atmospheric nitrogen, which is then utilized for protein synthesis (Mudryj et al. 2014), thus having high protein content. Protein constitutes approximately 20–40% of total nutrients in legumes, with lysine being the most abundant and methionine being the least abundant essential amino acid (Maphosa and Jideani 2017). In legumes, proteins content depends on variety, germination, and growing environment. Storage proteins constitute majority of legume proteins. Two major classes into which they are classified are albumins and globulins. According to Dahl et al. (2012), globulins are rich in glutamine, arginine, lysine, and aspartic acid amino acids and constitute more than 70% of total protein content of legumes. Because legume proteins are low in essential sulfur amino acids, they are inadequate protein source (Kouris-Blazos and Belski 2016). The low sulfur amino acid content in legumes is not entirely undesirable because it promotes calcium retention in body. The breakdown of sulfur-containing amino acids releases hydrogen ions, which demineralize bone and increase calcium excretion. As a result, the proteins in legumes may help the body retain calcium. Legume proteins assist in reducing the levels of low-density lipoproteins, which is crucial for the development of coronary heart diseases (Angeles et al. 2021).

Because of poor solubility and intermolecular disulfide linkages, legume proteins are low in digestibility. When a protein complexes with starch, the protein digestibility of legumes improves. Coupling of proteins with starch molecules results in formation of a hole in their structure; therefore, proteolytic enzymes are more accessible to catalyze proteolysis reaction (Ghumman et al. 2016).

12.3.2 Carbohydrates

Energy in diet primarily comes from carbohydrates content. Legumes are rich in carbohydrates (~50–65%) (Singh 2017) and mainly consist of monosaccharides, oligosaccharides like raffinose, and polysaccharides like starch (Ekanayake et al. 2000). Legumes starch has high retrogradation properties and varies in physicochemical and rheological properties. These properties reduce GI of starch by rendering it resistant to digestive enzymes (Singh 2011). Legumes are beneficial for reducing the risk of diabetes. It is well-known that foods with a low GI have ability to decrease blood sugar levels (Jenkins et al. 2012). Moreover, resistant starch, oligosaccharides, and dietary fiber, which constitute total carbohydrate content, facilitate bacterial growth in GI tract. Their fermentation favors butyric acid formation, which lowers the colon cancer risk. Legumes are also an important component of diets of patients suffering from celiac disease as they are devoid of glutens (Hosseini et al. 2018; Mlyneková et al. 2014). Grain legumes have fiber content ranging from 8.00% to 27.50%, of which 3.30–13.80% fiber is soluble. There are two forms of dietary fiber present in legumes: (i) water soluble fiber present in cotyledons and (ii) water insoluble fiber in seed coat. Dietary fiber-rich foods have been shown to reduce the incidence of various gastrointestinal disorders, including diverticulitis, gastroesophageal reflux disease, hemorrhoids and constipation, obesity, diabetes, hypertension, coronary heart disease, and stroke (Maphosa and Jideani 2017). Dietary fiber components from legumes are also used in industry. In the bakery, meat, extruded product, and beverage industries, they have been used as stabilizing agents, texturing agents, bulking ingredients, fat alternatives, and emulsion stabilizing agents.

12.3.3 Fat Content

Legumes are often low in fat and cholesterol. However, fat content is higher in some legumes such as peanuts (~45%), chickpeas (~15%), and soybeans (~47%) (Maphosa and Jideani 2017). Legume fat is primarily composed of mono- and polyunsaturated fatty acids (PUFA), with small amount of saturated fatty acids (Maphosa and Jideani 2017; Ofuya and Akhidue 2006). Polyunsaturated essential fatty acids are α-linolenic acid and α-linoleic acid, mainly present in legume fat. Since human body is unable to produce these PUFAs, they must be obtained from the diet.

12.3.4 Micronutrients

Legumes contain appreciable micronutrients content (vitamins and minerals). Some B group vitamins are more abundant than others vitamins, such as riboflavin, folic acid, and thiamine (Kouris-Blazos and Belski 2016; Ofuya and Akhidue 2006). They also contain considerable levels of vitamins K and E, which play crucial role in reducing risk of many diseases. For example, folic acid also reduces the incidence of neural tube abnormalities (Imbard et al. 2013). Legumes also contain good quantity of essential minerals, such as selenium, copper, chromium, iron, zinc, phosphorus, calcium, magnesium, and potassium (Brigide et al. 2014; Kouris-Blazos and Belski 2016). Minerals and vitamins are necessary for body to function properly. Legumes have lower bioavailability of iron than calcium and zinc. Consumption of legumes with vitamin C may improve the iron bioavailability (Hurrell and Egli 2010). Soybean has high calcium bioavailability, which is comparable to milk (Reinwald and Weaver 2010).

12.4 Nutraceutical Compounds in Legumes

Legumes are known to have high protein content (~20–40% protein) than cereals. They also contain appreciable level of carbohydrates, fiber, amino acids, and micronutrients. Legumes not only provide vital nutrients, but they also provide biologically active compounds having nutraceutical properties, such as antibacterial, antioxidant, anticarcinogenic, and angiotensin-converting enzyme (ACE) inhibitor activities. Raffinose family of oligosaccharides factors provide nondigestible carbohydrates stimulating ruminal microflora growth, thereby having prebiotic effects (Alles et al. 1999; Gupta and Abu-Ghannam 2012). However, flatulence is associated with oligosaccharide consumption in higher amount. Dietary fiber from legumes is beneficial for controlling blood glucose and cholesterol levels as well as for maintaining the colon health, which promotes efficient movement and excretion of waste from the body. The detailed account of different bioactive components present in legumes having nutraceutical values is presented in the following sections.

12.4.1 Bioactive Peptides and Proteins

Legumes contain biologically active proteins and peptides (Singh and Basu 2012). For example, angular in protein in *Vigna angularis*, α- and β-antifungal protein and unguilin in *Vigna unguiculata*, delandin in *Vigna umbellata,* and lectins in *Vigna sesquipedalis* (Katoch and Tripathi 2017). These proteins have anti-ACE enzyme activity with antimicrobial, antioxidant, anticancer, hypocholesterolemic, antidiabetic, and satiety effects (Chatterton et al. 2006; Seyler et al. 2007). Unguilin and α- and β-antifungal protein function as antidiabetic agents by blocking the actions of α- and β-glucosidase enzymes. Mice splenocytes responded mitogenically to ground bean lectins, which also reduce the viability of leukemia (L1210), hepatoma (HepG2), and leukemia (M1) cells. Lectin proteins also show antifungal activity against *Botrytis cinerea*, *Fusarium oxysporum*, *Mycosphaerella arachidicola*, *Colletotrichum gossypii,* and *Rhizoctonia solani* (Ye et al. 2000; Ye and Ng 2002).

12.4.2 Resistant Starch, Dietary Fiber, and Oligosaccharide Content

Legumes contain about ~50% and 65% carbohydrates primarily classified as starch, nonstarch polysaccharides, dietary fiber, and oligosaccharides (Bravo et al. 1998). Indigestible carbohydrates such as resistant starch are fermented by colonic bacteria to produce butyric acid, which has anti-colorectal cancer properties through preventing malignant alterations in intestinal cells (Whitehead et al. 1986). Additionally, resistant starch decreases stool pH (Malhotra 1982) and promotes fecal bulking (Cummings et al. 1992). Diets enriched with resistant starch are beneficial for diabetes management; therefore, legumes are particularly interesting in relation to diabetes management (Granfeldt et al. 1994). Short-chain fatty acids, such as acetic, propionic, and butyric acids, have prebiotic effects and are produced by lower gut bacteria in response to dietary fiber metabolism. While propionate functions as a natural HMG CoA reductase inhibitor and lowers cholesterol levels in stomach, propionic acid and butyric acid are metabolized in the liver to produce energy (Bueld et al. 1996). In addition to providing colon lining cells with a significant energy source, butyric acid has been proven to have anticancer properties that make it helpful in the treatment of colon cancer and ulcerative colitis. Dietary fiber lowers cholesterol level while simultaneously raising the level of short-chain fatty acids in the colon (Nishimura et al. 2000). According to Anderson et al. (2009), dietary fiber lowers blood pressure, improves blood glucose levels, and increases serum lipoprotein levels in diabetic people. Transient time of waste material in gastrointestinal tract is accelerated by insoluble fiber. As a result, the body is exposed to less harmful substances generated during digestion process. Dietary fiber can help in fighting weight loss as well. Although dietary fiber has no calories, its capacity to absorb water gives feeling of full stomach (Anderson et al. 2009).

The main component of total oligosaccharide content in legumes is Raffinose Family Oligosaccharides (RFO), which includes verbascose, stachyose, and raffinose. The growth of intestinal microflora, particularly *Bifidobacterium,* is stimulated by oligosaccharides, which have several positive health effects (Alles et al. 1999; Gupta and Abu-Ghannam 2012). With a faster rate of mineral absorption and a variety of bioactive characteristics, oligosaccharides are also helpful to human health due to their anticarcinogenic, antidiabetic, and anticardiovascular effects (Zhao 2021). However, consumption of oligosaccharide in higher amount is linked to flatulence production in gastrointestinal system.

12.4.3 Lipids, Phytosterols, and Phytoestrogens

Legumes seeds typically contain ~1–3% total lipids, out of which ~40–50% is neutral lipids, ~25–35% is phospholipids, and ~9% is glycolipids. Legume phospholipids have strong effects on serum lipoproteins levels (Anderson and Major 2002; Kirsten et al. 1993). Legumes also contain phytosterols, which are structurally related to cholesterol. Phytosterols possess a diverse range of biological activities. Diosgenin, campesterol, brassicasterol, sitosterol, stigmasterol, and guggulsterone are important phytosterols that have an impact on obesity (Racette et al. 2009). Increased dietary consumption of these compounds can also decrease blood glucose and LDL-C levels and prevent atherosclerosis (Tiwari et al. 2013). Since phytosterols have significant effect on intestinal genes and transcription factors expression, they play crucial role in regulation of cholesterol metabolism and transport in body.

Legumes also contain nonsteroidal phytochemicals known as phytoestrogens, which are structurally similar to gonadal estrogen hormone. Because of their polyphenolic nature, they have antioxidant properties that influence lipid and lipoprotein profiles, detoxify enzymes, and regulate steroid metabolism (Racette et al. 2009). They also interfere with the transport of calcium. They provide an alternative to hormone replacement therapy, which is better for cardiovascular system and may potentially lessen the symptoms associated with menopause.

12.4.4 Minerals and Vitamins

Legumes have appreciable amount of minerals and vitamins, essential to human health. Certain minerals such as selenium have antioxidative properties, and others function as cofactors in numerous enzyme reactions. Significant amounts of vitamins, including pyridoxine, thiamine, riboflavin, and folic acid, are also present and function as coenzymes in different enzymatic reactions. Legume contains traces of several vitamins such as vitamins C, K, and E. It is well recognized that vitamins C and E have antioxidant properties and prevent gastrointestinal tract from oxidizing vitamin A. They are also believed to keep cell membrane stability (Davies and Stewart 1987). Liver is the primary organ where vitamin K functions since it is essential for producing blood clotting proteins. Folic acid significantly decreases incidence of neural tube defects (NTDs), notably spina bifida, in infants. Because of its antioxidant properties, selenium displays a protective effect against lung, prostate, and colorectal cancers (Patterson and Levander 1997).

12.5 Nonnutritive Bioactive Compounds

Legumes contain a variety of compounds, which have positive as well as negative biological effects. A number of beneficial health effects, such as a decreased risk of intestinal diseases, coronary heart disease, and diabetes, have been reported for antinutrients, also known as antinutritional factors (or compounds) (ANFs or ANCs), at low concentrations (Duranti and Gius 1997). Beside nutritional significance, these compounds also hold significance in the plant defense (Katoch et al. 2016). Different nonnutritive bioactive compounds present in legumes have been presented under the following subheads.

12.5.1 Polyphenolic Content

Phenolic compounds protect individuals from oxidative damage by scavenging free radicals, establishing linkages with ions and preventing lipid peroxidation (Sundaram et al. 2013). Primary phenolic compound present in legumes include phenolic acids, flavonoids, anthocyanins, and tannins. They may occur in highly polymerized forms (tannins) or simple forms (phenolic acids). According to Kawsar et al. (2008), phenolic acids are often found as derivatives of caffeic and quinic acid. Phenolic acids are either esters of benzoic acid or cinnamic acid. Phenolic acids have been reported to have therapeutic advantages, such as platelet aggregation, detoxification of harmful chemicals, and inducing apoptosis in diseased cells (McDougall and Stewart 2005; Soobrattee et al. 2005). Moreover, these compounds are also linked to the hardness, astringency, flavor, color, and aroma of legume seeds.

Polyphenolic compounds possess antibacterial, antimutagenic, antiviral, anticancer, anti-inflammatory, and immunomodulatory properties (Aaby et al. 2004; Dos Santos et al. 2006; Lairon and Amiot 1999; Lule and Xia 2005; Perez 2003; Taguri et al. 2006). Polyphenols also protect against malignancies linked to hormones and cardiovascular disorders (Gupta and Abu-Ghannam 2012). Certain phenolic chemicals have been shown to have antiallergic and gastro-protective qualities such as flavonoids. Additionally, these dietary antioxidants may protect neural network from degeneration (Daniel 2003). High molecular weight polyphenolic compounds with multiple hydroxyl groups in their structure are known as tannins. They are able to form complexes with carbohydrates and proteins. Tannins are classified as hydrolyzable or condensed (nonhydrolyzable) based on the degree of condensation of the flavonoids (Singh et al. 2020). Proanthocyanidins (PACs), another name for condensed tannins, are polymeric and oligomeric flavonoids that, when heated to a high temperature, release anthocyanidins and catechins (Singh et al. 2020). Because tannins can attach to proteins, they assist in the removal of toxins from intestine. Tannins also contribute to oral hygiene by preventing the growth of germs that cause tooth decay (Niveditha and Sridhar 2012).

One of the important phenolic compounds in legumes is flavonoids that have significant influence of their color and flavor (Singh et al. 2017; Yang et al. 2018). In general, they have two aromatic rings joined by a bridge made of three carbons, which usually forms heterocyclic ring C. These compounds are low in molecular weight. Anthocyanins and anthoxanthins are two categories of flavonoids. The main flavonoids present in legume seeds are flavonols, flavanols, flavones, and anthocyanins. (Amarowicz and Pegg 2008). Flavonols and flavones can influence the color of the anthocyanins group (Wang et al. 2008). Acting as antioxidants, flavonoids protect body from a variety of diseases including aging, inflammation, cancer, autoimmune disorders, cataract, and arteriosclerosis (Katoch et al. 2023b).

Isoflavones have antioxidant activities and lower the risk of several diseases, such as osteoporosis, cancer, and cardiovascular disease. According to Kim (2021), menopausal symptoms can also be treated with isoflavones. By preventing LDL oxidation, natural phytoestrogens genistein and isoflavones daidzein can reduce the risk of atherosclerosis (Shashank et al. 2015). Daidzein and genistein have anticancer properties. In cancer cells, genistein helps in activation of apoptosis and angiogenesis. It decreases bioavailability of sex hormones. Daidzein stimulates human leukemia cells (HL-60) and B16 melanoma cells to differentiate. Additionally, daidzein helps in treatment of alcoholism by inhibiting ALDH-I enzyme. The diets including legume seeds have been linked to a lower incidence of cancer, according to epidemiological research (Peter et al. 1997).

12.5.2 Saponins

Saponins combine with steroidal or triterpenoid hydrophobic aglycone (sapogenin) and a carbohydrate moiety and form persistent, soap-like foams in aqueous solutions. Triterpenoid saponins are frequently found in legumes. Saponins content in chickpea, mung bean, pigeon pea, and rice bean varies from 0.05% to 0.23% (Jain et al. 2009). Katoch (2013) reported that saponin content of rice bean varies from 1.20 to 3.10 mg/100 g. Saponins' physiological activity may result from their ability to interact with mucosal cell membranes and their surface-active interactions with other digesting components. The latter can occasionally result in significant modifications in membrane biochemistry. These physiological interactions can inhibit the nutrient uptake (Katoch and Tripathi 2017).

Despite the fact that saponins lower the nutritive value of legumes, they have positive beneficial health effects such as antifungal, antiparasitic, hypocholesterolemic, immunomodulatory, expectorant, and vasoprotective activities (Sparg et al. 2004). Additionally, saponins also have anticancer properties in gut. Colon cancer is caused by metabolism of bile acid by intestinal microbes, which also produce secondary bile acids. By binding with bile acid and reducing its availability to microbes, saponins can help in stopping the colon from generating compounds that might cause cancer. Saponins increases the amount of cholesterol excreted in stools and consequently plays a significant function in human nutrition as they reduce the risk of heart disease and cholesterol levels in the body. Adjuvants based on saponins have a distinctive capability of inducing a cell-mediated immune response and formation of antibodies. Additionally, the adjuvant activity requires a minimal dosage (Oda et al. 2000). Because saponins stimulate interleukins and interferons synthesis, they have immunomodulatory effects and can be used as adjuvant in the production of vaccines (Sharma et al. 2020).

12.5.3 Lectins

Lectins are ubiquitously distributed proteins and are capable of cross-linking with carbohydrates and glycoconjugates preferentially and reversibly (Katoch and Tripathi 2021). Legume lectins are well-characterized family of plant lectins. Lectins are used for determining erythrocyte polyagglutinable conditions. They have shown to be helpful for investigating the structures of cell surfaces and associated membrane activities. Lectins are able to recognize changes in the structural profiles of glycans in certain disorders. These changes include increased fucosylation, sialylation, and branching patterns of complex carbohydrates. This characteristic makes lectins useful biomarker of diseases. Since lectins can recognize a variety of sugar structures, they are known to have antiviral, antifungal, and antibacterial activities (Katoch and Tripathi 2021).

The primary mechanism by which lectins exert antibacterial action is by their binding to extracellular glycans or the carbohydrate components of bacterial cell walls. Antibacterial lectins make strong interactions with tetrapeptide, *N*-acetyl-D-muramic acid (NAM), and *N*-acetyl-D-glucosamine (NAG) components present in the cell wall of gram-positive bacteria, as well as with lipopolysaccharides (LPS) present in gram-negative bacteria cell wall (Lagarda-Diaz et al. 2017; Qadir et al. 2013). This interaction inhibits host cytoplasm from penetrating the membrane and limits the ability of bacteria to interact with carbohydrates-binding proteins (Mishra et al. 2019). When lectins bind to fungal hyphae, they inhibit the development of fungus, decrease the absorption of nutrients, and delay the germination process of spores. In addition to this, lectin binding to the fungal cell wall causes hyphae to enlarge, cell content to vacuolate, and ultimately hyphal cell wall to lyse. The capacity of lectins to connect with complex glycans that are added as a result of posttranslational changes on viral envelope proteins, prevent the virus from interacting with host cells (Akkouh et al. 2015; Barton et al. 2014). It was first reported by Müller et al. (1988) that the D-mannose specific lectin from *Gerardiasa vaglia* inhibits HIV infection in H9 cells by inhibiting the development of syncytiums in HIV-1/human lymphocyte and HTLVIIIB/H9-Jurkat cell systems. According to Keyaerts et al. (2007) and Nascimento da Silva et al. (2020), lectins showing specificity in binding with mannose, glucose, and N-acetylglucosamine sugars are potential antiviral agent against coronaviruses. The hyacinth bean (*Lablab purpureus*) contains a glucose/mannose lectin called DLL-I (Flt3 receptor-interacting lectin (FRIL)). DLL-I lectin was recently isolated and investigated for antiviral properties against SARS-CoV-2 virus strain (hCoV-19/Taiwan/NTU04/2020) (Liu et al. 2020).

Cancer cells display altered glycan structures on their cell surfaces and this maybe used to develop lectin-based tools for cancer treatment and diagnostic studies (Brooks and Carter 2001; Konno et al. 2002). Legume lectins induce apoptosis in cancer cells. Lectins also induce cancer cells to produce reactive oxygen species (ROS) and undergo autophagic and apoptotic cell death when they bind to glycoconjugates produced on cancer cells (Mukhopadhyay et al. 2014; Panda et al. 2014). Furthermore, higher ROS levels in cancer cells trigger caspases and SEK/JNK pathways, which lead to apoptotic death of cancer cells (Polito et al. 2009). Tumor suppressor proteins (p38 and p53) are also expressed more often in cancer cells as a result of lectin binding with glycoconjugates (Hostanska et al. 2003; Lyu et al. 2002). Furthermore, lectin binding also alters signaling pathways that affect the expression of Bax, Fas, Bcl-XL and Bcl-2proteins; disturbs mitochondrial membrane potential; and triggers apoptosis in mitochondria via ROS-induced p38-p53 pathway (Lam and Ng 2010; Liu et al. 2009).

12.5.4 Phytic Acid

Phytic acid, also known as myo-inositol hexaphosphate or InsP6, plays a key role in storage of phosphates in seeds to be utilized as a source of energy during growth and development. Phytates, commonly referred to as salts of phytic acid, regulate various biological processes such as endocytosis, export of messenger RNA from nucleus to cytoplasm, chromatin remodeling, DNA repair, and maybe even hormone signaling (Kakati et al. 2010). In addition to its favorable effects in human diets due to its anticarcinogen properties, phytic acid also plays a major role in plant metabolism, stress, and disease resistance (Hussain et al. 2010). Additionally, it has been proposed that phytic acid is a component of the structure that may prevent HIV-1 viral genome transcription (Filikov and James 1998). Both endogenous and exogenous kinds of phytic acid may have hypoglycemic effects, thus play crucial role in the diabetes management. Phytic acid may also play a significant role in lowering serum levels of cholesterol and lipids and consequently lowering the risk of heart disease (Jenab and Thompson 2002). Studies conducted *in vitro* and *in vivo* have reported the strong anticancer (preventive as well as therapeutic) properties of phytic acid and inositol hexaphosphate (InsP6) (Shamsuddin 2002). The main component of the majority of inositol phosphates in cells is myo-inositol. Myo-inositol and InsP6 function together to inhibit progression of cancer. Myo-inositol has been proven to be an effective preventive measure against stomach, liver, and lung cancer (Nishino et al. 2000).

Phytic acid makes strong interactions with minerals, carbohydrates, charged proteins, amino acids, and enzymes (Figure 12.1). Because of their insoluble nature and difficulty in digestion, the resultant complexes usually have lower nutrient absorption potential. Protein and carbohydrates are broken down more slowly by phytic acid because it binds to digestive tract enzymes. The undigested starch travels to colon converted into short-chain fatty acids and have prebiotic effects. Additionally, polyvalent cations including iron, zinc, and calcium that are necessary for DNA synthesis and cell growth can chelate with phytic acid. Additionally, it affects gene expression, cell signaling cascades, and signal transduction (Jenab and Thompson 2002). Phytate can also assist in preventing the formation of cavities, plaque, and tartar in teeth by reducing the solubility of calcium, fluoride, and phosphate, the three main components of enamel that protect teeth from demineralization. Phytate is important in terms of nutrition and diabetes mellitus prevention and management. By selectively binding to serine-threonine phosphatases, it regulates calcium channel function and sustains insulin secretion. Insulin release is subsequently triggered by opening of intracellular calcium channels (Katoch and Tripathi 2017).

12.5.5 Enzyme Inhibitors

Legumes contain compounds that inhibit endogenous proteases, amylases, lipases, phosphatases, and glycosidases enzymes. Two types of serine protease inhibitors (PIs) are mainly present in legumes are Kunitz type (~20–25 kDa) and Bowman-Birk type (8–16 kDa). Because of their double-headed structure, these inhibitors can interact with two enzymes simultaneously. Due to disulfide linkage in their structure, these inhibitors display great stability over a broad pH and temperature range. From

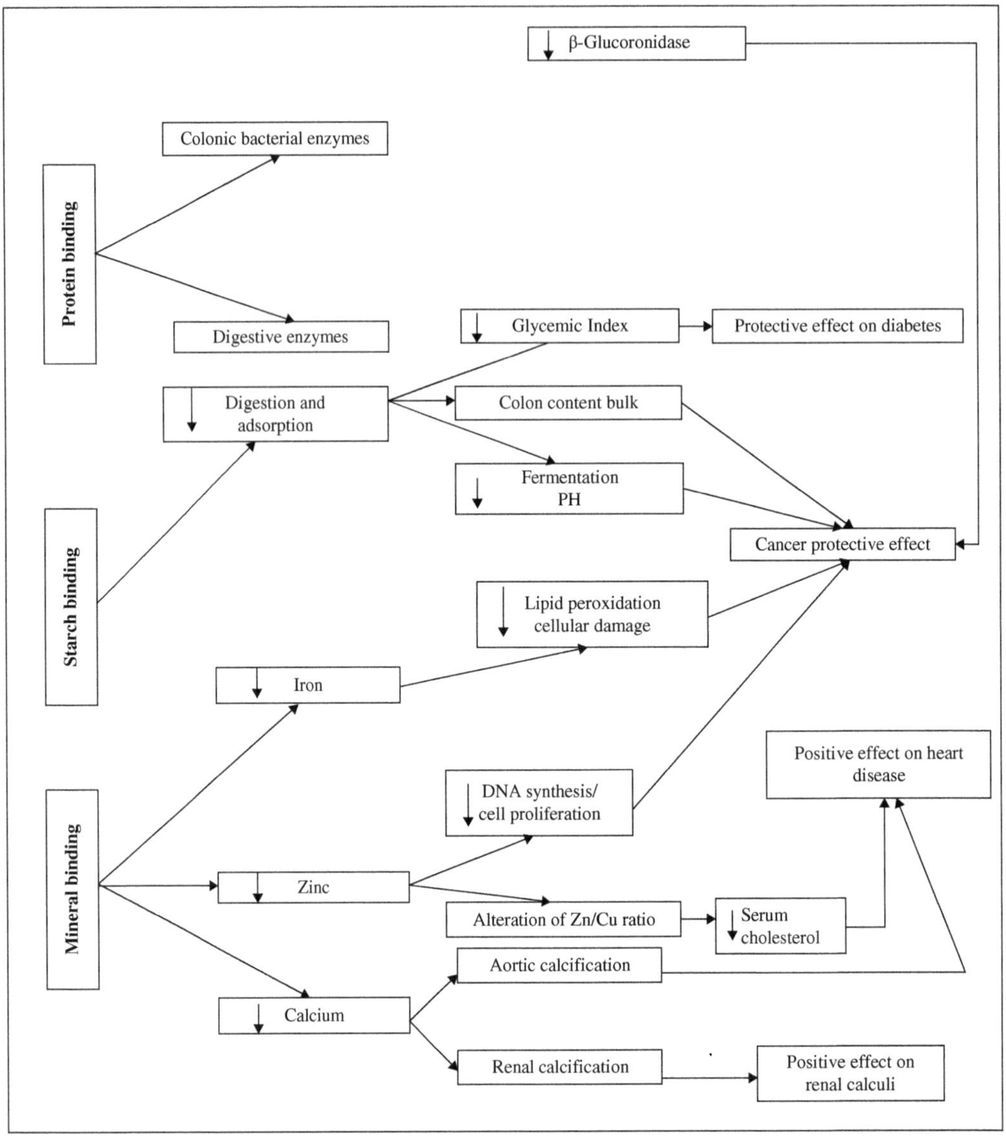

Figure 12.1 Nutraceutical potential of phytic acid. *Source:* Jenab and Thompson (2002); Katoch and Tripathi (2017).

a nutritional perspective, it is imperative to investigate serine PIs, such as trypsin and chymotrypsin (Katoch 2013), since their functions in the pathophysiology of numerous diseases, from cancer and inflammatory conditions to degenerative diseases, have been well established (Katoch et al. 2014).

a) ***PIs:*** PIs have been identified in many plant families and are especially abundant (~1–10% of total proteins) in storage organs (Katoch et al. 2014). Double-headed PIs called Bowman-Birk inhibitors (BBIs) can inhibit the enzymes chymotrypsin and trypsin. According to Kennedy (1998), antichymotryptic activity suppresses carcinogen-induced transformation more effectively than the antitryptic activity. BBIs show anti-inflammatory activity in addition to their anti-carcinogenic effects by inhibiting the protease that promotes inflammation (Ware et al. 1999). It has been demonstrated that the protein hydrolysates of peas, chickpeas, and mung bean exhibit ACE inhibitor activity. Legume ACE inhibitors have potential to improve cardiovascular health because ACE enzyme is essential in regulating blood pressure. Numerous epidemiological studies have identified inverse relationships between consumption of legumes and death rate from colon cancer as well as risks of gastric, pancreatic, and prostate cancer (Jain et al. 1999).

b) ***α-Amylase and α-glucosidase inhibitors:*** α-amylase inhibitors bind to the α-amylase enzyme isolated from porcine pancreas, which hydrolyzes starch into small oligosaccharides. This process delays the absorption of carbohydrates and lowers the blood sugar level (Gong et al. 2020). The most efficient remedy for hyperglycemia and its harmful physiological effects, especially in type 2 diabetes, has been demonstrated to be the use of α-glucosidase inhibitors (Hossain et al. 2020). According to Hossain et al. (2020), α-glucosidase inhibitors function by competitively inhibiting enzyme that hydrolyzes carbohydrates. This results in a delay in glucose absorption in small intestine, which in turn regulates hyperglycemia.

c) ***Tyrosinase inhibitor:*** an essential enzyme in the synthesis of melanin is tyrosinase. The hydroxylation of L-tyrosine to L-dopa and the oxidation of L-dopa to dopaquinone are the two distinct reactions in melanin synthesis pathway catalyzed by tyrosinase enzyme (Pillaiyar et al. 2017). Tyrosinase inhibitors like arbutin, hydroquinone, kojic acid, azelaic acid, L-ascorbic acid, ellagic acid, and tranexamic acid are important skin-whitening agents and have been used to treat dermatological hyperpigmentation (Pillaiyar et al. 2017).

The bioactive principles in legumes and their associated health benefits have been listed in Table 12.1.

Table 12.1 Bioactive components in legumes and their health benefits.

Sr. No.	Bioactive components	Beneficial health effects
1.	Protein (angularin, delandin, unguilin, Lunasin)	Anti–HIV activity, anticancer activity
2.	Minerals and vitamins	Antioxidant properties and lower incidence of cardiovascular diseases
3.	Fiber	Reducing the risk of colon cancer and cholesterol levels
4.	Oligosaccharides	Reducing the risk of colon cancer and prebiotic effects
5.	Saponins	Hypocholesterolaemic and anticarcinogenic activity
6.	Lectins	May aid in treating obesity and the formation of tumors Reduces mineral bioavailability, protects against oxidative DNA damage
7.	Phytates	Acts as anticarcinogen Antioxidant properties and protection from DNA damage
8.	Phenolic compounds	Reduces the risk of colon cancer, coronary heart disease, and menopause; inhibits the cyclooxygenase and angiotensin-converting enzyme (ACE)
9.	Flavonoids, isoflavones (phytoestrogens)	Act as antioxidant and anticarcinogenic Boost up bioavailability of NO
10.	Physterols	Protection against DNA damage, oxidative stress, heart diseases, metabolic disorders, and atherosclerosis
11.	Isoflavonoids (genistein and daidzein)	Protection against osteoporosis and cardiovascular diseases; inhibition of angiogenesis and cell cycle progression
12.	Protease inhibitors	Anticarcinogenic
13.	Amylase inhibitors	Reduces dietary starch utilization, beneficial for treating diabetes and obesity

Source: Katoch and Tripathi (2017); Singh and Basu (2012).

12.6 Nutraceutical Activities of Bioactive Compounds Present in Legumes

Different antioxidants are required by human body to counteract and avoid the adverse effects of free radicals produced by various metabolic processes. Free radicals may lead to harmful changes to lipids, proteins, and nucleic acids within cells. They can also act as a catalyst for various diseases, including cancer, diabetes, and heart disease (Katoch et al. 2023a). Natural endogenous antioxidants present in legumes include tannins, vitamin C, phenolic compounds, and tocopherols. In addition to their antioxidant properties, they can also function as metal chelators, reducing agents, and radical scavengers (Zhao et al. 2014). Antioxidant activity of legumes is positively correlated with their total phenolic content. Yao et al. (2012) reported a positive correlation between the total phenolic and flavonoid levels and antioxidant activity in rice bean seeds. Different processes like fermentation, germination, and heat processing have significant impacts on antioxidant properties of legumes. Antioxidant properties maybe improved after processing of seeds such as germination and fermentation. Different peptides and proteins from legumes also have antioxidant potential (Chen et al. 2017).

12.6.1 Anticancer Activities

Legumes contain phytochemicals that provide protection against cancer. Bioactive proteins and peptides, polyphenolic compounds and short-chain fatty acids, and other bioactive principles isolated from legume seeds possess nutraceutical properties, which minimize the risk of cancer and induce pro-apoptotic, antiproliferative, and anti-inflammatory effects (Rao et al. 2018). Numerous research investigations have shown strong anti-inflammatory properties of phenolic compounds and flavonoids are present in legumes. They regulate the levels of different inflammatory cytokines or mediators, like nitric oxide (NO), lipoxygenase (LOX), cyclooxygenase-2 (COX-2), tumor necrosis factor (TNF-α), nuclear factor-kappa (NF-κB), iNOS, and interleukins 1 (IL-1, 6, and 10) (Zhang et al. 2018). Numerous bioactivities have also been uncovered for legume peptides. It has been demonstrated that soybean saponins exert anticancer properties (Kerwin 2004). GABA and protein hydrolysates, two anti-inflammatory peptides derived from legumes, have anticancerous properties. A soybean-derived anticancer peptide called lunasin causes apoptosis and prevents chemical carcinogen-induced transformation of murine fibroblast cells into malignant foci (González-Montoya et al. 2017). ConA, a lectin from Jack bean, mediates p73-dependent apoptosis in p53-null cells by blocking Akt survival signaling and triggering Foxo-Bim signaling (Amin et al. (2007). Lectins induce ER stress and unfolded protein response (UPR) in tumor cells, resulting in upregulation of IRE-1α, CHOP, and caspase 12 enzyme (Chan et al. 2012). Legume lectins not only induce apoptosis but also use autophagy to exert anticancer activity (Bhutia et al. 2019).ConA lectin initiates autophagy in hepatoma cells via BNIP3-mitochondrial pathway, in which lectin permeability increases to mitochondrial surface as a result of mitochondrial depolarization. By modulating the PI3K/Akt/mTOR and MEK/ERK pathways, Con A also promotes autophagy in cancer cells (Roy et al. 2014). Furthermore, it has been revealed that ConA-mediated autophagy is significantly influenced by macrophage migration inhibitory factor (MIF) (Lai et al. 2015).

12.6.2 Protection from Cardiovascular Ailments

Cardiovascular disorders, such as stroke and coronary heart disease, are known to be linked with dyslipidemia, hypertension and type 2 diabetes. Legumes include a variety of bioactive substances that are important for preventing cardiovascular disease, including fiber, oligosaccharides, ACE, vitamins, and minerals. Venkidasamy et al. (2019) reported that legume phytosterols prevent intestinal absorption of cholesterol and subsequently decrease the levels of LDL-C and TAG content. In addition to this, they also increase HDL-C levels in blood, thus lower the risk of coronary heart diseases. Moreover, β-sterols have been shown to reduce blood pressure, serum cholesterol, and risk of coronary heart disease (Venkidasamy et al. 2019). In addition to improving body metabolism, these bioactive compounds also lower blood pressure, insulin resistance, serum cholesterol levels, and chronic inflammation. They also affect fibrinolysis and coagulation, which are crucial for formation of atherosclerotic plaques (Satija and Hu 2012). Legumes consumption has cardiovascular protective benefits as they reduce blood pressure and maintain lipids level in the body.

12.6.3 Antidiabetic Activities

Since legumes have low GI, they reduce the levels of lipids, insulin, and epididymal adipocyte volume in the blood plasma. Additionally, it has been proven that legumes regulate blood glucose levels, which helps in preventing diabetes. Furthermore, it has been documented that short-chain fatty acids and inhibition of α-amylase and α-glucosidase result in hypoglycemia and hypocholesterolemic effects (Singhal et al. 2014). Feeding Adzuki bean extracts to mice increases lipolysis, decreases the mice final body weight, and significantly lowers the level of liver lipids, total cholesterol, LDL-C, and serum triglyceride (Liu et al. 2017).

12.6.4 Hepato-protective Activities

Hepatotoxins stimulate the production of proinflammatory cytokines in liver cells. These cytokines trigger the generation of ROS that lead to inflammation and liver damage. Antioxidant and anti-inflammatory phytochemicals can prevent damage to liver cells (Mohd Ali et al. 2013). It has been reported that extracts from adzuki bean, black bean, chickpea, mung bean, and rice bean have similar hepato-protective effect. Mekky et al. (2016) reported substantial reduction in the pathological defects of liver cell structure in animals administrated with legume seeds extract.

12.7 Nutraceutical Uses of Legumes in Traditional Medicine System

The traditional Indian medicine system, Ayurveda, is one of the oldest systems of curing involving the use of herbal medicines. The two main texts of Ayurveda, the Charaka and Shushrutha Samhita, discuss numerous herbal cures for a wide range of ailments. It recommends a vegetarian diet, which can fuel our bodies and balance our metabolic processes.

According to Ayurveda, the foundation of healing is a healthy diet. Indian traditional meals are often considered to be balanced or part of the proper diet recommended by Ayurveda for numerous ailments. Indian traditional food is often referred to as "functional foods" since it contains a variety of body-healing chemical compounds that can be improved through different processing techniques including germination, fermentation, malting, etc. (Sarkar et al. 2015). Indian cuisine has its roots in the early Vedic era, when Aryans lived here. During this period, the initial concept for nutraceuticals was developed. Aryans believed that food served as the foundation for the cosmic moral cycle in addition to being necessary for life sustenance. These include cereals, pulses, etc. According to Ayurveda, beans are a crucial component of a balanced diet (Sarkar et al. 2015). In traditional Indian medicine, where nutrition is prioritized as part of treatment, lentils are specifically suggested as a convalescing food (Anand et al. 2017). More significantly, lentils in combination with rice were the main component of Aryan diet as they provide complimentary nutrients to rice. Khichadi (rice in combination with lentils) is a popular example of a diet that is extremely nourishing.

Mudga (green gram), which has a high iron and potassium content, is one of the most significant member of the samidhanyavarga. Modern food research has proven its cardiovascular and antiobesity benefits. Horse gram has antihelminthic properties and is well-known for its effects on cough, stomach glands, and urinary calculi. Ayurveda has detailed description of several pulses and their different biological functions. For numerous ailments, Ayurvedic doctors typically recommend inclusion of legumes in the diet (Nishteswar 2016; Sharma and Rao 2014). Legumes have a crucial role in the vegetarian diet since they support the structure of all tissues, particularly muscle. Legumes are suggested by Ayurvedic diet to be included in each and every meal of the day. Certain legumes, including mung bean, are highly valued in Ayurveda because they are easier to digest and more nutritious than other legumes. When cooked to a butter-soft consistency, mung bean can even be eaten by people with weak digestive systems, such as the elderly, sick, and newborns, among others. Herbal therapy in traditional Chinese medicine has long appreciated the nutraceutical uses of rice bean (*Vigna umbellate*). In Chinese dietetic therapy, rice bean has been utilized as an adjuvant in combination with other medicines. In the traditional Chinese medical system, a mixture of rice bean seeds and leaves extract is given orally to treat enuresis or frequent urination. To treat acute mastitis, traditional Chinese medicine combines rice bean seed, bitter wine, and powdered *Illicium anisatum* leaves. Coix Gruel (rice bean porridge) is consumed to ease constipation, reduce spleen swelling, relax mind, stimulate the heart, and revitalize the spleen. According to Wei et al. (2015), rice bean coix gruel is the best food for obese patients and pregnant women. Hemorrhoid bleeding can be treated using dry rice bean seeds, which have been soaked in vinegar. Rice bean paste reduces edema and decreases exudation. In the traditional Chinese medical system, rice bean is also used to treat varicella, acute lymphadenitis, boil sores, acute parotitis, and herpes zoster (Katoch et al. 2023b).

12.8 Legume-based Nutritionally Fortified Food

The development of highly consumed food products like milk or bread, which can be made even more nutraceutical by adding extracts from legume seeds has been the focus of research over the past decade (Alu'datt et al. 2017; Hernandez-Aguilar et al. 2020; McClements 2020). Sprouting lentil extracts combined with wheat flour have resulted in breads that are high in nutrients (Hernandez-Aguilar et al. 2020). Milk substitutes from legumes such as soybean are excellent for incorporating essential nutrients and nutraceuticals into an individual's diet. In such a scenario, it is simple to fortify meals with micronutrients, such as vitamin D, calcium, vitamin B12, and omega-3 fatty acids, by using legume milk substitutes such as soybean milk (McClements 2020). Replacing the egg yolk in mayonnaise preparation with proteins from legumes such as chickpeas, beans, or lupine could improve nutraceutical content in mayonnaise and turn it into a functional food (Alu'datt et al. 2017).

12.9 Conclusion

Legumes are staple foods that are crucial for global food and nutritional security. They are the rich source of proteins and have been considered as an alternative for animal proteins. In addition to protein, legumes also contain fiber, carbohydrates, amino acids, and appreciable levels of vitamins and minerals. Numerous bioactive compounds have been isolated and purified from legumes with nutraceutical properties. These compounds have positive impact on human health, as they prevent diseased conditions like cancer, diabetes, digestive tract disorders, cardiovascular diseases, and obesity. Because of many health advantages and prospective use as functional foods, the bioactive principles have attracted the attention of pharmaceutical industries. Underutilized food legumes such as rice bean have potential to produce nutraceutical foods. Due to the growing demand for protein and other essential nutrients worldwide and to reduce the risk of many diseases incidence, legumes are likely to play a more significant role in the diet of people in the near future.

References

Aaby, K., Hvattum, E., and Skrede, G. (2004). Analysis of flavonoids and other phenolic compounds using high-performance liquid chromatography with coulometric array detection: relationship to antioxidant activity. *Journal of Agricultural and Food Chemistry* 52: 4595–4603.

Akkouh, O., Ng, T.B., Singh, S.S. et al. (2015). Lectins with anti-HIV activity: a review. *Molecules* 20: 648–668.

Alles, M.S., Hartemink, R., Meyboom, S. et al. (1999). Effect of transgalactooligo-saccharides on the composition of the human intestinal microflora and on putative risk markers for colon cancer. *American Journal of Clinical Nutrition* 69: 980–991.

Alu'datt, M.H., Rababah, T., Alhamad, M.N. et al. (2017). Preparation of mayonnaise from extracted plant protein isolates of chickpea, broad bean and lupin flour: chemical, physiochemical, nutritional and therapeutic properties. *Journal of Food Science and Technology* 54 (6): 1395–1405.

Amarowicz, R. and Pegg, R.B. (2008). Legumes as a source of natural antioxidants. *European Journal of Lipid Science and Technology* 110 (10): 865–878.

Amin, A.R.M.R., Paul, R.K., Thakur, V.S., and Agarwal, M.L. (2007). A novel role for p73 in the regulation of Akt-Foxo1a-Bim signaling and apoptosis induced by the plant lectin, Concanavalin A. *Cancer Research* 67: 5617–5621.

Anand, P., Nathani, S., Rama Murthy, A., and Sharma, T. (2017). Nutritive and therapeutic potential of pulses in diabetes mellitus. *International Journal of Ayurveda Medical Sciences* 2 (2): 48–53.

Anderson, J.W., Baird, P., Davis, R.H. Jr. et al. (2009). Health benefits of dietary fiber. *Nutrition Reviews* 67 (4): 188–205.

Anderson, J.W. and Major, A.W. (2002). Pulses and lipaemia, short- and long-term effect: potential in the prevention of cardiovascular disease. *The British Journal of Nutrition* 88: 5263–3271.

Angeles, J.G.C., Villanueva, J.C., Uy, L.Y.C. et al. (2021). Legumes as functional food for cardiovascular disease. *Applied Sciences* 11: 5475. https://doi.org/10.3390/app11125475.

Barton, C., Kouokam, J.C., Lasnik, A.B. et al. (2014). Activity of and effect of subcutaneous treatment with the broad-spectrum antiviral lectin griffithsin in two laboratory rodent models. *Antimicrobial Agents and Chemotherapy* 58: 120–127.

Bhutia, S.K., Panda, P.K., Sinha, N. et al. (2019). Plant lectins in cancer therapeutics: targeting apoptosis and autophagy-dependent cell death. *Pharmacological Research* 144: 8–18.

Bravo, L., Siddhuraju, P., and Saura-Calixto, F. (1998). Effect of various processing methods on the in vitro starch digestibility and resistant starch content of Indian pulses. *Journal of Agricultural and Food Chemistry* 46: 4667–4674.

Brigide, P., Canniatt-Brazaca, S.G., and Silva, M.O. (2014). Nutritional characteristics of biofortified common beans. *Food Science and Technology* 34 (3): 493–500.

Brooks, S.A. and Carter, T.M. (2001). N-acetylgalactosamine, N-acetylglucosamine and sialic acid expression in primary breast cancer. *Acta Histochemica* 103: 37–51. https://doi.org/10.1078/0065-1281-00576.

Bueld, J.E., Bannenberg, G., and Netter, K.J. (1996). Effects of propionic acid and pravastatin on HMG-CoA reductase activity in relation to forestomach lesions in the rat. *Pharmacology & Toxicology* 78 (4): 229–234.

Chan, Y.S., Wong, J.H., Fang, E.F. et al. (2012). Isolation of a glucosamine binding leguminous lectin with mitogenic activity towards splenocytes and anti-proliferative activity towards tumor cells. *PLoS One* 7: e38961. https://doi.org/10.1371/journal.pone.0038961.

Chatterton, D.E.W., Smithers, G., Roupas, P., and Brodkorb, A. (2006). Bioactivity of β-lactoglobulin and α-lactalbumin–technological implications for processing. *International Dairy Journal* 16 (11): 1229–1240.

Chen, Z., Wang, J., Liu, W., and Chen, H. (2017). Physicochemical characterization, antioxidant and anticancer activities of proteins from four legume species. *Journal of Food Science and Technology* 54 (4): 964–972.

Cummings, J.H., Bingham, S., Heaton, K.W., and Eastwood, M.A. (1992). Faecal weight colon cancer risk, and dietary intake of non-starch polysaccharides (dietary fiber). *Gasteroenterology* 103: 1783–1789.

Dahl, W.J., Foster, L.M., and Tyler, R.T. (2012). Review of the health benefits of peas (*Pisum sativum* L.). *The British Journal of Nutrition* 108 (Suppl 1): S3–S10.

Daniel S (2003) An investigation into the neuro-protective prospective of curcumin. Electronic thesis collection TR 03-17. Rhodes University.

Davies, S. and Stewart, A. (1987). *The Drug-free Guide to Better Family Health Nutritional Medicine*, 543. Bungay: Richard Clay Ltd.

Dos Santos, M.D., Almeida, M.C., Lopes, N.P., and de Souza, G.E.P. (2006). Evaluation of the anti-inflammatory, analgesic and antipyretic activity of the natural polyphenol chlorogenic acid. *Biological & Pharmaceutical Bulletin* 29: 2236–2240.

Duranti, M. and Gius, C. (1997). Legume seeds: protein content and nutritional value. *Field Crops Research* 53: 31–45.

Ekanayake, S., Jansz, E.R., and Nair, B.M. (2000). Literature review of an underutilized legume: *Canavalia gladiata* L. *Plant Foods for Human Nutrition* 55: 305–321.

Filikov, A.V. and James, T.L. (1998). Structure-based design of ligands for protein basic domains: application to the hiv-1 tat protein. *Journal of Computer-Aided Molecular Design* 12: 229–240. https://doi.org/10.1023/A:1007949625270.

Ghumman, A., Kaur, A., and Singh, N. (2016). Functionality and digestibility of albumins and globulins from lentil and horse gram and their effect on starch rheology. *Food Hydrocolloids* 61: 843–850.

Gong, L., Feng, D., Wang, T. et al. (2020). Inhibitors of α-amylase and α-glucosidase: potential linkage for whole cereal foods on prevention of hyperglycemia. *Food Science & Nutrition* 8 (12): 6320–6337.

González-Montoya, M., Cano-Sampedro, E., and Mora-Escobedo, R. (2017). Bioactive peptides from legumes as anticancer therapeutic agents. *International Journal of Cancer and Clinical Research* 4: 81.

Granfeldt, Y., Liljeberg, H., Drews, A. et al. (1994). Glucose and insulin responses to barley products. Influence of food structure and amylose-amylopectin ratio. *The American Journal of Clinical Nutrition* 59: 1075–1082.

Gupta, S. and Abu-Ghannam, N. (2012). Probiotic fermentation of plant based products: possibilities and opportunities. *Critical Reviews in Food Science and Nutrition* 52: 183–199.

Hernandez-Aguilar, C., Domínguez-Pacheco, A., Palma Tenango, M. et al. (2020). Lentil sprouts: a nutraceutical alternative for the elaboration of bread. *Journal of Food Science and Technology* 57: 1817–1829.

Hossain, U., Das, A.K., Ghosh, S., and Sil, P.C. (2020). An overview on the role of bioactive α-glucosidase inhibitors in ameliorating diabetic complications. *Food and Chemical Toxicology* 145: 111738.

Hosseini, S.M., Soltanizadeh, N., Mirmoghtadaee, P. et al. (2018). Gluten-free products in celiac disease: nutritional and technological challenges and solutions. *Journal of Research in Medical Sciences* 23: 109.

Hostanska, K., Vuong, V., Rocha, S. et al. (2003). Recombinant mistletoe lectin induces p53-independent apoptosis in tumour cells and cooperates with ionising radiation. *British Journal of Cancer* 88: 1785–1792.

Hurrell, R. and Egli, I. (2010). Iron bioavailability and dietary reference values. *The American Journal of Clinical Nutrition* 91 (5): 1461S–1467S.

Hussain, I., Burhanuddin, M., and Bhuiyan, M.K.J. (2010). Evaluation of physiochemical properties of wheat and mung bean from Bangladesh. *International Journal of Food Safety* 12: 104–108.

Imbard, A., Benoist, J.-F., and Blom, H.J. (2013). Neural tube defects, folic acid and methylation. *International Journal of Environmental Research and Public Health* 10 (9): 4352–4389.

Jain, M.G., Hislop, G.T., Howe, G.R., and Ghadirian, P. (1999). Plant foods, antioxidants, and prostate cancer risk: findings from case-control studies in Canada. *Nutrition and Cancer* 34 (2): 173–184.

Jain, A.K., Kumar, S., and Panwar, J.D.S. (2009). Antinutritional factors and their detoxification in pulses—a review. *Agricultural Reviews* 30: 64–70.

Jenab, M. and Thompson, L.U. (2002). Role of phytic acid in cancer and other diseases. In: *Food Phytates* (ed. N.R. Reddy and S.K. Sathe), 225–248. Boca Raton: CRC Press.

Jenkins, D.J.A., Kendall, C.W.C., Augustin, L.S.A. et al. (2012). Effect of legumes as part of a low glycemic index diet on glycemic control and cardiovascular risk factors in type 2 diabetes mellitus: a randomized controlled trial. *Archives of Internal Medicine* 172 (21): 1653–1660.

Kakati, P., Deka, S.C., Kotoki, D., and Saikia, S. (2010). Effect of traditional methods of processing on the nutrient contents and some antinutritional factors in newly developed cultivars of green gram [*Vigna radiata* (L.) Wilezek] and black gram [*Vigna mungo* (L.) Hepper] of Assam, India. *Food Research Journal* 17: 377–384.

Katoch, R. (2013). Nutritional potential of rice bean (*Vigna umbellata*): an underutilized legume. *Journal of Food Science* 78: 8–16.

Katoch, R., Sanadya, S.K., Pathania, K., and Chaudhary, H.K. (2023a). Nutritional and nutraceutical potential of rice bean (*Vigna umbellata*)-a legume with hidden potential. *Frontiers in Nutrition* 10: 1126544. https://doi.org/10.3389/fnut.2023.1126544.

Katoch, R., Singh, S.K., Thakur, N. et al. (2014). Cloning, characterization, expression analysis and inhibition studies of a novel gene encoding Bowman–Birk type protease inhibitor from ricebean. *Gene* 546: 342–351.

Katoch, R. and Tripathi, A. (2017). Nutraceutical and pharmacological properties of *Vigna* species. *Indian Journal Of Agricultural Biochemistry* 30 (1): 10–20.

Katoch, R. and Tripathi, A. (2021). Research advances and prospects of legume lectins. *Journal of Biosciences* 46 (4): 104. https://doi.org/10.1007/s12038-021-00225-8.

Katoch, R., Tripathi, A., and Kiran (2023b). Rice bean (*Vigna umbellata*): a promising legume with unexplored potential. In: *Neglected and Underutilized Crops* (ed. M. Farooq and S. KHM), 647–680. Academic Press. https://doi.org/10.1016/B978-0-323-90537-4.00002-8.

Katoch, R., Tripathi, A., and Neelam, T.N. (2016). Current perspective of plant protection strategies using inhibitory proteins against insects. *Indian Journal of Agricultural Biochemistry* 29: 124. https://doi.org/10.5958/0974-4479.2016.00021.6.

Kawsar, S.M.A., Huq, E., Nahar, N., and Ozeki, Y. (2008). Identification and quantification of phenolic acids in *Macrotyloma uniflorum* by reversed phase-HPLC. *American Journal of Plant Physiology* 3 (4): 165–172.

Kennedy, A.R. (1998). Chemopreventive agents: protease inhibitors. *Pharmacology & Therapeutics* 78: 167–209.

Kerwin, S. (2004). Soy saponins and the anticancer effects of soybeans and soy-based foods. *Current Medicinal Chemistry. Anti-Cancer Agents* 4 (3): 263–272.

Keyaerts, E., Vijgen, L., Pannecouque, C. et al. (2007). Plant lectins are potent inhibitors of corona viruses by interfering with two targets in the viral replication cycle. *Antiviral Research* 75: 179–187. https://doi.org/10.1016/j.antiviral.2007.03.003.

Kim, I.S. (2021). Current perspectives on the beneficial effects of soybean isoflavones and their metabolites for humans. *Antioxidants* 10: 1064.

Kirsten, R., Heintz, B., Nelson, K. et al. (1993). Polyenylphosphatidylcholine improves the lipoprotein profile in diabetic patients. *International Journal of Clinical Pharmacology and Therapeutics* 32: 53–56.

Konno, A., Hoshino, Y., Terashima, S. et al. (2002). Carbohydrate expression profile of colorectal cancer cells is relevant to metastatic pattern and prognosis. *Clinical & Experimental Metastasis* 19: 61–70.

Kouris-Blazos, A. and Belski, R. (2016). Health benefits of legumes and pulses with a focus on Australian sweet lupins. *Asia Pacific Journal of Clinical Nutrition* 25: 1–17.

Kutoš, T., Golob, T., Kač, M., and Plestenjak, A. (2003). Dietary fibre content of dry and processed beans. *Food Chemistry* 80: 231–235.

Lagarda-Diaz, I., Guzman-Partida, A.M., and Vazquez-Moreno, L. (2017). Legume lectins: proteins with diverse applications. *International Journal of Molecular Sciences* 18: 1242. https://doi.org/10.3390/ijms18061242.

Lai, Y.C., Chuang, Y.C., Chang, C.P., and Yeh, T.M. (2015). Macrophage migration inhibitory factor has a permissive role in concanavalin A-induced cell death of human hepatoma cells through autophagy. *Cell Death & Disease* 6: e2008. https://doi.org/10.1038/cddis.2015.349.

Lairon, D. and Amiot, M.J. (1999). Flavonoids in food and natural antioxidants in wine. *Current Opinion in Lipidology* 10: 23–28.

Lam, S.K. and Ng, T.B. (2010). First report of a haemagglutinin-induced apoptotic pathway in breastcancer cells. *Bioscience Reports* 30: 307–317.

Liu, B., Cheng, Y., Zhang, B. et al. (2009). *Polygonatum cyrtonema* lectin induces apoptosis and autophagy in human melanoma A375 cells through a mitochondria-mediated ROS-p38-p53 pathway. *Cancer Letters* 275: 54–60.

Liu, Y.M., Shahed-Al-Mahmud, M., Chen, X. et al. (2020). A carbohydrate-binding protein from the edible Lablab beans effectively blocks the infections of influenza viruses and SARS-CoV-2. *Cell Reports* 32: 108016. https://doi.org/10.1016/j.celrep.2020.108016.

Liu, R., Zheng, Y., Cai, Z., and Xu, B. (2017). Saponins and flavonoids from Adzuki bean (*Vigna angularis* L.) ameliorate high-fat diet-induced obesity in ICR mice. *Frontiers in Pharmacology* 8: 687. https://doi.org/10.3389/fphar.2017.00687.

Lule, S.U. and Xia, W. (2005). Food phenolics, pros and coin: a review. *Food Review International* 21: 367–388.

Lyu, S.Y., Choi, S.H., and Park, W.B. (2002). Korean mistletoe lectin-induced apoptosis in hepatocarcinoma cells is associated with inhibition of telomerasevia mitochondrial controlled pathway independent of p53. *Archives of Pharmacal Research* 25. https://doi.org/10.1007/BF02975269.

Malhotra, S.L. (1982). Faecal urobilinogen levels and pH of stools in population groups with different incidence of cancer of the colon and their possible role in its etiology. *Journal of the Royal Society of Medicine* 75: 709–714.

Maphosa, Y. and Jideani, V.A. (2017). The role of legumes in human nutrition. *INTECH.* https://doi.org/10.5772/intechopen.69127.

Martín-Cabrejas, M.A. (2019). *Legumes: An Overview. Legumes: Nutritional Quality, Processing and Potential Health Benefits*, 1–18. Croydon: CPI Group (UK) Ltd.

McClements, D.J. (2020). Development of next-generation nutritionally fortified plant-based milk substitutes: structural design principles. *Food* 9 (4): 421.

McDougall, G.J. and Stewart, D. (2005). The inhibitory effects of berry polyphenols on digestive enzymes. *BioFactors* 23: 189–195.

Mekky, R.H., Fayed, M.R., El-Gindi, M.R. et al. (2016). Hepatoprotective effect and chemical assessment of a selected Egyptian Chickpea cultivar. *Frontiers in Pharmacology* 7. https://doi.org/10.3389/fphar.2016.00344.

Mishra, A., Behura, A., Mawatwal, S. et al. (2019). Structure-function and application of plant lectins in disease biology and immunity. *Food and Chemical Toxicology* 110827. https://doi.org/10.1016/j.fct.2019.110827.

Mlyneková, Z., Chrenková, M., and Formelová, Z. (2014). Cereals and legumes in nutrition of people with celiac disease. *International Journal of Celiac Disease* 2 (3): 105–109.

Mohd Ali, N., Mohd Yusof, H., Long, K. et al. (2013, 2013). Antioxidant and hepatoprotective effect of aqueous extract of germinated and fermented mung bean on ethanol-mediated liver damage. *BioMed Research International* 693613. https://doi.org/10.1155/2013/693613.

Mudryj, A.N., Yu, N., and Aukema, H.M. (2014). Nutritional and health benefits of pulses. *Applied Physiology, Nutrition, and Metabolism* 39 (11): 1197–1204.

Mukhopadhyay, S., Panda, P., Das, D. et al. (2014). Abrus agglutinin suppresses human hepatocellular carcinoma *in vitro* and *in vivo* by inducing caspase-mediated cell death. *Acta Pharmacologica Sinica* 35: 814–824.

Müller, W.E., Renneisen, K., Kreuter, M.H. et al. (1988). The D-mannose-specific lectin from *Gerardia savaglia* blocks binding of human immunodeficiency virus type I to H9 cells and human lymphocytes *in vitro*. *Journal of Acquired Immune Deficiency Syndromes* 1: 453–458.

Nascimento da Silva, L.C., JSP, M., Oliveira, W.F. et al. (2020, 2020). Exploring lectin–glycan interactions to combat COVID-19: lessons acquired from other enveloped viruses. *Glycobiology* cwaa099. https://doi.org/10.1093/glycob/cwaa099.

Nishimura, N., Kiriyama, T.Y., and Kiriyama, S. (2000). Plasma cholesterol-lowering effect on rats of dietary fiber extracted from immature plants. *Bioscience, Biotechnology, and Biochemistry* 64: 2543–2551.

Nishino, H., Tokuda, H., Satomi, Y. et al. (2000). Cancer chemoprevention by phytochemicals and their related compounds. *Asian Pacific Journal of Cancer Prevention* 1 (1): 49–55.

Nishteswar, K. (2016). Ayurvedic concept of food and nutrition. *Journal of Nutrition & Food Sciences* 6 (4): 1–8.

Niveditha, V.R. and Sridhar, K.R. (2012). Antioxidant activity of raw, cooked and *Rhizopus oligosporus* fermented beans of canavalia of coastal sand dunes of Southwest India. *Journal of Food Science and Technology* 51 (11): 3253–3260.

Oda, K., Matsuda, H., Murakami, T. et al. (2000). Adjuvant and haemolytic activities of 47 saponins derived from medicinal and food plants. *Biological Chemistry* 381 (1): 67–74.

Ofuya, Z. and Akhidue, V. (2006). The role of pulses in human nutrition: a review. *Journal of Applied Sciences and Environmental Management* 9 (3). https://doi.org/10.4314/jasem.v9i3.17361.

Osorio-Díaz, P., Bello-Pérez, L.A., Sáyago-Ayerdi, S.G. et al. (2003). Effect of processing and storage time on in vitro digestibility and resistant starch content of two bean (*Phaseolus vulgaris* L.) varieties. *Journal of the Science of Food and Agriculture* 83: 1283–1288.

Panda, P.K., Mukhopadhyay, S., Behera, B. et al. (2014). Antitumor effect of soybean lectin mediated through reactive oxygen species-dependent pathway. *Life Sciences* 111 (1–2): 27–35.

Patterson, B.H. and Levander, O.A. (1997). Naturally occurring selenium compounds in cancer chemoprevention trials: a workshop summary. *Cancer Epidemiology, Biomarkers & Prevention* 6: 63–69.

Perez, R.M. (2003). Antiviral activity of compounds isolated from plants. *Pharmaceutical Biology* 41: 107–157.

Peter, B., Kaufman, J.A., Duke, B.H. et al. (1997). A comparative survey of leguminous plants as sources of the isoflavones, genistein and daidzein: implications for human nutrition and health. *The Journal of Alternative and Complementary Medicine* 3: 7–12.

Pham, H.H. and Luan, N. (2021). Nutraceutical properties of legume seeds: phytochemical compounds. In: *Legumes Research*, vol. 2 (ed. J.C. Jimenez-Lopez and A. Clemente). IntechOpen. https://doi.org/10.5772/intechopen.100171.

Pillaiyar, T., Manickam, M., and Namasivayam, V. (2017). Skin whitening agents: medicinal chemistry perspective of tyrosinase inhibitors. *Journal of Enzyme Inhibition and Medicinal Chemistry* 32 (1): 403–425.

Polito, L., Bortolotti, M., Farini, V. et al. (2009). Saporin induces multiple death pathways in lymphoma cells with different intensity and timing as compared to ricin. *The International Journal of Biochemistry & Cell Biology* 41: 1055–1061.

Priego-Poyato, S., Rodrigo-García, M., Escudero-Feliu, J. et al. (2021). Current advances research in nutraceutical compounds of legumes, pseudocereals and cereals. In: *Grain and Seed Proteins Functionality* (ed. J.C. Jimenez-Lopez). IntechOpen. https://doi.org/10.5772/intechopen.97782.

Qadir, S., Wani, I.H., Rafiq, S. et al. (2013). Evaluation of antimicrobial activity of a lectin isolated and purified from *Indigofera Heterantha*. *Advances in Bioscience and Biotechnology* 4: 999–1006. https://doi.org/10.4236/abb.2013.411133.

Racette, S.B., Spearie, C.A., Phillips, K.M. et al. (2009). Phytosterol-deficient and high-phytosterol diets developed for controlled feeding studies. *Journal of the American Dietetic Association* 109 (12): 2043–2051.

Rao, S., Chinkwo, K.A., Santhakumar, A.B., and Blanchard, C.L. (2018). Inhibitory effects of pulse bioactive compounds on cancer development pathways. *Diseases* 6 (3): 72. https://doi.org/10.3390/diseases6030072.

Reinwald, S. and Weaver, C.M. (2010). Soy components vs. whole soy: are we betting our bones on a long shot? *The Journal of Nutrition* 140 (12): 2312S–2317S.

Roy, B., Pattanaik, A.K., Das, J. et al. (2014). Role of PI3K/Akt/mTOR and MEK/ERK pathway in Concanavalin A induced autophagy in HeLa cells. *Chemico-Biological Interactions* 210: 96–102.

Sarkar, P., Lohith Kumar, D.H., Dhumal, C. et al. (2015). Traditional and ayurvedic foods of Indian origin. *Journal of Ethnic Foods* 2 (3): 97–109.

Satija, A. and Hu, F.B. (2012). Cardiovascular benefits of dietary fiber. *Current Atherosclerosis Reports* 14 (6): 505–514.

Seyler, J.E., Wildman, R.E.C., and Layman, D.K. (2007). Protein as a functional food ingredient for weight loss and maintaining body composition. In: *Handbook of Nutraceuticals and Functional Foods* (ed. R.E.C. Wildman), 391–407. Boco Raton: CRC Press.

Shamsuddin, A.K. (2002). Anticancer function of phytic acid. *International Journal of Food Science and Technology* 37: 769–782.

Sharma, R., Palanisamy, A., Dhama, K. et al. (2020). Exploring the possible use of saponin adjuvants in COVID-19 vaccine. *Human Vaccines & Immunotherapeutics* 16 (12): 2944–2953.

Sharma, P. and Rao, M.V. (2014). Shimbhi dhanya varga (group of legumes and pulses): a preventive and curative perspective. *International Ayurvedic Medical Journal* 2 (6): 11.

Shashank, A., Tidke, D., Ramakrishna, S. et al. (2015). Nutraceutical potential of soybean: review. *Asian Journal of Clinical Nutrition* 7: 22–32.

Singh, B., Singh, J.P., Kaur, A., and Singh, N. (2017). Phenolic composition and antioxidant potential of grain legume seeds: a review. *Food Research International* 101: 1–16.

Singh, J. and Basu, P. (2012). Non-nutritive bioactive compounds in pulses and their impact on human health: an overview. *Food and Nutrition Sciences* 03: 1664–1672.

Singh, J.P., Singh, B., and Kaur, A. (2020). Bioactive compounds of legume seeds. In: *Phytochemistry Bioactive Compounds in Underutilized Vegetables and Legumes*, 1–21. Cham: Springer.

Singh, N. (2011). Functional and physicochemical properties of pulse starch. In: *Pulse Foods Processing, Quality and Nutraceutical Applications* (ed. B.K. Tiwari, A. Gowen, and B. McKenna), 87–112. https://doi.org/10.1016/B978-0-12-818184-3.00005-2.

Singh, N. (2017). Pulses: an overview. *Journal of Food Science and Technology* 54 (4): 853–857.

Singhal, P., Kaushik, G., and Mathur, P. (2014). Antidiabetic potential of commonly consumed legumes: a review. *Critical Reviews in Food Science and Nutrition* 54 (5): 655–672.

Soobrattee, M.A., Neergheen, V.S., Luximon-Ramma, A. et al. (2005). Phenolics as potential antioxidant therapeutic agents: mechanism and actions. *Mutation Research* 579: 200–213.

Sparg, S.G., Light, M.E., and van Staden, J. (2004). Biological activities and distribution of plant saponins. *Journal of Ethnopharmacology* 94 (2–3): 219–243.

Sundaram, U., Marimuthu, M., Anupama, V., and Gurumoorthi, P. (2013). Comparative antioxidant quality evaluation of underutilized/less common south Indian legumes. *International Journal of Pharma and Bio Sciences* 4: 117–126.

Taguri, T., Tanaka, T., and Kouno, I. (2006). Antibacterial spectrum of plant polyphenols and extracts depending upon hydroxyphenyl structure. *Biological & Pharmaceutical Bulletin* 29: 2226–2235.

Tiwari, A.T., Abhinay, B., Babu, K.S. et al. (2013). Pigeon pea seed husks as potent natural resource of antioxidant and anti hyperglycaemic activity. *International Journal of Green Pharmacy* 7: 252–257.

Venkidasamy, B., Selvaraj, D., Nile, A. et al. (2019). Indian pulses: a review on nutritional, functional and biochemical properties with future perspectives. *Trends in Food Science and Technology* 88. https://doi.org/10.1016/j.tifs.2019.03.012.

Wang, M.L., Gillaspie, A.G., Morris, J.B. et al. (2008). Flavonoid content in different legume germplasm seeds quantified by HPLC. *Plant Genetic Resources* 6 (1): 62–69.

Ware, J.H., Wan, X.S., Newberne, P., and Kennedy, A.R. (1999). Bowman-Birk inhibitor concentrate reduces colon inflammation in mice with dextran sulfate sodium-induced ulcerative colitis. *Digestive Diseases and Sciences* 44 (5): 986–990.

Wei, Y., Yan, J., Long, F., and Lu, G. (2015). *Vigna umbellata* (Thunb.) Ohwi et Ohashi or *Vigna angularis* (Willd.) Ohwi et Ohashi (Chixiaodou, rice bean). In: *Dietary Chinese Herbs: Chemistry, Pharmacology and Clinical Evidence* (ed. Y. Liu, Z. Wang, and J. Zhang), 551–560. New York: Springer.

Whitehead, R.G., Young, G.P., and Bhathal, P.S. (1986). Effect of short chain fatty acids on new human colon carcinoma cell line (LIM 1215). *Gut* 27: 1457–1463.

Yang, Q.Q., Gan, R.Y., Ge, Y.Y. et al. (2018). Polyphenols in common beans (*Phaseolus vulgaris* L.): chemistry, analysis, and factors affecting composition. *Comprehensive Reviews in Food Science and Food Safety* 17 (6): 1518–1539.

Yao, Y., Wang, C.L.-X., Wang, S., and Ren, G. (2012). Major phenolic compounds, antioxidant capacity and antidiabetic potential of rice bean (*Vigna umbellata* L.) in China. *International Journal of Molecular Sciences* 13: 2707–2716.

Ye, X.Y. and Ng, T.B. (2002). Isolation of a new cyclophilin-like protein from chickpeas with mitogenic, antifungal and anti-HIV-1 reverse transcriptase activities. *Life Sciences* 70 (10): 1129–1138.

Ye, X.Y., Wang, H.X., and Ng, T.B. (2000). Structurally dissimilar proteins with antiviral and antifungal potency from cowpea (*Vigna unguiculata*) seeds. *Life Sciences* 67 (26): 3199–3207.

Zhang, B., Peng, H., Deng, Z., and Tsao, R. (2018). Phytochemicals of lentil (*Lens culinaris*) and their antioxidant and anti-inflammatory effects. *Journal of Food Bioactives* 1 (1): 93–103. https://doi.org/10.31665/JFB.2018.1128.

Zhao, C. (2021). Oligosaccharides in food. In: *Handbook of Dietary Phytochemicals* (ed. J. Xiao, S.D. Sarker, and Y. Asakawa). Singapore: Springer. https://doi.org/10.1007/978-981-15-4148-3_32.

Zhao, Y., Du, S., Wang, H., and Cai, M. (2014). *In vitro* antioxidant activity of extracts from common legumes. *Food Chemistry* 152: 462–466.

13

Nutraceutical Potential of Herbal Beverages

Anoma Chandrasekara[1], Sashya Diyapaththugama[1], and Fereidoon Shahidi[2]

[1] *Department of Applied Nutrition, Wayamba University of Sri Lanka, Makandura, Gonawila, Sri Lanka*
[2] *Department of Biochemistry, Memorial University of Newfoundland, St. John's, Newfoundland, Canada*

13.1 Introduction

Herbal beverages are popular among consumers who seek alternative complementary methods for the management of noncommunicable diseases (NCDs) and overall health and wellness. In addition to hydration as their major function, there are increasing alternative uses of herbal beverages, such as for slimming, antiaging, immunity boosting, mood fixing, and cosmetic purposes among others. They are an integral part of the food culture of some countries such as China, India, Korea, Malaysia, Indonesia, and Sri Lanka where traditional medical systems are established and are widely practiced for centuries. Herbal beverages are presented in several modalities, namely ready-to-serve bottled or canned drinks, infusions, and as dried products in bulk or tea bags form. They have entered into the beverage market besides traditional commonly used tea, coffee, and cocoa, which are also produced from different plant parts (Chandrasekara and Shahidi 2018).

Herbal beverages provide a myriad of bioactive compounds, namely carotenoids, alkaloids, phenolic acids, flavonoids, polyacetylenes, saponins, coumarins, and terpenoids among others in addition to nutrients (Chandrasekara and Shahidi 2018; Peter and Babu 2012; Shahidi et al. 2017; Shahidi et al. 2018). They compounds are provide a number of physiological benefits, including antioxidative, antimicrobial, anti-inflammatory, antiallergic, antithrombotic, antimutagenic, anticarcinogenic effects, and antiaging properties (Chandrasekara and Shahidi 2018; Shahidi et al. 2018; Shahidi and Ambigaipalan 2015; Upadhyay and Singh 2021, 2023). The aim of this chapter is to provide a review of herbal beverages, biomolecules therein and constituent nutraceutical potential in *in vitro*, and *in vivo* systems.

13.1.1 Herbal Beverages

Herbal beverages are generally infusions of plant parts and they are not composed of *Carmellia sinesis,* which is considered the second most important beverage next to water in the world. Herbal beverages are also described as herbal tisanes, infusions, or botanical infusions to avoid confusion with true traditional teas, which are manufactured from buds and leaves of *C. sinesis* and are

Herbal Nutraceuticals: Products and Processes, First Edition. Edited by Santosh Kumar Upadhyay and Sudhir Pratap Singh.

submitted to different processing conditions to produce a wide variety of teas including black tea, green tea, oolong tea, and white tea. Different morphological fresh or dried plant parts, such as leaves, stems, bark, peels, roots, seeds, fruits, buds, and flowers, as well as whole plants of varied maturity stages are used for the preparation of herbal beverages. For instance, among Asian herbal beverages *Aerva lanata (Polpala)*, *Aegle marmelos* (bael), *Centella asiatica* (Indian Pennywort), *Cassia auriculata (Ranawara)*, *Coriander sativum* (coriander), *Hemidesmus indicus (Iramusu)* are popular in Sri Lanka. Further, new additions of herbal beverage preparations such as *Moringa oleifera* leaf (Moringa), *Hibiscus sinensis* flower (Hibiscus), *Nyctanthes arbortristis* flower (Night Jasmine), *Osbeckia octandra* flower and leaf (*Heen Bovitiya*), *Psidium guajava* leaf (Guava), *Annona muricata* leaf (Soursop), and *Clitoria ternatea* flower (Butterfly pea) are gaining popularity among health-conscious consumers (Figure 13.1).

The bael beverage is prepared using dried leaves, buds, and flowers or immature fruits whereas the *Ranawara* beverage is prepared using dried buds and flowers (Chandrasekara et al. 2018; Chandrasekara et al. 2020, 2021). The whole of dried herb of *Polpala,* Indian Pennywort, and *Iramusu* is generally used for beverage preparation. Depending on the mode of preparation, when using the source plant material for consumption, herbal beverages can be infusions (by steeping or maceration) or decoctions (by boiling) (Chandrasekara and Shahidi 2018). Herbal beverages can also be derived from a single part of the herb or a combination of different parts of a variety of herbs. A special decoction that is a combination of dried coriander seeds and ginger (*Zingiber officinale*) rhizome is a popular traditional herbal beverage used by individuals at all ages in Sri Lanka, for immunity boosting and reducing muscle pain during common cold and fever.

Figure 13.1 Herbal beverages of flower infusions: (a) night jasmine; (b) hibiscus; (c) butterfly pea; (d) sesbania. *Source:* Anoma Chandrasekara (Chapter contributor).

Furthermore, additional processing and combination of different herbs give herbal beverages with a variety of taste. For instance, dried coriander seeds, ginger rhizome, and long pepper (*Piper longum*) fruits are roasted together and are ground to a coarse powder, which is used to prepare an infusion for improving immunity and relief of common cold and cough.

Herbal beverages are composed of unique flavors, specific aromas, and vibrant colors adding variety to a balanced diet. A majority of herbal beverages are naturally free from detectable levels of caffeine. However, beverages like yerba mate contains caffeine (Rzasa-Duran et al. 2022). Country-specific agencies assigned by the government regulates the marketing of herbal beverages ensuring the safety of its populations. The US Food and Drug administration regulate herbal beverages under dietary supplements (FDA 2022). In Canada, herbal beverages are placed among natural health products (Health Canada 2023). Table 13.1 presents popular herbal beverages consumed by populations in different regions of the world to promote their health and wellness in addition to managing disease conditions.

Table 13.1 Herbal beverages consumed in different parts of the world.

Region	Local name	Scientific name	Plant part	Health benefits
Asia	Bael, bengal quince, *Beli*	*Aegle marmelos*	Dried leaves, buds, flowers, immature fruits, bark	Anticancer, antidyslipidaemia, antihyperglycemic, antidiabetic, anti-inflammatory
	Tanner's Cassia, *Avartaki*, *Ranawara*	*Cassia auriculata*	Dried leaves, flowers	Antioxidant, antihyperglycemic
	Polpala	*Aerva lanata*	Dried whole plant	Antihyperglycemic, anti-inflammatory
	Indian Sarasaparilla, *Iramusu*	*Hemidesmus indicus*	Dried whole plant	Antioxidant, antithrobotic, antiplatelet aggregation
	Pegaga, Indian pennywort, *Gotukola*	*Centella asiatica*	Dried whole plant	Anticancer, increase the activity of antioxidant enzymes
	Coriander, *Koththmalli*	*Coriandum sativum*	Dried fruits	Antioxidant, antibacterial, hypocholesterolemic, anticancer, anti-inflammatory
	Ginger, *Inguru*	*Zingiber officinale*	Rhizome	Anti-inflammatory, hypoglycemic
	Cinnamon, *Kurundu*		Bark, leaves	
	Jackfruit	*Artocarpus heterophyllus*	Mature dried leaves	Hypoglycemic, hypolipidemic, antioxidant, antibacterial
	Rang Jued	*Tunbergia laurifolia*	Leaves, stems	Antioxidant, detoxification anticancer, anti-inflammatory
	Heen Bovitiya	*Osbeckia octandra*	Flowers, leaves, young stems	Hypoglycemic, antioxidant, anticancer, anti-inflammatory, hypolipidemic, hepatoprotective, antiangiogenic
	Moringa	*Moringa olifera*	Leaf, tender stem	Antihyperglycemic, antidyslipidemic, analgesic, antiulcer, antihypertensive, radioprotective, antioxidant, immune-modulatory

(*Continued*)

Table 13.1 (Continued)

Region	Local name	Scientific name	Plant part	Health benefits
Africa	Rooibos	*Aspalathus linearis*		Antioxidant, anticancer
	Borututu	*Cochlospermum angolensis*	Roots	Anticancer, hepatoprotective, antioxidant
	Honey bush	*Cyclopia subternata Vogel*	Leaves, stems	Antioxidant, anticancer
South America	Yerba mate	*Ilex paraguariensis*		Hypocholesterolemic, hepatoprotective, cardiovascular protective, detoxification, cognitive functions, anticancer
	Sage, Adacayi, Minchi	*Salvia officinalis*		Antioxidant, anti-inflammatory
	Rosehips	*Rosa canina*	Fruits	Anti-inflammatory
	Acai berry	*Euterpe oleracea Mart.*	Fruits	Hypocholesterolemic, hepatoprotective, cardio protective, gastroprotective, neuroprotective, anticancer, antioxidant, anti-inflammatory
Europe	Chamomile	*Matricaria chamomilla*; *Chamaemelum nobile*	Dried flowers	Antioxidant, hypocholesterolemic, anticancer, anti-inflammatory
	Heart leaved mooseed	*Tinospora cordifolia*	Stem, roots	Antidiabetic, anti-inflammatory
	Dagcayi	*Sideritis condensate*	Dried aerial parts	Antioxidant, antibacterial
	Peppermint	*Mentha piperita*	Dried leaves	Antioxidant, antitumor

13.2 Bioactive Compounds in Herbal Beverages

13.2.1 Phenolic Compounds

Phenolic compounds are the major group of nonnutrient bioactive compounds found in herbal beverages. The molecules are consisted of an aromatic ring attached to one or more hydroxyl groups (Shahidi and Naczk 2004). Phenolics are derived secondary metabolites of plants from a number of biosynthetic precursors, namely acetate, pyruvate, and aromatic amino acids, as well as acetyl coenzyme A and malonyl coenzyme A. Phenolic acids, coumarins, flavonoids, and tannins are examples for compounds commonly found in herbal beverages (Chandrasekara and Shahidi 2018).

13.2.2 Phenolic Acids

Hydroxybenzoic acids and hydroxycinnamic acids are the two main groups of phenolic acids found in herbal beverages (Chandrasekara and Shahidi 2018). Major hydroxybenzoic acids (C_6–C_1) include *p*-hydroxybenzoic, gallic, vanillic, syringic, and protocatechuic acids, among others. The constituent hydroxycinnamic acids commonly found in herbal beverages include *p*-coumaric, caffeic, ferulic, and sinapic acids (Chandrasekara 2019).

13.2.3 Flavonoids

Flavonoids are common constituent bioactive compounds in herbal beverages and are responsible for number of health benefits reported in addition to color, taste, and mouth feel (Chandrasekara and Shahidi 2018). The precursor of flavonoids is a phenylpropanoid compound, which is condensed with three molecules of malonyl coenzyme A. The end products of the reaction are chalcones and the reaction is catalyzed by chalcone synthase. Furthermore, resulting chalcones get cyclized under acidic conditions and form different subclasses of flavonoids, namely flavones, flavonols, flavonones, flavononols, isoflavones, anthocyanidins, and flavanols (Shahidi and Naczk 2004). Flavones and flavonols are generally found as aglycones in foods. The common flavonoids reported in the human diet include flavonols (quercetin, kaempferol, and myricetin), flavones (luteolin, apigenin, and chrysin), flavanols (catechin, epicatechin, epigallocatechin, epicatechin gallate, and epigallocatechin gallate), flavanones (narigenin, hesperitin, and eriodictyol), anthocyanidins (cyanidin, malvidin, peonidin, petunidin, pelargonidin, and dephinidin), and isoflavones (genistein, daidzein, and glycitein) (Shahidi and Naczk 2004).

13.2.4 Tannins

Tannins are a group of compounds with diverse structures, which exhibit several beneficial bioactivities such as prevention of lipid peroxidation and superoxide formation with free radical scavenging activity. These are either hydrolysable or condensed, the latter is also known as proanthocyanidin. Tannins have the ability to bind and precipitate proteins, thereby reducing the protein digestion. Tannins may also be present in the form of complex tannins (Shahidi and Naczk 2004).

13.2.5 Coumarins

Coumarins are lactones of *cis-O*-hydroxycinnamic acid derivatives, which exist in the free form or as glycosides. Three types of coumarins, namely simple coumarins, furanocoumarins (psoralens), and pyranocoumarins are found in foods (Shahidi and Naczk 2004). Cinnamon is the main source of coumarins in the human diet (Blahova and Svobodova 2012).

13.2.6 Terpenes

Terpenes and terpenoid derivatives are derived from isoprene (2-methylbutadiene) units and are secondary metabolites in plants (Gershenzon and Kreis 1999). The C_5H_8 isoprene units produce different classes of terpenoids by polymerization and include hemiterpenes consisting of a single C_5 isoprene unit, monoterpenes (C_{10}), sesquiterpenes (C_{15}), diterpenes (C_{20}), sesterterpenes (C_{25}), triterpenes (C_{30}), carotenoids (C_{40}), and polyterpenes consisting of long chains of many isoprene units. Sterols and triterpenes, which are accumulated as glycosides (saponins) in plants, belong to triterpene group of compounds. Saponins have surface-active properties and are glycosylated (aglycone named sapogenin) secondary metabolites (Liu and Henkel 2002).

13.2.7 Carotenoids

Carotenoids are pigments possessing provitamin A and antioxidant activities with yellow, orange, and red colors. They are hydrocarbons containing a 40-carbon atom skeleton of eight isoprene units. The structures may be cyclized at one or both ends and contain a varying number of hydrogen atoms; the addition of oxygen-containing functional groups forms xanthophylls. Common carotenes include *β*-carotene, α-carotene, and lycopene, whereas xanthophylls include lutein, zeaxanthine, cryptoxanthin, canthaxanthin, and astaxanthin in the diet. *β*-Carotene, α-carotene, and cryptoxanthin are prominent due to their provitamin A activity (Young and Lowe 2018).

13.2.8 Polyacetylenes

Polyacetylenes are comprised of carbon-carbon triple bond or alkynyl functional group (Minto and Blacklock 2008). Apiaceae and Araliaceae families are known to contain plants with aliphatic C_{17}-polyacetylenes of the falcarinol type such as falcarinol and falcarindiol. Several herbs such as American ginseng root (*Panax quinquefolium*), *Peucedanum praeruptorum*, *Echinacea pallida*, *Bupleurum spinosum*, and *Atractylodes lancea* contain polyacetylenes (Christensen et al. 2006; Pellati et al. 2006).

13.3 Bioactivities of Herbal Beverages

Herbal beverages have been used as social drinks for centuries by people in different countries of the world and are known natural remedies (Table 13.1). However, the increasing lifespan along with a number of chronic ailments, such as obesity, cardio vascular diseases, cancers, and type 2 diabetes associated with unhealthy diet and sedentary lifestyle, increases the demand for complementary and alternative medicine (CAM) treatments, which include herbal beverages among other dietary supplements. Due to the interactions among bioactive compounds, herbal beverages that are natural formulations provide better activities than those of isolated bioactive components (Caesar and Cech 2019). Globalization expands the availability of herbal beverages in every region of the world despite of their cultural specificity and availability of source plants.

13.4 Antioxidant Properties of Herbal Beverages

Beverages prepared from herbs can be used as a component of a balanced diet. They may provide a number of beneficial effects attenuating oxidative stress (Cabrera et al. 2006). They contain several bioactives including polyphenolic compounds with effective antioxidant properties. Oxidants are formed during usual aerobic metabolism and the body consists of internal antioxidant mechanisms, either enzymatic or nonenzymatic, to provide defenses against their harmful actions to the body. However, the presence of pathophysiological conditions or inadequate intake of compounds providing antioxidant defense in the body may lead to excess formation of oxidants, thus resulting in an imbalance between oxidants and antioxidants, and eventual oxidative stress. Oxidant species, namely hydrogen peroxide, superoxide anion, and hydroxyl radicals, are generated during the reduction of molecular oxygen and are reactive oxygen species (ROS). Other reactive species include singlet oxygen, ozone, hypochlorous acid, nitric oxide, peroxynitrite, and lipid peroxyl radicals. Sources of oxidative stress can be internal or external. Internal oxidative stress is mainly caused by peroxysomes, detoxifying enzymes belonging to P450 enzyme system, xanthine oxidase, and the nicotinamide adenine dinucleotide (NADPH) oxidase complex (Sosa et al. 2013). Radiation, which includes x-irradiation, ultraviolet, ultrasound and microwave, and dietary compounds are the external sources of oxidants (Sies 1997). Hydroxyl radical being one of the most potent oxidants when produced in the body can initiate lipid peroxidation, breaking DNA single or double strands, and oxidation of any other bioorganic molecule such as proteins (McCord 2000). DNA mutations can lead to aging and carcinogenesis while alterations in the protein structure change its functionality. Lipid peroxidation by oxidative stress increases cell permeability and cell death (Sosa et al. 2013). By damaging the intracellular organic biomolecules such as proteins, lipids, DNA, and RNA, oxidative stress leads to etiology of aging and several diseases in humans (Sosa et al. 2013). They include inflammatory diseases, such as arthritis, vasculitis, glomerulonephritis, systemic lupus erythematosus, and adult respiratory distress syndrome, as well as ischemic diseases such as cardiovascular diseases, diabetes, and cancers among others. In addition, autoimmune diseases and neurodegenerative disorders such as multiple sclerosis, Alzheimer's disease, Parkinson disease, amyotrophic lateral sclerosis, and muscular dystrophy are also included (McCord 2000). Bioactive compounds of herbal beverages that exert antioxidant activities are responsible for attenuating a myriad of diseases and damages caused by oxidative stress.

13.5 Herbal Beverages in Asia

13.5.1 Bael

A. marmelos, belonging to the family *Rutaceae* is known by several names (Table 13.1). Bael is a tropical fruit bearing indigenous plant in the Southeast Asian region and is found in countries such as Sri Lanka, India, Pakistan, Thailand, Bangladesh, and Myanmar. Dried flowers, buds, leaves, and immature fruit are mainly used for beverage preparation (Figure 13.2). Bioactivities, such as anti-inflammatory, anticancer, antibacterial, antiulcer, antimalarial, radioprotective, analgesic, antiviral, antidyslipidemic, and antidiabetic activities, are reported from extracts, decoctions, or infusions of different parts of bael tree (Agarwal et al. 2009, 2010; Arul et al. 2005; Badam et al. 2002; Dhankhar et al. 2011; Jagetia et al. 2004; Jagetia and Venkatesh 2007; Kamalakkannan and Prince 2003, 2005; Narender et al. 2007; Rajadurai and Prince 2005; Upadhya et al. 2004). Extracts of dried bael flowers and buds have been reported to contain a number of phenolic acids, such as vanillic, *p*-coumaric, chlorogenic, caffeic, and gentisic acids. Furthermore, flavonoids, mainly catechin, and quercetin were reported (Chandrasekara et al. 2018). In addition, carotenoids, coumarins, alkaloids, and tannins are reported in different parts of bael tree (Baliga et al. 2011; Dhankhar et al. 2011; Maity et al. 2009; Narendhirakannan et al. 2006). Bioactives such as marmelosin, marmelide, luvangetin, aurapten, psoralen, aegelin, scoparone, scopoletin, tannic acid, xanthotoxol, β-sitosterol, and

(a) (b) (c)

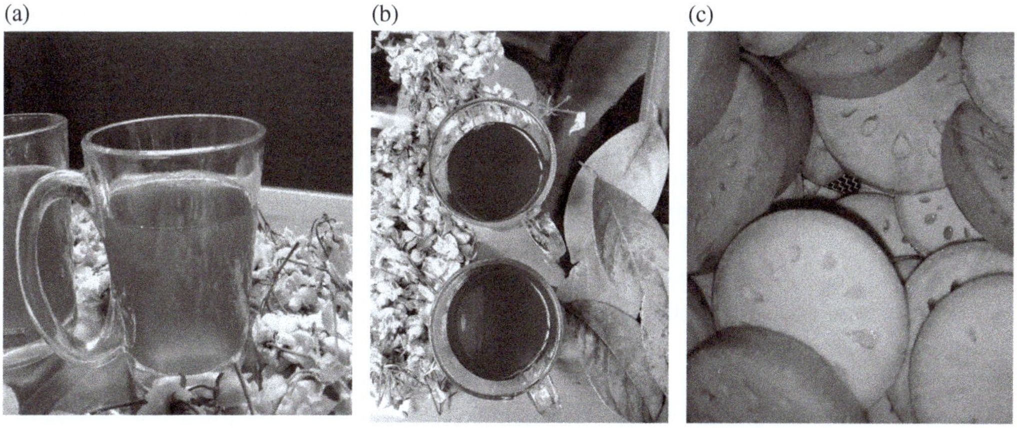

Figure 13.2 Herbal beverages: (a) dried Cassia flowers and buds; (b) dried Cassia flowers and mature jackfruit leaves; (c) immature bael fruit cuts prepared for drying. *Source:* Anoma Chandrasekara (Chapter contributor).

tannin are reported in bael fruit (Kamalakkannan and Prince 2005; Maity et al. 2009). In addition, ellagic acid, protocatechuic acid derivatives, and resveratrol derivatives have been identified in bael fruit extracts (Chandrasekara et al. 2020). In the same study, Chandrasekara et al. (2020) showed that consumption of 250 ml of bale fruit beverage improved the plasma antioxidant capacity of healthy subjects by 75% after 120 minute compared to the baseline.

13.5.2 Indian Pennywort

C. asiatica is commonly used by Asian populations and is popular as a memory booster. The whole dried herb is used as the only ingredient or in combination with other herbal components such as garlic bulbs, coriander seeds, or ginger rhizomes. *C. asiatica* is a rich source of alkaloids, terpenoids phenolic acids, polyacetylenes, flavonoids, and saponins that improve the action of antioxidant enzymes, namely catalase, superoxide dismutase, and glutathione peroxidase (Minto and Blacklock 2008; Singh et al. 2012; Kunjumon et al. 2022). The principal triterpenoids identified in *C. asiatica* include asiatic acid, madecassic acid, asiaticoside, and madecassoside and are known as major centellosides (Bonfill et al. 2005). The neuroprotective, memory enhancing, wound healing, and cardioprotective properties of centellosides have been reported (Belwal et al. 2019; Razali et al. 2019; Sun et al. 2020). Furthermore, a polyacetylene compound, cadiyenol identified from the areal part of the plant exhibited cell apoptosis by 63% in mouse lymphoma cells (Govindan et al. 2007).

13.5.3 Tanner's Cassia

C auriculata L (*Ranawara*), commonly known as *Tanner's Cassia,* is abundantly grown in India and Sri Lanka and is a popular traditional herbal beverage. Use of *Tanner's Cassia* for therapeutic purposes in diabetes, conjunctive, and ophthalmic conditions has been reported (Gupta et al. 2009; Kalaivani and Umamaheswari 2008; Abesundara et al. 2004; Latha and Pari 2003a, 2003b). Chandrasekara et al. (2021) reported phenolic acid, ellagic acid and the flavonoids gallocatechin, and derivatives of isorhamnetin, myricetin, and phloridzin in *Cassia* flowers. In addition, dimeric gallocatechin, and trimeric propelargonidin have been identified as major proanthocyanidins in aqueous extracts of *Cassia* flowers. Several flavonoids, such as kaempferol-3-*O*-rutiniside, kaempferol, quercetin, and luteolin have also been identified from the arial parts of *C. auriculata* (Juan-Badaturugae et al. 2011). Herbal beverages prepared using dried Cassia flowers and buds reported reduced plasma glucose level and enhanced plasma antioxidant capacity at the end of 120 minute postingestion (Chandrasekara et al. 2021).

13.5.4 Jackfruit

Artocarpus heterophyllus (jackfruit) is a tropical tree belonging to the family *Moraceae* and is abundantly found in Asia, Africa, and in some regions of South America (Wang et al. 2017). Parts such as leaves, bark, roots of the tree have been used in folklore medicine to treat ailments (Prakash et al. 2009). Flavones, such as artocarpin, artocarpesin, artocarpetin, artocarpanone, norartocarpetin, isoartocarpin, cylcoartocarpin, dihydromorin, cynomacurin, morin, oxydihydroartocarpesin, and cycloartinone have been reported from different morphological parts of *A. heterophyllus* (Rama Rao et al. 1973). Leaves contain phenolic compounds, terpenoids, coumarin, alkaloids, chalcones, stibenoids, lignans, steroids, and saponins (Thapa et al. 2016; Wang et al. 2017). The leaf extracts have demonstrated hypoglycemic, hypolipidemic, antioxidative, and antibacterial effects (Thapa et al. 2016; Biworo et al. 2015; Swami et al. 2012; Omar et al. 2011; Loizzo et al. 2010; Chandrika et al. 2006). In hot water extracts of matured leaves, ellagic acid, and a derivative of monomeric flavonol, epigallocatechin were identified (Chandrasekara et al. 2021). Furthermore, a herbal beverage produced by decoction of dried leaves exhibited lower plasma glucose concentration compared to the control of 50 g of glucose in healthy adults (Chandrasekara et al. 2021). Jackfruit leaf herbal beverage increases plasma antioxidant capacity of participants, 120 minute postingestion. The high content of epigallocatechin derivative present in Jackfruit leaves could be responsible for the reported favorable health effects (Chandrasekara et al. 2021). Furthermore, herbal beverage prepared as a blend of dried Jackfruit leaves and Cassia flowers and buds showed the highest increment of plasma antioxidant capacity suggesting a synergistic effect of ingredients (Chandrasekara et al. 2021).

13.5.5 Rang Jued

Tunbergia laurifolia (*Rang Jued*) is a popular herb in Thailand where the leaves and stems are processed to prepare herbal beverages. Leaves contain several phenolic compounds, such as apigenin, caffeic acid, catechin, rosmarinic acid, rutin, and quercetin (Essiedu et al. 2023; Junsi and Siripongvutikorn 2016; Ruangpayungsak et al. 2018). Several bioactivities such as anti-inflammatory, detoxification, antioxidant, and anticancer potentials of *T. laurifolia* have been reported (Junsi et al. 2017).

13.5.6 Heen Bovitiya

O. octandra (*Heen Bovitiya*) is an endemic plant in Sri Lanka. The leaves, young stems, and flowers are used for the preparation of herbal beverages. In Ayurveda medicine, leaves, roots, and bark are used for treating diabetes mellitus, hepatitis, jaundice, hyperlipidemia, hemorrhoids, and ascites. Furthermore, leaf extract of *O. octandra* has shown beneficial effects against the liver damage induced by galactosamine, paracetamol, and tert-butyl hydroperoxide (Thabrew et al. 1995a, 1995b). Kim et al. (2022) recently demonstrated that leaf extract of *O. octandra* repressed the proliferation of oral squamous cell carcinoma (OSCC) cells through several mechanisms such as interruption of DNA replication among others. Furthermore, the extract of leaf triggered the apoptotic response via caspase 3 activation in OSCC cells. In addition, Bogahawaththa et al. (2021) reported hepatoprotective effect of *O. octandra* leaf in a wistar rat model of thioacetomide-induced liver cirrhosis. They further established strong anti-angiogenic and hepatic anti-fibrotic effects of *O. octandra in vivo*.

13.5.7 Moringa

Moringa leaf and tender stem infusions have been popular recently due to a number of bioactivities, namely antihyperglycemic, antidyslipidemic, analgesic, antiulcer, antihypertensive, radioprotective, and antioxidant, as well as immune-modulatory actions that were reported (Stohs and Hartman 2015). *M. oleifera,* known as moringa and drum stick tree, is a perennial plant found indigenously in India and widespread in many tropical and subtropical countries such as Sri Lanka, Africa, South and Central America, Mexico, Hawaii, Southeast Asia, and the southern part of Japan. Moringa leaves and fruits are used as vegetables, and leaves and bark have been used in traditional medicines. The antioxidant activities of moringa leaf infusions have been reported. Chumark et al. (2008) showed free radical scavenging ability in 1,1-diphenyl-2-picrylhydrazyl (DPPH) radical and inhibition of Cu^{2+}-induced low-density lipoprotein oxidation in vitro of hot water extract of Moringa leaf. Further, Moringa leaf and stem herbal beverages are good sources of natural antioxidants and, specially, constituents phenolic acids, chlorogenic, and caffeic acids are effective on superoxide anion radical scavenging activity (Sugahara et al. 2018).

13.6 Tropical Fruit Herbal Beverages

Novel herbal beverages are emerging based on tropical fruits such as Wood apple (*Limonia acidissima)*, Java plum (*Syzygium cumini)*, June plum *(Spondias dulcis)*, and Carambola *(Averrhoa carambola)* due to their unique fragrance, aroma, taste, therapeutic properties, and lower content of caffeine (Pękal et al. 2011) (Figure 13.3).

13.6.1 Wood Apple

Wood apple is rich in coumarins including 6-methoxy-7-geranyloxy coumarin, demethylsuberosin, bergapten, osthenol, isopimpinellin, and psoralen. Flavonoids, phenolic acids, and quinones are 3,5,6-trimethoxy-flavone-7-O-β-D-glucopyranoside, gallocatechin, saponarin, vitexin, gallic acid, 4-hydroxybenzoic acid, and 2,6-dimethoxy benzoquinone have been isolated from the fruit (Masud Parvez and Sarker 2021). Wood apple has shown antimicrobial, anti-diabetes, and anti-inflammatory activities (Pandey et al. 2014). In addition, anti-ulcerative colitis, gastroprotective, antidiarrheal, hepatoprotective, and inhibition of lipid peroxidation are therapeutic properties have been reported (Sharma et al. 2021).

13.6.2 Java Plum

Java plum trees grow in tropical countries, such as Sri Lanka, India, Bangladesh, Burma, Nepal, Pakistan, and Indonesia, and are a rich source of gallic acid and anthocyanins such as delphinidin-3-gentiobioside, malvidin-3-laminaribioside, petunidin-3-gentiobioside, cyanidin diglycoside, petunidin, and malvidin (Ayyanar and Subash-Babu 2012). The fruits and seeds of Java plum have been shown to

Figure 13.3 Herbal beverages prepared from dried tropical fruits: (a) Carambola; (b) wood apple; (c) June plum; (d) Java plum. *Source:* Anoma Chandrasekara (Chapter contributor).

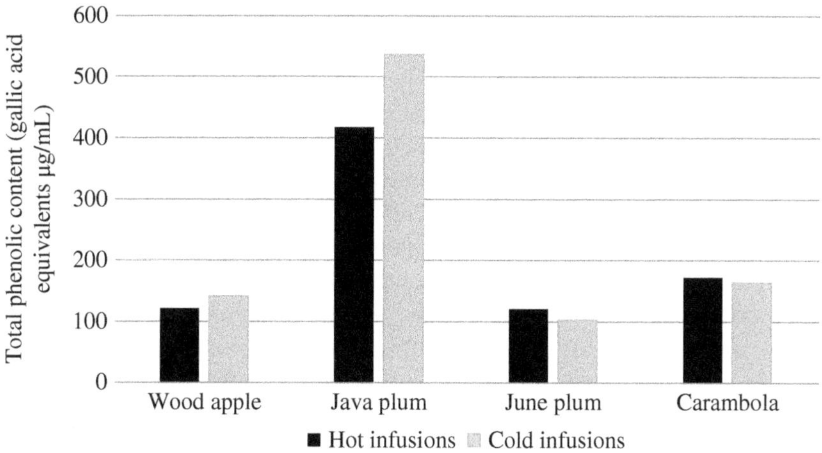

Figure 13.4 Total phenolic content of hot and cold infusions of wood apple, Java plum, June plum, and Carambola dried fruits. *Source:* Anoma Chandrasekara (Chapter contributor).

exhibit health benefits such as hypoglycemic, anti-fertility, antihyperlipidemic, anorexigenic, gastroprotective, anti-ulcerogenic, antimicrobial, anti-inflammatory, antibacterial, anti-HIV, antileishmanial, and antifungal, and antidiarrheal effects (Ayyanar and Subash-Babu 2012; Chagas et al. 2019; Ravi et al. 2004).

13.6.3 June Plum

June plum is a tropical tree that originated in Melanesia and spread to Sri Lanka, India, Indonesia, Malaysia, and many other tropical nations (Jayarathna et al. 2020). June plum contains flavonoids, tannins, and terpenoids (Islam et al. 2013). Parts of plant are used to treat sore throat, skin inflammation, itchiness, internal ulceration, poor sight, and eye infections. Their antidiabetic, antimicrobial, cytotoxic, and thrombolytic potentials have been reported (Sinan et al. 2021).

13.6.4 Carambola

Carambola is a tree found in different countries of the world and is extensively grown in Southeast Asia (Muthu et al. 2016). Carambola is rich in saponins, flavonoids, alkaloids, tannins, and pyrogallic steroids. Other phytochemicals such as anthocyanin and anthocyanidin, chalcones and aurones, leucoanthocyanidins, catechins, and triterpenoids are also present in Carambola (Khoo et al. 2010; Lakmal et al. 2021; Muthu et al. 2016). The health benefits reported for Carambola include anti-diabetes, anti-inflammatory, anti-infective, antiulcer, antitumor, and immune-boosting effects (Lakmal et al. 2021). A recent study showed that total phenolic content of hot and cold infusions of wood apple, Java plum, June plum, and Carambola dried fruits ranged from 108 to 536 gallic acid equivalents µg/mL of infusions of dry material of fruit in 200 mL of water (Figure 13.4) (Udani et al. 2024).

13.7 Herbal Beverages in Africa

Aspalathus linearis (Rooibos) is a shrub native to South Africa and is used to prepare a herbal beverage rich in polyphenolic antioxidants but with no caffeine and very little tannins. It is reported to have therapeutic effects on insomnia, allergies, asthma, and nervous breakdowns (Joubert et al. 2008).

Rooibos is rich in polyphenols and a rare source of the dietary dihydrochalcones, namely aspalathin, and nothofagin (McKay and Blumberg 2006, 2007). Decoction of honey bush, an African herbal beverage is used as a restorative and as an expectorant in chronic catarrh and pulmonary tuberculosis (McKay and Blumberg 2007). Major constituent polyphenols in honey bush are xanthone mangiferin and the flavonones, hesperitin and isokuranetin (McKay and Blumberg 2007). Rooibos and honey bush teas have demonstrated potent antioxidant and antimutagenic activities *in vitro*. Canda et al. (2014) showed that Rooibos herbal beverages demonstrate enhanced antioxidant status of the liver in an induced oxidative stress situation of an animal model. Furthermore, Rooibos showed cardioprotective effects by number of modes. Chrysoeriol is a flavone, present at low levels in Rooibos is a potential agent in preventing and treating vascular diseases in humans. Moreover, chrysoeriol is able to inhibit the migration of smooth muscle cells inside the aorta, a key cause of atherosclerosis (Aboulaghras et al. 2022; Cha et al. 2009). In addition, Rooibos demonstrated angiotensin-converting enzyme (ACE) inhibitory activities and reduced several biomarkers associated with cardiovascular disease (Marnewick 2010; Persson et al. 2010).

13.8 Herbal Beverages in South America

Yerba mate (*Ilex paraguariensis*) is a caffeine containing herbal beverage consumed by native people in South America as a social and medicinal beverage for centuries. Several health promoting properties such as central nervous system stimulating effects, hepatoprotective, diuretic, detoxification, precognitive functions, hypocholesterolemic, anticancer, and weight loss properties have been reported for yerba mate (Heck and de Mejia 2007; Rzasa-Duran et al. 2022). Furthermore, several studies have demonstrated anti-inflammatory, antimutagenic, antioxidant, and lipid lowering effects (Vanderjagt et al. 2002; Lobato et al. 2009; Bracesco et al. 2011; Gao et al. 2013, Rzasa-Duran et al. 2022).

Several bioactive compounds in yerba mate have been identified to exert beneficial health effects and these include caffeoyl derivatives such as caffeic acid, mono-, and dicaffeoylquinic acids, methylxanthines (caffeine and theobromine), and flavonoids, namely rutin, quercetin, and kaempferol (Isobella et al. 2010). Rzasa-Duran et al. (2022) reported the content of neochloro genic acid, chlorogenic acid, isochlorogenic acid, and cryptochlorogenic acid in yerba mate. The content of the quantitatively dominant neochlorogenic acid ranged from 1.24 to 39.03 mg/g of dry weight of sample. Arcari et al. (2013) reported the potency of bioactive compounds of yerba mate in regulation of the gene expression associated with adipogenesis. Yerba mate extract down-regulated the expression of genes Creb-1 and C/EBPα and up-regulated the expression of genes such as Dlk1, Gata2, Gata3, Klf2, Lrp5, Pparγ2, Sfrp1, Tcf7l2, Wnt10b, and Wnt3 that related to the control of adipogenesis (Arcari et al. 2013).

Açaí berry herbal beverage has recently become popular in many regions of the world due to a number of health benefits demonstrated. The açaí palm (*Euterpe oleracea Mart.*) is in the family Arecaceae and is native to tropical Central and South America (Fornari Laurindo et al. 2023). It has been used for centuries as a dietary plant with many other utilities in Amazon region in South America but its recent introduction to world market has boomed global demand for açaí berry (de Lima Yamaguchi et al. 2015). Açaí berry is a rich source of phenolic compounds and includes phenolic acids, anthocyanins, and proanthocyanidins in addition to lignans (ALNasser et al. 2022). Several phenolic acids, namely protocatechuic, *p*-hydroxybenzoic, ferulic, vanillic, caffeic, and syringic have been identified in açaí berry pulp (Schauss et al. 2006). In addition, the main flavonoid constituents of açaí berry, anthocyanins, include cyanidin-3-sambubioside, cyanidin 3-glucoside, cyanidin3-rutinoside, peonidin-3-rutinoside, pelargonidin-3-glucoside, and delphinidin-3-glucoside (Fornari

Laurindo et al. 2023). Other flavonoids reported are homoorientin, orientin, taxifolin, isovitexin, and scoparin among others. Several in vitro and in vivo studies reported antioxidant and anti-inflammatory actions of açaí berry, thus render a myriad of cardioprotective, gastroprotective, hepatoprotective, neuroprotective, renoprotective, antilipidemic, antidiabetic, and antineoplastic health benefits (Fornari Laurindo et al. 2023).

13.9 Herbal Beverages in Europe

Popularity of herbal beverages across Europe has expanded steadily due to their functional properties. Chamomile belonging family Asteraceae is composed of a number of varieties, such as *Matricaria chamomilla, Chamomilla recutita,* and *Chamaemelum nobile.* Chamomile tea is a popular herbal beverage in Europe. This beverage is prepared using dried flower heads as the single ingredient. The main compounds found of infusions include phenolic compounds, which consisted of flavonoids, such as apigenin, quercetin, patuletin, and luteolin (Mckay and Blumberg 2006). The main compounds of the constituent essential oil are the terpenoids alpha-bisabolol and azulenes, including chamazulene. Chamomile has demonstrated moderate level of antioxidant and antimicrobial activities, and significant antiplatelet activity *in vitro.* In addition, *in vivo* studies have shown the potent anti-inflammatory, antimutagenic, and cholesterol-lowering activities of chamomile (Mckay and Blumberg 2006). A number of bioactivities, namely antimicrobial, hepatoprotective, anticancer, anti-inflammatory, antioxidant, hypocholesterolemic, antiparasitic, antiaging, analgesic, antiallergic, and antihypertensive properties of chamomile, have been reported (Lee and Shibamoto 2002; Babenko and Shakhova 2006; Sah et al. 2022). Chamomile beverage has been used as a treatment for inflammatory diseases. Water-soluble components of chamomile include phenolic acids, flavonoids, and coumarins. Major flavonoids reported are apigenin, quercetin, patuletin, and luteolin (Sah et al. 2022). The chamomile intervention inhibited the release of LPS-induced prostaglandin $E(_2)$ in RAW 264.7 macrophages *in vitro.* This was due to the inhibition of COX-2 enzyme activity by chamomile extracts. Furthermore, chamomile caused reduction in LPS-induced COX-2 mRNA and protein expression, without affecting COX-1 expression (Srivastava and Gupta 2009).

Peppermint herbal beverage, brewed from the leaves of *Mentha* species, are used in combination with others or singly to give a calming effect. There are multiple food uses of *Mentha* species in food, beverages, and confectionary. Among health benefits reported *in vitro,* antibacterial activity against a range of pathogenic bacteria and antioxidant activity are predominant among others. Antioxidant potential of peppermint beverages further render several health benefits such as anticancer, anti-obesity, anti-inflammatory, antidiabetic, and cardioprotective effects (Tafrihi et al. 2021). Several phenolic compounds were reported in the leaves of *Mentha piperita,* and rosmarinic acid and caffeic acid were among phenolic acids, whereas luteolin, gallocatechin, epigallocatechin gallate, and catechins were among flavonoids in addition to menthone, isomenthone, hexadecanoic acid, *cis*-carveol, carvone, and limonene (Tafrihi et al. 2021).

13.10 Conclusions

Herbal beverages play a major role in disease risk reduction and management of NCDs. Specific characteristic flavors, aromas, and colors of herbal beverages add variety to the diet adding nutrients and nonnutrient bioactive compounds. An array of herbal beverages prepared from different morphological parts of plants belonging to several botanical families are used in the world.

Although bioactive potential of some herbal beverages is established through in vitro, in vivo, and ex vivo systems, further research is warranted to investigate the safety of frequent consumption of constituent bioactive compounds that execute health benefits and their mode(s) of actions in disease risk attenuation and health promotion.

References

Abesundara, K.J., Matsui, T., and Matsumoto, K. (2004). Alpha-Glucosidase inhibitory activity of some Sri Lanka plant extracts, one of which, Cassia auriculata, exerts a strong antihyperglycemic effect in rats comparable to the therapeutic drug acarbose. *Journal of Agricultural Food Chemistry* 52: 2541–2545.

Aboulaghras, S., Sahib, N., Bakrim, S. et al. (2022). Health benefits and pharmacological aspects of chrysoeriol. *Pharmaceuticals* 15: 973.

Agarwal, R., Gupta, S.K., Srivastava, S. et al. (2009). Intraocular pressure-lowering activity of topical application of *Aegle marmelos* fruit extract in experimental animal models. *Ophthalmic Research* 42: 112–116.

Agrawal, A., Verma, P., and Goyal, P. (2010). Chemomodulatory effects of *Aegle marmelos* against DMBA-induced skin tumorigenesis in Swiss albino mice. *Asian Pacific Journal of Cancer Prevention* 11: 1311–1314.

ALNasser, M.N., Mellor, I.R., and Carter, W.G. (2022). A preliminary assessment of the nutraceutical potential of Acai berry (Euterpe sp.) as a potential natural treatment for Alzheimer's disease. *Molecules* 27: 4891.

Arcari, D.P., Santos, J.C., Gambero, A., and Ribeiro, M.L. (2013). The in vitro and in vivo effects of yerba mate (*Ilex paraguariensis*) extract on adipogenesis. *Food Chemistry* 141: 809–815.

Arul, V., Miyazak, S., and Dhananjayan, R. (2005). Studies on the anti-inflammatory, antipyretic and analgesic properties of the leaves of *Aegle marmelos* Correa. *Journal of Ethnopharmacology* 96: 159–163.

Ayyanar, M. and Subash-Babu, P. (2012). *Syzygium cumini* (L.) skeels: a review of its phytochemical constituents and traditional uses. *Asian Pacific Journal of Tropical Biomedicine* 2: 240–246.

Babenko, N.A. and Shakhova, E.G. (2006). Effects of *Chamomilla recutita* flavonoids on age-related liver sphingolipid turnover in rats. *Experimental Gerontology* 41: 32–39.

Badam, L., Bedekar, S., Joshi, S.K.B., and SP. (2002). In vitro antiviral activity of bael (*Aegle marmelos* Corr) upon human coxsackieviruses B1-B6. *The Journal of Communicable Diseases* 34: 88–99.

Baliga, M.S., Bhat, H.P., Joseph, N., and Fazal, F. (2011). Phytochemistry and medicinal uses of the bael fruit (*Aegle marmelos* Correa): a concise review. *Food Research International* 11: 1768–1775.

Belwal, T., Andola, H.C., Atanassova, M.S. et al. (2019). Gotu Kola (*Centella asiatica*). In: *Nonvitamin and Nonmineral Nutritional Supplements* (ed. S.M. Nabavi and A.S. Silva), 265–275. Academic Press.

Biworo, A., Tanjung, E., Iskandar, K., and Suhartono, E. (2015). Antidiabetic and antioxidant activity of jackfruit (*Artocarpus heterophyllus*) extract. *Journal of Medical and Bioengineering* 4 (4): 318–323.

Blahova, J. and Svobodova, Z. (2012). Assessment of coumarin levels in ground cinnamon available in the Czech retail market. *Scientific World Journal* 2012: 263851. https://doi.org/10.1100/2012/263851.

Bogahawaththa, S., Kodithuwakku, S.P., Wijesundera, K.K. et al. (2021). Anti-fibrotic and anti-angiogenic activities of *Osbeckia octandra* leaf extracts in thioacetamide-induced experimental liver cirrhosis. *Molecules* 26: 4836.

Bonfill, M., Mangas, S., Cusido, R.M. et al. (2005). Identification of triterpenoid compounds of *Centella asiatica* by thin-layer chromatography and mass spectrometry. *Biomedical Chromatography* 20: 151–153.

Bracesco, N., Sanchez, A.G., Contreras, V. et al. (2011). Recent advances on *Ilex paraguariensis* research: minireview. *Journal of Ethnopharmacology* 136: 378–384.

Cabrera, C., Artacho, R., and Giménez, R. (2006). Beneficial effects of green tea-a review. *Journal of the American College of Nutrition* 25: 79–99.

Caesar, L.K. and Cech, N.B. (2019). Synergy and antagonism in natural product extracts: when 1 + 1 does not equal 2. *Natural Product Reports* 36: 869–888.

Canda, B.D., Oguntibeju, O.O., and Marnewick, J.L. (2014). Effects of consumption of Rooibos (*Aspalathus linearis*) and a Rooibos-derived commercial supplement on hepatic tissue injury by tert-Butyl hydroperoxide in Wistar rats. *Oxidative Medicine and Cellular Longevity* 2014: 716832. https://doi.org/10.1155/2014/716832.

Cha, B., Shi, W.L., Yonezawa, T. et al. (2009). An inhibitory effect of chrysoeriol on platelet-derived growth factor (PDGF)-induced proliferation and PDGF receptor signaling in human aortic smooth muscle cells. *Journal of Pharmacological Sciences* 110: 105–110.

Chagas, V.T., Coelho, R.M.R.D.S., Gaspar, R.S. et al. (2019). Protective effects of a polyphenol-rich extract from *Syzygium cumini* (L.) skeels leaf on oxidative stress-induced diabetic rats. *Oxidative Medicine and Cellular Longevity* 9: 5785798.

Chandrasekara, A. (2019). Phenolic Acids. In: *Encyclopedia of Food Chemistry* (ed. P. Varelis, L. Melton, and F. Shahidi) Reference Work, 535–545. Elsevier.

Chandrasekara, A. and Shahidi, F. (2018). Herbal beverages: bioactive compounds and their role in disease risk reduction—a review. *Journal of Traditional and Complementary Medicine* 8: 451–458.

Chandrasekara, A., Daugelaite, J., and Shahidi, F. (2018). DNA scission and LDL cholesterol oxidation inhibition and antioxidant activities of bael (*Aegle marmelos*) flower extracts. *Journal of Traditional and Complementary Medicine* 8: 428–435.

Chandrasekara, A., Januka, T., Kumari, D. et al. (2020). Phenolic antioxidants of bael fruit herbal tea and effects on postprandial glycemia and plasma antioxidant status in healthy adults. *Journal of Food Bioactives* 11: 75–83.

Chandrasekara, A., Herath, A., Abeynayake, R. et al. (2021). Synergies of herbal teas on improved plasma antioxidant capacity and postprandial hypoglycemic response in healthy adults. *Journal of Food Bioactives* 13.

Chandrika, U.G., Wedage, W.S., Wickramasinghe, S.M.D.N., and Fernando, W.S. (2006). Hypoglycaemic action of the flavonoid fraction of Artocarpus heterophyllus leaf. *African Journal of Traditional Complementary and Alternative Medicines* 3: 42–50.

Christensen, L.P., Jensen, M., and Kidmose, U. (2006). Simultaneous determination of ginsenosides and polyacetelyns in American ginseng root (*Panax quinquefolium* L.) by high performance liquid chromatography. *Journal of Agricultural and Food Chemistry* 54: 8995–9003.

Chumark, P., Khunawat, P., Sanvarinda, Y. et al. (2008). The in vitro and ex vivo antioxidant properties, hypolipidae mic and antiatherosclerotic activities of water extract of *Moringa oleifera* Lam. leaves. *Journal of Ethnopharmacology* 116: 439–446.

Dhankhar, S., Ruhil, S., Balhara, M. et al. (2011). *Aegle marmelos* (Linn.) correa: a potential source of phytomedicine. *Journal of Medicinal Plants Research* 5: 1497–1507.

Essiedu, J.A., Gonu, H., Adadi, P., and Withayagiat, U. (2023). Polyphenols and antioxidant activity of *Thunbergia laurifolia* infused tea under drying conditions. *Journal of Food Quality* 2023: doi:10.1155/2023/5046880.

FDA (2022) US Food and Drug Administration. https://www.fda.gov/food/information-consumers-using-dietary-supplements/supplement-your-knowledge (accessed 1 December 2023).

Fornari Laurindo, L., Barbalho, S.M., Guiguer, E.L. et al. (2023). Açaí (*Euterpe oleracea Mart.*) in health and disease: a critical Review. *Nutrients* 15: 989. https://doi.org/10.3390/nu15040989.
Gao, H., Long, Y., Jiang, X. et al. (2013). Beneficial effects of yerba mate tea (*Ilex paraguariensis*) on hyperlipidemia in high-fat-fed hamsters. *Experimental Gerontology* 48: 572–578.
Gershenzon, J. and Kreis, W. (1999). Biosynthesis of monoterpenes, sesquiterpenes, diterpenes, sterols, cardiac glycosides and steroid saponins. In: *Biochemistry of Plant Secondary Metabolites, Annual Plant Reviews* (ed. M. Wink), 222–299. Sheffield: Sheffield Academic Press.
Govindan, G., Sambandan, T.G., Govindan, M. et al. (2007). A bioactive polyacetylene compound isolated from *Centella asiatica*. *Planta Medica* 73: 597–599.
Gupta, S., Sharma, S.B., Bansal, S.K., and Prabhu, K.M. (2009). Antihyperglycemic and hypolipidemic activity of aqueous extract of *Cassia auriculata* L. leaves in experimental diabetes. *Journal of Ethnopharmacology* 123: 499–503.
Health Canada(2023). Natural health product regulation in Canada: natural health product inspection program. https://www.canada.ca/en/health-canada/services/drugs-health-products/natural-non-prescription/regulation/inspection-program.html (accessed 1 December 2023).
Heck, C.I. and de Mejia, E.G. (2007). Yerba mate tea (*Ilex paraguariensis*): a comprehensive review on chemistry, health implications, and technological considerations. *Journal of Food Science* 72: R138–R151.
Islam, S.M.A., Ahmed, K.T., Manik, M.K. et al. (2013). A comparative study of the antioxidant, antimicrobial, cytotoxic and thrombolytic potential of the fruits and leaves of *Spondias dulcis*. *Asian Pacific Journal of Tropical Biomedicine* 3 (9): 682–691.
Isobella, S., Cogoi, L., Lopez, P. et al. (2010). Study of the bioactive compounds variation during yerba mate (*Ilex paraguariensis*) processing. *Food Chemistry* 122: 695–699.
Jagetia, G.C. and Venkatesh, P. (2007). Inhibition of radiation-induced clastogenicity by *Aegle marmelos* (L.) correa in mice bone marrow exposed to different doses of gamma-radiation. *Human & Experimental Toxicology* 26: 111–124.
Jagetia, G., Venkatesh, P., and Baliga, M. (2004). Evaluation of the radioprotective effect of bael leaf (*Aegle marmelos*) extract in mice. *International Journal of Radiation Biology* 80: 281–290.
Jayarathna, P.L., Jayawardena, J.A.E., and Vanniarachchy, M.P.G. (2020). Identification of physical, chemical properties and flavor profile of *Spondias dulcis* in three maturity stages. *IRJAES* 5 (1): 208–211.
Joubert, E., Gelderblom, W.C.A., Louw, A., and de Beer, D. (2008). South African herbal teas: *Aspalathus linearis, yclopia* spp. and *Athrixia phylicoides*-A review. *Journal of Ethnopharmacology* 119: 376–412.
Juan-Badaturugae, M., Habtemariam, S., and Thomas, M.J.K. (2011). Antioxidant compounds from a South Asian beverage and medicinal plant, *Cassia auriculata*. *Food Chemistry* 125: 221–225.
Junsi, M. and Siripongvutikorn, S. (2016). *Tunbergia laurifolia*, a traditional herbal tea of Tailand: botanical, chemical composition, biological properties and processing influence. *International Food Research Journal* 23: 923.
Junsi, M., Siripongvutikorn, S., Yupanqui, C.T., and Usawakesmanee, W. (2017). Phenolic and flavonoid compounds in aqueous extracts of *Tunbergia laurifolia* leaves and their effect on the toxicity of the carbamate insecticide methomyl to murine macrophage cells. *FFHD* 7: 529–544.
Kalaivani, A. and Umamaheswari, A. (2008). Anti-hyperglycemic and antioxidant properties of *Cassia auriculata* leaves and flowers on alloxan induced diabetic rats. *Pharmacology* 1: 204–217.
Kamalakkannan, N. and Prince, P.S. (2003). Hypoglycaemic effect of water extracts of *Aegle marmelos* fruits in streptozotocin diabetic rats. *Journal of Ethnopharmacology* 87: 207–210.
Kamalakkannan, N. and Prince, P.S. (2005). The effect of *Aegle marmelos* fruit extract in streptozotocin diabetes: a histopathological study. *Journal of Herbal Pharmacotherapy* 5: 87–96.

Khoo, H.E., Prasad, K.N., Kong, K.W. et al. (2010). A review on underutilized tropical fruits in Malaysia. *Guangxi Academy of Agricultural Sciences* 41 (7): 698–702.

Kim, J.Y., Kim, J., Ratnayake Bandara, B.M. et al. (2022). Leaf extract of *Osbeckia octandra* induces apoptosis in oral squamous cell carcinoma cells. *BMC Complementary Medicine and Therapies* 22: 20.

Kunjumon, R., Johnson, A.J., and Baby, S. (2022). *Centella asiatica*: secondary metabolites, biological activities and biomass sources. *Phytomedicine Plus* 2: 100176.

Lakmal, K., Yasawardene, P., Jayarajah, U., and Seneviratne, S.L. (2021). Nutritional and medicinal properties of star fruit (*Averrhoa carambola*): a review. *Food Science & Nutrition* 9 (3): 1810–1823.

Latha, M. and Pari, L. (2003a). Antihyperglycaemic effect of *Cassia auriculata* in experimental diabetes and its effects on key metabolic enzymes involved in carbohydrate metabolism. *Clinical and Experimental Pharmacology & Physiology* 30: 38–43.

Latha, M. and Pari, L. (2003b). Preventive effects of *Cassia auriculata* L. flowers on brain lipid peroxidation in rats treated with stretozocin. *Molecular and Cellular Biochemistry* 243: 23–28.

Lee, K.G. and Shibamoto, T. (2002). Determination of antioxidant potential of volatile extracts isolated from various herbs and spices. *Journal of Agricultural and Food Chemistry* 50: 4947–4952.

de Lima Yamaguchi, K.K., Pereira, L.F.R., Lamarão, C.V. et al. (2015). Amazon acai: chemistry and biological activities: a review. *Food Chemistry* 179: 137–151.

Liu, J. and Henkel, T. (2002). Traditional Chinese medicine (TCM): are polyphenols and saponins the key ingredients triggering biological activities? *Current Medicinal Chemistry* 9: 1483–1485.

Lobato, R., Matsumoto, T., Mendonça, S. et al. (2009). Effects of maté tea intake on ex vivo LDL peroxidation induced by three different pathways. *Nutrients* 1: 18–29.

Loizzo, M.R., Tundis, R., Chandrika, U.G. et al. (2010). Antioxidant and antibacterial activities on foodborne pathogens of *Artocarpus heterophyllus* Lam. (Moraceae) leaves extracts. *Journal of Food Science* 75 (5): 291M–295M.

Maity, P., Hansda, D., Bandyopadhyay, U., and Mishra, D.K. (2009). Biological activities of crude extracts and chemical constituents of bael, *Aegle marmelos* (L.) Corr. *Indian Journal of Experimental Biology* 47: 849–861.

Marnewick, J.L. (2010). Rooibos and honeybush: recent advances in chemistry, biological activity and pharmacognosy. In: *African Natural Plant Products: New Discoveries and Challenges in Chemistry and Quality*, ACS Symposium Series, vol. 1021 (ed. H.R. Juliani, J.E. Simon, and C.T. Ho), 277–294. Washington DC: American Chemical Society.

Masud Parvez, G.M. and Sarker, R.K. (2021). Pharmacological potential of wood apple (*Limonia acidissima*): a review. *IJMFM&AP* 7 (2): 40–47.

McCord, J.M. (2000). The evolution of free radicals and oxidative stress. *The American Journal of Medicine* 108 (8): 652–659.

McKay, D.L. and Blumberg, J.B. (2006). A review of the bioactivity and potential health benefits of chamomile tea (*Matricaria recutita* L.). *Phytotherapy Research* 20: 519–530.

McKay, D.L. and Blumberg, J.B. (2007). A review of the bioactivity of South African herbal teas: rooibos (*Aspalathus linearis*) and honeybush (*Cyclopia intermedia*). *Phytotherapy Research* 21: 1–16.

Minto, R.E. and Blacklock, B.J. (2008). Biosynthesis and function of polyacetylenes and allied natural products. *Progress in Lipid Research* 47: 233–306.

Muthu, N., Lee, S.Y., Phua, K.K., and Bhore, S.J. (2016). Nutritional, medicinal and toxicological attributes of star-fruits (*Averrhoa carambola* L.): a review. *Bioinformation* 12 (12): 420–424.

Narender, T., Shweta, S., Tiwari, P. et al. (2007). Antihyperglycemic and antidyslipidemic agent from *Aegle marmelos*. *Bioorganic & Medicinal Chemistry Letters* 17: 1808–1811.

Narendhirakannan, R.T., Subramanian, S., and Kandaswamy, M. (2006). Biochemical evaluation of antidiabetogenic properties of some commonly used Indian plants on streptozotocin-induced diabetes in experimental rats. *Clinical and Experimental Pharmacology & Physiology* 33: 1150–1157.

Omar, H.S., El-Beshbishy, H.A., Moussa, Z. et al. (2011). Antioxidant activity of *Artocarpus heterophyllus* Lam. (jackfruit) leaf extracts: remarkable attenuations of hyperglycemia and hyperlipidemia in streptozotocin-diabetic rats. *Scientific World Journal* 11: 788–800.

Pandey, S., Satpathy, G., and Gupta, R.K. (2014). Evaluation of nutritional, phytochemical, antioxidant and antibacterial activity of exotic fruit *Limonia acidissima*. *JPP* 3 (2): 81–88.

Pękal, A., Dróżdż, P., Biesaga, M., and Pyrzynska, K. (2011). Evaluation of the antioxidant properties of fruit and flavoured black teas. *European Journal of Nutrition* 50: 681–688.

Pellati, F., Calo, S., Benvenuti, S. et al. (2006). Isolation and structure elucidation of cytotoxic polyacetylenes and polyene from *Echinacea pallida*. *Phytochemistry* 67: 1359–1364.

Persson, I.A., Persson, K., Hägg, S., and Andersson, R.G.G. (2010). Effects of green tea, black tea and rooibos tea on angiotensin-converting enzyme and nitric oxide in healthy volunteers. *Public Health Nutrition* 3: 730–737.

Peter, K.V. and Babu, K.N. (2012). Introduction to herbs and spices: medicinal uses and sustainable production. In: *Handbook of Herbs and Spices* (ed. K.V. Peter), 1–16. Elsevier.

Prakash, O., Kumar, R., Mishra, A., and Gupta, R. (2009). *Artocarpus heterophyllus* (Jackfruit): an overview. *Pharmacognosy Reviews* 3 (6): 353–358.

Rajadurai, M. and Prince, P.S. (2005). Comparative effects of *Aegle marmelos* extract and alpha tocopherol on serum lipids, lipid peroxides and cardiac enzyme levels in rats with isoproterenol-induced myocardial infarction. *Singapore Medical Journal* 46: 78–81.

Rama Rao, A.V., Varadan, M., and Venkataraman. (1973). Colouring matter of the *A. hetetophyllus*. *Indian Journal of Chemistry* 11: 298–299.

Ravi, K., Ramachandran, B., and Subramanian, S. (2004). Protective effect of *Eugenia jambolana* seed kernel on tissue antioxidants in streptozotocin-induced diabetic rats. *Biological and Pharmaceutical Bulletin* 27 (8): 1212–1217.

Razali, N.N.M., Ng, C.T., and Fong, L.Y. (2019). Cardiovascular protective effects of Centella asiatica and its triterpenes: a review. *Planta Medica* 85: 1203–1215.

Ruangpayungsak, N., Sithisarn, P., and Rojsanga, P. (2018). High performance liquid chromatography fngerprinting and chemometric analysis of antioxidant quality of *Tunbergia laurifolia* leaves. *Journal of Liquid Chromatography & Related Technologies* 41: 713–721.

Rzasa-Duran, E., Kryczyk-Poprawa, A., Sułkowska-Ziaja, K. et al. (2022). Yerba mate as a source of elements and bioactive compounds with antioxidant activity. *Antioxidants* 11: 371.

Sah, A., Naseef, P.P., Kuruniyan, M.S. et al. (2022). A comprehensive study of therapeutic applications of Chamomile. *Pharmaceuticals (Basel).* 15: 1284.

Schauss, A.G., Wu, X., Prior, R.L. et al. (2006). Antioxidant capacity and other bioactivities of the freeze-dried Amazonian palm berry, Euterpe oleraceae mart. (acai). Journal of Agricultural. *Food Chemistry* 54: 8604–8610.

Shahidi, F. and Ambigaipalan, P. (2015). Phenolics and polyphenolics in foods, beverages and spices: antioxidant activity and health effects—a review. *Journal of Functional Foods* 18: 820–897.

Shahidi, F. and Naczk, M. (2004). *Phenolics in Food and Nutraceuticals*. Boca Raton: CRC Press.

Shahidi, F., Ambigapalan, P., and Chandrasekara, A. (2017). Recent advances in phytochemicals in fruits and vegetables. In: *Fruit and Vegetable Phytochemicals: Chemistry and Human Health*, 2ee (ed. E.M. Yahia), 1323–1356. UK: John Wiley & Sons, Ltd.

Shahidi, F., Chandrasekara, A., and Zhong, Y. (2018). Bioactive phytochemicals in vegetables. In: *Handbook of Vegetables and Vegetable Processing*, 2ee (ed. M. Siddiq and M.A. Uebersax), 181–222. UK: John Wiley & Sons Ltd.

Sharma, P., Gothalwal, R., and Tenguria, R.K. (2021). Phytochemical screening of fruit and stem bark extracts of *Limonia acidissima* (L.). *Journal of Advanced Scientific Research* 12: 326–329.

Sies, H. (1997). Oxidative stress: oxidants and antioxidants. *Experimental Physiology* 26 (82): 291–295.

Sinan, K.I., Zengin, G., Zheleva-Dimitrova, D. et al. (2021). Exploring the chemical profiles and biological values of two spondias species (*S. dulcis* and *S. mombin*): valuable sources of bioactive natural products. *Antioxidants* 10 (11): 1771.

Singh, D., Gupta, R., and Saraf, S.A. (2012). Herbs-are they safe enough? an overview. *Critical Reviews in Food Science and Nutrition* 52: 876–898.

Sosa, V., Moliné, T., Somoza, R. et al. (2013). Oxidative stress and cancer: an overview. *Ageing Research Reviews* 31: 376–390.

Srivastava, J.K. and Gupta, S. (2009). Health promoting benefits of chamomile in the elderly population. In: *Complementary and Alternative Therapies in the Aging Population* (ed. R. Watson), 135–158. Elsevier Inc., Academic Press.

Stohs, S.J. and Hartman, M.J. (2015). Review of the safety and efficacy of *Moringa oleifera*. *Phytotherapy Research* 29: 796–804.

Sugahara, S., Chiyo, A., Fukuoka, K. et al. (2018). Unique antioxidant effects of herbal leaf tea and stem tea from *Moringa oleifera* L. especially on superoxide anion radical generation systems. *Bioscience, Biotechnology, and Biochemistry* 82: 1973–1984.

Sun, B., Wu, L., Wu, Y. et al. (2020). Therapeutic potential of *Centella asiatica* and its triterpenes: a review. *Frontiers in Pharmacology* 11: 568032.

Swami, S.B., Thakor, N.J., Haldankar, P.M., and Kalse, S.B. (2012). Jackfruit and its many functional components as related to human health: a review. *Comprehensive Reviews in Food Science and Food Safety* 11 (6): 565–576.

Tafrihi, M., Imran, M., Tufail, T. et al. (2021). The wonderful activities of the genus mentha: not only antioxidant properties. *Molecules* 26: 1118.

Thabrew, M.I., Hughes, R.D., Gove, C.D. et al. (1995a). Protective effects of *Osbeckia octandra* against paracetamol-induced liver injury. *Xenobiotica* 25: 1009–1017.

Thabrew, M.I., Gove, C.D., Hughes, R.D. et al. (1995b). Protective effects of *Osbeckia octandra* against galactosamine and tert-butyl hydroperoxide induced hepatocyte damage. *Journal of Ethnopharmacology* 49: 69–76.

Thapa, N., Thapa, P., Bhandari, J. et al. (2016). Study of phytochemical, antioxidant and antimicrobial activity of *Artocarpus heterophyllus*. *Nepal Journal of Biotechnology* 4: 47–53.

Udani WDL, Sahabandu M, Chandrasekara A. (2024). Effect of hot and cold brewing on phenolic content, antioxidant and enzyme inhibitory activities of novel fruit teas available in Sri Lanka. *Proceedings of the Annual Scientific Sessions of the Nutrition Society of Sri Lanka*, Hotel Ramada, Colombo, Sri Lanka (27–28 January), p. 52.

Upadhya, S., Shanbhag, K.K., Sunethea, G. et al. (2004). A study of hypoglycemic and antioxidant activity of *Aegle marmelos* in alloxan induced diabetic rats. *Indian Journal of Physiology and Pharmacology* 48: 476–480.

Upadhyay, S.K. and Singh, S.P. (2021). *Bioprospecting of Plant Biodiversity for Industrial Molecules*. Wiley https://doi.org/10.1002/9781119718017.

Upadhyay, S.K. and Singh, S.P. (2023). *Plants as Bioreactors for Industrial Molecules*. Wiley https://doi.org/10.1002/9781119875116.

VanderJagt, T.J., Ghattas, R., VanderJagt, D.J. et al. (2002). Comparison of the total antioxidant content of 30 widely used medicinal plants of New Mexico. *Life Sciences* 70: 1035–1040.

Wang, X.L., Di, X.X., Shen, T. et al. (2017). New phenolic compounds from the leaves of *Artocarpus heterophyllus*. *Chinese Chemical Letters* 28: 37–40.

Young, A.J. and Lowe, G.L. (2018). Carotenoids—antioxidant properties. *Antioxidants* 7: 28.

14

Nutraceuticals Present in Edible Oils

Rupasree Mukhopadhyay[1] *and Shruti Kabra*[2]

[1] *Department of Genetics and Biotechnology, Telangana Mahila Viswavidyalayam, Koti, Hyderabad, Telangana, India*
[2] *Department of Food and Nutrition, Telangana Mahila Viswavidyalayam, Koti, Hyderabad, Telangana, India*

14.1 Introduction

"Nutraceutical" is a term used as an amalgam of the two words "nutrition" and "pharmaceutical." It was coined by DeFelice (1989) to emphasize the role of food or part of food, which possesses pharmaceutical activity. Nutraceuticals are mainly defined as a group of food-derived products with diverse medicinal and health benefits (DeFelice 1995). In addition to the enhancement of nutritional values of foodstuff, these products also help to safeguard and treat cancer, osteoporosis, arthritis, retinopathy, cardiovascular diseases, aging, and also many neurological disorders like Alzheimer's disease (Maddi et al. 2007; Radhika et al. 2011; Taylor 2004). Due to the presence of bioactive components like flavonoids, phenols, phytosterols, etc., various nutraceuticals are often known for their neutralizing effects on free radicals, boosting the immune system, regulating serum cholesterol levels, and maintaining overall wellness of health.In fact, it has been established that nutraceuticals form an integral part of human diet as they provide all the essential nutrients. A diverse range of plant-derived bioactive compounds called phytochemicals have the potentiality to naturally modulate the metabolic processes to aid in drug discovery and remedy for human diseases like cancer, cholesterolemia, and immunological disorders (Moiseeva et al. 2007; Thomford et al. 2018). These nutrient-rich molecules are available in various forms like whole foods dietary supplements and isolated compounds, which clearly distinguishes the nutraceuticals from functional foods (Santana-Gálvez et al. 2019). Nutraceuticals include bioactive components from natural origin, while functional foods functional foods have nutritious components that imparts well-being and safeguards from diseases (Shahidi 2009). Hence, a food is considered as functional food only if it contains nutraceutical components in it.

Generally, nutraceuticals are classified and characterized by their chemical nature, food availability, and the mode of action of the biologically active component in it (Figure 14.1). Traditional nutraceuticals are the compounds directly obtained from nature and can be conveniently used without any modification of their original form. Such compounds are widely subcategorized into phytochemicals, probiotic bacteria, enzymes, chemical compounds, nutrients, and plant-derived nutrients (Ali et al. 2019; Dixit et al. 2021a,b; Singh and Upadhyay 2021, 2023; Upadhyay and Singh 2021, 2023).The nontraditional nutraceuticals include fortified and recombinant compounds. In current days,

Herbal Nutraceuticals: Products and Processes, First Edition. Edited by Santosh Kumar Upadhyay and Sudhir Pratap Singh.

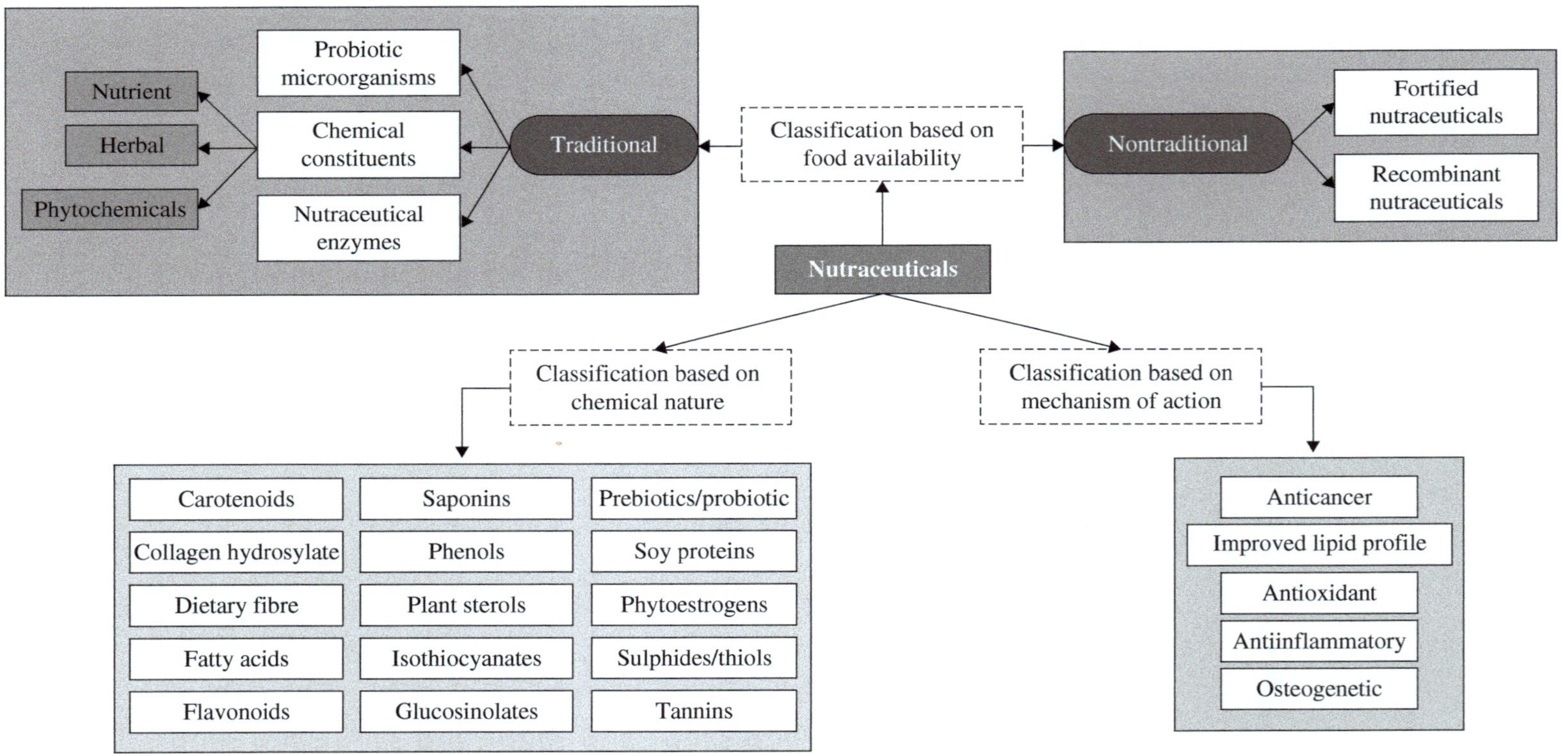

Figure 14.1 Classification of nutraceuticals based on food availability, chemical nature, and mechanism of action.

Table 14.1 List of naturally occurring nutraceuticals from food items.

Food product	Nutraceutical
Cruciferous vegetables	Glucosinolates, isothiocyanates, indoles, carotenoids
Eggs	Ovophospholipids, fatty acids including ALA, docosa hexanoic acid (DHA), biopeptides, vitamins
Fish oil	EPA, DHA
Grapes (skin)	Red wine resveratrol
Kale, spinach, corn	Citrus lutein, zeaxanthin
Onion	Allyl garlic sulfur compounds
Soy and other leguminous plants	Protein, isoflavones, protease inhibitors,phytosterols, saponins, phenolic acids, phytic acid
Tea, berries, red wine	Polyphenolic compounds, catechins, anthocyanins
Vegetable oils (linseed, evening primrose, amaranth)	Lignins, fatty acids, α-linolenic acid, squalene
Wheat bran	B-glucan cereal products, conjugated linoleic acid (CLA)

Source: Adapted from Sawicka et al. (2020).

numerous strategies of breeding and biotechnology are adapted to enhance the nutritional value of food and to improve human health (Gupta et al. 2010; Upadhyay 2021, 2024). The natural/conventional nutraceuticals comprise of herbal components, single nutrients, supplements, functional foods, and bioactive phytochemicals (Sawicka et al. 2020). The biologically active phytochemicals form a major group of nutraceuticals including phenolic compounds, polyphenols, flavonoids, anthocyanins, ellagic acid, resveratrol, isoflavones, flavanones, etc. (Sakthinathan and Nandhini 2017).

Usually, nutraceuticals are found naturally in common food items. These compounds may support the health and general well-being of an individual, besides supporting the physical performance and functioning of gastrointestinal system, cardiovascular system, and immune system. A generalized list of naturally occurring nutraceuticals in food items is given in Table 14.1. The evaluation of nutritional components of food items and their effect on human health is a continuous process.Nutraceuticals are available in isolated form as powders, capsules, tablets, and syrups. However, the natural sources of nutraceuticals are better preferred over chemicals due to their fewer side effects and better results. Nutraceuticals offer promising benefits to human beings in various capacities ranging from improving the nutritional value of diet, providing natural harmless medication and overcoming nutrient deficiencies. In addition to improving good health, nutraceuticals introduce good eating habits and enhance the lifestyle. Therefore, nutraceuticals are considered as an imperative and integral part of proactive medicine in recent times, which not only boosts the quality of life and health status, but also helps in avoiding diseases decreasing the cost of time-taking therapies.

14.2 Edible Oils and Their Uses

Edible oils from plants, animal, or microbial origin form an important part of culinary science, as it enriches the recipes and improves the palatability of food. Plant sources are the major sources for oil production in the world as compared to that of animal and marine sources (Gunstone 2002). The global vegetable oil production accounts for 83.7%, which provide nutraceuticals in appreciable amounts as the main sources are sunflower,palm, rapeseed, and soybean (Sharma et al. 2011).Edible oils mostly contain fatty acid glycerides obtained from vegetable sources, apart from small quantities of other lipids like phosphatides, unsaponifiable constituents, and free fatty acids. Some of the commonly used edible oils extracted from vegetables are described in the following sections.

14.2.1 Coconut Oil

Coconut oil is being used as a food and cooking oil since thousands of years. This oil is obtained from dried coconut meal with about 57–75% recovery. Coconut oil mainly contains medium-chain fatty acids, including caprylic acid, capric acid, and lauric acid that forms the triglyceride molecules and imparts antimicrobial properties. Unlike other oils, it is poor in unsaturated fatty acids (9%), but has a high proportion of saturated fatty acids (SFAs) (90%). The chemical composition of coconut oil indicates that it is the highest natural source of lauric acid (about 49.5%), while it contains a very low concentration of cholesterol (about 0–14 ppm) (Rathinam et al. 2002). Oleochemicals and linoleum acid are present at only 9% concentrations. Lauric acid along with its derivative monolaurin make up to approximately 50% of the fat-derived lipid of coconut (Boateng et al. 2016). The high lauric acid content present in coconut oil undergoes conversion into monolaurium within the human body that aides in killing viruses, bacteria, and protozoa (Selvaraj et al. 2020). In general, coconut oil finds applications in the manufacturing of soap, vanaspati, margarine, pharmaceutical preparation, confectionery and cookie filling, and cooking avial and banana chips due to its characteristic flavor. The recent globally accepted version, virgin coconut oil is known for its natural aroma and nutrient profile, which is preserved as it is extracted by natural means without using any chemical refining, bleaching, or deodorizing. The natural aroma is also due to the fresh, mature endosperm of the coconut, which is used for the extraction of oil. In its pure, pristine raw form, it is enriched with the natural form of vitamin E and has very low fatty acid content (0.1%), which makes it apt for human consumption as a nutraceutical (Selvaraj et al. 2020). Coconut oil tends to impart immunity to small babies against many diseases due to the structural similarity of the medium-chain fatty acids to the fats present in mother's milk. Virgin coconut oil possesses antioxidant, anti-inflammatory, and antimicrobial properties, and it elevates high-density lipoprotein (HDL) levels, which aid in protecting against cardiovascular disorders. As a nutraceutical, it increases the bioavailability of amino acids and micronutrients by enhancing its absorption, thereby enhancing the nutritive value of food (Selvaraj et al. 2020).

14.2.2 Cottonseed Oil

Cottonseed oil is obtained from the fresh seeds of cotton plants by cold-pressed process. In current days, this oil has gained popularity due to excessive saturated fat contents like oleic, linoleic, and palmitic acid, which are beneficial for hair and skin. Cottonseed oil contains about 50% of necessary polyunsaturated linoleic acid, which is much higher than other traditional cooking oils (about 30% in sesame, olive, safflower, and coconut oil). This plays a very important role in humans as it prevents the thickening of coronary veins (Waheed et al. 2010). It is consumed as a salad or cooking oil while sometimes it is used as shortening. Its neutral flavor doesnot overpower other products, and with a high smoke point of approximately 420°F, it is well-suited for use in restaurants and highly processed foods. The presence of the abundant natural antioxidant compound, alpha-tocopherol in cottonseed oil, prolongs the shelf life of products and frying time for many foods (Sekhar and Rao 2011). It can be used for cooking and baking at home. It has properties that prevent inflammation, free radical formation, cancer,help in wound healing, and decreasing cholesterol levels (Riaz et al. 2021). The unsaturated fatty acids in cottonseed oil may help lower LDL and increase HDL levels, which improve blood pressure and reduce the incidence of heart disease and stroke (Fattore and Massa 2018; Mahesar et al. 2017).

14.2.3 Groundnut/Peanut Oil

In India, groundnut oil accounts for about 50% of the total edible oil produced annually. Extracted from shelled and crushed peanuts through techniques, such as hydraulic pressing, expeller pressing, and solvent extraction, this oil is clear amber colored and is predominantly used as a cooking and salad oil (Akhtar et al. 2014). There are three types of groundnut oil, i.e. refined, gourmet, and 100% peanut oil. The refined peanut oil is a highly stable, allergen-, cholesterol-, and trans-fat-free cooking oil with a pleasing flavor. It has minimal saturated fats and a high smoke point of 229.4°C, which makes it a popular oil for deep-fat frying applications (List 2016).Gourmet peanut oil contains notable levels of vitamin E and phytosterols, while 100% peanut oil is highly aromatic and is usually blended with other edible oils (Peanut Institute 2011). The fatty acids in groundnut oil include oleic acid (52%), linoleic acid (LA) (32%), and palmitic acid, while stearic acid, arachidic acid, eicosenoic acid, behenic acid, and linoceric acid are present in minute quantities (Ozcan 2010). It also contains natural nutraceutical compounds like vitamin E, antioxidants, squalene, phytosterols, and *p*-coumaric acid for health benefits. High oleic acid in peanut oil has a positive effect on various risk factors for cardiovascular diseases, such as blood pressure, lipid profiles, and glucose metabolism. Groundnut oil contains resveratrol, which shows potential anti-inflammatory activity along with nutraceutical and anticancer applications (Chang et al. 2006; Djoko et al. 2007; Kang et al. 2010). The presence of phytochemicals like campesterol, β-sitosterol, resveratrol, and sigmasterol imparts anticancerous properties to this oil (Awad et al. 2000; Lopes et al. 2011). It also lowers the perils of cardiac disorders and Alzheimer's disease (Chen et al. 2005; Stephens et al. 2010).

14.2.4 Olive Oil

Olive oil is derived from the fruit in contrast to other oils that are derived from seeds of plants. The main components in olive oil include 55–83% monounsaturated fatty acids (MUFAs) like oleic and palmitoleic acid, 4–20% polyunsaturated fatty acids (PUFAs) like linoleic and linolenic acid, and 8–14% SFAs like arachidonic, myristic, stearic, and palmitic acid (Ghanbari Shendi et al. 2018). Olive oil contains high-value components such as phenolic compounds, α-, β-, and γ-tocopherols, sterols, vitamins E and A, lignans, and antioxidants. Olive oil caters to numerous health benefits including cancer chemopreventive activity, lowering blood pressure, reducing incidence of myocardial infarction, anti-inflammatory activity, and treating cardiovascular disorders (Cicerale et al. 2010; Ghanbari Shendi 2019; Juan et al. 2006; López-Biedma et al. 2016b; Miles et al. 2005; Owen et al. 2003). Tocopherols and several tocotrienols present in olive oil bring about protection of cells in response to oxidative stress (Beltrán et al. 2010). Evident health benefits are observed by the use of extra virgin olive oil (EVOO) for mothers in prenatal periods, children, adolescents, and especially the elderly and in hospital diets (Frankel 2011; Salazar et al. 2017).

14.2.5 Palm Oil

Palm oil derived from *Plammesocarpis* used worldwide for cooking, frying oil, in margarines, shortening, vanaspati, and fat blends (Gunstone 2005). The agrifood industry is a major consumer of palm oil and its derivatives especially in the sectors of industrial baked goods, chocolate products, ice creams, confectionery, and even diet meal substitutes. It comprises an equal balance of unsaturated fatty acids and SFAs, with 44% palmitic acid, 5% stearic acid, 40% oleic acid, and

10% linoleic and linolenic acid. Palm oil is abundant in minor components such as carotenoids, tocopherols, tocotrienols, vitamin E, sterols, phospholipids, squalene, aliphatic hydrocarbons, and aliphatic alcohols (Mancini et al. 2015). In crude state, the high carotenoid content in palm oil boosts vitamin A levels in the blood, thereby exerting preventive effects on certain eye conditions. The antioxidant properties of palm oil are attributed to the carotenoids and vitamin E present in it. Additionally, tocopherols and tocotrienols also enhance the oxidative stability of red palm oils. Serving as antioxidants, they extend the shelf life of food products, stabilize oils and fats, and reduce free radical damage (Al-Saqer et al. 2004). It also plays a role in aging and cancer prevention (OngandGoh 2002). Vitamin E, particularly tocotrienols are also reported to suppress the synthesis of cholesterol in liver (Edem 2002). Although palm oil has been reported to have a relatively higher saturation level as compared to many other vegetable oils, its triglyceride profile and other nutritionally beneficial compounds make it a healthy alternative option for human consumption.

14.2.6 Rapeseed or Mustard Oil

Rapeseed oil derived from *Brassica napus* is the third most commonly produced vegetable oil worldwide (Lomascolo et al. 2012).Owing to its low levels of saturated fats and high content of MUFAs and PUFAs (especially oleic and α-linolenic fatty acids), it is considered to be classified among the healthiest vegetable oils (Wroniak and Rękas 2016). Mustard (*Brassica nigra* L.) seeds have a high fat content of about 47% and oil extracted from it has a characteristic sharp taste and odor. It has high amount of erucic acid and LA (Grygier 2022; Lietzow 2021). The peculiar odor of mustard oil is by virtue of allyl isothiocyanate. Due to its spicy and hot flavor, mustard oil is mostly used for cooking and making pickles in Asian countries. Rapeseed/mustard oil is rich in unsaturated fats and bioactive compounds, such as phenolic acids, phytosterols, diglycerides, flavones, vitamin E, and flavonols, which contribute to its health-promoting properties (Mishra and Manchanda 2012). Rapeseed oil exhibits beneficial effects on diabetes and boosts lipid metabolism. The major antioxidant of mustard oil is glucosinolate, which accounts for antibiotic, fungicidal, and cancer prevention properties. Additionally, the high amount of alphalinolenic acid helps to control the high cholesterol level and heart diseases (Kaur et al. 2019).

14.2.7 Ricebran Oil

Ricebran oil, obtained from the bran of rice grains after the hulling and dehusking of paddy, has gained importance as a healthy oil. The health benefits of this oil is attributed to the balanced fatty acid composition containing about 41% MUFA (oleic acid), 37% PUFA (omega-3, omega-6 fatty acids), and 22% SFAs (mainly palmitic, myristic, and stearic acid) (Riantong and Jorg 2021). It also contains high amounts of unsaponifiable components like γ-oryzanol (1.5–1.8%), vitamin E, tocotrienols (0.07%), squalene, and small quantities of unbound phytosterols like γ-sitosterol, campesterol, and stigmasterol (Pali 2013). Because of its high smoke point of 255°C, ricebran oil is considered ideal for deep frying as well as for pan or stir frying (Rohman 2014). It is good for baking and grilling and also used for salad dressings as it gets easily emulsified. It is nonallergenic, less absorbed, and has no trans-fats (Sharma and Das 2013). It is a natural source of oryzanol and vitamin E, and exhibits antioxidant property in addition to cholesterol-lowering, anti-inflammatory, antithrombotic, anticancer, and antidiabetic effects (Lai et al. 2019). Due to its nutritional benefits, it has become the oil of choice, which improves the overall immunity of human beings.

14.2.8 Sesame Oil

Sesame oil, processed out from the seeds of sesame, consists of various fatty acids (14% saturated, 39% MUFA, and 46% PUFA) and nonfat antioxidants, including sesamin, sesamol, sesamolin, and tocopherol (Pal et al. 2010). It is highly stable and rarely turns rancid in hot climates. Sesame oil finds use in salad dressings, cooking oil, and also for various medicinal purposes. It is also added into pickles for its flavor and preservative property. Sesame oil is an excellent source of PUFAs including omega-3, omega-6, and omega-9 fatty acids, which are essential for the growth and development and help in averting and treating of chronic diseases such as coronary heart disease, hypertension, cancers, diabetes, inflammation, and arthritis (Orsavova et al. 2015; Shivhare and Satsangee 2012). It also contains significant amounts of lignans (sesamin, episesamin, and sesamolin) that are responsible for antihypertensive antibacterial, anti-inflammatory, and antioxidant properties (Duan et al. 2023; Langyan et al. 2022). The sesaminol glucosides and tocopherol also help in increasing stability and gives a longer shelflife to sesame oil (Suja et al. 2004). In Ayurveda, this oil is used for treating diseases such as hepatitis, diabetes, and migraines.

14.2.9 Soybean Oil

Soybean oil obtained from raw bean by solvent extraction is the world's leading oil, both in terms of production and consumption. It is used in vanaspati, cooking oil, mayonnaise, spreads, baby foods, and cake mixes. Soybean oil is known for its lipid profile comprising of 58% PUFAs, 24% MUFAs, and 16% saturated fats (Chuffa et al. 2014). The high amount of unsaturated fatty acids includes α-linolenic acid (omega-3 acid), linoleic, γ-linolenic, and arachidonic acid (omega-6 acid), and oleic acid (omega-9 acid) (Nikolic et al. 2009; Olguin et al. 2003). It also contains various phytochemicals with nutraceutical properties like vitamin E (antioxidant), phytosterols, principally B-sitosterol (prevent cholesterol absorption and decrease blood LDLcholesterol levels), and vitamin K (for bones and neuronal protection) (Chuffa et al. 2014).

14.2.10 Sunflower Oil

Sunflower oil is second to soybean oil in production of edible oils in the world (Nandha et al. 2014) containing 20% oleic acid, 69% linoleic acid, and 11% SFAs (Khan et al. 2015).It is used as a cooking oil, salad oil, and fordeep frying because of its light, clean, odorless taste, high smoke point (232°C), and low level of saturated fats (RadhaMadhavi et al. 2010). It contains more vitamin E than any other vegetable oils, and is rich in vitamins A and D that makes it beneficial for human consumption. Sunflower oil has an increased content of oleic and LA that aids in reducing cholesterol levels and reduction in cardiac diseases (Chowdhury et al. 2007), while high levels of phytosterols control cholesterol and reduce risk of cancer (Phillips et al. 2005).

Research on different edible oils indicates that different vegetable oils provide a rich source of nutraceuticals vital for human health and nutrition that include phytosterols and phytostanols, carotenoids, triacylglycerols, lignans, oryzanols, policosanol, lecithin, tocopherols and tocotrienols, and free fatty acids.The MUFAs and PUFAs present in the oils aids in combating coronary heart diseases, cancer, depression, and diabetes. Additional phytochemicals like tocopherols (natural antioxidants), phytosterols (reduce cholesterol), and other phenolic compounds help to reduce cholesterol and prevent aging, cataract formation, oxidative stress, cancer, cardiovascular diseases, and modulate immune responses (Karuna and Prasad 2015; Leung et al. 2001). Hence, it becomes very important to identify and extract the nutraceuticals present in important edible oils and explore their therapeutic advantages for control and treatment of human diseases.

14.3 Important Nutraceuticals Present in Edible Oils

Vegetable oil-based nutraceuticals such as carotenoids, essential fatty acids (EFAs), phytosterols, tocopherols, tocotrienols, phenolics, and lignans not only add value to the oil quality, but also provide various health benefits (Dhara and Chakrabarti 2020). Hence, it is essential to explore new sources of oil with balanced nutraceutical profiles for utility in food as well as to alternatively serve as a feedstock for biofuel production to address the increasing global demands. Elaborate traditional oil-refining processes result in the loss of important nutritional components, posing a big challenge to the processor to enhance the quality of edible oils retaining all the nutraceuticals in their intact form. Though a variety of nutraceuticals are now commercially available in the market, vegetable fats and oils provide double benefits as they provide both EFAs and nutraceuticals (Karuna and Prasad 2015). Apart from being a source of concentrated energy, edible oils also have a profound impact on health when consumed from different sources and in required quantities because of the nutraceuticals present in them. Therefore, a detailed discussion about the common nutraceuticals found in edible oils along with their health benefits would provide an opportunity to choose the perfect oil in a regular balanced diet.

14.3.1 Essential Fatty Acids

The chemical structure of oils and fats consists of a triglyceride structure wherein three fatty acid molecules are esterified to a glycerol backbone. Different kinds of oils are defined by the varied types of fatty acids present in them and the composition of fatty acids plays a prominent role in assessing the nutritional significance of the oil. Usually, fatty acids are categorized as SFAs, which lack any unsaturation, MUFAs containing a single unsaturation, and PUFAsshowing more than one unsaturation (Figure 14.2) (Vergallo 2020).Among these fatty acids, PUFA finds more importance as a nutraceutical. Oils with elevated levels of PUFAs, like linoleic acid (an omega-6 fatty acid) and alpha linolenic acid (ALA, an omega-3 fatty acid), show better prospects for human intake. The nomenclature of these fatty acids is described by the position of the first double bond with regard to the methyl end of the fatty acid molecule (also called the omega ω end) (Hulbert et al. 2005). Eventually, omega-3 fatty acid gets its name when the first double bond is present on the third carbon atom from methylend based on the presence of the first double bond in the third carbon atom from the methyl end, while the location of double bond in the sixth and ninth carbon from the methyl endgives namesto omega-6 and omega-9 fatty acids, respectively (Kaur et al. 2014).

EFAs cannot be synthesized in human system and need to be exogenously supplemented in the food. The EFAs are viewed as nutraceuticals because of their essential role in crucial biochemical pathways concerning the concentration of lipoproteins, biological membrane fluidity, receptors, and enzyme activity, eicosanoids production, blood pressure regulation, and finally, the mineral metabolism. They act as antiarrhythmic, antiatherogenic, anti-inflammatory, antithrombotic, and hypolipidemic compounds that help to prevent cardiovascular diseases, cancer, diabetes, and osteoporosis (Flachs et al. 2005; Hu et al. 2001; Mišurcová et al. 2011; Mobraten et al. 2013; Weiss et al. 2005).

Among the EFAs present in edible oils, LAcontributes to more than 50% of the lipid content. The intake of adequate amounts of LA through oils shows a decrease in plasma low-density lipoprotein cholesterol (LDL-c), while replacement of 5% SFAs with ω6-type PUFAs reducedLDL-c by up to 10% (Vergallo 2020). The content of LA present in different vegetable oils vary considerably depending on geographical and seasonal variations viz. safflower (60–75%), poppy (70–72%), sunflower (60–68%), hemp (54–56%), walnut (50–53%), soybean (43–56%), cottonseed (42–52%), sesame (40–50%), and corn (34–62%) oils. ALA is amply found in flaxseed oil (35–55%), hemp (15–25%), canola (9–11%), and soybean (6–8%) (Karuna and Prasad 2015). Table 14.2 indicates the

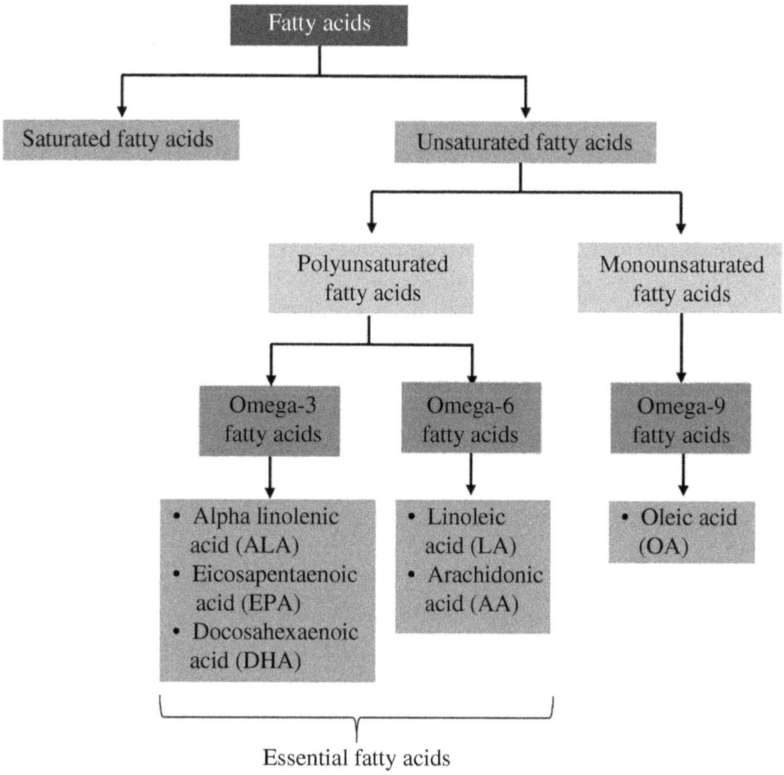

Figure 14.2 Fatty acids present in edible vegetable oils.

Table 14.2 Essential fatty acids (g/100g) in some vegetable oils.

Vegetable oil	Energy (Kcal)	Total lipid (g)	LA (g)	ALA (g)	Total saturated fat
Canola	884	100	18.6	9.14	7.37
Corn	900	100	53.5	1.16	13.0
Cottonseed	884	100	51.9	0.20	25.9
Grapeseed	884	100	69.6	0.10	9.60
Olive	884	100	9.76	0.76	13.8
Groundnut	884	100	32.0	0.00	16.9
Safflower	884	100	12.7	0.10	7.54
Sesame	884	100	41.3	0.30	14.2
Soybean	884	100	51.0	6.79	15.7
Sunflower	884	100	65.7	0.00	10.3
Walnut	884	100	52.9	10.4	9.10

Source: U.S.D.A (2020).

composition of EFAs in some edible oils as accounted by U.S.D.A (2020). These EFAs help in the growth and development of the brain, stimulation of skin and hair growth, the maintenance of reproductive function, the health of bones, and the regulation of metabolism (Karuna and Prasad 2015). Dietary ALA (ω-3) plays explicit role in controlling high blood cholesterol and BP, strengthening the immune system, safeguarding from cancer, and diminishing infertility in males (Rajaram 2014). The nonenzymatic oxidation of ALA produces phytoprostanes (isoprostane analogs) with anti-inflammatory properties. Thus, ALA-enriched oil in diet tends to reduce the oxidative stress and inflammation during myocardial infarction (Mazzocchi et al. 2021). Although both LA and ALA show significant health benefits, the dietary intake of these fatty acids should be adequately balanced (at a ratio of 0.3:1 LA:ALA) for proper functioning of physiological processes (Karuna and Prasad 2015). The highest available MUFA, oleic acid has the ability to bring about a significant reduction in mortality risk from cardiovascular events and stroke (Marangoni et al. 2020).

LA and ALA also serve as precursors of other EFAs like arachidonic acid (AA, ω-6), gamma linoleic acid (GLA, ω-6), eicosapentaenoic acid (EPA, omega-3), and docosahexaenoic acid (DHA, ω-3) (Jyoti Kumari and Avtar 2023). GLA is naturally found in oils of hempseed (6–8%), evening primrose (8–12%), borage (25%), and black currant seed oil (Firouzi 2017). On consumption, within the human body, GLA is chemically modified into dihomo-ϒ-linoleic acid (DGLA), which serves as a precursor to the anti-inflammatory metabolites, prostaglandin PGE1 and 15-OH-DGLA. A daily dosage of 1.4g of GLA brings critical improvements in rheumatoid arthritis (Joseph et al. 2016). DGLA is enzymatically converted to AA, which is beneficial for reducing inflammation and is recommended in limited dosages for babies, pregnant ladies, and adults to combat intestinal and blood flukes, maintain cell membrane fluidity and control ion channels (Tallima and El Ridi 2018). EPA and DHA, abundant in marine and microbial oils, are EFAs formed from ALA and serve as anti-inflammatory agents, greatly regulate the growth and development of fetus, and help in retina health and brain functioning (Dunstan et al. 2007; Etschmann et al. 2007; Neff et al. 2011). In addition to these, the medium-chain triglycerides (MCT) found in substantial amounts in coconut and palm kernel oil are reported to have good nutraceutical properties. Since MCT can be easily broken down and absorbed into the blood stream, it gives improved energy utilization and causes weight loss (Mumme and Stonehouse 2015; Onge and Jones 2002; Papamandjaris et al. 1998).MCTs also have noteworthy effects on muscle building, muscle strength, and diabetes control (Abe et al. 2016; Han et al. 2007).

14.3.2 Vitamin E: Tocopherols and Tocotrienols

Vitamin E is made up of eight isomers of the sub categories tocopherols (side chain with saturated phytyl unit) and tocotrienols (side chain with unsaturated isoprenyl unit), which are collectively known as tocols. Each subgroup consists of four isoforms viz. α (5,7,8-trimethyltocol), β (5,8-dimethyltocol), γ (7,8-dimethyltocol), and δ (8-methyltocol) (Morya et al. 2022). These isoforms vary with each other in their chemical structure (Figure 14.3) and the difference is mainly owing to the position and number of methyl groups (Sen et al. 2006; Shahidi and De Camargo 2016).

Usually, edible oils contain less than 0.3% of tocopherols and tocotrienols.The principal tocol-rich oils include barley, oat germ, and wheat germ oil (Kumar and Krishna 2013; Moreau et al. 2007; Peterson 1995). Tocols have a prominent role in preventing oxidation of MUFAs and PUFAs and are present in higher concentrations in the highly unsaturated vegetable oils (Shahidi and Shukla 1996). Tocopherols are antioxidant compounds, mostly oil-soluble and primarily present in olive, soybean, sunflower, and maize oils. Vegetable oils contain predominantly α-, γ-, and δ-tocopherols in more abundance, compared to the β-isomer, which is less common and present in wheat germ oil. The edible oils like sunflower, olive, wheatgerm, almond, and groundnut contain more of α-tocopherols, while the oils like soybean, cottonseed, canola, corn, linseed, camelina,and walnut have more amount of γ-tocopherols (Camargo et al. 2012; Caporaso et al. 2015; Grilo et al. 2014; Karmowski et al. 2015; Schwartz et al. 2008). The other isoform

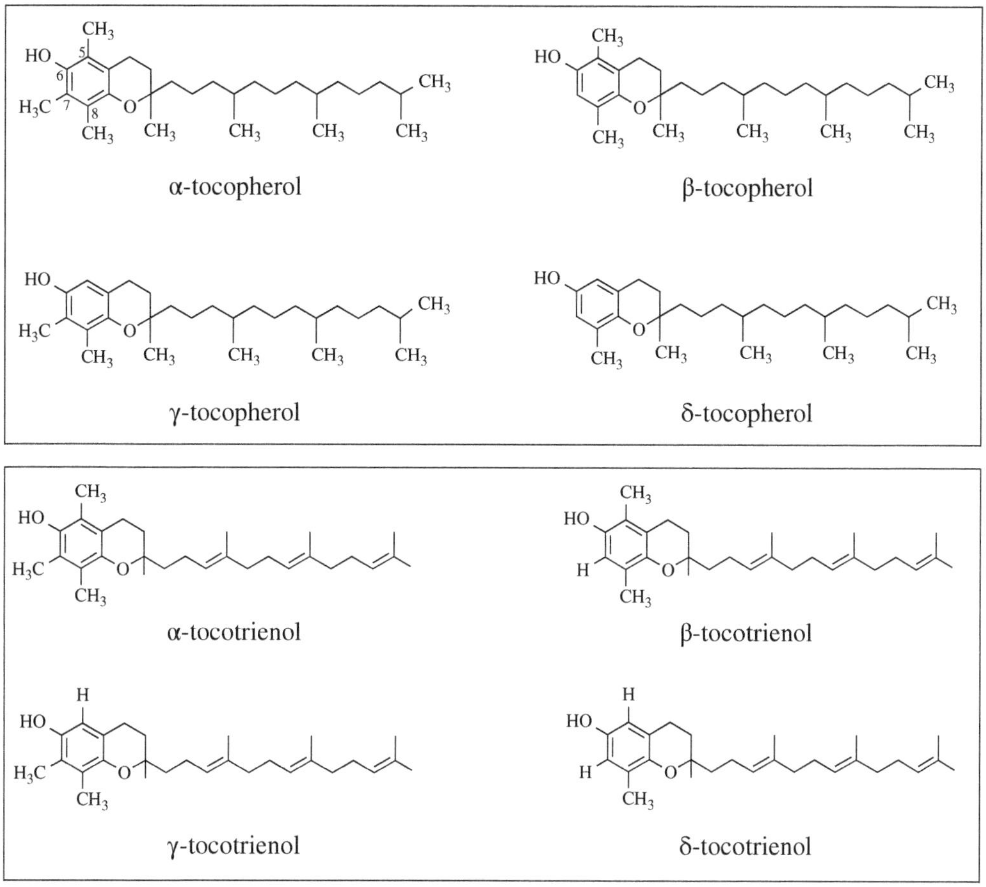

Figure 14.3 The chemical structure of isoforms of tocopherols and tocotrienols (α, β, γ, δ).

δ-tocopherol is present in castor and soybean oil, and in meager amounts in wheat germ oil (Aggarwal et al. 2010). The constitution of vitamin E in palm oil shows presence of about 70% tocotrienols and the rest 30% tocopherols (Karmowski et al. 2015; Sundram et al. 2003). The refining process of vegetable oils generates a tocopherol mixture as a by-product, which comprises of γ-tocopherol (58%), δ-tocopherol (24%), α-tocopherol (13%), and β-tocopherol (0.5%) (Ju et al. 2009; Lee et al. 2009). Tocopherols are usually found in the lipid areas of the cells and tissue imparting significant lipid-soluble antioxidant properties to be used in food. The lipid-rich domains of cells include the membranes of mitochondria, lipoproteins (like LDL-c), and fat depots of the cells. Tocotrienols are predominantly found in palm, rice bran, and annatto oils (Ng et al. 2004; Rukmini and Raghuram 1991; Zou and Akoh 2015). The α-tocotrienol is a powerful lipid-soluble antioxidant about 40–60 times more effective than the tocopherols. Palm oil is the most abundant source of α-tocotrienol and also contains the γ and δ homologs (Dauqan et al. 2011). Some other oils that serve as natural sources of tocotrienols are barley, cocoa butter, coconut oil, soybean, and wheat germ oil. Research has indicated that addition of tocopherols and tocotrienols resulted in an increase in the stability against oxidation and the shelf life of various oils and fats (Damirchi 2010; Huang et al. 1994). Tocotrienol-rich fractions have been seen to enhance antioxidant capability and oxidative stabilities of different oils and food formulations (Zou and Akoh 2015). These unique properties of tocotrienols were applied variedly in food, pharmaceutical, and cosmetic industries (Burton et al. 1998; Let et al. 2007).

Vitamin E serves as a health-beneficiary nutraceutical in oils that also plays a role in delaying and treating many age-related ailments (Serbinova et al. 1991). It protects the cell membranes from damage due to free radicals, thereby decreasing risks of cardiovascular disorders and Alzheimer's disease (Karuna and Prasad 2015; Schauss et al. 2013; Sylvester and Theriault 2003). Several health issues like anemia, neurological disorders, weak immune system, and vision impairment may arise due to vitamin E deficiency (Rizvi et al. 2014). Different types of tocopherols are reported to have antitumor, antioxidant, and hypocholesterolemic properties, have significant impact on cancer in colorectal, prostate, and lungs, and are also used in anti-obesity formulations (Morya et al. 2022; Sylvester 2011). Tocotrienols contain three unsaturated bonds in their carbon side chain and one chiral center that enables them to penetrate tissues with saturated fatty layers more effectively than tocopherols and thus, promotes efficient metabolic functions (Suzuki et al. 1993). Tocotrienols have been demonstrated to possess anti-inflammatory and antiangiogenic properties and are used as anticancer and antitumor compounds for the treatment of breast cancer (Sylvester 2011; Sylvester and Theriault 2003). Nevertheless, there is significant potential to improvise the role of tocotrienol both as anticancer or anti-obesity agent (Wong and Lindsay 2013).

14.3.3 Phytosterols and Phytostanols

Phytosterols and phytostanols make up for a significant part of the unsaponifiable matter and is mainly found in various vegetable oils and fats. Structurally, phytosterols are triterpene analogs of cholesterol and bear the structure comprising of a four-ringed steroid nucleus containing a methyl or ethyl group at C-24 position and occasionally an additional double bond on a side chain at C-22 position (Bai et al. 2021). Majority of the 4-methyl phytosterols possess double bonds present between the C-5 and C-6 within the ring system and are, therefore, given the name Δ5-phytosterols. There is yet another group of familiar desmethylsterols that have a double bond located between C-7 and C-8, instead of C-5 and C-6, and hence known as Δ7-phytosterols (Müller and Bracher 2015). The steroidal moiety at the Δ-5 position is saturated in case of phytostanols, also termed as stanols (Moreau et al. 2018). Edible vegetable oils happen to be the main source of routine consumption of phytosterols, with more prevalence in rapeseed/canola (22.9%), sesame, soybean, corn, and ricebran oil (Bai et al. 2021; Yang et al. 2019). Plant sterols lack any flavor or taste and are heat-stable, which gives them the advantage as antipolymerization agents in vegetable oils used for frying purposes. The vital components of phytosterols, namely, β-sitosterols, stigmasterols, Δ 5-avenasterols campesterols, and brassicasterols (Figure 14.4), are found in oils like wheatgerm, corn, rapeseed, soybean, ricebran,palm, flaxseed, olive, peanut, and sesame (Dutta 2004; Verleyen et al. 2002). In vegetable oils, phytosterols occur as free sterols or steroyl esters (Fassbender et al. 2008; Weihrauch and Gardner 1978). Stanol esters are often derived by hydrogenation of phytosterols from tall oil (Miettinen et al. 1996). These ester forms offer an easier incorporation into food. In certain instances, phytostanol and phytosterol esters are used as fat substitutes because the phytostanol/sterol moiety present in the ester molecule fails to contribute to any energy to the body. Thus, phytostanol esters have the ability to alter the fatty acid composition of a fat blend thereby replacing the hard fat found in margarines and spreads.

Phytosterols and phytostanols present in edible oils play a critical role in reducing the levels of serum cholesterol and also act against certain types of cancers (Damirchi 2010; Gylling et al. 2014; Moreau et al. 2018). Owing to their structural similarity, the phytosterols act as cholesterol-lowering agents by competing with the cholesterol within the body, to form micelles in the lumen of intestines and inhibits the absorption of cholesterol (Jones et al. 2000; Wester 2000). Plant sterols and stanols play a prominent role in regulating fluidity and permeability of membranes, modulating the signal transduction pathways, and activity of enzymes bound to membranes. They also form the precursors of bile acids and steroid hormones. Other health-promoting effects exhibited by

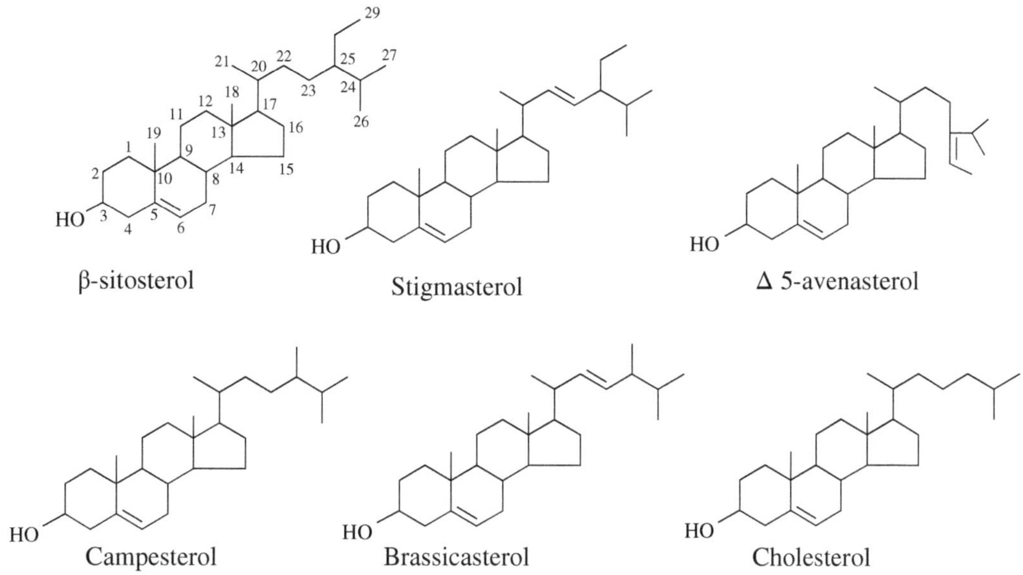

Figure 14.4 The chemical structures of phytosterols in comparison to cholesterol.

phytosterols include immunomodulatory (Bouic 2001), anti-inflammatory (Othmanand Moghadasian 2011), anticancer (Woyengo et al. 2009) and anticardiovascular morbidity (Rong et al. 2016) effects. The first commercial margarine containing plant-based sterols was introduced to the European market in 1997. Phytostanols are virtually unabsorbable, a characteristic that attributes to its hypocholesterolemic property (Moreau et al. 2002; Miettinen et al. 1995). Studies indicated that the daily intake of 2–3 g of phytosterol and/or phytostanol may bring down the levels of total cholesterol and LDL by 10%, thereby reducing the vulnerability of heart disease by about 25% (Katan et al. 2003; Law 2000). In 1995, plant stanol esters were introduced into margarine as dietary adjuncts to reduce the levels of cholesterol in the body. Phytosterol or stanol esters were shown to lower down the cholesterol absorption and LDL cholesterol levels by 10–15% (Weber et al. 2002). By decreasing the absorption of carotenoids, they also reduce the risk of cancers and heart diseases.

14.3.4 Oryzanol

Oryzanol (γ-oryzanol) was first discovered in ricebran approximately 60 years back (Kaneko and Tsuchiya 1954).Initially, γ-oryzanol, extracted as a significant component (1–3%) from the oil of rice bran, was presumed to be only one component, but was later discovered to be a mixture constituting ferulic acid (4-hydroxy-3-methoxy cinnamic acid), various phytosterol esters, and triterpene alcohols (Rogers et al. 1993; Kim et al. 2013). Three main components that are included in 80% γ-oryzanol are cycloartenylferulate, 24-methylenecycloartanyl ferulate, and campesterylferulate (Figure 14.5) (Patel and Naik 2004).

As an antioxidant compound, the most important health benefits of γ-oryzanol include its ability to lower serum and plasma cholesterol, reduce cholesterol absorption, and use in treatment of hyperlipidaemia (Saji et al. 2019; Zavoshy et al. 2012). Cycloartenol (structurally similar to cholesterol) is reported to compete with the binding sites and sequestrate cholesterol and thus used in treating hyperlipidaemia (Bumrungpert et al. 2019; Karuna and Prasad 2015; Yu et al. 2007). The ferulic acid ester present in rice exhibits biological effects like antiaging, antidiabetes, anticholesterol actions, anti-inflammation, and relieves symptoms associated with menopause (Khalid et al. 2017; Punia et al. 2021; Rigo et al. 2014). It is also used to improve insulin

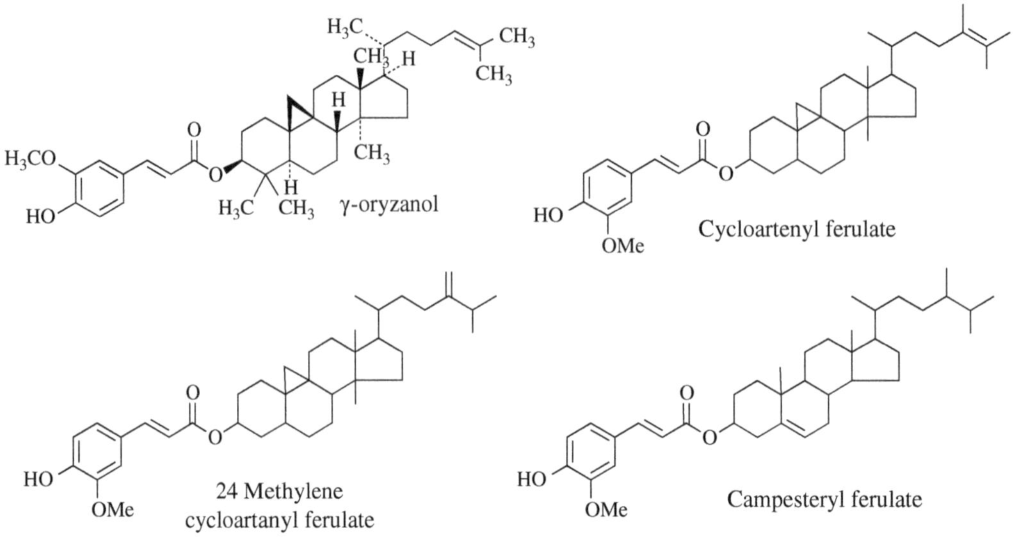

Figure 14.5 Chemical structures of γ-oryzanol and its components viz. cycloartenylferulate, 24-methylene cycloartanylferulate, and campesterylferulate.

sensitivity,stimulates immunity, modulates the secretion of pituitary glands, regulates platelet aggregation, and inhibits gastric acid secretion(Cheng et al. 2010; Dhara and Chakrabarti 2020; Park et al. 2017). γ-Oryzanol is a hydrophobic compound insoluble in water and exhibits low absorption, poor bioavailability, and a high antioxidant activity. It is demonstrated as a powerful inhibitor for iron-driven hydroxyl radical formation (Suh et al. 2007; Kim et al. 2013). The use of oryzanol in protecting skin from UV radiation and delaying aging effects on skin are well reported (Noboru and Yusho 1970). Some studies also established the role of γ-oryzanol in hindering the production of free radicals in muscle tissue, which in turn decreases fatigue and exhaustion of muscles, thereby increasing the growth and endurance of muscle mass (Eslami et al. 2014; Fry et al. 1997; Rosenbloom et al. 1992).

14.3.5 Squalene

Squalene is an isoprenoid intermediate, formed by six isoprene units assembled into a triterpene ($C_{30}H_{50}$) that gives this compound its lipid character (Figure 14.6). This polyunsaturated hydrocarbon molecule serves as a physiological precursor compound in plants or animals and humans utilized for the biosynthesis of photosterol/cholesterol (Huang et al. 2009). The six double bonds present between the carbon atoms (C=C) imparts high levels of unsaturation into the squalene molecule, because of which it gets easily oxidized on exposure to atmospheric oxygen (Lopez et al. 2014). Oxidized squalene shows reduced stability and decreased nutritional value. The extreme unsaturation in linear chain limits its production at cellular level inside the body, making it essential to be supplemented through squalene-rich diets (Sumi et al. 2018).

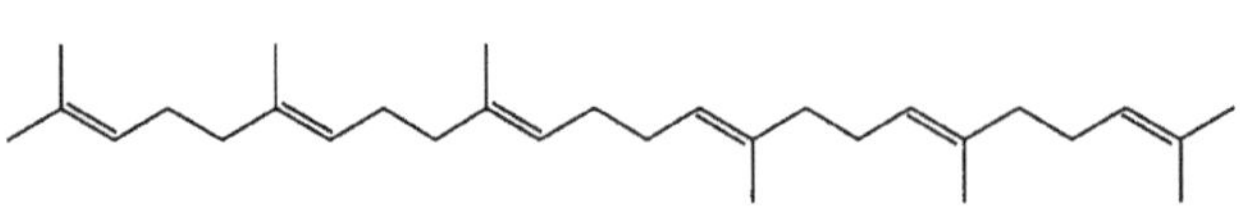

Figure 14.6 The chemical structure ofsqualene (2,6,10,15,19,23-hexamethyl-6,6,10,14,18,20-tetracosahexane).

Although the liver oil of deep-sea sharks is considered to be the primary source of squalene (about 40% of its body weight), it is also found in reasonable quantities in different vegetable oils. Significant amounts of squalene are present in olive oil, amaranth oil, palm oil, wheat germ oil, grapeseed, peanut, soybean, and a very meager amount(1%) in ricebran oil (Reddyand Couvreur 2009; Tikekar et al. 2008). The varying levels of squalene in plant oils are strictly correlated to the source, geographical location, and season/period of growth, along with the method of extraction employed (Popa et al. 2014). Among plant sources, squalene was first found in the olive oil (Thorbjarnarsonand Drummond 1935), which has eventually become the second highest natural source of squalene. In fact, a prescribed level of squalene content in olive oil is a mark for the EVOO, which greatly improves the stability of the oil during the frying process (Tsimidou 2010). The concentration of squalene assessed by Frega et al. (1992) in various edible oils showed values of 564 mg/100 g in olive oil, 27.9 mg/100 g in hazelnuts oil, 27.4 mg/100 g in peanut and corn oil, 14.1 mg/100 g in grape seed oil, and 9.9 mg/100 g in soybean oil, while another study in Europe reported the concentration of squalene in vegetable oilsranging 1.7–4.6 g/kg in olive oil, 0.1–0.17 g/kg in corn oil, 0.03–0.2 g/kg in soybean oil, and 0–0.19 g/kg in sunflower oil (Grigoriadou et al. 1990; Naziri et al. 2011).

Squalene finds importance as a nutraceutical in oils due to its pharmacological, cosmetic, and nutritional properties. Squalene extracted from vegetal sources are much better preferred over that from shark liver oil in the cosmetic as well as pharmaceutical industry because of its light consistency, nongreasy texture, high stability, excellent spreadability, colorless, and odorless properties (Popa et al. 2014). Research conducted in the Mediterranean area showed that a diet that is rich in olive oil component, decreases the risk of Coronary Heart Disease (CHD) and high cholesterol (Chan et al. 1996; Strandberg et al. 1990). It was observed in research studies that squalene absorption occurs faster than cholesterol in the circulatory system, which is then deposited in adipose tissue or muscles (EsraGüneş et al. 2013; Reddy and Couvreur 2009; Relas et al. 2001). Hence, it contributes significantly to weight loss and cholesterol control (Grande et al. 2018).An *in vivo* study showed that squalene significantly reduces triglycerides, fatty acid, and cholesterol levels in the plasma and tissues of heart. It has been shown to possess antioxidant, antilipidemic, and membrane-stabilizing properties as well (Farvin et al. 2006). Further, it was observed that an increased dietary intake of squalene lowers the levels of farnesyl pyrophosphate resulting in deceased tumor growth and better chemoprotection (Newmark 1999; Rao et al. 1998). A squalene-rich diet of about 2–5 g/day lowers the risk of cancers and other degenerative diseases (Kelly 1999; Owen et al. 2000; Owen et al. 2004; Smith 2000). The effective use of squalene was seen in developing vaccines against hepatitis B and C, HIV, malaria, and H_1N_1 viruses (Hagan et al. 1997; Lippi et al. 2010). More recently, it was also proven to play a vital role in development of COVID-19 vaccines (Chakrabarti and Dhara 2020; Dhara and Chakrabarti 2020).

Squalene acts as a natural antioxidant molecule by quenching singlet oxygen and protects skin cells from oxidative damage of UV light and also prevents premature aging of the skin. Hence, it finds tremendous applications in the cosmetics industry, especially in skincare products (Wołosik et al. 2013). It hydrates the skin and aids in regeneration of healthy skin cells. It acts as an antioxidant agent, humectant, and emollient and is used to treat many skin disorders including seborrhea disorders, acne, dermatitis, and psoriasis (Saint-Leger et al. 1986; Wołosik et al. 2013). Additionally, it protects the biomembranes of immune cells to strengthen immunity and also helps in wound healing (Karuna and Prasad 2015). Recently, an emulsion containing squalene and other surfactants was formulated, which was found to be effective as an immune system stimulant. Due to its nonpolar nature, squalene tends to show an affinity toward nonionized drugs and can be used to purify the effects of xenobiotic substances in human body (Kelly 1999).

14.3.6 Carotenoids

Carotenoids are isoprenoid group of molecules that are fat-soluble and form a portion of the unsaponifiable portion of edible oils. Most of the carotenoids are tetraterpenoid (C40) molecules, made of eight units of isoprenoids arranged into a linear and symmetrical molecule, which has a reverse order in the center. The basic structure is cyclic in nature and can be modified by hydrogenation, dehydrogenation, oxidation, and cyclization to give different derivatives. Carotenoids are classified into: the carotenes (linear hydrocarbons with cyclization at one or both ends, e.g. α-carotene, β-carotene) and the xanthophylls (oxygenated derivatives of carotenes, e.g. lutein, neoxanthin, zeaxanthin, violaxanthin) (Botella-Pavía and Rodríguez-Concepción 2006). These pigments attain their high levels of chemical reactivity due to the easy isomerization and oxidation of their conjugated double bonds (Oliver and Palou 2000). The chemical structures of some important carotenoids are illustrated in Figure 14.7. Crude palm oil and palm kernel oils serve as the richest sources of carotenoids (at concentration of about 500–700 ppm), while trace amounts of carotenoids (concentrations less than 100 ppm), particularly β-carotene, are also present in vegetable oils like corn oil, cottonseed oil,groundnut oil, rapeseed oil, soybean oil, sunflower oil, and olive oil (Demchenko 1969; Froehling et al. 1972; Harold and Purcell 1967; Jacobsberg 1974; Mosquera, M.l., Fernandez, J.G. 1986; Ong and Tee 1992; Zonta and Stancher 1987). The major carotenoids present in palm oil include β-carotene, δ-carotene, phytofluene, lycopene, and β-zeacarotene (May 1994). During the processes of refining, bleaching and deodorization, majority of the carotenoid pigments are removed, resulting in destabilization of the oils. To overcome this problem and photosensitized oxidation of the oils, β-carotene is usually incorporated into edible vegetable oils, which extend their shelf life and storage properties (Karuna and Prasad 2015).

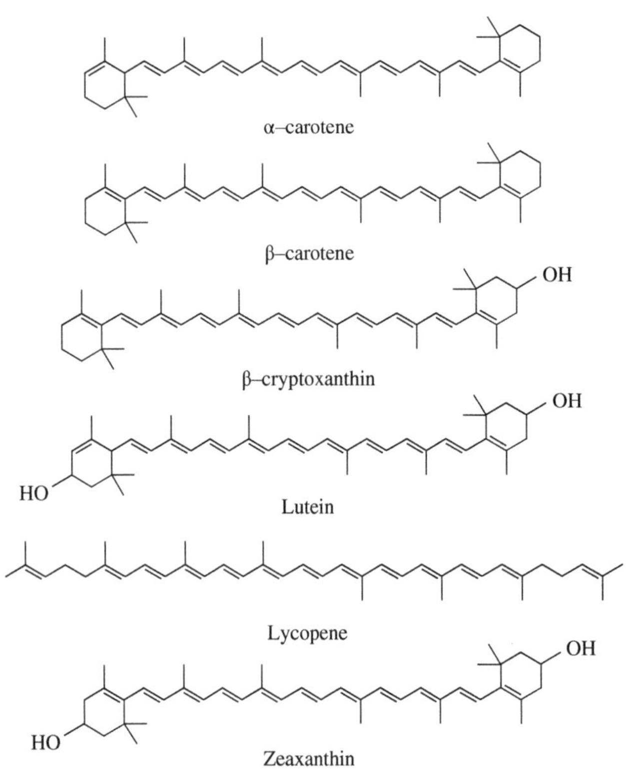

Figure 14.7 The chemical structure of some carotenoids.

Among the carotenoids, β-carotene is more important to human health as it is the precursor of vitamin A. The chemical conversion of β-carotene into retinol is catalyzed by the enzyme carotene deoxygenase, which is present in the liver and mucosal portion of human intestine. Normally, the rate of this conversion is very slow, but dispersion in a lipid medium enhances the absorption of carotenoids effectively (Demchenko 1969).Thus, the carotenoids are blended with edible oils that have potential in functional food and nutraceutical market, as recommended by WHO (Zanqui et al. 2021). Carotenoids offer numerous health benefits in the form of nutraceuticals. It enhances immunity, lowers the risk of degenerative diseases, and shows radicalscavenging, immune-stimulating, and antioxidant properties, along with protective effects against UV exposure and cardiovascular diseases (Mezzomo et al. 2015). These functions can be attributed to the long conjugated double bonds present in all carotenoids. Some carotenoids like α-carotene, β-carotene, and lycopene exhibit anti-obesity/hypolipidemic and anticancer properties (Hughes 2001; Tan and Norhaizan 2019).

14.3.7 Lecithin (Phospholipids)

Phospholipids form an important dietary component of many vegetable oils. A mixture of phospholipids isolated from fats and oils is generally termed as lecithin. The color of these compounds ranges from pale tan to deep reddish-brown and the consistency may vary from fluid to plastic strong (Vergallo 2020). It combines well with oils, fats and water. Because of this amphiphilic nature, lecithin is used as an effective emulsifier to improve the texture, flavor and quality of foods like flavored chocolates, readily soluble cocoa powders, bread rolls, salad dressings, soft-spread margarines and crispy pasta (Belayneh et al. 2018; Cabezas et al. 2012). Lecithins derived from vegetable sources are utilized as emulsifiers in food and cosmetic industry owing to their lower toxicity and high nutritive values (Nguyen et al. 2014). Physiologically, lecithin stabilizes the cell membranes, stimulates metabolic activity in cells and acts as antioxidant. It is a gum substance, obtained either by degumming crude gum in processed vegetable oils, or by enzymatic degumming where hydrolysis of lecithin results in cleavage of a single fatty acid group from the moiety of lecithin converting it into a water-soluble product, lysolecithin (Ceci et al. 2008; Gupta 2017). Lecithins are made up primarily of fatty acids, glycerine, phosphoric acid and choline, mainly comprising of four types of phospholipids viz. phosphatidylcholine (PC), phosphatidylethanolamine (PE), phosphatidylinositol (PI), and phosphatidic acid (PA) (Figure 14.8) (Bergfeld et al. 2015; Law 2000; Nieuwenhuyzen and Tomas 2008).

Lecithin is present in many vegetable oils like corn, sunflower, soybean, mustard, cottonseed, and ricebran. Among all the edible oils, the highest lecithin percentage is seen in soybean oil, while the lowest in sesame oil (Table 14.3). Apart from soybean, ricebran oil, mustard oil, sunflower oil, and cottonseed oil also contain a good number of phospholipids. The percentage of phosphatidylcholine is found to be the highest in most of the oils, followed by phosphatidylethanolamine (Table 14.3).

Besides their applications in food and cosmetics sector, lecithins are widely used in the pharmaceutical industry due to their high nutritional content, capacity to stabilize emulsions and ability to form liposomes. Phosphatidylcholine (PC), which is a significant compound required for acetylcholine biosynthesis, is considered as one of the most nutritionally significant constituents of lecithin. It not only exhibits nutraceutical effects by promoting healthy metabolism, reducing cholesterol levels and apoptosis,but is also associated with memory loss diseases like Alzheimer's (Duric et al. 2012; Küllenberg et al. 2012; Smith et al. 2016; Station 2008). Noticably, when cells are deprived of PC, it causes liver damage and apoptosis (Miller 2002). The potential uses of PC in several pharmaceutical formulations is also reported (Dacaranhe and Terao 2001).

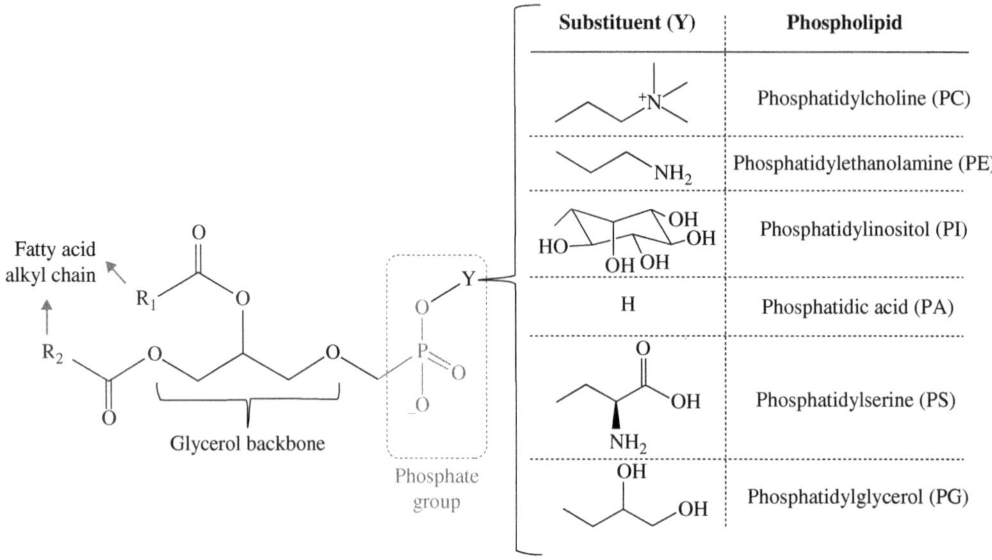

Figure 14.8 The chemical structures of different phospholipid lecithins. *Source:* Adapted from Alhajj et al. (2020).

Table 14.3 The phospholipid composition of lecithins sourced from vegetable oils.

		Phospholipid composition of lecithins (%)					
Vegetable oil	**Lecithin content**	**PC**	**PE**	**PI**	**PA**	**PS**	**PG**
Soybean	60.0–81.9	10.0–55.3	6.5–34.0	1.7–41.0	0.2–12.0	0.2–6.0	1.0–2.0
Sunflower	38.9–43.1	14.0–64.0	4.8–24.0	11.0–35.7	1.3–13.0	1.0–4.0	0.9–1.0
Mustard	71.3–71.3	16.7–74.1	6.5–64.9	5.3–33.0	2.4–28.0	1.0–4.0	—
Corn	57.5–68.1	30.0–68.1	3.0–13.9	14.5–19.8	9.0–9.4	1.0	1.0–1.4
Rice bran	33.0–33.6	3.2–38.0	11.8–33.2	5.8–19.7	2.5–22.6	—	8.6–8.6
Cottonseed	—	23.2–35.9	13.5–20.1	13.4–13.4	8.8	2.4–26.0	7.6
Peanut	—	23.0–39.4	8.0–18.8	17.0–22.2	2.0–6.6	—	—
Sesame	0.7–3.0	39.0	10.0–19.0	13.0–14.0	27.0–30.0	—	—
Wheat germ	—	40.0–60.0	9.0–15.0	13.0–20.0	—	—	—

PC, phosphatidylcholine; PE, phosphatidylethanolamine; PI, phosphatidylinositol; PA, phosphatidic acid; PS, phosphatidylserine; PG, phosphatidylglycerol.
Source: List (2015), Guiotto et al. (2015), Jala and Prasad (2015); Robert et al. (2020); Sun et al. (2018), Szuhaj et al. (2020), Wang (2015), Ambrosewicz-Walacik et al. (2015), Serna-Saldivar et al. (2015), Cansell et al. (2017), and Dunford (2009).

Phosphatidylethanolamine (PE), a precursor of PC, influences the topology of cell membranes, promotes cell fusion, mitochondrial biogenesis, oxidative phosphorylations and ensures proper liver function (Calzada et al. 2016; Li et al. 2006; van der Veen et al. 2014). PA functions as a signaling lipid in cell membranes and produces other lipids during resistance training (Escalante et al. 2016; Hoffman et al. 2012; Joy et al. 2014), while phosphatidylserine (PS) is required for the maintenance of healthy cell membranes of neurons and myelin sheath and also helps in signal transmission of cerebral cortex as PS constitutes about 13–15% of phospholipids in the cerebral cortex of humans (Glade and Smith 2015; Kim et al. 2014). Phosphatidylglycerol (PG), on the other hand, is used for treatment of respiratory distress syndrome and plays a role in maintaining proper pulmonary performance (Furse 2017; Hallman et al. 1985). The long-chain phospholipids possess the ability to form liposomes and are often used in drug-delivery systems (van Hoogevest and Wendel 2014). Lecithins are also associated with capacity of reducing LDL cholesterol and promoting HDL cholesterol syntheses, thereby decreasing the risk of cardiovascular diseases (Lin et al. 2004). Abnormal phospholipid metabolism is also associated with cancer (Dobrzyńska et al. 2005).

14.3.8 Phenolic Compounds

Phenolic compounds, generally produced by secondary metabolism in plants are mainly classified into five categories based on their properties: (i) Phenolic acids (e.g. hydroxycinnamic acids and hydroxybenzioic acids), (ii) Flavonoids (e.g. anthocyanins and anthocyanidins, flavonols, flavones and isoflavones), (iii) Stilbenes, (iv) Quinones and (v) Lignans and tannins. Phenolic compounds form one of the important constituents of vegetable oils, which impart antioxidant properties and play an important role in maintaining the overall quality and the thermal stability of edible oils along with extending their shelf life (Dimitrios 2006; Zeb 2021). They are also linked to the sensory properties (color and taste) of food and serve preservative functions. The vegetable oils containing reasonable amounts of phenolic compounds include olive, sesame and groundnut oils. Specific phenolic compounds were found in different edible oils viz.lignans (sesamin, sesamol, and sesmolin) in sesame oil, isoflavones in soybean oil, sinapic acid derivatives in rapeseed oil, and phenolic alcohols in olive oil (Jimenez-Lopezet al. 2021; Puupponen-Pimiä et al. 2001; Zhang et al. 2022). The phenolic compounds differ in different oils. For instance, the phenolic compounds in soybean oil mainly include isoflavones (viz. glycitein, daidzein, genistein, and their glucosides), phenolic acids (like caffeic acid, chlorogenic acid, ferulic acid, p-coumaric acid, syringic acid, salicylic acid, vanillic acid, protocatechuic acid, etc.), and anthocyanins (Król-Grzymała and Amarowicz 2020). Rapeseed contains important phenolic compounds in the form of sinapine, sinapic acid, and sinapolyglucoside (Wang et al. 2021). Olive oil contains oleuropein and its hydrolytic products, hydroxytyrosol and tyrosol, as some of the predominant phenolic compounds (Mancebo-Campos et al. 2014). High amounts of phenolic compounds are present in sesame oil, which include phenolic acids (like *p*-coumaric acid, ferulic acid, and *p*-hydroxybenzoic acid), flavonoids, and lignans (Chen et al. 2020; Shi et al. 2017).

The chemical structure of phenolic compounds shows at least one benzene ring containing one or more hydroxyl groups in it (Lund 2021) and ranges from small molecules to macromolecules in different oilseeds (Król-Grzymała and Amarowicz 2020; Ma et al. 2014). This kind of unique chemical structures of the phenolic compounds confer them with unparalleled functional bioactivity like antidiabetic, antioxidant, anticancer, anti-inflammatory, antimicrobial, anti-obesity, antiviral, cardioprotective, neuroprotective,hypoglycemic, and hypolipidemic effects (Alu'datt et al. 2013; López-Fernández et al. 2020). The phenolic compounds present in edible oils contribute to its oxidative stability and serve as antioxidants to reduce oxidative stress impacts on health like cardiovascular disorders, cancer, aging, diabetes, free radical scavenging, modulation of the immune response,and neurodegenerative diseases (Dilini and Terrence 2013; Li et al. 2014; Lucas 2020; Matías-Pérez et al. 2023). Phenolic compounds also possess strong antithrombotic, antimicrobial, antidiabetic, and antiatherogenic effects (Awuchiet al. 2022).

Lignans and isoflavones are common phenolic compounds which are found in very low quantities in some oils. Taking into consideration the nutritional and health benefits of these compounds, they can be placed under the nutraceuticals.

14.3.8.1 Lignans

Lignans are basically a class of polyphenolic compounds, consisting of dimers of phenylpropane units, largely found in oilseeds like sesame, flaxseed, and pumpkin. They serve a variety of physiological roles, including boosting antioxidant capacity and γ-tocopherol bioavailability, lowering blood lipid and AA levels, and function as anti-inflammatory agents. While there are no suggestive daily intake recommendations for lignans, some studies indicate an improvement of degenerative diseases based on dietary intake of lignans (Rodríguez-García et al. 2019).Lignans are not much prevalent in the commonly used edible oils. However, a few edible oils like sesame, flaxseed, and virgin olive oil contain specific lignans in variable quantities (Aludatt et al. 2017). Sesame oil contains upto 3.38–11.53 mg/g of lignin content, which makes it suitable for edible purpose (Shi et al. 2017). The main aglycon lignans found in sesame oil include the fat-soluble sesamin and sesamolin, while it also contains trace amounts of minor aglycons like sesaminol, sesamol, sesamolinol, matairesinol, lariciresinol, pinoresinol, and episesamin (Dar and Arumugam 2013; Eldin et al. 2011; Katsuzaki et al. 1992). Among different cultivated varieties of sesame, the black sesame seeds contain greater amounts of sesamin and sesamolin (1.98–9.41 mg/g and 1.06–3.35 mg/g respectively) in comparison to the white variety (Shi et al. 2017). Sesamin is more abundantly present in sesame oil than the other components and is found associated with lipid and glucose metabolism. It acts as an anti-inflammatory agent by inhibiting the Δ5-desaturase, helps in scavenging free radicals, and plays a major function in hypertension management by decreasing the LDL and increasing HDL levels (Hirata et al. 1996; Penalvo et al. 2006; Visavadiya and Narasimhacharya 2008). Sesamolin and sesaminol tend to inhibit lipid peroxidation of membranes by acting as antioxidants. Sesamin and sesamolin act as anticancerous, antihypertensive, and hypo-cholesterolemic agents (Dar and Arumugam 2013). Sesame lignans are also utilized for hormone therapy for postmenopausal women due to their estrogen-antagonist effect (Prisna et al. 2011). Virgin olive oil contains the natural lignans, (+)-pinoresinol, and 1-acetoxypinoresinol, which exhibit important biological and pharmacological properties like lowering the incidence of atherosclerosis, cardiac diseases, and cancers (López-Biedma et al. 2016a; Medina-Remón et al. 2017). Flaxseed lignans exhibit antioxidant activity along with acting as estrogenic/antiestrogenic compounds against various cancers (Adlercreutz et al. 1992; Hutchins and Slavin 2003; Kitts et al. 1999; Prasad 1997; Prasad 2000; Rickard and Thompson 1997; Waters and Knowler 1982). The major component of flaxseed lignin, secoisolariciresinoldiglucoside (SDG), reduces the growth of cancerous tumors, apart from being a good antioxidant (Touré and Xueming 2010).

14.3.8.2 Isoflavones

Oils and fats have another group of phenolic compounds, the isoflavones, also known as "phytoestrogen" because of their resemblance to chemical structure of estrogen (Kurzer and Xu 1997). Isoflavones are naturally available in soybean and seeds of legumes like beans, lentils, etc. (Gacek 2014). In fact, soybean oil has been considered as one of the richest sources of isoflavones with antioxidant properties, mainly the components namely genistein, genistin, daidzein, glycitein, and glycitin (Wang et al. 2013). Other nutraceutical properties of isoflavones include their influence on relief of postmenopausal vasomotor symptoms, postmenopausal nutrition, delay in osteoporosis, reduction of inflammation, and cardiovascular diseases (Agostoni et al. 2012; Ahsan and Mallick 2017; Dalaiset al. 2003; Mace et al. 2019; Taku et al. 2011).

Hence, different nutraceuticals with significant health benefits, present in various edible oils, decide the nutritional quality of the oil, which is vital for well-being of the human body. Evidently, edible vegetable oils provide almost 50% of the total dietary fat (approximately 39 g/day) and the major contributors to this are the three oils, rapeseed (12.8%), groundnut (7.4%), and soybean (6.5%) (Ma and Zhao 2014). A comparative study on the fatty acid contents of different vegetable

Table 14.4 Evaluation of edible oils based on nutraceuticals.

Edible oil	SFA	AFA	MUFA	LA	ALA	Vit E	PS	Total score
Rapeseed oil (low EC)	1.32	0.76	12.28	3.75	1.51	12.18	114.03	6
Soybean oil	5.00	2.01	5.54	9.88	1.25	18.6	63.42	5
Peanut oil	3.68	2.12	8.50	6.56	0.02	8.4	51.94	4
Sunflower oil	2.18	1.08	6.04	10.40	0.04	10.9	74.45	5
Flaxseed oil	1.62	0.94	3.74	2.83	10.71	7.50	116.98	6
Olive oil	2.70	2.01	15.02	1.20	0.11	4.40	54.02	3
Palm oil	8.78	8.01	8.24	2.05	0.04	3.05	3.05	2
Corn oil	2.80	2.35	5.84	9.89	0.11	10.20	206.41	5
Camellia oil	1.90	1.55	15.76	1.30	0.00	6.50	23.58	3
Sesame oil	2.78	1.63	7.56	8.34	0.06	13.71	62.34	4

SFA, saturated fatty acids; AFA, atherosclerotic fatty acids; MUFA, monounsaturated fatty acids; LA, linoleic acid; ALA, alfalinolenic acid; Vit E, vitamin E; PS, phytosterols; EC, erucic acid.
Source: Adapted from Zhao et al. (2021).

oils suggested that oleic acid and linoleic acid comprise the highest percentages in most of the oils. However, maintaining a prolonged diet pattern that is highly enriched with oleic and linoleic acid results in a higher intake of omega-6 fatty acids, disturbing the balance between the n-6 and n-3 PUFAs (Yang et al. 2013; Zhao et al. 2021). As the proportion and content of fatty acid intake has a prominent impact on human health, calculated consumption of an appropriate amount is of utmost importance to maintain normal fat metabolism in the body, thereby avoiding the risks of hypertension, hyperlipidemia, and other associated diseases (Meng et al. 2009). A comparison of different vegetable oils showed the highest content of vitamin E in soybean oil (Wen et al. 2016), while highest sterol contents in corn oil and rapeseed oil (Tu et al. 2018). Table 14.4 evaluates the content of important nutraceuticals in different edible oils, which are graded based on the nutritional quality (Zhao et al. 2021). The scores of edible oils in the table are indicative of the fact that none of the edible oils can meet with the complete balanced dietary requirements of nutraceuticals when used individually. Hence, it is always recommended to use the vegetable oils in commercial blended forms in combinations so as to include the major nutraceuticals, thereby providing maximal benefits as per the body demands.

14.4 Conclusion

Our health is determined by a well-balanced nutritional diet in which oils and fats play a vital role. Various food preparations utilize different edible oils. Unlike the myth of causing negative health impacts, oils help in maintaining a healthy body, if consumed in proper quantities and proportions. Due to presence of essential nutrients and bioactive compounds called nutraceuticals, the edible oils provide proper nutrition and also aid in the prevention and treatmentof various diseases (Nasri et al. 2014). The nutraceuticals like EFAs, tocopherols and tocotrienols, phytosterols and stanols, oryzanol, etc., are essential for general health maintenance. Besides their antioxidant properties, nutraceuticals lower the cholesterol levels, reduce hypertension, combat heart diseases, help in treating cancers, apoptosis, inflammation, immunological and neurological disorders. Oil-based nutraceuticals are in high demand in food, cosmetic, and pharmaceutical industries, as they tend to increase the shelf life of food items, delay aging and enhance nutritive quality of food items. Currently, increased awareness about fitness and health created by social media is encouraging people to lead healthier lifestyles, eat nutritious food, and exercise more. Selection of edible oils from vegetable origin containing balanced fatty acids, high content of vitamin E, higher levels of antioxidants, polyphenols and nutrients in the daily diet is advisable. However, different oils contain different categories of nutraceuticals in variable quantities and proportions. Thus, inclusion of different oils with different nutraceutical components into the diet provides a perfect balance in nutrient intake, thereby providing an alternative therapy against nutritional deficiencies and diseases. Nanoformulations of oils fortified with nutraceutical components in recent years have opened up prospective avenues for general health management using dietary fats and oils.The advancement in research and technology, exploiting better extraction techniques, and biotechnological intervention for nutraceutical overexpression in oils would enhance the nutritive quality and therapeutic value of edible oils further to solve general health issues in the modern world without medical intervention.

References

Abe, S., Ezaki, O., and Suzuki, M. (2016). Medium-chain triglycerides in combination with leucine and vitamin d increase muscle strength and function in frail elderly adults in a randomized controlled trial. *Journal of Nutrition* 146 (5): 1017–1026.

Adlercreutz, H., Mousavi, Y., Clark, J. et al. (1992). Dietary phytoestrogens and cancer: in vitro and in vivo. *Journal of Steroid Biochemistry and Molecular Biology* 41: 331–337.

Aggarwal, B.B., Sundaram, C., Prasad, S., and Kannappan, R. (2010). Tocotrienols, the vitamin E of the 21st century: its potential against cancer and other chronic diseases. *Biochemical Pharmacology* 80: 1613–1631.

Agostoni, C., Bresson, J., Tait, S. et al. (2012). Scientific opinion on the substantiation of health claims related to soy isoflavones and maintenance of bone mineral density (ID 1655) and reduction of vasomotor symptoms associated with menopause. *EFSAJ* 10: 2847.

Ahsan, M. and Mallick, A.K. (2017). The effect of soy isoflavones on the menopause rating scale scoring in perimenopausal and postmenopausal women: a pilot study. *Journal of Clinical and Diagnostic Research* 11: 13–16.

Akhtar, S., Khalid, N., Ahmed, I. et al. (2014). Physicochemical characteristics, functional properties, and nutritional benefits of peanut oil: areview. *Critical Reviews in Food Science and Nutrition* 54: 1562–1575.

Alhajj, M.J., Montero, N., Yarce, C.J., and Salamanca, C.H. (2020). Lecithins from vegetable, land, and marine animal sources and their potential applications for cosmetic, food, and pharmaceutical sectors. *Cosmetics* 7 (4): 87.

Ali, A., Ahmad, U., Akhtar, J. et al. (2019). Engineered nano scale formulation strategies to augment efficiency of nutraceuticals. *Journal of Functional Foods* 62: 103554.

Al-Saqer, J.M., Sidhu, J.S., Al-Hooti, S.N. et al. (2004). Developing functional foods using redpalmolein. IV. Tocopherols and tocotrienols. *Food Chemistry* 85: 579–583.

Alu'datt, M.H., Rababah, T., Ereifej, K., and Alli, I. (2013). Distribution, antioxidant and characterisation of phenolic compounds in soybeans, flaxseed and olives. *Food Chemistry* 139: 93–99.

Aludatt, M.H., Rababah, T., Alhamad, M.N. et al. (2017). A review of phenolic compounds in oil-bearing plants: distribution, identification and occurrence of phenolic compounds. *Food Chemistry* 218: 99–106.

Ambrosewicz-Walacik, M., Tańska, M., and Rotkiewicz, D. (2015). Phospholipids of rapeseeds and rapeseed oils: factors determining their content and technological significance—areview. *Food Review International* 31: 385–400.

Awad, A.B., Chan, K.C., Downie, A.C., and Fink, C.S. (2000). Peanuts as a source of b-sitosterol, a sterol with anticancer properties. *Nutrition and Cancer* 36: 238–241.

Awuchi, C.G., Chukwu, C.N., Iyiola, A.O. et al. (2022). Bioactive compounds and therapeutics from fish: revisiting their suitability in functional foods to enhance human wellbeing. *BioMed Research International* 2022: 3661818–3661866.

Bai, G., Chuanguo, M., and Chen, X. (2021). Phytosterols in edible oil: distribution, analysis and variation during processing. *Grain & Oil Science and Technology* 4 (1): 33–44.

Belayneh, H.D., Wehling, R.L., Cahoon, E., and Ciftci, O.N. (2018). Lipid composition and emulsifying properties of *Camelina sativa* seed lecithin. *Food Chemistry* 242: 139–146.

Beltrán, G., Jiménez, A., Del Río, C. et al. (2010). Variability of vitamin E in virgin olive oil by agronomical and genetic factors. *Journal of Food Composition and Analysis* 23: 633–639.

Bergfeld, W.F., Belsito, D.V., Hill, R.A. et al. (2015). *Safety Assessment of Lecithin and Other Phosphoglycerides as used in Cosmetics*. Washington: Cosmetic Ingredient Review.

Boateng, L., Ansong, R., Owusu, W.B., and Steiner-Asiedu, M. (2016). Coconut oil and palm oil's role in nutrition, health and national development: a review. *Ghana Medical Journal* 50 (3): 189–196.

Botella-Pavía, P. and Rodríguez-Concepción, M. (2006). Carotenoid biotechnology in plants for nutritionally improved foods. *PhysiologiaPlantarum* 126 (3): 369–381.

Bouic, P.J. (2001). The role of phytosterols and phytosterolins in immune modulation: a review of the past 10 years. *Current Opinion in Clinical Nutrition and Metabolic Care* 4: 471–475.

Bumrungpert, A., Chongsuwat, R., Phosat, C., and Butacnum, A. (2019). Rice bran oil containing γ-oryzanol improves lipid profiles and antioxidant status in hyperlipidemic subjects: a randomized double-blind controlled trial. *Journal of Alternative and Complementary Medicine* 25 (3): 353–358.

Burton, G.W., Traber, M.G., Acuff, R.V. et al. (1998). Human plasma and tissue alpha-tocopherol concentrations in response to supplementation with deuterated natural and synthetic vitamin E. *The American Journal of Clinical Nutrition* 67 (4): 669–684.

Cabezas, D.M., Madoery, R., Diehl, B.W.K., and Tomás, M.C. (2012). Emulsifying properties of different modified sunflower lecithins. *Journal of the American Oil Chemists' Society* 89: 355–361.

Calzada, E., Onguka, O., and Claypool, S.M. (2016). Phosphatidyl ethanolamine metabolism in health and disease. *International Review of Cell and Molecular Biology* 321: 29–88.

Camargo, A.C., Souza Vieira, T.M., Arce, M.A.R. et al. (2012). Gamma radiation induced oxidation and tocopherols decrease in in-shell, peeled and blanched peanuts. *International Journal of Molecular Sciences* 13 (3): 2827–2845.

Cansell, M., Bardeau, T., Morvan, E. et al. (2017). Phospholipid profiles of oleaginous pressed cakes using NMR and gas chromatography. *Journal of the American Oil Chemists' Society* 94: 1219–1223.

Caporaso, N., Savarese, M., Paduano, A. et al. (2015). Nutritional quality assessment of extra virgin olive oil from the Italian retail market: do natural antioxidants satisfy EFSA health claims? *Journal of Food Composition and Analysis* 40: 154–162.

Ceci, L.N., Constenla, D.T., and Crapiste, G.H. (2008). Oil recovery and lecithin production using water degumming sludge of crude soybean oils. *Journal of the Science of Food and Agriculture* 88: 2460–2466.

Chakrabarti, P. and Dhara, O. (2020). Lipid based nutraceuticals and nanoformulations: emerging applications in pharmaceuticals and cosmetics industries. *Molecules* 26 (18): 5510. https://doi.org/10.30954/ndp.bio.2020.3.

Chan, P., Tomilsonm, B., Leemm, C.B., and Lee, Y.S. (1996). Effect and safety of low dose prevastatin and squalene alone and in combination in elderly patients with hypercholesterolemia. *Journal of Clinical Pharmacology* 36: 422–427.

Chang, J.C., Lai, Y.H., Djoko, B. et al. (2006). Biosynthesis enhancement and antioxidant andanti-inflammatory activities of peanut (*Arachishypogaea* L.) arachidin-1, arachidin-3, and isopentadienylresveratrol. *Journal of Agricultural and Food Chemistry* 54: 10281–10287.

Chen, Y., Lin, H., Lin, M. et al. (2020). Effect of roasting and in vitro digestion on phenolic profiles and antioxidant activity of water-soluble extracts from sesame. *Food and Chemical Toxicology* 139: 111239–111248.

Chen, J., Zhou, Y., Mueller-Steiner, S. et al. (2005). SIRT1 protects against microglia-dependent amyloid-b toxicity through inhibiting NF-kB signaling. *The Journal of Biological Chemistry* 280: 40363–40374.

Cheng, H.H., Ma, C.Y., Chou, T.W. et al. (2010). Gamma-oryzanol ameliorates insulin resistance and hyperlipidemia in rats with streptozotocin/nicotinamide-induced type 2 diabetes. *International Journal for Vitamin and Nutrition Research* 80 (1): 45–53.

Chowdhury, K., Banu, L.A., Khan, S., and Latif, A. (2007). Studies on the fatty acid compositionof edible oil. *Bangladesh Journal of Scientific and Industrial Research* 42: 311–316.

Chuffa, L.G., Rocha, V.F., Silva, D., and Franco, D. (2014). Soybean seed oil: nutritional composition, healthy benefits and commercial applications. In: *Seed Oil: Biological Properties, Health Benefits and Commercial Applications*, 1ee (ed. A. Varnham), 1–54. USA: Nova Science Publishers.

Cicerale, S., Lucas, L., and Keast, R. (2010). Biological activities of phenolic compounds present in virgin olive oil. *International Journal of Molecular Sciences* 11: 458–479.

Dacaranhe, C.D. and Terao, J. (2001). A unique antioxidant activity of phosphatidylserine on iron-induced lipid peroxidation of phospholipid bilayers. *Lipids* 36 (10): 1105–1110.

Dalais, F.S., Ebeling, P.R., Kotsopoulos, D. et al. (2003). The effects of soy protein containing isoflavones on lipids and indices of bone resorption in postmenopausal women. *Clinical Endocrinology* 58 (6): 704–709.

Damirchi, S. (2010). Minor compounds of olive oil: phytosterols and tocopherols. In: *Olive Oil and Health* (ed. J.D. Corrigan), 141–168. New York: Nova Science Publishers.

Dar, A.A. and Arumugam, N. (2013). Lignans of sesame: purification methods, biological activities and biosynthesis-a review. *Bioorganic Chemistry* 50: 1–10.

Dauqan, E., Sani, H.A., Abdullah, A. et al. (2011). Vitamin E and beta carotene composition in four different vegetable oils. *American Journal of Applied Sciences* 8 (5): 407–412.

DeFelice, S.L. (1989). *The Nutraceutical Revolution: Fueling a Powerful, New International Market*. Como: Harvard University Advanced Management Program in Biomedical Research and Development.

DeFelice, S.L. (1995). The nutraceutical revolution: its impact on food industry R&D. *Trends in Food Science and Technology* 6: 59–61.

Demchenko, A.L. (1969). Levels of carotenes and tocopherols in barley oil. *IzvVysshChebZavedPishchTekhnol* 5: 18–20.

Dhara, O. and Chakrabarti, P.P. (2020). Oils and fats-largely unexplored source of nutraceuticals. *Novel Techniques in Nutrition & Food Science* 5 (1) NTNF. 000603: 394–405.

Dilini, B. and Terrence, M. (2013). Antioxidant activity and total phenolic content of sesame (*Sesamumindicum* L.) seedoil. *Tropical Agricultural Research* 24: 296–302.

Dimitrios, B. (2006). Sources of natural phenolic antioxidants. *Trends in Food Science and Technology* 17: 505–512.

Dixit, S., Shukla, A., Singh, V., and Upadhyay, S.K. (2021a). Bioprospecting of natural compounds for industrial and medical applications: current scenario and bottleneck. In: *Bioprospecting of Plant Biodiversity for Industrial Molecules*. Wiley; Portico. https://doi.org/10.1002/9781119718017.ch3.

Dixit, S., Shukla, A., Singh, V., and Upadhyay, S.K. (2021b). Engineering of plant metabolic pathway for nutritional improvement: recent advances and challenges. In: *Genome Engineering for Crop Improvement*. Wiley; Portico. https://doi.org/10.1002/9781119672425.ch20.

Djoko, B., Robin, Y.Y.C., Shee, J.J., and Liu, Y.W. (2007). Characterization of immunological activities of peanut stilbenoids, arachidin-1, piceatannol, and resveratrol on lipopolysaccharide-induced inflammation of RAW 264.7 macrophages. *Journal of Agricultural and Food Chemistry* 55: 2376–2383.

Dobrzyńska, I., Petelska, B., Sulkowski, S., and Figaszewski, Z. (2005). Changes in electric charge and phospholipids composition in human colorectal cancer cells. *Molecular and Cellular Biochemistry* 276 (1-2): 113–119.

Duan, Y., Miao, H., Ju, M. et al. (2023). Nutraceutomics of the ancient oilseed crop sesame (*Sesamum indicum* L.). In: *Compendium of Crop Genome Designing for Nutraceuticals* (ed. C. Kole), 471–501. Singapore: Springer.

Dunford, N.T. (2009). Wheat germ oil. In: *Gourmet and Health-Promoting Specialty Oils* (ed. R.A. Moreau and A.B.T.-G. Kamal-Eldin), 359–376. Amsterdam: Elsevier.

Dunstan, J.A., Mitoulas, L.R., Dixon, G. et al. (2007). The effects of fish oil supplementation in pregnancy on breast milk fatty acid composition over the course of lactation: a randomized controlled trial. *Pediatric Research* 62 (6): 689–694.

Duric, M., Sivanesan, S., and Bakovic, M. (2012). Phosphatidylcholine functional foods and nutraceuticals: a potential approach to prevent non-alcoholic fatty liver disease. *European Journal of Lipid Science and Technology* 114: 389–398.

Dutta, P.C. (2004). *Phytosterols as Functional Food Components and Nutraceuticals*. New York: CRC Press.

Edem, D.O. (2002). Palm oil: biochemical, physiological, nutritional, hematological, and toxicological aspects: a review. *Plant Foods for Human Nutrition* 57: 319–341.

Eldin, A., Moazzami, A., and Washi, S. (2011). Sesame seed lignans: potent physiological modulators and possible ingredients in functional foods & nutraceuticals. *Recent Patents on Food, Nutrition & Agriculture* 3 (1): 17–29.

Escalante, G., Alencar, M., Haddock, B., and Harvey, P. (2016). The effects of phosphatidic acid supplementation on strength, body composition, muscular endurance, power, agility, and vertical jump in resistance trained men. *Journal of the International Society of Sports Nutrition* 13: 24.

Eslami, S., Esa, N.M., Marandi, S.M. et al. (2014). Effects of gamma oryzanolsupplementation on anthropometric measurements & muscular strength in healthy malesfollowing chronic resistance training. *The Indian Journal of Medical Research* 139 (6): 857–863.

Esra Güneş, F., ÜniversitesiSağlıkBilimleriFakültesi, M, Diyetetik Bölümü, B., and Türkiye, İ. (2013). Medical use of squalene as a natural antioxidant derleme/review. *Journal of Marmara University Institute of Health Sciences* 33 (44): 220–228.

Etschmann, S.K., Shadid, R., Campoy, C. et al. (2007). Effects of fish-oil and folate supplementation of pregnant women on maternal and fetal plasma concentrations of docosahexaenoic acid and eicosapentaenoic acid: a European randomized multicenter trial. *The American Journal of Clinical Nutrition* 85 (5): 1392–1400.

Farvin, K.H.S., Anandan, R., Hari Senthil Kumar, S. et al. (2006). Cardioprotective effect of squalene on lipid profile in isoprenaline-induced myocardial infarction in rats. *Journal of Medicinal Food* 9 (4): 531–536.

Fassbender, K., Lütjohann, D., Dik, M.G. et al. (2008). Moderately elevated plant sterol levels are associated with reduced cardiovascular risk—the LASA study. *Atherosclerosis* 196: 283–288.

Fattore, E. and Massa, E. (2018). Dietary fats and cardiovascular health: a summary of the scientific evidence and current debate. *International Journal of Food Sciences and Nutrition* 69 (8): 916–927.

Firouzi, S.R. (2017). Herbal oil supplement with hot-nature diet for multiple sclerosis. In: *Nutrition and Lifestyle in Neurological Autoimmune Diseases* (ed. R.R. Watson and W.D.S. Killgore). Academic Press.

Flachs, P., Horakova, O., Brauner, P. et al. (2005). Polyunsaturated fatty acids of marine origin upregulate mitochondrial biogenesis and induce β-oxidation in white fat. *Diabetologia* 48: 2365–2375.

Frankel, E.N. (2011). Nutritional and biological properties of extra virgin olive oil. *Journal of Agricultural and Food Chemistry* 59: 785–792.

Frega, N., Bocci, F., and Lercker, G. (1992). Direct gas chromatographic analysis of the unsaponifiable fraction of different oils with a polar capillary column. *Journal of the American Oil Chemists' Society* 69: 447–450.

Froehling, P.E., Bosch, G., and Boekenoogen, H.A. (1972). Fatty acid composition of carotenoid esters in soybean and rapeseed oils. *Lipids* 7: 447–449.

Fry, A.C., Bonner, E., Lewis, D.L. et al. (1997). The effects of gamma-oryzanolsupplementa-tion during resistance exercise training. *International Journal of Sport Nutrition* 7: 318–329.

Furse, S. (2017). Is phosphatidylglycerol essential for terrestrial life? *Journal of Chemical Biology* 2017 (10): 1–9.

Gacek, M. (2014). Soy and legume seeds as sources of isoflavones: selected individual determinants of their consumption in a group of perimenopausal women. *Menopausal Review* 13 (1): 27–31.

Ghanbari Shendi, E. (2019). Olive oil: nutraceutical and pharmaceutical food. *International Journal of Medical Reviews* 6 (1): 28–30.

Ghanbari Shendi, E., Sivri, O.D., Ozkaya, M.T., and Ustunel, N.F. (2018). Changes occurring in chemical composition and oxidative stability of virgin olive oil during storage. *Oilseeds and fats, Crops and Lipids* 25 (6): 1–8.

Glade, M.J. and Smith, K. (2015). Phosphatidylserine and the human brain. *Nutrition* 31: 781–786.

Grande, M.A.L., Gorinstein, S., Rangel, E. et al. (2018). Plant sources, extractionmethods, and uses of squalene. *International Journal of Agronomy* 1–13. https://doi.org/10.1155/2018/1829160.

Grigoriadou, D., Androulaki, A., Psomiadou, E., and Tsimidou, M.Z. (1990). Solid phase extraction in the analysis of squalene and tocopherols in olive oil. *Food Chemistry* 105 (2): 675–680.

Grilo, E.C., Costa, P.N., Gurgel, C.S.S. et al. (2014). Alpha-tocopherol and gamma-tocopherol concentration in vegetable oils. *Food Science and Technology* 34 (2): 379–385.

Grygier, A. (2022). Mustard seeds as a bioactive component of food. *Food Review International* 1–14.

Guiotto, E.N., Tomás, M.C., and Diehl, B.W.K. (2015). Sunflower lecithin. In: *Polar Lipids*, 57–75. Amsterdam: Elsevier.

Gunstone, F.D. (2002). Production and trade of vegetable oils. In: *Vegetable Oils in Food Technology: Composition, Properties and Uses*, 1ee, 1–24. Boca Raton: Blackwell Publishing Ltd.

Gunstone, F.D. (2005). Vegetable oils. In: *Baileys Industrial Oil and Fat Products* (ed. F. Shadini), 228–229. USA: John Wiley and Sons.

Gupta, M.K. (2017). Chapter 4—degumming. In: *Practical Guide to Vegetable Oil Processing*, 2ee (ed. M.K. Gupta), 41–78. AOCS Press.

Gupta, S., Chauhan, D., Mehla, K. et al. (2010). An overview ofnutraceuticals: current scenario. *Journal of Basic and Clinical Pharmacy* 1: 55–62.

Gylling, H., Plat, J., Turley, S. et al. (2014). Plant sterols and plant stanols in the management of dyslipidaemia and prevention of cardiovascular disease. *Atherosclerosis* 232 (2): 346–360.

Hagan, D.T., Ott, G.S., and Nest, G.V. (1997). Recent advances in vaccine adjuvants: the development of MF59 emulsion and polymeric microparticles. *Molecular Medicine Today* 3 (2): 69–75.

Hallman, M., Enhorning, G., and Possmayer, F. (1985). Composition and surface activity of normal and phosphatidylglycerol-deficient lung surfactant. *Pediatric Research* 19: 286–292.

Han, J.R., Deng, B., Sun, J. et al. (2007). Effects of dietary medium-chain triglyceride on weight loss and insulin sensitivity in a group of moderately overweight free-living type 2 diabetic Chinese subjects. *Metabolism* 56 (7): 985–991.

Harold, E.P. and Purcell, A.K. (1967). Carotenoid pigments of peanut oil. *Journal of the American Oil Chemists Society* 44 (5): 328–330.

Hirata, F., Fujita, K., Ishikura, Y. et al. (1996). Hypercholesterolemic effect of sesame lignan in human. *Atherosclerosis* 122 (1): 135–136.

Hoffman, J.R., Stout, J.R., Williams, D.R. et al. (2012). Efficacy of phosphatidic acid ingestion on lean body mass, muscle thickness and strength gains in resistance-trained men. *Journal of the International Society of Sports Nutrition* 9: 47.

Hu, F.B., Manson, J.A.E., and Willet, W.C. (2001). Types of dietary fat and risk of coronary heart disease: a critical review. *Journal of the American College of Nutrition* 20: 5–19.

Huang, S.W., Frankel, E.N., and German, J.B. (1994). Antioxidant activity of - and -tocopherols in bulk oils and in oil-in-water emulsions. *Journal of Agricultural and Food Chemistry* 42: 2108–2114.

Huang, Z.R., Lin, Y.K., and Fang, J.Y. (2009). Biological and pharmacological activities of squalene and related compounds: potential uses in cosmetic dermatology. *Molecules* 14: 540–554.

Hughes, D.A. (2001). Dietary carotenoids and human immune function. *Nutrition* 17: 823–827. https://doi.org/10.1016/S0899-9007(01)00638-4.

Hulbert, A.J., Turner, N., Storlien, L.H., and Else, P.L. (2005). Dietary fats and membrane function: implications for metabolism and disease. *Biological Reviews of the Cambridge Philosophical Society* 80 (1): 155–169.

Hutchins, A.M. and Slavin, J.L. (2003). Effects of flaxseed on sex hormone metabolism. In: *Flaxseed in Human Nutrition*, 2ee (ed. L.U. Thompson and S.C. Cunnane), 126–149. Champaign: AOCS Press.

Jacobsberg, B. (1974). Palm oil characteristics and quality. In: *Proceedings of the 1st Choo Yuen May 7 MARDI Workshop on Oil Palm Technology. Kuala Lumpur, Malaysia* (ed. O.S. Chai and A. Awalludin), 48–70. Malaysia Agricultural Research and Development Insitute (MARDI).

Jala, R.C.R. and Prasad, R.B.N. (2015). Rice bran lecithin: compositional, nutritional, and functional characteristics. In: *Polar Lipids*, 35–55. Amsterdam: Elsevier.

Jimenez-Lopez, C., Pereira, A.G., Lourenço-Lopes, C. et al. (2021). Main bioactive phenolic compounds in marine algae and their mechanisms of action supporting potential health benefits. *Food Chemistry* 341: 128262.

Jones, P.J., Raeini-Sarjaz, M., Ntanios, F.Y. et al. (2000). Modulation of plasma lipid levels and cholesterol kinetics by phytosterol versus phytostanol esters. *Journal of Lipid Research* 41: 697–705.

Joseph, E., Pizzorno, N.D., Michael, T. et al. (2016). Rheumatoid arthritis. In: *The Clinician's Handbook of Natural Medicine*, 3ee, 875–893. Elsevier.

Joy, J.M., Gundermann, D.M., Lowery, R.P. et al. (2014). Phosphatidic acid enhances mTOR signaling and resistance exercise induced hypertrophy. *Nutrition and Metabolism* 11: 29.

Ju, J., Hao, X., Lee, M.J. et al. (2009). A gamma-tocopherol-rich mixture of tocopherols inhibits colon inflammation and carcinogenesis in azoxymethane and dextran sulfate sodium-treated mice. *Cancer Prevention Research (Philadelphia, Pa.)* 2: 143–152.

Juan, M.E., Wenzel, U., Ruiz-Gutierrez, V. et al. (2006). Olive fruit extracts inhibit proliferation and induce apoptosis in HT-29 human colon cancer cells. *The Journal of Nutrition* 136 (10): 2553–2557.

Jyoti Kumari, N. and Avtar, R. (2023). Nutraceuticals in vegetable oils. *Vigyan Varta* 4 (9): 85–87.

Kaneko, R. and Tsuchiya, T. (1954). New compound in rice bran and germ oils. *Journal of the Chemical Society of Japan* 57: 526–529.

Kang, L., Heng, W., Yuan, A. et al. (2010). Resveratrol modulates adipokine expression and improves insulin sensitivity in adipocytes: relative to inhibition of inflammatory responses. *Biochimie* 92: 789–796.

Karmowski, J., Hintze, V., Kschonsek, J. et al. (2015). Antioxidant activities of tocopherols/tocotrienols and lipophilic antioxidant capacity of wheat, vegetable oils, milk and milk cream by using photochemiluminescence. *Food Chemistry* 175: 593–600.

Karuna, M.S.L. and Prasad, R.B.N. (2015). Vegetable oil-based nutraceuticals. In: *Plant Biology and Biotechnology* (ed. B. Bahadur, M. Venkat Rajam, L. Sahijram, et al.), 793–812. New Delhi: Springer.

Katan, M.B., Grundy, S.M., Jones, P. et al. (2003). Efficacy and safety of plant stanolsand sterols in the management of blood cholesterol levels. *Mayo Clinic Proceedings* 78: 965–978.

Katsuzaki, H., Kawasumi, M., Kawakishi, S., and Osawa, T. (1992). Structure of novel antioxidantivelignan glucosides isolated from sesame seed. *Bioscience, Biotechnology, and Biochemistry* 56 (12): 2087–2088.

Kaur, N., Chugh, V., and Gupta, A.K. (2014). Essential fatty acids as functional components of foods-a review. *Journal of Food Science and Technology* 51 (10): 2289–2303. https://doi.org/10.1007/s13197-012-0677-0.

Kaur, R., Sharma, A., Rani, R. et al. (2019). Medicinal qualities of mustard oil and its role in human health against chronic diseases: areview. *Asian Journal of Dairy and Food Research* 38 (2): 98–104.

Kelly, G.S. (1999). Squalene and its potential clinical uses. *Alternative Medicine Review* 44 (11): 29–36.

Khalid, N., Kobayashi, I., Neves, M.A. et al. (2017). Encapsulation of β sitosterol plus γ oryzanol in O/W emulsions: formulation characteristics and stability evaluation with microchannel emulsification. *Food and Bioproducts Processing* 102: 222–232.

Khan, S., Choudhary, S., Pandey, A. et al. (2015). Sunflower oil: efficient oil source for human consumption. *Emergent Life Sciences Research* 1: 1–3.

Kim, H.Y., Huang, B.X., and Spector, A.A. (2014). Phosphatidylserine in the brain: metabolism and function. *Progress in Lipid Research* 56: 1–18.

Kim, M.J., Park, J., Kim, J. et al. (2013). Effects of heat treatment and visible light exposure on the oxidative stability of rice bran and of rice bran oil. *Food Science and Biotechnology* 22: 1–6.

Kitts, D.D., Yuan, Y.V., Wijewickreme, A.N., and Thompson, L.U. (1999). Antioxidant activity of the flaxseed lignansecoisolariciresinoldiglycoside and its mammalian lignan metabolites enterodiol and enterolactone. *Molecular and Cellular Biochemistry* 202: 91–100.

Król-Grzymała, A. and Amarowicz, R. (2020). Phenolic compounds of soybean seeds from two European countries and their antioxidant properties. *Molecules* 25: 2075–2085.

Küllenberg, D., Taylor, L.A., Schneider, M., and Massing, U. (2012). Health effects of dietary phospholipids. *Lipids in Health and Disease* 11: 3.

Kumar, G.S. and Krishna, A.G.G. (2013). Studies on the nutraceutical's composition of wheat derived oils wheat bran oil and wheat germ oil. *Journal of Food Science and Technology* 52 (2): 1145–1151.

Kurzer, M.S. and Xu, X. (1997). Dietary phytoestrogens. *Annual Review of Nutrition* 17: 353–381.

Lai, O.M., Jacoby, J.J., Leong, W.F. et al. (2019). Nutritional studies of rice bran oil. In: *Rice Bran and Rice Bran Oil*, 1ste (ed. L.Z. Cheong and X. Xu), 19–54. USA: Academic Press and AOCS.

Langyan, S., Yadava, P., Sharma, S. et al. (2022). Food and nutraceutical functions of sesame oil: an underutilized crop for nutritional and health benefits. *Food Chemistry* 389: 132990.

Law, M. (2000). Plant sterol and stanol margarines and health. *British Medical Journal* 320: 861–864. https://doi.org/10.1136/bmj.320.7238.861.

Lee, H.J., Ju, J., Paul, S. et al. (2009). Mixed tocopherols prevent mammary tumorigenesis by inhibiting estrogen action and activating PPAR-gamma. *Clinical Cancer Research* 15: 4242–4249.

Let, M.B., Jacobsen, C., and Meyer, A.S. (2007). Ascorbyl palmitate, gamma-tocopherol, and EDTA affect lipid oxidation in fish oil enriched salad dressing differently. *Journal of Agricultural and Food Chemistry* 55 (6): 2369–2375.

Leung, K.S., Galano, J.M., Oger, C. et al. (2001). Enrichment of alpha-linolenic acid in rodent diet reduced oxidative stress and inflammation during myocardial infarction. *Free Radical Biology & Medicine* 162: 53–64.

Li, Z., Agellon, L.B., Allen, T.M. et al. (2006). The ratio of phosphatidylcholine to phosphatidylethanolamine influences membrane integrity and steatohepatitis. *Cell Metabolism* 3: 321–331.

Li, A., Li, S., Zhang, Y.J. et al. (2014). Resources and biological activities of naturalpolyphenols. *Nutrients* 6: 6020–6047.

Lietzow, J. (2021). Biologically active compounds in mustard seeds: atoxicological perspective. *Food* 10: 2089.

Lin, Y., Meijer, G.W., Vermeer, M.A., and Trautwein, E.A. (2004). Soy protein enhances the cholesterol-lowering effect of plant sterol esters in cholesterol-fed hamsters. *Journal of Nutrition* 134 (1): 143–148.

Lippi, G., Targher, G., and Franchini, M. (2010). Vaccination, squalene and anti-squalene antibodies: facts or fiction? *European Journal of Internal Medicine* 21 (2): 70–73.

List, G.R. (2015). Soybean lecithin: food, industrial uses, and other applications. In: *Polar Lipids*, 1–33. Amsterdam: Elsevier.

List, G.R. (2016). Processing and food uses of peanut oil and protein. In: *Peanuts* (ed. H.T. Stalker and R.F. Wilson), 405–428. USA: AOCS Press.

Lomascolo, A., Uzan-Boukhris, E., Sigoillot, J.C., and Fine, F. (2012). Rapeseed and sunflower meal: areview on biotechnology status and challenges. *Applied Microbiology and Biotechnology* 95: 1105–1114.

Lopes, R.M., Agostini-Costa, T.D.S., Gimenes, M.A., and Silveira, D. (2011). Chemical composition and biological activities of Arachis species. *Journal of Agricultural and Food Chemistry* 59: 4321–4320.

Lopez, S., Bermudez, B., Montserrat-de la Paz, S. et al. (2014). Membrane composition and dynamics: a target of bioactive virgin olive oil constituents. *Biochimica et BiophysicaActa–Biomembranes* 1838 (6): 1638–1656.

López-Biedma, A., Sánchez-Quesada, C., Beltrán, G. et al. (2016b). Phytoestrogen (+)-pinoresinol exerts antitumor activity in breast cancer cells with different oestrogen receptor statuses. *BMC Complementary and Alternative Medicine* 16 (1): 350.

López-Biedma, A., Sánchez-Quesada, C., Delgado-Rodríguez, M., and Gaforio, J.J. (2016a). The biological activities of natural lignans from olives and virgin olive oils: *a review*. *Journal of Functional Foods* 26: 36–47.

López-Fernández, O., Domínguez, R., Pateiro, M. et al. (2020). Determination of polyphenols using liquid chromatography-tandem mass spectrometrytechnique (LC-MS/MS): a review. *Antioxidants* 9: 479–505.

Lucas, E.W. (2020). Handbook of cereal science and technology, revised and expanded. In: *Oilseeds and Oil-Bearing Materials 337 When Experimentally Fortifying Breads with Vegetable Food* (ed. K. Karel). Boca Raton: CRC Press.

Lund, M.N. (2021). Reactions of plant polyphenols in foods: impact of molecular structure. *Trends in Food Science and Technology* 112: 241–251.

Ma, Y., Kosińska-Cagnazzo, A., Kerr, W.L. et al. (2014). Separationand characterization of soluble esterified and glycoside-bound phenolc compounds in dry-blanched peanut skins by liquid chromatography-electrospray ionization mass spectrometry. *Journal of Agricultural and Food Chemistry* 62: 11488–11504.

Ma, G. S., and Zhao, L.Y. (2014). Report on the 2010-2013 China National Nutrition and Health Surveillance; Proceedings of the China Nutrition Research and Development Symposium; Beijing, China. 6–13. Mace, T.A., Ware, M.B., King, S.A. et al. (2019). Soy isoflavones and their metabolites modulate cytokine-induced natural killer cell function. *Scientific Reports* 9: 5068.

Maddi, V.S., Aragade, P.D., Digge, V.G., and Nitalikar, M.N. (2007). Importance of nutraceuticals in health management. *Pharmacognosy Reviews* 1 (2): 377.

Mahesar, S.A., Shah, S.N., Shirazi, S.T.H. et al. (2017). Outcome of refining on the physicochemical properties of cottonseed oil. *Pakistan Journal of Analytical & Environmental Chemistry* 18 (2): 105–111.

Mancebo-Campos, V., Salvador, M.D., and Fregapane, G. (2014). Antioxidant capacity of individual and combined virgin olive oil minor compounds evaluated at mild temperature (25 and 40°C) as compared to accelerated and antiradical assays. *Food Chemistry* 150: 374–381.

Mancini, A., Imperlini, E., Nigro, E. et al. (2015). Biological and nutritional properties of palm oil and palmitic acid: effects on health. *Molecules* 20 (9): 17339–17361.

Marangoni, F., Agostoni, C., Borghi, C. et al. (2020). Dietary linoleic acid and human health: focus on cardiovascular and cardiometabolic effects. *Atherosclerosis* 292: 90–98.

Matías-Pérez, D., Sánchez-Medina, M.A., Hernández-Bautista, E., and García-Montalvo, I.A. (2023). The consumption of vegetable oils and their possible participation as a neuroprotector. *Turkish Journal of Biochemistry* 48 (2): 142–144.

May, C.Y. (1994). Palm oil carotenoids. *Food and Nutrition Bulletin* 15 (2): 1–8.

Mazzocchi, A., De Cosmi, V., Risé, P. et al. (2021). Bioactive compounds in edible oils and their role in oxidative stress and inflammation. *Frontiers in Physiology* 12: 659551.

Medina-Remón, A., Casas, R., Tressserra-Rimbau, A. et al. (2017). Polyphenol intake from a Mediterranean diet decreases inflammatory biomarkers related to atherosclerosis: asubstudy of the PREDIMED trial. *British Journal of Clinical Pharmacology* 83: 114–128.

Meng, L.P., Zhang, J., Wang, C.R. et al. (2009). Analysis of dietary fatty acid intake status of Chinese middle-aged and elderly residents. *Food Nutrition Chinese* 10: 58–61.

Mezzomo, N., Tenfen, L., Farias, M.S. et al. (2015). Evidence of anti-obesity and mixed hypolipidemic effects of extracts from pink shrimp (*Penaeusbrasiliensis*and *Penaeuspaulensis*) processing residue. *Journal of Supercritical Fluids* 96: 252–261.

Miettinen, T.A., Puska, P., Gylling, H. et al. (1995). Reduction of serum cholesterol with sitostanol-ester margarine in a mildly hypercholesterolemic population. *The New England Journal of Medicine* 333: 1308–1312. https://doi.org/10.1056/NEJM199511163332002.

Miettinen, T., Vanhanen, H., and Wester, I. (1996). *Use of a Stanol Fatty Acid Ester for Reducing Serum Cholesterol Level*. US Patent.

Miles, E.A., Zoubouli, P., and Calder, P.C. (2005). Differential anti-inflammatory effects of phenolic compounds from extra virgin olive oil identified in human whole blood cultures. *Nutrition* 21 (3): 389–394.

Miller, D.L. (2002). Health benefits of lecithin and choline. *Cereal Foods World* 47 (5): 178.

Mishra, S. and Manchanda, S.C. (2012). Cooking oils for heart health. *Journal of Preventive Cardiology* 1 (3): 3–131.

Mišurcová, L., Ambrožová, J.V., and Samek, D. (2011). Seaweed lipids as nutraceuticals. *Advances in Food and Nutrition Research* 64: 339–355.

Mobraten, K., Haug, T.M., Kleiveland, C.R., and Lea, T. (2013). Omega-3 and omega-6 PUFAs induce the same GPR120-mediated signaling events, but with different kinetics and intensity in Caco-2 cells. *Lipids in Health and Disease* 12: 101–107.

Moiseeva, E.P., Almeida, G.M., Jones, G.D., and Manson, M.M. (2007). Extended treatment with physiologic concentrations of dietary phytochemicals results in altered gene expression, reduced growth, and apoptosis of cancer cells. *Molecular Cancer Therapeutics* 6: 3071–3079.

Moreau, R.A., Nyström, L., Whitaker, B.D. et al. (2018). Phytosterols and their derivatives: structural diversity, distribution, metabolism, analysis, and health-promoting uses. *Progress in Lipid Research* 70: 35–61.

Moreau, R.A., Wayns, K.E., Flores, R.A., and Hicks, K.B. (2007). Tocopherols and tocotrienols in barley oil prepared from germ and other fractions from scarification and sieving of hulless barley. *Cereal Chemistry Journal* 84 (6): 587–592.

Moreau, R.A., Whitaker, B., and Hicks, K. (2002). Phytosterols, phytostanols, and their conjugates in foods: Structural diversity, quantitative analysis, and health-promoting uses. *Progress in Lipid Research.* 41: 457–500.

Morya, S., Menaa, F., Jiménez-López, C. et al. (2022). Nutraceutical and pharmaceutical behavior of bioactive compounds of miracle oilseeds: an overview. *Food* 11 (13): 1824.

Mosquera, M.l., Fernandez, J.G. (1986). Composition and evaluation of chlorophylls and carotenoids during the development and maturation of the olive tree fruits. *Grasas y Aceites (Seville)* 37: 33742.

Müller, C. and Bracher, F. (2015). Determination by GC-IT/MS of phytosterols in herbal medicinal products for the treatment of lower urinary tract symptoms and food products marketed in Europe. *Planta Medica* 81: 613–620.

Mumme, K. and Stonehouse, W. (2015). Effects of medium-chain triglycerides on weight loss and body composition: a meta-analysis of randomized controlled trials. *Journal of the Academy of Nutrition and Dietetics* 115 (2): 249–263.

Nandha, R.H., Singh, K., and Garg and Rani S. (2014). Therapeutic potential of sunflower seeds: an overview. *International Journal of Research and Development in Pharmacy & Life Sciences* 3: 967–972.

Nasri, H., Baradaran, A., Shirzad, H., and Kopaei, M.R. (2014). New concepts in nutraceuticals as alternative for pharmaceuticals. *International Journal of Preventive Medicine* 5 (12): 1487–1499.

Naziri, E., Mantzouridou, F., and Tsimidou, M.Z. (2011). Squalene resources and uses point to the potential of biotechnology. *Lipid Technology* 23 (12): 270–273.

Neff, L.M., Culiner, J., Rundles, S.C. et al. (2011). Algal docosahexaenoic acid affects plasma lipoprotein particle size distribution in overweight and obese adults. *The Journal of Nutrition* 141 (2): 207–213.

Newmark, H.L. (1999). Squalene, olive oil, and cancer risk: review and hypothesis. *Annals of the New York Academy of Sciences* 889 (1): 193–203.

Ng, M.H., Choo, Y.M., Ma, A.N. et al. (2004). Separation of vitamin E (tocopherol, tocotrienol, and tocomonoenol) in palm oil. *Lipids* 39 (10): 1031–1035.

Nguyen, M.T., Van de Walle, D., Petit, C. et al. (2014). Mapping the chemical variability of vege, able lecithins. *Journal of the American Oil Chemists' Society* 91: 1093–1101.

Nieuwenhuyzen, W.V. and Tomas, M.C. (2008). Update on vegetable lecithin and phospholipid technologies. *European Journal of Lipid Science & Technology* 110: 472–486.

Nikolic, N.C., Cakic, S.M., Novakovic, S.M. et al. (2009). Effect of extraction techniques on yield and composition of soybean oil. *Macedonian Journal of Chemical and Chemical Engineering* 28: 173–179.

Noboru, K. and Yusho, T. (1970). *Oryzanol Containing Cosmetics*, vol. 70, 32078. *Japanese Patent.*

Olguin, M.C., Hisano, N., D'Ottavio, E.A. et al. (2003). Nutritional and antinutritional aspect of an Argentinean soy flour assessed on weanling rats. *Journal of Food Composition and Analysis* 16: 441–448.

Oliver, J. and Palou, A. (2000). Chromatographic determination of carotenoids in foods. *Journal of Chromatography A* 881 (1-2): 543–555.

Ong, A.S. and Goh, S.H. (2002). Palm oil: a healthful and cost-effective dietary component. *Food and Nutrition Bulletin* 23: 11–22.

Ong, A.S.H. and Tee, E.S. (1992). Natural sources of carotenoids from plants and oils. In: *Methods in enzymology. Carotenoids*, vol. 213 (ed. L. Packer), 142–167. San Diego: Academic Press.

Onge, M.P. and Jones, P.J. (2002). Physiological effects of medium-chain triglycerides: potential agents in the prevention of obesity. *Journal of Nutrition* 132 (3): 329–332.

Orsavova, J., Misurcova, L., Ambrozova, J.V. et al. (2015). Fatty acids composition of vegetable oils and its contribution to dietary energy intake and dependence of cardiovascular mortality on dietary intake of fatty acids. *International Journal of Molecular Sciences* 16 (6): 12871–12890.

Othman, R.A. and Moghadasian, M.H. (2011). Beyond cholesterol-lowering effects of plant sterols: clinical andexperimental evidence of anti-inflammatory properties. *Nutrition Reviews* 69: 371–382.
Owen, R.W., Giacosa, A., Hull, W.E. et al. (2000). Olive-oil consumption and health: the possible role of antioxidants. *The Lancet Oncology* 1: 107–112.
Owen, R.W., Haubner, R., Mier, W. et al. (2003). Isolation, structure elucidation and antioxidant potential of the major phenolic and flavonoid compounds in brined olive drupes. *Food and Chemical Toxicology* 41 (5): 703–717.
Owen, R.W., Haubner, R., Würtele, G. et al. (2004). Olives and olive oil in cancer prevention. *European Journal of Cancer* 13 (4): 319–326.
Ozcan, M.M. (2010). Some nutritional characteristics of kernel and oil of peanut (*Arachishypogaea* L.). *Journal of Oleo Science* 59: 1–5.
Pal, A., Khanum, F., and Bawa, A. (2010). Nutritional, medicinal and industrial uses of sesame (*Sesamumindicum* L.) seeds—an overview. *Agriculturae Conspectus Scientificus* 75 (4): 159–168.
Pali, V. (2013). Rice bran oil-unique gift of nature: a review. *Agricultural Reviews* 34: 288–294.
Papamandjaris, A.A., MacDougall, D.E., and Jones, P.J. (1998). Medium chain fatty acid metabolism and energy expenditure: obesity treatment implications. *Life Sciences* 62 (14): 1203–1215.
Park, H.Y., Lee, K.W., and Choi, H.D. (2017). Rice bran constituents: immunomodulatory and therapeutic activities. *Food & Function* 8 (3): 935–943.
Patel, M. and Naik, S. (2004). Gamma-oryzanol from rice bran oil-A review. *Journal of Scientific and Industrial Research* 63.
Peanut Institute (2011). Types of peanut oil. http://www.peanut-institute.org/health-and-nutrition(Accessed October 20).
Penalvo, J.L., Hopia, A., and Adlercreutz, H. (2006). Effect of sesamin on serum cholesterol and triglycerides level in LDL-receptor deficient mice. *European Journal of Nutrition* 45: 439–444.
Peterson, D.M. (1995). Oat tocols-concentration and stability in oat products and distribution within the kernel. *Chemistry Journal* 72 (1): 21–24.
Phillips, K.M., Ruggio, D.M., and Ashraf, K.M. (2005). Phytosterol composition of nuts and seeds commonly consumed in the United States. *Journal of Agriculture Food* 53 (24): 9436–9445.
Popa, I., Băbeanu, N., Niă, S., and Popa, O. (2014). Squalene-natural resources and applications. *Farmacia* 62: 840–862.
Prasad, K. (1997). Hydroxyl radical-scavenging property of secoisolariciresinoldiglucoside (SDG) isolated from flaxseed. *Molecular and Cellular Biochemistry* 168: 117–123.
Prasad, K. (2000). Antioxidant activity of secoisolariciresinoldiglucoside-derived metabolites, secoisolariciresinol, enterodiol, and enterolactone. *International Journal of Angiology* 9: 220–225.
Prisna, P., Apinya, T., Nuchanart, R. et al. (2011). Estrogenic activities of sesame lignans and their metabolites on human breast cancer cells. *Journal of Agricultural and Food Chemistry* 59 (1): 212–221.
Punia, S., Kumar, M., Siroha, A.K., and Purewal, S.S. (2021). Rice bran oil: emerging trends in extraction, health benefit, and its industrial application. *Rice Science* 28 (3): 217–232.
Puupponen-Pimiä, R., Nohynek, L., and Meier, C. (2001). Antimicrobial properties of phenolic compounds from berries. *Journal of Applied Microbiology* 90: 494–507.
Radhamadhavi, B., Kanakadurga Devi, N., Saimrudula, B., and Nagendra Babu, R. (2010). The importance of biodegradable bio-oil sunflower. *International Journal of Pharm Tech Research* 2 (3): 1913–1915.
Radhika, P.R., Singh, R.B., and Sivakumar, T. (2011). Nutraceuticals: an area of tremendous scope. *International Journal of Research in Ayurveda and Pharmacy* 2 (2): 410–415.
Rajaram, S. (2014). Health benefits of plant-derived α-linolenic acid. *The American Journal of Clinical Nutrition* 100 (suppl): 443S–448S.
Rao, C., Newmark, H., and Reddy, B. (1998). Chemopreventive effect of squalene on colon cancer. *Carcinogenesis* 19 (2): 287–290.

Rathinam, P., Sreekumar, P., and Nanda Kumar, T.B. (2002). *Coconut Products*, 29–32. *Coconut Development Board Publication.*

Reddy, L.H. and Couvreur, P. (2009). Squalene: a natural triterpene for use in disease management and therapy. *Advanced Drug Delivery Reviews* 61 (15): 1412–1426.

Relas, H., Gylling, H., and Miettinen, T. (2001). Fate of intravenously administered squalene and plant sterols in human subjects. *Journal of Lipid Research* 42 (6): 988–994.

Riantong, S. and Jorg, J.J. (2021). Nutrition and applications of rice bran oil: a mini-overview. *Science and Technology of Cereals, Oils and Foods* 29 (3): 47–53.

Riaz, T., Iqbal, M., Mahmood, S. et al. (2021). Cottonseed oil: a review of extraction techniques, physicochemical, functional, and nutritional properties. *Critical Reviews in Food Science and Nutrition* 63: 1–19.

Rickard, S.E. and Thompson, L.U. (1997). Phytoestrogens and lignans: effects on reproduction and chronic disease. In: *Antinutrients and Phytochemicals in Foods* (ed. F. Shahidi), 273–293. New York: Oxford University Press.

Rigo, L.A., Pohlmann, A.R., Guterres, S.S., and Beck, R.C.R. (2014). Rice bran oil: benefits to health and applications in pharmaceutical formulations. In: *Wheat and Rice in Disease Prevention and Health: Benefits, Risks and Mechanisms of Whole Grains in Health Promotion* (ed. R.R. Watson, V. Preedy, and S. Zibadi), 311–322. Academic Press.

Rizvi, S., Raza, S.T., Ahmed, F. et al. (2014). The role of vitamin e in human health and some diseases. *Sultan Qaboos University Medical Journal* 14 (2): e157–e165.

Robert, C., Couëdelo, L., Vaysse, C., and Michalski, M.-C. (2020). Vegetable lecithins: a review of their compositional diversity, impact on lipid metabolism and potential in cardiometabolic disease prevention. *Biochimie* 169: 121–132.

Rodríguez-García, C., Sánchez-Quesada, C., Toledo, E. et al. (2019). Naturally lignan-rich foods: adietary tool for health promotion? *Molecules* 24: 917.

Rogers, E.J., Rice, S.M., Nicolosi, R.J. et al. (1993). Identification and quantitation of γ-oryzanol components and simultaneous assessment of tocols in rice bran oil. *Journal of the American Oil Chemists' Society* 70 (3): 301–307.

Rohman, A. (2014). Rice bran oil's role in health and cooking. In: *Wheat and Rice in Disease Prevention and Health. Benefits, Risks and Mechanisms of Whole Grains in Health Promotion* (ed. R.R. Watson, V.R. Preedy, and S. Zibadi), 481–490. Amsterdam: Elsevier.

Rong, S., Xu, R., and Li, W.F. (2016). Phytosterols and dementia. *Plant Food for Human Nutrition* 71: 1–8.

Rosenbloom, C., Millard Stafford, M., and Lathrop, J. (1992). Contempo-rary ergogenic aids used bystrength/power athletes. *Journal of the Academy of Nutrition and Dietetics* 92: 1264–1265.

Rukmini, C. and Raghuram, T.C. (1991). Nutritional and biochemical aspects of the hypolipidemic action of rice bran oil-a review. *Journal of the American College of Nutrition* 10 (6): 593–601.

Saint-Leger, D., Bague, A., Cohen, E., and Lchivot, M. (1986). A possible role for squalene in thepathogenesis of acne. I. In vitro study of squalene oxidation. *British Journal of Dermatology* 114 (5): 535–542.

Saji, N., Francis, N., Schwarz, L.J. et al. (2019). Rice bran derivedbioactive compounds modulate risk factors of cardiovascular disease and type 2 diabetesmellitus: an updated review. *Nutrients* 11: 2736.

Sakthinathan, B. and Nandhini, U.D. (2017). Phytochemicals—anutraceutical source of vegetables. *Chemical Science Review and Letters* 6 (24): 2133–2137.

Salazar, D., López-Cortés, I., and Salazar-García, D. (2017). Olive oil: composition and health benefits. In: *Olive Oil. Sensory Characteristics, Composition and Importance for Human Health* (ed. T. Fritjof and B. Henning), 1–37. New York: Nova Science Publishers.

Santana-Gálvez, J., Cisneros-Zevallos, L., and Jacobo-Velázquez, D.A. (2019). A practical guide for designing effective nutraceutical combinations in the form of foods, beverages, and dietary supplements against chronic degenerative diseases. *Trends in Food Science and Technology* 88: 179–193.

Sawicka, B., Ziarati, P., Krochmal-Marczak, B., and Skiba, D. (2020). Nutraceuticals in food and pharmacy: areview. *Journal of Agronomy* 74: 7–31.

Schauss, A.G., Endres, J.R., and Clewell, A. (2013). Safety of unsaturated vitamin E tocotrienols and their isomers. In: *Tocotrienols: Vitamin E Beyond Tocopherols*, 2ee (ed. R.R. Watson, V.R. Preedy, et al.), 17–36. Boca Raton: CRC Press.

Schwartz, H., Ollilainen, V., Piironen, V., and Lampi, A.M. (2008). Tocopherol, tocotrienol and plant sterol contents of vegetable oils and industrial fats. *Journal of Food Composition and Analysis* 21 (2): 152–161.

Sekhar, S.C. and Rao, B. (2011). Cottonseed oil as health oil. *Pertanika Journal of Tropical Agricultural Science* 34 (1): 17–24.

Selvaraj, K.S., Bharathi, A., Sivakumar, V. et al. (2020). Virgin coconut oil: a nutraceutical and therapeutic food. *International Journal of Chemical Studies* 8: 662–663.

Sen, C.K., Khanna, S., and Roy, S. (2006). Tocotrienols: vitamin E beyond tocopherols. *Life Sciences* 78: 2088–2098.

Serbinova, E., Kagan, V., Han, D., and Packer, L. (1991). Free radical recycling and intramembrane mobility in the antioxidant properties of alpha-tocopherol and alpha-tocotrienol. *Free Radical Biology & Medicine* 10 (5): 263–275.

Serna-Saldivar, S.O., Gutiérrez-Uribe, J.A., and García-Lara, S. (2015). Phytochemical profiles and nutraceutical properties of corn and wheat tortillas. In: *Tortillas*, 65–96. Amsterdam: Elsevier.

Shahidi, F. (2009). Nutraceuticals and functional foods: whole versus processed foods. *Trends in Food Science and Technology* 20: 376–387.

Shahidi, F. and De Camargo, A.C. (2016). Tocopherols and tocotrienols in common and emerging dietary sources: occurrence, applications, and health benefits. *International Journal of Molecular Sciences* 17 (10): 1745.

Shahidi, F. and Shukla, V.K.S. (1996). Nontriacylglycerol constituents of fats, oils. *International News on Fats, Oils and Related Materials* 7: 1227–1232.

Sharma, S. and Das, A. (2013). Rice bran oil: a cooking medium with health benefits. In: *CRRI Technology Bulletin 61*(1st Revision) (ed. B.N. Sarangi, G.A.K. Kumar, and S. Dalal). ICAR.

Sharma, M., Mondal, A.K., and Gupta, S.K. (2011). Production and trade of major world oil crops. In: *Technological Innovations in Major World Oil Crops*, 1–15. New York, NY, USA: Springer.

Shi, L.K., Liu, R.J., Jin, Q.Z., and Wang, X.G. (2017). The contents of lignans in sesame seeds and commercial sesame oils of China. *Journal of the American Oil Chemists' Society* 94: 1035–1044.

Shivhare, N. and Satsangee, N. (2012). Wonders of sesame: nutraceutical uses and health benefits. In: *Chemistry of Phytopotentials: Health, Energy and Environmental Perspectives* (ed. L. Khemani, M. Srivastava, and S. Srivastava), 63–68. Berlin, Heidelberg: Springer.

Singh, S.P. and Upadhyay, S.K. (Éds.) (2021). *Bioprospecting of Microorganism-Based Industrial Molecules*. Wiley. https://doi.org/10.1002/9781119717317.

Singh, S.P. and Upadhyay, S.K. (Éds.) (2023). *Microbial Bioreactors for Industrial Molecules*. Wiley. https://doi.org/10.1002/9781119874096.

Smith, T.J. (2000). Squalene: potential chemopreventive agent. *Expert Opinion on Investigational Drugs* 9 (8): 1841–1848.

Smith, R.E., Rouchotas, P., and Fritz, H. (2016). Lecithin (phosphatidylcholine): healthy dietary supplement or dangerous toxin? *Natural Products Journal* 6 (4): 242–249.

Station, A.R. (2008). Impact of dietary phospholipids on human health. *ALP Science* 524: 1–15.

Stephens, A.M., Dean, L.L., Davis, J.P. et al. (2010). Peanuts, peanut oil, and fat free peanut flour reduced cardiovascular disease risk factors and the development of atherosclerosis in Syrian golden hamsters. *Journal of Food Science* 75: 116–122.

Strandberg, T.E., Tilvis, R.S., and Miettinen, T.A. (1990). Metabolic variables of cholesterol during squalene feeding in humans: comparison with cholestyramine treatment. *Journal of Lipid Research* 31: 1637–1643.

Suh, M.H., Yoo, S.H., and Lee, H.G. (2007). Antioxidative activity and structural stability of microencapsulated γ oryzanol in heat treated lards. *Food Chemistry* 100 (3): 1065–1070.

Suja, K.P., Jayalekshmy, A., and Arumughan, C. (2004). Free radical scavenging behavior of antioxidant compounds of sesame (*Sesamumindicum* L.) in DPPH• system. *Journal of Agricultural and Food Chemistry* 52: 912–915.

Sumi, E.R., Anandan, R., Rajesh, R. et al. (2018). Nutraceutical and therapeutic applications of squalene. *Fishery Technology* 55: 229–237.

Sun, N., Chen, J., Wang, D., and Lin, S. (2018). Advance in food-derived phospholipids: sources, molecular species and structure as well as their biological activities. *Trends in Food Science and Technology* 80: 199–211.

Sundram, K., Sambathamurth, I.R., and Tan, Y.A. (2003). Palm fruit chemistry and nutrition. *Asia Pacific Journal of Clinical Nutrition* 12: 355–362.

Suzuki, Y.J., Tsuchiya, M., Wassall, S.R. et al. (1993). Structural and dynamic membrane properties of alpha-tocopherol and alpha-tocotrienol: implication to the molecular mechanism of their antioxidant potency. *Biochemistry* 32 (40): 10692–10699.

Sylvester, P.W. (2011). Synergistic anticancer effects of combined γ-tocotrienol with statin or receptor tyrosine kinase inhibitor treatment. *Genes & Nutrition* 7 (1): 63–74.

Sylvester, P. and Theriault, A. (2003). Role of tocotrienols in the prevention of cardiovascular disease and breast cancer. *Current Topics in Nutraceutical Research* 1 (2): 121–136.

Szuhaj, B.F., Yeo, J., and Shahidi, F. (2020). Lecithins. In: *Bailey's Industrial Oil and Fat Products*, 1–86. Hoboken: Wiley.

Taku, K., Melby, M.K., Nishi, N. et al. (2011). Soy isoflavones for osteoporosis: an evidence-based approach. *Maturitas* 70 (4): 333–338.

Tallima, H. and El Ridi, R. (2018). Arachidonic acid: physiological roles and potential health benefits-A review. *Journal of Advanced Research* 11: 33–41.

Tan, B.L. and Norhaizan, M.E. (2019). Carotenoids: how effective are they to prevent age-related diseases? *Molecules* 24: 1801.

Taylor, C.L. (2004). Regulatory frameworks for functional foods and dietary supplements. *Nutrition Reviews* 62 (2): 55–59.

Thomford, N.E., Senthebane, D.A., Rowe, A. et al. (2018). Natural products for drug discovery in the 21st century: innovations for novel drug discovery. *International Journal of Molecular Sciences* 19: E1578.

Thorbjarnarson, T. and Drummond, J.C. (1935). Occurrence of an unsaturated hydrocarbon in olive oil. *Analyst* 60 (706): 23–29.

Tikekar, R.V., Ludescher, R.D., and Karwe, M.V. (2008). Processing stability of squalene in amaranthand antioxidant potential of amaranth extract. *Journal of Agricultural and Food Chemistry* 56: 10675–10678.

Touré, A. and Xueming, X. (2010). Flaxseed lignans: source, biosynthesis, metabolism, antioxidant activity, bio-active components, and health benefits. *Comprehensive Reviews in Food Science and Food Safety* 9 (3): 261–269.

Tsimidou, M. (2010). Squalene and tocopherols in olive oil: importance and methods of analysis. In: *Olives and Olive Oil in Health and Disease Prevention* (ed. V.R. Preedy and R.R. Watson), 561–567. New York: Elsevier Inc.

Tu, R.Y., Wu, G.H., Luo, R.C. et al. (2018). Analysis of phytosterols in edible vegetable oils on the market in Beijing. *Capital Journal of Public Health* 12: 84–88.

Upadhyay, S.K. (Éd.). (2021). *Genome Engineering for Crop Improvement*. Wiley. https://doi.org/10.1002/9781119672425.

Upadhyay, S.K. (Éd.). (2024). *Applications of Genome Engineering in Plants*. Wiley. https://doi.org/10.1002/9781394183913.

Upadhyay, S.K. and Singh, S.P. (Éds.) (2021). *Bioprospecting of Plant Biodiversity for Industrial Molecules*. Wiley. https://doi.org/10.1002/9781119718017.

Upadhyay, S.K. and Singh, S.P. (Éds.) (2023). *Plants as Bioreactors for Industrial Molecules*. Wiley. https://doi.org/10.1002/9781119875116.

Van der Veen, J.N., Lingrell, S., Da Silva, R.P. et al. (2014). The concentration of phosphatidylethanolamine in mitochondria can modulate ATP production and glucose metabolism in mice. *Diabetes* 63: 2620–2630.

Van Hoogevest, P. and Wendel, A. (2014). The use of natural and synthetic phospholipids as pharmaceutical excipients. *European Journal of Lipid Science and Technology* 116: 1088–1107.

Vergallo, C. (2020). Nutraceutical vegetable oil nano formulations for prevention and management of diseases. *Nanomaterials* 10 (6): 1232.

Verleyen, T., Forcades, M., Verhe, R. et al. (2002). Analysis of free and esterified sterols in vegetable oils. *Journal of the American Oil Chemists' Society* 79: 117–122.

Visavadiya, N.P. and Narasimhacharya, A.V.R.L. (2008). Sesame as a hypocholesteraemic and antioxidant dietary component. *Food and Chemical Toxicology* 46 (6): 1889–1895.

Waheed, A., Rasool, G., and Asghar, A. (2010). Effect of interesterified palm and cottonseed oil blends on cookie quality. *Agriculture and Biology Journal of North American* 1: 402–406.

Wang, T. (2015). Soybean: processing. In: *Encyclopedia of Food Grains*, 2ee. Amsterdam: Elsevier.

Wang, Q., Ge, X., Tian, X. et al. (2013). Soy isoflavone: the multipurpose phytochemical (Review). *Biomedical Reports* 1 (5): 697–701.

Wang, W., Yang, B., Li, W. et al. (2021). Effects of steam explosion pretreatmenton the bioactive components and characteristics of rapeseed and rapeseed products. *LWT—Food Science and Technology* 143: 111172–111181.

Waters, A.P. and Knowler, J.T. (1982). Effect of a lignan (HPMF) on RNA synthesis in the rat uterus. *Reproduction, Fertility and Development* 66: 379–381.

Weber, N., Weitkamp, P., and Mukherjee, K.D. (2002). Cholesterol-lowering food additives: lipase-catalysed preparation of phytosterol and phytostanol esters. *Food Research International* 35 (2-3): 177–181.

Weihrauch, J.L. and Gardner, J.M. (1978). Sterol content of foods of plant origin. *Journal of the American Dietetic Association* 73: 39–47.

Weiss, L.A., Barrett-Connor, E., and Von Mühlen, D. (2005). Ratio of n-6 to n-3 fatty acids and bone mineral density in older adults: the rancho bernardo study. *The American Journal of Clinical Nutrition* 81: 934–938.

Wen, Y.Q., Liu, Y. L., Liu, Y., Xu, Y. H., Wang, L.Y. (2016). In study on the oxidation stability and the content of vitamin E in edible vegetable oil (oil branch of Chinese cereals and oils association), *Paper collection of the 25th annual academic meeting and product exhibition meeting of oil branch of China Grain and Oil Society*.

Wester, I. (2000). Cholesterol-lowering effect of plant sterols. *European Journal of Lipid Science and Technology* 102 (1): 37–44.

Wołosik, K., Knaś, M., Zalewska, A., and Niczyporuk, M. (2013). The importance and perspective of plant-based squalene in cosmetology. *Journal of Cosmetic Science* 64 (1): 19–65.

Wong, W.Y. and Lindsay, B. (2013). Tocotrienols as possible treatments for obesity. In: *Tocotrienols: Vitamin E Beyond Tocopherols*, 2ee (ed. T. Barrie, R.R. Watson, and V.R. Preedy), 195–208. Boca Raton: Taylor & Francis (CRC Press).

Woyengo, T.A., Ramprasath, V.R., and Jones, P.J.H. (2009). Anticancer effects of phytosterols. *European Journal of Clinical Nutrition* 63: 813–820.

Wroniak, M. and Rękas, A. (2016). Nutritional value of cold-pressed rapeseed oil during long term storage as influenced by the type of packaging material, exposure to light &oxygen and storage temperature. *Journal of Food Science and Technology* 53: 1338–1347.

Yang, C.Y., Liu, X.M., and Che, Z.Y. (2013). Determination of fatty acid profiles in fifteen kinds of edible vegetable oil by gas chromatography-mass spectrometry. *Food Science* 34: 211–214.

Yang, R., Xue, L., Zhang, L. et al. (2019). Phytosterol contents of edible oils and their contributions to estimated phytosterol intake in the Chinese diet. *Food* 8 (8): 334.

Yu, S., Nehus, Z.T., Badger, T.M., and Fang, N. (2007). Quantification of vitamin E and gamma-oryzanol components in rice germ and bran. *Journal of Agricultural and Food Chemistry* 55 (18): 7308–7313.

Zanqui, A.B., Barros, T.V., Barão, C.E. et al. (2021). Production of blends of edible oil and carrot carotenoids using compressed propane: enhancement of stability and nutritional characteristics. *The Journal of Supercritical Fluids* 171: 105–189.

Zavoshy, R., Noroozi, M., and Jahanihashemi, H. (2012). Effect of low-calorie diet with rice bran oil on cardiovascular risk factors in hyperlipidemic patients. *Journal of Research in Medical Sciences* 17 (7): 626–631.

Zeb, A. (2021). A comprehensive review on different classes of polyphenolic compounds present in edible oils. *Food Research International (Ottawa Ont.)* 143: 110312.

Zhang, Y., Xiao, H., Lv, X. et al. (2022). Comprehensive review of composition distribution and advances in profiling of phenolic compounds in oilseeds. *Frontiers in Nutrition* 9: 1044871. https: //www.frontiersin.org/articles/10.3389/fnut.2022.1044871.

Zhao, X., Xiang, X., Huang, J. et al. (2021). Studying the evaluation model of the nutritional quality of edible vegetable oil based on dietary nutrient reference intake. *ACSOmega* 6 (10): 6691–6698.

Zonta, F. and Stancher, B. (1987). Carotenoid content and the quality of olive oils. *RivistaItalianaDelleSostanze Grasse* 64: 53–55.

Zou, L. and Akoh, C.C. (2015). Antioxidant activities of annatto and palm tocotrienol-rich fractions in fish oil and structured lipid-based infant formula emulsion. *Food Chemistry* 168: 504–511.

15

Nutraceuticals from Conventional and Nonconventional Fruits

Karishma Sebastian, Manjula Belagihalli Siddalingappa, Jeevitha Devaraju, and Panneer Selvam Ranchana

Division of Horticulture, School of Agricultural Sciences, Karunya Institute of Technology and Sciences, Coimbatore, India

15.1 Introduction

A food or its component thereof, which offers a health or medical advantage to the body, including the avoiding or curing of disease, is referred to as a nutraceutical (Ishak et al. 2017). Combining the concepts "nutrition" with "pharmaceutical," Chairman of the Foundation for Innovation in Medicine, Stephen Defelice, created the name "nutraceutical" (Biesalski 2001). According to the Canadian Health Ministry, a nutraceutical is a substance that has been extracted or refined from food, is typically sold in a therapeutic form, and has been shown to have advantages for both the body's physiological processes and the prevention of chronic illnesses (Pandey et al. 2010). Nutraceuticals can help you return to nature in a way that is both therapeutic and protective. There is a growing appreciation and understanding of the connection between eating more fruits and vegetables and being healthier, these days. A diet heavy in fruits and vegetables has been linked to a lower incidence of chronic illnesses due to their high mineral and vitamin content and the abundance of bioactive substances like anti-inflammatory molecules and antioxidants (Donno et al. 2013). Mahima et al. (2013) reported fruits and vegetables as the least expensive and most accessible source of nutrients such as dietary fiber, protein, carbohydrates, vitamins, minerals, and essential amino acids, which lowers the risk of obesity, metabolic disorders, and cardiovascular disease. Studies have demonstrated the significant potential of biologically active molecules found in plant-based foods, especially phytochemicals, to modify various processes involved in the onset of diseases, including cancer, diabetes, cardiovascular disease, Alzheimer's disease, pulmonary disorders, and other degenerative illnesses (Dixit et al. 2021; Upadhyay and Singh 2021, 2023).

In this chapter, nutraceuticals from conventional and nonconventional fruits are discussed in detail.

Herbal Nutraceuticals: Products and Processes, First Edition. Edited by Santosh Kumar Upadhyay and Sudhir Pratap Singh.

15.2 Apple

As a member of the Rosaceae family, apples (*Malus domestica* Borkh.) are one of the most significant and extensively grown temperate fruits worldwide. Apples are an excellent source of plant nutrients, especially phenolic compounds and other dietary carbohydrates. Apple phenolics are plant-based compounds that protect the cell from the damaging effects of free radicals and inhibit low-density lipoprotein oxidation, making them powerful antioxidants. Flavonols (quercetin, kaempferol, and rutin), dihydrochalcones (phloretin and phloridzin), flavan-3-ols (epicatechin and procyanidins), and phenolic acids (caffeic acid and coumaric acid) are the structural types of polyphenols found in apples (Rana and Bhushan 2016). Additionally, studies have linked the consumption of apples with favorable effects on bone health, weight management, pulmonary function, aging and cognitive decline, and gastrointestinal health (Hyson 2011).

15.3 Peach

Peach fruit (*Prunus persica*) is known for its significant level of phytochemicals such as carotenoids, vitamins, phenolic compounds, organic acids, and volatiles. It has been found that the phenolic chemicals found in peaches, specifically quercetin, catechins, and cyanidin derivatives, are important for their antibacterial, anti-inflammatory, and antioxidant properties (Bento et al. 2022). These compounds have a multitude of positive impacts on cells by scavenging free radicals and acting in cell signaling pathways.

15.4 Plum

Japanese plum (*Prunus salicina* Lindl.) fruits are found to be rich in sugars, organic acids, and phenolic compounds. It is believed that plums' health advantages are partly attributed to the antioxidant activity of phenolic chemicals (Crozier et al. 2009).

15.5 Apricot

Apricot (*Prunus armeniaca* L.) is a nutrient-dense fruit with several health advantages as it is high in bioactive chemicals. The fruit is considered a functional food due to its high content of bioactive phytochemicals such as carotenoids, flavonoids, phenolics, and antioxidants (Bartolini et al. 2015). Major carotenoids in apricot fruits include β-carotene, γ-carotene, lycopene, β-cryptoxanthin, phytoene, phytofluene, and lutein (Ayour et al. 2016). These fruits contain the following flavonols: glycoside rutin, resveratrol, quercetin, and vanillin; the phenolic components include salicylic acid, procatechin, gallic acid, ferulic acid, caffeic acid, 4-aminobenzoic acid, and chlorogenic acid (Sochor et al. 2010).

15.6 Sweet Cherry

Sweet cherries (*Prunus avium* L.) are popular fruits due to their flavor, color, nutritional value, and health benefits. It is a nutrient-dense food that contains considerable levels of key nutrients and bioactive dietary components such as potassium, vitamin C, carotenoids, polyphenols, and fiber (McCune et al. 2010). Furthermore, cherries are high in tryptophan, serotonin, and melatonin (Garrido et al. 2012). According to Kelley et al. (2018), cyanidin-3-glucoside and cyanidine-3-rutinoside are the major anthocyanins, and hydroxycinnamates and flavin-3-ols are the phenolic compounds present in cherry. Ferretti et al. (2010) reported that the anti-inflammatory and antioxidant capabilities of cherry fruit are aided by vitamins E and C, carotenoids, polyphenols, and melatonin.

15.7 Mango

Mango (*Mangifera indica* L.), which belongs to Anacadiaceae family, is a highly prized fruit with a powerful aroma, vibrant peel pigmentation, excellent taste, and great nutritional value due to its high levels of minerals, beta-carotene, and vitamin C.

Rymbai et al. (2013) identified mango pulp as a dietary source of various bioactive compounds viz., polyphenols, ascorbic acid, carotenoids, and lupeol. As per Kim et al. (2007), gallic acid and hydrolyzable tannins constitute approximately 98% of total polyphenols. Additional polyphenols found in mango pulp comprise gallo-tannins, phenolic acids, flavonoids, and xanthones. These polyphenols work together to enhance the fruit's antioxidant capacity. It has been discovered that these antioxidants can stop the development of multiple ailments, including high-density lipoprotein cholesterol, triglycerides, and coronary atherosclerosis, which is brought on by raised cholesterol.

Ajila et al. (2007) reported ascorbic acid as the major vitamin in mango pulp and Chena et al. (2004) identified β-carotene as the dominant carotenoid in mango, which possesses nutraceutical function. Mango pulp contains a pentacyclic triterpene called luteol, which has been shown to have a variety of pharmacological effects, including potent antioxidant, antimutagenic, and anti-inflammatory effects (Saleem et al. 2004). Additionally, research revealed that lupeol prevented DNA strand breaks caused by 7, 12-dimethylbenz(a)anthracene (Nigam et al. 2007), which decreased the occurrence of tumors, their body burden, and caused a significant postponement in tumor emergence (Saleem et al. 2004).

According to Rao et al. (2023), higher plants and mangoes contain a glucosyl xanthone called mangiferin, which has anticancer, antioxidant, antimutagenic, antidiabetic, antibacterial, and anti-inflammatory effects.

15.8 Banana

Banana fruits (*Musa* spp.) have two potent nutraceuticals viz., flavonoids and fructans. Flavonoids are antioxidant phytonutrients and 100 g edible portion of banana fruit contains 60 mg flavonoids, and the major flavonoids present in it are gallocatechin, catechin, and epicatechin. By way of low-density lipoprotein reduction, free radical scavenging, metal ion sequestration, and thrombic tendency reduction, flavonoids were found to protect against coronary heart disease.

Fructans, the second nutraceutical present in banana fruit is a soluble but indigestible nonstarch polysaccharide that was found to provide dietary fiber. 100 g of edible portion of banana fruit contains approximately 2.7 g dietary fiber and 0.3 g fructans. Fructans protect the interior intestinal lining from bacteria which may otherwise lead to bowel inflammation, and they also help promote normal bowel movement. In the intestine, fiber reduces triglyceride buildup and enhances calcium and magnesium absorption (Vaganan et al. 1998).

15.9 Citrus

Citrus fruits are rich in flavonoids and vitamin C, and their unique scent is ascribed to limonoids and flavonoids in the peel. The highest dietary sources of citrus bioflavonoid hesperidin, a flavanone glycoside, are tangelos and *Citrus sinensis*. Lemon and orange peels and their membrane regions have higher hesperidin contents. Hemorrhoids and vein dysfunction are treated with hesperidin (Garg et al. 2001).

According to Aschoff et al. (2015), numerous phytochemicals are found in oranges, including flavonoids (naringenin), carotenoids (beta-carotene, lutein, and beta-cryptoxanthin), and volatile organic compounds (terpenes, alcohols, esters, aldehydes, and ketones) that give oranges their scent. Limonoids, which are abundant in citrus fruits, were found to have antifungal, antioxidant, anticancer, and antibacterial qualities that improve human health (Russo et al. 2016). Strong antioxidant and anticancer effects have been demonstrated for naringin and its aglycone naringenin, which are mostly found in grapefruits (Alam et al. 2014).

15.10 Grapes

Grapes (*Vitis* spp.) are one of the oldest fruit crops that humans have domesticated. According to Palai (2023), grapes have health-promoting properties because of their phytochemicals, which include phenolic acids, stilbenes, flavanols, flavanones, flavones, flavan-3-ols, isoflavonoids, anthocyanins, and proanthocyanidins. Grapes have antiatherosclerotic, anti-inflammatory, immunomodulatory, antiplatelet, antibacterial, antiallergic, and antimutagenic properties that promote good health (Georgiev et al. 2014). Consuming grapes lowers the chance of getting chronic illnesses like malignancies, diabetes, and neurological, cardiovascular, gastrointestinal, and respiratory conditions (Lal et al. 2002).

15.11 Guava

The tropical fruit guava (*Psidium guajava*), belonging to the Myrtaceae family, is abundant in essential nutrients. The fruit, sometimes known as "super-fruit," easily falls into the category of new functional foods due to its distinct flavor, taste, and health-promoting properties. Vitamin C and minerals including calcium, phosphorus, and iron are abundant in guava fruits (Nanjundaswamy et al. 1964). Joseph and Priya (2011) reported that essential oils found in guava fruits are said to offer antinociceptive, insecticidal, repellent, anticancer, and anti-inflammatory properties. Guava is a good source of lycopene; 100 g of pink guava fruit has 5204 g of lycopene, which is about twice as much as tomatoes. According to Weng (2010), lycopene in pink guavas protects the skin from UV radiation and may help prevent prostate cancer.

15.12 Papaya

Carica papaya L., papaya fruit, is widely renowned around the world, with tropical and subtropical countries producing most of it. Its pulp has folate, pantothenic acid, potassium, magnesium, dietary fibers, and vitamins A, C, and E. The elements of this composition, according to

Santana et al. (2019), have positive effects on the circulatory system, shielding it against cardiovascular diseases and preventing damage from free radicals. Additionally, it has been suggested that it aids in the medical management of diabetes and lowers cholesterol. According to Manosroi et al. (2014), papain, a proteolytic enzyme found in papaya, possesses gelationlytic and antioxidant qualities.

15.13 Pineapple

The pineapple, or *Ananas comosus*, is a tropical fruit with a unique aroma and flavor due to the complex combinations and minute quantities of volatile compounds it contains. According to Sahu et al. (2017), the primary bioactive components of pineapple are flavonoids, ascorbic acid, β-carotene, and phenolic compounds. It is also found to be a rich source of β-glucan, proteoglycan, lectin, polysaccharides, triterpenoids, dietary fiber, lentinan, steroids, glycopeptides, terpenes, saponins, xanthones, coumarins, and alkaloids. It is regarded as one of the most beneficial fruits for the synthesis of value-added compounds like antioxidants, organic acids, bromelain, and phenolic compounds due to its physicochemical makeup and nutritional worth.

Bromelein, an active principle found in pineapple, belongs to the proteolytic enzyme group. Treatments for angina pectoris, bronchitis, sinusitis, surgical trauma, thrombophlebitis, wound debridement, and enhanced drug absorption—particularly for antibiotics—are among the conditions for which bromelain is beneficial. According to Pavan et al. (2012), it also relieves diarrhea, osteoarthritis, and a number of cardiovascular conditions.

15.14 Sapota

The evergreen tropical tree sapota (*Manilkara achras* Forb.), sometimes referred to as sapodilla, is said to have several health benefits for humans due to its high ascorbic acid and phenolic compound content. It is high in nutrients (sugars, amino acids, protein, and acids), minerals (calcium, potassium, and iron), and bioactive compounds, the majority of which are ellagitannins, gallotannins, flavonoids (anthocyanins and flavanols) depsides, and phenolic acid (Bangar et al. 2022). Because a 100 g fresh sample of sapota fruit has an antioxidant capacity equivalent to 3000 mg L^{-1} ascorbic acid, it is considered a high antioxidant source (Shui et al. 2004). The fruit's antioxidant activity is also greatly enhanced by polyphenolics, which are compounds that comprise basic blocks of gallocatechin, catechin, or both (Lim 2013).

15.15 Annona

Custard apple (*Annona squamosa*) of Annonaceae family is a potential source of nutraceuticals. Bhardwaj et al. (2014) claim that it is an excellent source of minerals, phenolic compounds, and naturally occurring antioxidants.

Annonaceous acetogenins are the active component responsible for the antitumor activities of annonaceous fruits (Dutta et al. 2017). Annonaceous acetogenins are considered to block mammalian mitochondrial NADH-ubiquinone reductase (Complex I) and cause cell death in the stomach. According to Han et al. (2015), annonaceous acetogenin can trigger cancer cell death via apoptosis, proposing a potential cancer treatment.

15.16 Pomegranate

Middle Eastern by origin, pomegranates (*Punica granatum*) are becoming increasingly popular as functional foods and sources of nutraceuticals because of their high polyphenolic content, particularly punicalagins, punicalins, gallagic acid, and ellagic acid. These molecules are converted to ellagic acid and urolithins during digestion, implying that the bioactive chemicals that provide in vivo antioxidant action may not be the same as those found in whole food. Pomegranate's health benefits can also be influenced by anthocyanins and the unique fatty acid content of the seed oil (Johanningsmeier and Harris 2011).

15.17 Dates

It has been discovered that the fruits of the date palm (*Phoenix dactylifera*), which are abundant in energy, vitamins, minerals, and easily absorbed sugars, help the body swiftly replenish and regenerate, especially during fasting. In addition to minerals like potassium, phosphorus, magnesium, and calcium, date palm fruits have been shown to contain phenolic compounds, anthocyanins, carotenoids, and sterols, among other phytochemicals and phytonutrients. Al-Alawi et al. (2017) identified zeaxanthin, β-cryptoxanthin, lycopene, lutein, and β-carotene as the carotenoids present in Khalas variety of date palm fruits from the United Arab Emirates. It was discovered that these fruits' flavonoids and phenolic components had antioxidant and free radical-scavenging qualities (Al-Okbi 2022).

15.18 Avocado

The nutrient-dense *Persea americana* avocado fruit has a single, enormous seed encircled by a delicious, creamy, smooth-textured edible portion and a thick, rough skin that turns purple-black when ripe. It has dietary fiber, magnesium, potassium, vitamin A, vitamin B6, vitamin C, vitamin E, vitamin K1, folate, niacin, pantothenic acid, riboflavin, lutein/zeaxanthin, choline, and phytosterols (Hasler and Brown 2009). Avocados have a medium-low calorie density due to their high water content (72%) and low dietary fiber content (6.8%). Studies have indicated that the potassium and lutein found in avocados may support healthy blood pressure and control oxidative/inflammatory stress. Avocado fruits have also been shown in studies to have heart-healthy benefits on blood lipid profiles (Colquhoun et al. 1992). Avocados' substantial amount of unsaturated fatty acids (polyunsaturated and monounsaturated fatty acids) and low content of saturated fatty acids are mostly to blame for this. Avocado oil rich in monounsaturated fatty acids in a water-based matrix seems to enhance the absorption of nutrients and phytochemicals while reducing the flavor and texture of the dietary fiber (Unlu et al. 2005). Natural phytosterols and dietary fiber in this fruit also play a secondary cholesterol-lowering role (Dreher and Davenport 2013).

15.19 Aonla

Emblica officinalis Gaertn or *Phyllanthus emblica* Linn, commonly known as Indian gooseberry, or amla, is the most significant medicinal plant used in Southeast Asian traditional folk medicine systems, including ayurveda. Nearly all the vegetative and reproductive portions of aonla are abundant in various nutraceuticals, with iron, calcium, carotene, niacin, phosphorous, riboflavin, and thiamine among the most important. Amla is used to cure a wide range of conditions in traditional medicine, including hyperacidity, anemia, eye irritation, diarrhea, nervine debility, jaundice, leucorrhea, liver problems, anomalies in the urine, and cough. According to scientific research, amla has the following benefits: it is antiviral, antifungal, antibacterial, antimutagenic, antioxidant, antidiabetic, antiulcerogenic, anti-inflammatory, hypolipidemic, immunomodulatory, antitussive, analgesic, antipyretic, adaptogenic, antiatherogenic, neutralizes snake venom, antidiarrheal, antiatherosclerotic, neuroprotective, nephroprotective, hepatoprotective, and wound healing abilities (Baliga et al. 2019).

15.20 Strawberry

Due to its low calorie content, easily digestible sugar content, and plenty of bioactive phytochemicals such as vitamin C and other antioxidant components like phenolic acids, polyphenols, and flavonoids, strawberry fruits (*Fragaria × ananassa*) can play a significant role in a person's diet.

Strawberry phenolic compounds include hydroxycinnamic acids (*p*-cumaric) and hydroxybenzoic acids (gallic and ellagic acids). Anthocyanins (pelargonidin-3-glycoside pigment), flavones, flavan-3-ols (catechins), flavonols (quercetin, kaempferol, and myricetin), flavanones, and hydrolyzable tannins (ellagitannins) are the structural classes of flavonoids (Zebrowska et al. 2016) present in the fruit. According to Ikram et al. (2019), strawberry phenolic compounds, contrary to vitamin C, are responsible for antioxidant capabilities, and these dietary antioxidants inhibit or minimize the potential cancer-inducing oxidative damage. The study also pointed out that strawberry flavonoids can also chelate metal ions, thereby preventing the formation of reactive oxygen species.

15.21 Kiwifruit

The kiwifruit, or *Actinidia deliciosa*, is a fruit rich in nutrients and has a high vitamin C content. It also includes potassium, folate, vitamin E, and dietary fiber, along with a range of bioactive substances including enzymes, phytonutrients, and antioxidants that support various bodily processes and functions (Richardson et al. 2018).

One of the most prevalent micronutrient problems globally today is low iron status, which is associated with several detrimental health effects. Beck et al. (2010) reported that kiwifruit's high vitamin C content could help boost iron absorption.

15.22 Rambutan

Nephelium lappaceum L., often known as rambutan, is a tropical plant native to Southeast Asia and a member of the Sapindaceae family. Its peel, seed, and pulp contain a variety of bioactive components and its fruit is abundant in antioxidants, dietary fiber, vitamins, and minerals. As per Afzaal et al. (2023), the fruit of the rambutan tree, which has high nutritional and bioactive contents, has anticancer, antiallergic, antiobesity, antidiabetic, antimicrobial, antihypercholesterolemic, and antihyperglycemic effects.

15.23 Mangosteen

Mangosteen (*Garcinia mangostana*) of the Guttiferae family is a tropical evergreen tree with enormous potential for use as a source of nutraceuticals. In Southeast Asia, the plant's many parts have long been used as medicine to treat urinary tract infections, diarrhea, and skin conditions (Morton 1987). It has been reported that the presence of hydroxy citric acid (HCA) contributes to the noticeable tart taste of mangosteen. As per Hayamizu et al. (2003), HCA is regarded as an appetite suppressor that can help prevent and lessen the accumulation of visceral fat.

Mangosteen pericarp has twenty times the antioxidant activity and ten times the amount of phenolic compounds of the edible white portions, according to Lal et al. (2002). The phenolic chemicals found in the pericarp of mangosteen include gallocatechin, epigallocatechin, epicatechin, catechin, epiafzelechin, and afzelechin (Fu et al. 2007), which have oxygen radical scavenging capacities and according to Jung et al. (2006), xanthone group identified in mangosteen pericarp includes gartanin, garcinone E, garcinone D, α-, β-, and γ-mangostins, tovophyllin, smeathxanthone, mangostinone, 1-isomangostin, eudraxanthone G,1,5-dihydroxy-2-(3- methylbut-2-enyl)-3-methoxy- and 1,7-dihydroxy-2-(3-methylbut-2-enyl) -3-methoxyxanthone.

15.24 Litchi

It has long been recognized that litchi (*Litchi chinensis*) provides medicinal benefits for a variety of ailments. Litchi contains several different biologically functional compounds namely flavonoids, anthocyanins, phenolics, sesquiterpenes, triterpenes, and lignans (Yao et al. 2021). The antioxidant and radioprotective qualities of litchi fruit were found to be notably abundant and a robust relationship was found by Saxena et al. (2011) between the phenolic and flavonoid contents and the antioxidant parameters.

15.25 Longan

Longan (*Dimocarpus longan* Lour.), a member of the Sapindaceae family, is used in traditional Chinese medicine. The arilloid fruit pulp is found to be a rich source of potassium, which is necessary for the healthy operation of muscles and neurons. Additionally, the fruit pulp of longan contains minerals such as iron, calcium, phosphorous, and magnesium as well as vitamins like vitamin C, riboflavin, thiamin, and niacin (Lal et al. 2020).

A considerable number of bioactive compounds, including phenolic acids, polysaccharides, and flavonoids, which have antiviral, antibacterial, antioxidant, anti-inflammatory, and anticarcinogenic properties, have been discovered in the pericarp of longan fruits (Yang et al. 2008). Its pulp has been utilized traditionally in Chinese medicine in Asian nations for a long time to cure neurological pain and swelling, palpitations, amnesia, lengthen life, soothe nerves, reduce sleeplessness, and boost blood metabolism (Li et al. 2015; Yi et al. 2012).

15.26 Blueberry

The blueberry, or *Vaccinium* sp., is a fruit crop in the Ericaceae family that yields fruit with an indigo hue. Apart from their flavor, blueberries are mostly known for their decent amount of vitamin C. An average of 100 g of blueberries contains 10 mg of ascorbic acid, which is one-third of the daily required amount (Prior et al. 1998). It is also thought to be a storehouse of beneficial phytochemicals. Numerous health benefits of this fruit have been attributed to the presence of phenolic acids, which include *p*-coumaric, chlorogenic, caffeic, ferulic, and cinnamic acids, and flavonoids, which include anthocyanidins. Retinal damage, diabetes, obesity, stomach, liver, heart, and kidney inflammation, tumors, microbial infections, cognitive decline, and bone loss have all been shown to be alleviated by berries (Patel 2014).

Working memory gains and a further reduction in depressed symptoms were observed in adult individuals with mild cognitive impairment who took blueberry juice supplements for three months (Krikorian et al. 2010). These berries are especially high in the flavonoid subclass known as anthocyanidins, which have the ability to penetrate the blood–brain barrier and cluster in areas related to learning and memory.

15.27 Raspberry

One of the oldest fruits, the raspberry (*Rubus idaeus* L.), is a member of the Rosaceae family having been utilized for millennia for human nourishment and folk medicine. According to Rao and Snyder (2010), these fruits are abundant in bioactive chemical compounds viz, ellagitannins, anthocyanins, and ascorbic acid that have a high antioxidant capacity and can help prevent chronic human diseases. Apart from these biochemical compounds, raspberry fruits are good sources of pelargonidins, cyanidins, catechins, quercetin, kaempferol, gallic acid, and salicylic acid (Mazur et al. 2014). According to Spencer (2010), the components of berries have the potential to guard against damage caused by reactive oxygen species, which has been connected to the advancement of brain conditions like Alzheimer's illness.

15.28 Walnut

The most widespread tree nut in the world is the walnut (*Juglans regia* L.). The nut's nutritional value is linked to the seed (kernel) and is a nutrient-dense food due to its high oil content. The composition of walnut oil is dominated by unsaturated fatty acids (mostly linoleic, with low levels of oleic and linolenic acids) and it also contains phenols, sterols, squalene, melatonin, vitamins, and minerals. Phenolic compounds, which are abundant in the seed coat but are poorly extracted with the oil, were discovered to have potent antioxidant capabilities. Anderson et al. (2001) reported that walnut lowers cholesterol, oxidative stresses induced by free radicals, and harmful inflammations that damage health.

15.29 Fig

Figs (*Ficus carica*, L.) are a kind of deciduous tree or shrub native to Southwest Asia and the Middle East and are thought to be one of the earliest domesticated plants. According to Sandhu et al. (2023), figs are rich in bioactive compounds such as carotenoids, phenolic acids, flavonoids, and tocopherols. These substances have been utilized for centuries in traditional medicine to treat inflammatory, respiratory, gastrointestinal, metabolic, and cardiovascular conditions. Arvaniti et al. (2019) report that polyphenolic compounds present in figs have demonstrated antioxidant activity in vitro through their ability to regulate reactive oxygen and nitrogen species, which are continuously created during regular processes of metabolism.

15.30 Olive

The evergreen olive tree (*Olea europaea*, Oleaceae), which grows in the Mediterranean region's temperate climate, yields olive fruit. One of the mainstays of the Mediterranean diet, table olives are a unique Greek commodity. It has been used to identify sources of phenolic chemicals with significant biological characteristics (Boskou 2010). According to Romero (2004), tyrosol, hydroxytyrosol, and oleanolic acid are the three main phenolic chemicals found in table olives. Benavente-Garcıa et al. (2000) demonstrated that oleuropein, another phenolic chemical in this fruit, can effectively combat a variety of parasites, viruses, yeasts, fungi, molds, and retroviruses.

Olive oil, another core component of the Mediterranean diet, has been associated with the primary prevention of multiple long-term illnesses, including cancer, diabetes, and cardiovascular disease. It is a good source of polyphenols and monounsaturated fatty acids, among other bionutraceuticals. Many antiatherogenic, anticancer, and antidiabetic properties of olive oil are attributed to components including hydroxytyrosol, tyrosol, oleocanthal, and oleuropein, mainly because of their antioxidation and anti-inflammatory properties (Alkhatib 2020).

15.31 Jamun

Jamun (*Syzygium cumini*) also known as black plum, is an excellent source of bioactive components. According to Rizvi et al. (2022), jamun peel, pulp, and seed are all high in antioxidants, polyphenols, flavonoids, minerals, vitamins, and phytochemicals. Several research have confirmed jamun's antioxidative (Swami et al. 2012), antidiabetic (Helmstadter 2008), and anti-inflammation (Sagrawat et al. 2006) properties in the treatment of metabolic syndrome. Grover et al. (2000) and Sisodia and Bhatnagar (2009) reported the stomachic, diuretic, hepatoprotective, tonic, and hematinic properties of jamun due to the presence of active principles viz., jamboline, a pale yellow essential oil, and ellagic acid.

15.32 Passionfruit

Passifloraceae is a family of plants that are grown throughout the world in tropical and subtropical climates. Higher concentrations of vitamin C, total phenolic compounds, total carotenoids, and antioxidant activity have all been reported in *Passiflora edulis* (Barbosa Santos et al. 2021). Consequently, it has been reported that pulp from passion fruit is an excellent source of bioactive substances including saponins as well as secondary metabolites like phenols, carotenoids, flavonoids, and tannins (Dos Reis et al. 2018).

15.33 Phalsa

The underutilized native minor berry known as Phalsa (*Grewia asiatica* L.) is a little fruit in the Tiliaceae family with significant nutritional and therapeutic value. Along with a variety of bioactive substances, it is a good supplier of amino acids, proteins, vitamins A and C, minerals, phosphorus, and iron. Vitamin C, sugars, acids, phenols, alkaloids, tannins, flavonoids, saponins, glycosides, steroids, and mucilaginous components were all confirmed to be present in phalsa fruit by Gupta et al. (2006).

According to Singh et al. (2009), the fruit's deep purple color is mostly caused by the presence of anthocyanins, which are significant bioactive substances. Phalsa fruits have a well-established credibility for treating a wide range of illnesses, including fever, blood, respiratory, and cardiac issues. They also have properties that make them antioxidant, antidiabetic, anticancer, antirheumatic, antipyretic, hypoglycemic, radioprotective, anti-inflammatory, antiulcer, antimalarial, and hepatoprotective (Ray and Bala 2019).

15.34 Acerola

Malpighia emarginata, or acerola, is a plant that is rich in ascorbic acid and a variety of other phytonutrients, including flavonoids, anthocyanins, phenolics, and carotenoids. Scientists and pharmaceutical companies have become increasingly interested in this fruit in the last few years. The fruit possesses an astounding amount of ascorbic acid (around 50–100 times more than orange or lemon fruit) at 1500–4500 mg/100 g. The fruit is abundant in phytonutrients, has a high capacity for antioxidants, and has a number of interesting biofunctional properties, such as the potential to whiten skin and reverse multidrug resistance (Prakash and Baskaran 2018). Mezadri et al. (2008) report that the fruit is rich in flavonoids, carotenoids, phenolics, and anthocyanins and according to De Assis et al. (2000), the fruit additionally has pro-vitamin A, vitamins B1 and B2, niacin, albumin, calcium, iron, and phosphorus.

15.35 Mulberry

Mulberry (*Morus* sp.), which is grown all over the world under various agroecological conditions, is the food source for silkworms (*Bombyx mori*). Mulberry fruits have long been utilized in China, as well as in Asia and Africa, for culinary and medicinal purposes. The fruits of the mulberry tree are found to be rich in bioactive phytochemicals with strong antioxidant effects (Bhattacharjya et al. 2021). These elements, individually or in combination, have synergistic or direct effects on diabetes, inflammation, tumors, hepatic disorders, immunomodulation, hyperlipidemia, brain damage, and chronic diseases. Zhang et al. (2018) identified mulberry fruits as a storehouse of polyphenols which include flavonoids (anthocyanins and flavonols) and phenolic acids (benzoic acid and hydroxycinnamic acid).

As people age, their bodies produce less melatonin, a neurohormone that controls their circadian rhythm. A recent study demonstrated that people may successfully absorb and excrete plant melatonin, which has anti-free radical and growth-promoting properties. Therefore, melatonin deficits can be overcome, and human health is improved by eating plant-available melatonin through food supplements. High concentrations of melatonin, which has been connected to anti-inflammatory and antioxidant properties, have been discovered in mulberry fruit (Yuan and Zhao 2017).

15.36 Prickly Pear

Due to its pleasant organoleptic qualities, the prickly pear (*Opuntia ficus-indica* (L.) Mill.), which is native to Mexico, has been introduced to various areas of the globe, including Australia, Africa, and the Mediterranean basin. The Cactaceae family, which includes the prickly pear, is home to about 1500 different species of cacti. Additionally, it is becoming more well-known to consumers because of the profusion of bioactive substances like carotenoids, polyphenols, vitamins, polyunsaturated fatty acids, and sterols. These substances are linked to important biological and functional activities like anti-inflammatory, antioxidant, and hypoglycemic qualities (Albuquerque et al. 2020). A fascinating source of phytochemicals, prickly pears can be used to create innovative products with health-enhancing characteristics, including functional foods, natural additives, or dietary supplements, in the food, cosmetics, or pharmaceutical industries.

15.37 Sea Buckthorn

Sea buckthorns (*Hippophae rhamnoides* L.), which belong to Elaeagnaceae family, are deciduous shrubs that are highly resilient and can tolerate temperatures as low as −43°C during winters. Despite being highly acidic, the berries are still quite nutrient-dense and packed with necessary amino acids and proteins. Additionally, they contain mineral elements namely calcium, phosphorus, iron, and particularly potassium, which is the most abundant element overall (Bal et al. 2011). Furthermore, these fruits are rich in vitamins such as tocopherols, vitamin C, and carotenoids, including lycopene, zeaxanthin, and β-carotene (Gao et al. 2000). Additional vitamins found in berries include folic acid, B1, B2 and K (Bekker and Glushenkova 2001).

Because Hippophae berries contain a high concentration of naturally occurring antioxidants, they also have some of the greatest antioxidant activity among therapeutic plants. Ascorbic acid is their main antioxidant (Rosch et al. 2003). Isorhamnetin is the flavonoid that is present in the greatest amount followed by isorham-netin-3-O-13-D-glucoside, rutin, ustzagalin, quercetin, myricetin, and kaempferol (Wang et al. 2011).

For almost a millennium, Tibetan traditional medicine has been aware of the potential benefits of Hippophae in treating cardiovascular disorders. Together with the unsaturated fatty acids in oils, the flavonoids found in the various parts of Hippophae may improve heart health (Zeb 2006), avert heart attacks (Eccleston et al. 2002), and lessen the symptoms of diabetes (Wang et al. 2011). Consuming sea buckthorn may have these advantages through reducing blood sugar, eliminating free radicals (Suomela et al. 2006), lessening low-density lipoproteins vulnerability to oxidation (Eccleston et al. 2002), and having antihypertensive effects (Wang et al. 2011).

15.38 Bael

Tropical and subtropical regions of Southeast Asia are home to large cultivation of bael (*Aegle marmelos* (L.) Correa), a member of the Rutaceae family. Bael fruit has been a prominent area of research in recent years, owing to its high nutritional value (carbohydrates, proteins, minerals, and vitamins) and the presence of numerous phytochemicals that contribute to its nutraceutical potential. These phytochemicals are alkaloids, flavonoids, phenolic acids (protocatechuic acid, gallic acid, and ellagic acid), carotenoids, pectin, tannins, coumarins, and terpenoids (Sharma et al. 2022). Rajan et al. (2011) reported that bael fruit extracts of both alcoholic and aqueous origin have great antioxidant potential.

Nutraceuticals made from fruits are becoming increasingly popular due to the recognition that they can significantly improve health and the general understanding of their benefits in day-to-day living. Details of reported nutraceuticals from different fruits are given below (Table 15.1).

Table 15.1 Various nutraceuticals present in fruits and their potential health benefits.

Fruits	Active principle	Beneficial health effects	Reference
Apple	Quercetin, catechin, gallic acid, chlorogenic acid, phloretin	Chronic diseases (CVD, asthma, and cancer)	Boyer and Liu (2004)
Plum	Amygdalin, emulsin, malic acid	Laxative, emollient, antioxidant properties, prevent atherosclerosis	Shivhare et al. (2011)
Sweet cherry	Anthocyanins (cyanidin 3-glucoside)	Antioxidant activity	Tsuda et al. (1994)
Mango	Gallic acid	Diuretic, tonic, anthelmintic, are rich in iron, thus helpful in anemia, and beta-carotene helps enhance the immune system	Wauthoz et al. (2007)
Banana	Vitamin A, tannin, minerals viz., Ca, Mg, P, S, Fe, and Cu	Demulcent, laxative, emollient, smoothens the stomach	Imam and Akter (2011)
Lime, lemon	Geraniol, citral, hesperidin glucosides	Juice is good for asthma, headaches, pneumonia, and arthritis. Blood purifier and a mild diuretic	Bertuzzi et al. (2013)
Sour orange, bitter orange	d–linalool, hesperidin glucoside	Rich in vitamin C and beta-carotene possesses anticancer properties. Helpful in asthma attacks, bronchitis, and atherosclerosis	Peixoto et al. (2012)
Grapes	Flavonoids, anthocyanins, tannins	Against cardiovascular diseases	Inti and Faoro (2006)
Guava	Eugenol, essential oil, minerals	Laxative, antibacterial property against diarrhea-causing organisms	Gutierrez et al. (2008)
Papaya	Papain, papayic acid, carpaine, carposide	Antibacterial, antiviral, anti-inflammatory, and antipyretic potential	Singh and Ali (2011)
Pineapple	Bromelin	Antidiabetic, antioxidant, hyperlipedemic properties	Khan et al. (2011)
Custard apple	Acrid	Enhances immunity, helpful in prevention of cancer, antitumor, and antibacterial property	Pandey and Barve (2011)
Pomegranate	Carotene, thiamine, riboflavin, nicotinic acid, vit. C, pectin	Lowers blood pressure and functions as an antioxidant. anti-inflammatory and protective against breast cancer	Jurenka (2008)
Aonla	Essential oil, emblic acid, phyllanthin	Possesses antiviral, antimutagenic, antiallergic, and antibacterial properties	Anibijuwon and Udeze (2009)
Strawberry	Kaempferol, gallic acid, ellagic acid, chlorogenic acid	Reduced rates of cancer, inflammation, hypertension, and cardiovascular disease-related deaths	Basu et al. (2014)
Mangosteen	Mangosim, mangostin	Astringent, febrifuge	Obolskiy et al. (2009)
Blueberry	Chlorogenic acid, secoisolariciresinol, pterostilbene	Improvements in working memory and reduced depressive symptoms	Krikorian et al. (2010)
Raspberry	Ellagic acid, quercetin, gallic acid, cyanidins, pelargonidins, catechins, kaempferol, salicylic acid	Protect against reactive oxygen species damage and reduce inflammatory mediators	Miller and Shukitt-Hale (2012)
Jamun	Jamboline, a pale yellow essential oil, ellagic acid	Stomachic, diuretic, hepatoprotective, tonic, haematinic	Sisodia and Bhatnagar (2009)
Bael	Mucilage, pectin, tannin, volatile oil, bitter principle	Astringent, laxative, stomachic	Sharma et al. (2011)

15.39 Conclusions

Health-protective functional and nutraceutical foods are gaining popularity in the nutrition industry nowadays because they support overall health while minimizing medication expenditures. Numerous studies have demonstrated the unmistakable link between eating fruits and health advantages like boosting immunity, lowering oxidative damage to cells, and preventing the development of cancer. These characteristics are linked to a variety of chemicals that scavenge free radicals, including vitamins, phenolic compounds, terpenoids, and nitrogen compounds all of which have potent antioxidant capabilities. These nutraceuticals found in fruits, if properly utilized, have the potential to increase the economic relevance of the fruits. Investigating the possible nutraceutical properties of fruits is crucial given the rapidly growing nutraceutical market.

References

Afzaal, M., Saeed, F., Bibi, M. et al. (2023). Nutritional, pharmaceutical, and functional aspects of rambutan in industrial perspective: an updated review. *Food Science and Nutrition* 11 (7): 3675–3685.

Ajila, C.M., Bhat, S.G., and Rao, P.U.J.S. (2007). Valuable components of raw and ripe peels from two Indian mango varieties. *Food Chemistry* 102: 1006–1011.

Alam, M.A., Subhan, N., Rahman, M.M. et al. (2014). Effect of citrus flavonoids, naringin and naringenin, on metabolic syndrome and their mechanisms of action. *Advances in Nutrition* 5 (4): 404–417.

Al-Alawi, R.A., Al-Mashiqri, J.H., Al-Nadabi, J.S. et al. (2017). Date palm tree (*Phoenix dactylifera* L.): natural products and therapeutic options. *Frontier in Plant Science.* 8: 845.

Albuquerque, T.G., Pereira, P., Silva, M.A. et al. (2020). Prickly pear. In: *Nutritional Composition and Antioxidant Properties of Fruits and Vegetables* (ed. A.K. Jaiswal), 709–728. Elsevier.

Alkhatib, A. (2020). Olive oil nutraceuticals and chronic disease prevention: more than an offshoot of the Mediterranean diet. In: *The Mediterranean Diet : an evidence-based approach (2nd ed.).* (ed. V.R. Preedy and R. Watson), 363–370. Elsevier.

Al-Okbi, S.Y. (2022). Date palm as source of nutraceuticals for health promotion: a review. *Current Nutrition Reports* 11 (4): 574–591.

Anderson, K.J., Teuber, S.S., Gobeille, A. et al. (2001). Walnut polyphenolics inhibit in vitro human plasma and LDL oxidation. *The Journal of Nutrition* 131 (11): 2837–2842.

Anibijuwon, I.I. and Udeze, A.O. (2009). Antimicrobial activity of *Carica papaya* (pawpaw leaf) on some pathogenic organisms of clinical origin from southwestern Nigeria. *Ethnobotanical Leaflets* 13: 850–864.

Arvaniti, O.S., Samaras, Y., Gatidou, G. et al. (2019). Review on fresh and dried figs: Chemical analysis and occurrence of phytochemical compounds, antioxidant capacity and health effects. *Food Research International* 119: 244–267.

Aschoff, J.K., Kaufmann, S., Kalkan, O. et al. (2015). In vitro bioaccessibility of carotenoids, flavonoids, and vitamin C from differently processed oranges and orange juices [*Citrus sinensis* (L.) Osbeck]. *Journal of Agricultural and Food Chemistry* 63 (2): 578–587.

Ayour, J., Sagar, M., Alfeddy, M.N. et al. (2016). Evolution of pigments and their relationship with skin color based on ripening in fruits of different Moroccan genotypes of apricots (*Prunus armeniaca* L.). *Scientia Horticulturae* 207: 168–175.

Bal, L.M., Meda, V., Naik, S.N. et al. (2011). Sea buckthorn berries: a potential source of valuable nutrients for nutraceuticals and cosmoceuticals. *Food Research International* 44 (7): 1718–1727.

Baliga, M.S., Shivashankara, A.R., Thilakchand, K.R. et al. (2019). Hepatoprotective effects of the Indian Gooseberry (*Emblica officinalis* Gaertn): a revisit. In: *Dietary Interventions in Liver Disease* (ed. R.R. Watson and V.R. Preedy), 193–201. Academic Press.

Bangar, S.P., Sharma, N., Kaur, H. et al. (2022). A review of sapodilla (*Manilkara zapota*) in human nutrition, health, and industrial applications. *Trends in Food Science and Technology* 127: 319–334.

Barbosa Santos, T., De Araujo, F.P., Neto, A.F. et al. (2021). Phytochemical compounds and antioxidant activity of the pulp of two brazilian passion fruit species: *Passiflora cincinnata* Mast. and *Passiflora edulis* Sims. *International Journal of Fruit Science* 21 (1): 255–269.

Bartolini, S., Leccese, A., and Viti, R. (2015). Quality and antioxidant properties of apricot fruits at ready-to-eat: Influence of the weather conditions under Mediterranean coastal area. *Journal Food Process Technology* 7 (1): 1–6.

Basu, A., Nguyen, A., Betts, N.M. et al. (2014). Strawberry as a functional food: an evidence-based review. *Critical Reviews in Food Science and Nutrition* 54 (6): 790–806.

Beck, K., Conlon, C.A., Kruger, R. et al. (2010). Gold kiwifruit consumed with an iron-fortified breakfast cereal meal improves iron status in women with low iron stores: a 16-week randomised controlled trial. *British Journal of Nutrition* 105 (1): 101–109.

Bekker, N.P. and Glushenkova, A.I. (2001). Components of certain species of the Elaeagnaceae family. *Chemistry of Natural Compounds* 37: 97–116.

Benavente-Garcıa, O., Castillo, J., Lorente, J. et al. (2000). Antioxidant activity of phenolics extracted from *Olea europaea* L. leaves. *Food Chemistry* 68 (4): 457–462.

Bento, C., Goncalves, A.C., Silva, B. et al. (2022). Peach (*Prunus persica*): phytochemicals and health benefits. *Food Reviews International* 38 (8): 1703–1734.

Bertuzzi, G., Tirrillin, B., Angelini, P., and Venanzoni, R. (2013). Antioxidant action of citrus limonum essential oil on skin. *Europeon Journal of Medicinal Plants* 3 (1): 1–9.

Bhardwaj, A., Satpathy, G., and Gupta, R.K. (2014). Preliminary screening of nutraceutical potential of *Annona squamosa*, an underutilized exotic fruit of India and its use as a valuable source in functional foods. *Journal of Pharmacognosy and Phytochemistry* 3 (2): 172–180.

Bhattacharjya, D., Sadat, A., Dam, P. et al. (2021). Current concepts and prospects of mulberry fruits for nutraceutical and medicinal benefits. *Current Opinion in Food Science* 40: 121–135.

Biesalski, H.K. (2001). Nutraceuticals: the link between nutrition and medicine. In: *Nutraceuticals in Health and Disease Prevention* (ed. K. Kramer, P.P. Hoppe, and L. Packer), 1–26. New York: Marcel Dekker Inc.

Boskou, G. (2010). Antioxidant capacity and phenolic profile of table olives from the Greek market. In: *Olives and Olive Oil in Health and Disease Prevention* (ed. V.R. Preedy and R.R. Watson), 925–934. Academic Press.

Boyer, J. and Liu, R.H. (2004). Apple phytochemicals and their health benefits. *Nutrition Journal* 3: 1–15.

Chena, J.P., Tai, C.Y., and Chen, B.H. (2004). Improved liquid chromatographic method for determination of carotenoids in Taiwanese mango (*Mangifera indica* L.). *Journal of Chromatography* 1054: 261–268.

Colquhoun, D.M., Moores, D., Somerset, S.M. et al. (1992). Comparison of the effects on lipoproteins and apolipoproteins of a diet high in monounsaturated fatty acids, enriched with avocado, and a high-carbohydrate diet. *The American Journal of Clinical Nutrition* 56 (4): 671–677.

Crozier, A., Jaganath, I.B., and Clifford, M.N. (2009). Dietary phenolics: chemistry, bioavailability and effects on health. *Natural Product Reports* 26 (8): 1001–1043.

De Assis, S.A., Lima, D.C., and de Faria Oliveira, O.M.M. (2000). Acerola's pectin methylesterase: studies of heat inactivation. *Food Chemistry* 71 (4): 465–467.

Dixit, S., Shukla, A., Singh, V., and Upadhyay, S.K. (2021). Bioprospecting of natural compounds for industrial and medical applications: current scenario and bottleneck. In: *Bioprospecting of Plant Biodiversity for Industrial Molecules*. Wiley; Portico. https://doi.org/10.1002/9781119718017.ch3.

Donno, D., Cavanna, M., Beccaro, G.L. et al. (2013). Currants and strawberries as bioactive compound sources: determination of antioxidant profiles with HPLCDAD/MS. *Journal of Applied Botany and Food Quality* 86 (1): 1–10.

Dos Reis, L.C.R., Facco, E.M.P., Salvador, M. et al. (2018). Antioxidant potential and physicochemical characterization of yellow, purple and orange passion fruit. *Journal of Food Science and Technology* 55: 2679–2691.

Dreher, M.L. and Davenport, A.J. (2013). Hass avocado composition and potential health effects. *Critical Reviews in Food Science and Nutrition* 53 (7): 738–750.

Dutta, F., Marak, M.C., and Meena, A.K. (2017). Nutraceuticals from major fruit crops. *International Journal of Agriculture Innovations and Research* 5 (6): 1–5.

Eccleston, C., Baoru, Y., Tahvonen, R. et al. (2002). Effects of an antioxidant-rich juice (sea buckthorn) on risk factors for coronary heart disease in humans. *The Journal of Nutritional Biochemistry* 13 (6): 346–354.

Ferretti, G., Bacchetti, T., Belleggia, A. et al. (2010). Cherry antioxidants: from farm to table. *Molecules* 15 (10): 6993–7005.

Fu, C., Loo, A.E.K., Chia, F.P.P. et al. (2007). Oligomeric proanthocyanidins from mangosteen pericarps. *Journal of Agricultural and Food Chemistry* 55 (19): 7689–7694.

Gao, X., Ohlander, M., Jeppsson, N. et al. (2000). Changes in antioxidant effects and their relationship to phytonutrients in fruits of sea buckthorn (*Hippophae rhamnoides* L.) during maturation. *Journal of Agricultural and Food Chemistry* 48 (5): 1485–1490.

Garg, A., Garg, S., Zaneveld, L.J.D. et al. (2001). Chemistry and pharmacology of the citrus bioflavonoid hesperidin. *Phytotherapy Research* 15 (8): 655–669.

Garrido, M., Espino, J., Toribio-Delgado, A.F. et al. (2012). A jerte valley cherry-based product as a supply of tryptophan. *International Journal of Tryptophan Research* 5: 9–14.

Georgiev, V., Ananga, A., and Tsolova, V. (2014). Recent advances and uses of grape flavonoids as nutraceuticals. *Nutrients* 6 (1): 391–415.

Grover, J.K., Vats, V., and Rathi, S.S. (2000). Anti-hyperglycemic effect of *Eugenia jambolona* and *Tinospora cordifolia* in experimental diabetes and their effectson key metabolic enzymes involved in carbohydrate metabolism. *Journal of Ethnopharmacology* 73 (3): 461–470.

Gupta, M.K., Sharma, P.K., Ansari, S.H. et al. (2006). Pharmacognostical evaluation of *Grewia asiatica* fruits. *International Journal of Plant Sciences* 1 (2): 249–251.

Gutierrez, R.M.P., Mitchell, S., and Solis, R.V. (2008). *Psidium guajava*. A review of its traditional uses, phytochemistry and pharmacology. *Journal of Ethnopharmacology* 117: 1–27.

Han, B., Wang, T.D., Shen, S.M. et al. (2015). Annonaceous acetogenin mimic AA005 induces cancer cell death via apoptosis inducing factor through a caspase-3-independent mechanism. *BMC Cancer* 15 (1): 1–11.

Hasler, C.M. and Brown, A.C. (2009). Position of the American Dietetic Association: functional foods. *Journal of the American Dietetic Association* 109 (4): 735–746.

Hayamizu, K., Ishii, Y., Kaneko, I. et al. (2003). Effects of *Garcinia cambogia* (hydroxycitric acid) on visceral fat accumulation: a double-blind, randomized, placebo-controlled trial. *Current Therapeutic Research* 64 (8): 551–567.

Helmstadter, A. (2008). *Syzygium cumini* (L.) Skeels (Myrtaceae) against diabetes–125 years of research. *Die Pharmazie-An International. Journal of Pharmaceutical Sciences* 63 (2): 91–101.

Hyson, D.A. (2011). A comprehensive review of apples and apple components and their relationship to human health. *Advances in Nutrition* 2 (5): 408–420.

Ikram, S., Abassi, U.A., and Khalid, N. (2019). Strawberry (*Fragaria ananassa* Duch): phytochemicals, nutraceuticals and health benefits. A brief review. *World Journal of Biology and Biotechnology* 4 (3): 25–34.

Imam, M.Z. and Akter, S. (2011). *Musa paradisiaca* L. and *Musa sapientum* L.: a phytochemical and pharmacological review. *Journal of Applied Pharmaceutical Science* 1: 14–20.

Inti, M. and Faoro, F. (2006). Grape phytochemicals: a bouquet of old and new nutraceuticals for human health. *Medical Hypotheses* 67 (4): 833–838.

Ishak, K.A., Annuar, M.M., and Ahmad, N. (2017). Nano-delivery systems for nutraceutical application. In: *Nanotechnology Applications in Food* (ed. A.E. Oprea and A.M. Grumezescu), 179–202. Academic Press.

Johanningsmeier, S.D. and Harris, G.K. (2011). Pomegranate as a functional food and nutraceutical source. *Annual Review of Food Science and Technology* 2: 181–201.

Joseph, B. and Priya, R.M. (2011). Phytochemical and biopharmaceutical aspects of *Psidium guajava* (L.) essential oil: a review. *Research Journal of Medicinal Plants* 5 (4): 432–442.

Jung, H.A., Su, B.N., Keller, W.J. et al. (2006). Antioxidant xanthones from the pericarp of *Garcinia mangostana* (Mangosteen). *Journal of Agricultural and Food Chemistry* 54 (6): 2077–2082.

Jurenka, J.S. (2008). Therapeutic application of pomegranate (*Prunus granatum* L.) a review. *Alternative Medicine Review* 13 (2): 128–144.

Kelley, D.S., Adkins, Y., and Laugero, K.D. (2018). A review of the health benefits of cherries. *Nutrients* 10 (3): 368.

Khan, S.M.D., Ravikumar, V., and Neelima, K. (2011). Pharmacological intervention of the fruit of plant *Ananas comosus* acting as wound healing agent in various animal models. *International Journal Pharmacy and Technology* 3: 1807–1824.

Kim, Y., Brecht, K.J., and Talcott, S.T. (2007). Anti-oxidant phytochemical and fruit quality changes in mango (*Mangifera indica* L.) following hot-water immersion and controlled atmosphere storage. *Food Chemistry* 10: 1016.

Krikorian, R., Shidler, M.D., Nash, T.A. et al. (2010). Blueberry supplementation improves memory in older adults. *Journal of Agricultural and Food Chemistry* 58 (7): 3996–4000.

Lal, N., Chandola, J.C., Kumar, A. et al. (2002). Pharmaceutical and nutraceutical values of fruit crops for human health: an overview. *Agricultural Mechanization in Asia, Africa and Latin America* 53 (9): 9631–9644.

Lal, N., Sahu, N., Jayswal, D.K. et al. (2020). Traditional, medicinal and nutraceutical values of minor fruit: longan. *Current Journal of Applied Science and Technology* 39 (41): 59–70.

Li, L., Xu, J., Mu, Y. et al. (2015). Chemical characterization and anti-hyperglycaemic effects of polyphenol enriched longan (*Dimocarpus longan* Lour.) pericarp extracts. *Journal of Functional Foods* 13: 314–322.

Lim, T.K. (2013). *Manilkara zapota*. In: *Edible Medicinal and Non-medicinal Plants*, vol. 6 (ed. T.K. Lim), 110–118. Springer.

Mahima, M., Verma, A.K., Tiwari, R. et al. (2013). Nutraceuticals from fruits and vegetables at a glance: a review. *Journal of Biological Sciences* 13 (2): 38–47.

Manosroi, A., Chankhampan, C., Pattamapun, K. et al. (2014). Antioxidant and gelatinolytic activities of papain from papaya latex and bromelain from pineapple fruits. *Chiang Mai Journal of Science* 41 (3): 635–648.

Mazur, S.P., Nes, A.N., Wold, A.B. et al. (2014). Quality and chemical composition of ten red raspberry (*Rubus idaeus* L.) genotypes during three harvest seasons. *Food Chemistry* 160: 233–240.

McCune, L.M., Kubota, C., Stendell-Hollis, N.R. et al. (2010). Cherries and health: a review. *Critical Reviews in Food Science and Nutrition* 51 (1): 1–12.

Mezadri, T., Villaño, D., Fernández-Pachón, M.S. et al. (2008). Antioxidant compounds and antioxidant activity in acerola (*Malpighia emarginata* DC.) fruits and derivatives. *Journal of Food Composition and Analysis* 21 (4): 282–290.

Miller, M.G. and Shukitt-Hale, B. (2012). Berry fruit enhances beneficial signaling in the brain. *Journal of Agricultural and Food Chemistry* 60 (23): 5709–5715.

Morton, F.J. (1987). Fruits of warm climates. *Purdue University* 301–304.

Nanjundaswamy, A.M., Lakshminarayana, S., and Siddappa, G.S. (1964). Preparation and preservation of guava juice. *Indian Food Packer* 18 (4): 13–17.

Nigam, N., Prasad, S., and Shukla, Y. (2007). Preventive effects of lupeol on DMBA-induced DNA alkylation. *Food Chemistry and Toxicology* 10: 1016.

Obolskiy, D., Pischel, I., Siriwatanametanon, N. et al. (2009). *Garcinia mangostana* L.: a phytochemical and pharmacological review. *Phytotheraphy Research* 23 (8): 1047–1065.

Palai, S. (2023). The role of grapes in nutraceuticals and functional foods. In: *Fruits and Their Roles in Nutraceuticals and Functional Foods*, 300–334. CRC Press.

Pandey, M., Verma, R.K., and Saraf, S.A. (2010). Nutraceuticals: new era of medicine and health. *Asian Journal of Pharmaceutical and Clinical Research* 3 (1): 11–15.

Pandey, N. and Barve, D. (2011). Phytochemical and pharmacological review on *Annona squamosa* Linn. *International Journal of Research in Pharmaceutical and Biomedical Sciences* 2 (4): 1404–1412.

Patel, S. (2014). Blueberry as functional food and dietary supplement: the natural way to ensure holistic health. *Mediterranean Journal of Nutrition and Metabolism* 7 (2): 133–143.

Pavan, R., Jain, S., and Kumar, A. (2012). Properties and therapeutic application of bromelain: a review. *Biotechnology Research International* 2012: 1–6.

Peixoto, J.S., Comar, J.F., Moreira, C.T. et al. (2012). Effects of *Citrus aurantium* (bitter orange) fruit extracts and p-synephrine on metabolic fluxes in the rat liver. *Molecules* 17 (5): 5854–5869.

Prakash, A. and Baskaran, R. (2018). Acerola, an untapped functional superfruit: a review on latest frontiers. *Journal of Food Science and Technology* 55: 3373–3384.

Prior, R.L., Cao, G., Martin, A. et al. (1998). Antioxidant capacity as influenced by total phenolic and anthocyanin content, maturity, and variety of Vaccinium species. *Journal of Agricultural and Food Chemistry* 46 (7): 2686–2693.

Rajan, S., Gokila, M., Jency, P. et al. (2011). Antioxidant and phytochemical properties of *Aegle marmelos* fruit pulp. *International Journal of Current Pharmaceutical Research* 3 (2): 65–70.

Rana, S. and Bhushan, S. (2016). Apple phenolics as nutraceuticals: assessment, analysis and application. *Journal of Food Science and Technology* 53: 1727–1738.

Rao, A.V. and Snyder, D.M. (2010). Raspberries and human health: a review. *Journal of Agricultural and Food Chemistry* 58 (7): 3871–3883.

Rao, U.M., Dudekula, J.B., Bhatt, S. et al. (2023). Role of phytopharmaceuticals in inflammatory disorders. *Phytopharmaceuticals and Herbal Drugs* 433–451.

Ray, A.B. and Bala, K. (2019). Bioactive compounds and health benefits of phalsa: an underutilized fruit. *Food Bioactives: Functionality and Applications in Human Health* 109: 793–795.

Richardson, D.P., Ansell, J., and Drummond, L.N. (2018). The nutritional and health attributes of kiwifruit: a review. *European Journal of Nutrition* 57: 2659–2676.

Rizvi, M.K., Rabail, R., Munir, S. et al. (2022). Astounding health benefits of jamun (*Syzygium cumini*) toward metabolic syndrome. *Molecules* 27 (21): 7184.

Romero, L.M. (2004). Physiological stress in ecology: lessons from biomedical research. *Trends in Ecology and Evolution* 19 (5): 249–255.

Rosch, D., Bergmann, M., Knorr, D. et al. (2003). Structure—antioxidant efficiency relationships of phenolic compounds and their contribution to the antioxidant activity of sea buckthorn juice. *Journal of Agricultural and Food Chemistry* 51 (15): 4233–4239.

Russo, M., Arigò, A., Calabrò, M.L. et al. (2016). Bergamot (*Citrus bergamia* Risso) as a source of nutraceuticals: limonoids and flavonoids. *Journal of Functional Foods* 20: 10–19.

Rymbai, H., Srivastav, M., Sharma, R.R. et al. (2013). Bio-active compounds in mango (*Mangifera indica* L.) and their roles in human health and plant defence—a review. *The Journal of Horticultural Science and Biotechnology* 88 (4): 369–379.

Sagrawat, H., Mann, A.S., and Kharya, M.D. (2006). Pharmacological potential of *Eugenia jambolana*: a review. *Pharmacognosy Magazine* 2 (6): 96–105.

Sahu, P., Sushma, D., and Lal, N. (2017). Nutraceuticals profiling of queen and king varieties of pineapple (*Ananas comosus*) (Pineapple). *International Journal of Chemical Studies* 5 (3): 25–31.

Saleem, M., Afaq, F., Adhami, V.M. et al. (2004). Lupeol modulates NF-(kappa) B and PI3K/Akt pathways and inhibits skin cancer in CD-1 mice. *Oncogene* 23: 5203–5214.

Sandhu, A.K., Islam, M., Edirisinghe, I. et al. (2023). Phytochemical composition and health benefits of figs (Fresh and Dried): a review of literature from 2000 to 2022. *Nutrients* 15 (11): 2623.

Santana, L.F., Inada, A.C., Espirito Santo, B.L.S.D. et al. (2019). Nutraceutical potential of *Carica papaya* in metabolic syndrome. *Nutrients* 11 (7): 1608.

Saxena, S., Hajare, S.N., More, V. et al. (2011). Antioxidant and radioprotective properties of commercially grown litchi (*Litchi chinensis*) from India. *Food Chemistry* 126 (1): 39–45.

Sharma, G.N., Dubey, S.K., Sharma, P. et al. (2011). Medicinal values of bael (*Aegle marmelos* (L.) Corr.): a review. *International Journal of Current Pharmaceutical Review and Research* 1: 12–22.

Sharma, N., Radha, K., M. et al. (2022). *Aegle marmelos* (L.) Correa: an underutilized fruit with high nutraceutical values: a review. *International Journal of Molecular Sciences* 23 (18): 10889.

Shivhare, Y., Yadav, S., and Soni, P. (2011). Estimation of antihelminthic potential of *Prunus domestica*. *International Journal of Drug Discovery and Herbal Research* 1: 63–64.

Shui, G., Wong, S.P., and Leong, L.P. (2004). Characterization of antioxidants and change ofantioxidant levels during storage of *Manilkara zapota* L. *Journal of Agricultural and Food Chemistry* 52: 7834–7841.

Singh, D., Chaudhary, M., Chauhan, P.S. et al. (2009). Value addition to forest produce for nutrition and livelihood. *Indian Forester* 135 (9): 1271–1284.

Singh, O. and Ali, M. (2011). Phytochemical and antifungal profiles of the seeds of *Carica papaya* L. *Indian Journal of Pharmaceutical Sciences* 73 (4): 447–451.

Sisodia, S.S. and Bhatnagar, M. (2009). Hepatoprotective activity of *Eugenia jambolana* Lam. In carbon tetra-chloride treated rats. *Indian Journal of Pharmacology* 41 (1): 23–27.

Sochor, J., Zitka, O., Skutkova, H. et al. (2010). Content of phenolic compounds and antioxidant capacity in fruits of apricot genotypes. *Molecules* 15 (9): 6285–6305.

Spencer, J.P. (2010). The impact of fruit flavonoids on memory and cognition. *British Journal of Nutrition* 104 (3): S40–S47.

Suomela, J.P., Ahotupa, M., Yang, B. et al. (2006). Absorption of flavonols derived from sea buckthorn (*Hippophae rhamnoides* L.) and their effect on emerging risk factors for cardiovascular disease in humans. *Journal of Agricultural and Food Chemistry* 54 (19): 7364–7369.

Swami, S.B., Thakor, N.S.J., Patil, M.M. et al. (2012). Jamun (*Syzygium cumini* L.): a review of its food and medicinal uses. *Food and Nutrition Sciences* 1100–1117.

Tsuda, T., Watanabe, M., Ohshima, K. et al. (1994). Antioxidative activity of the anthocyanin pigments cyanidin 3-O-. beta.-D-glucoside and cyanidin. *Journal of Agricultural and Food Chemistry* 42 (11): 2407–2410.

Unlu, N.Z., Bohn, T., Clinton, S.K. et al. (2005). Carotenoid absorption from salad and salsa by humans is enhanced by the addition of avocado or avocado oil. *The Journal of Nutrition* 135 (3): 431–436.

Upadhyay, S.K. and Singh, S.P. (Eds.) (2021). *Bioprospecting of Plant Biodiversity for Industrial Molecules*. Wiley. https://doi.org/10.1002/9781119718017.

Upadhyay, S.K. and Singh, S.P. (Éds.) (2023). *Plants as Bioreactors for Industrial Molecules*. Wiley. https://doi.org/10.1002/9781119875116.

Vaganan, M.M., Ravi, I., and Mustaffa, M.M. (1998). *Banana: A Nutritive and Nutraceuticals—Rich Fruit*. Thiruchirapalli: National Research Centre for Banana.

Wang, B., Lin, L., Ni, Q. et al. (2011). *Hippophae rhamnoides* Linn. for treatment of diabetes mellitus: a review. *Journal of Medicinal Plants Research* 5 (13): 2599–2607.

Wauthoz, N., Balde, A., Balde, E.S. et al. (2007). Ethnopharmacology of *Mangifera indica* L. bark and pharmacological studies of its main C-glucosylxanthone, mangiferin. *International Journal of Biomedical and Pharmaceutical Sciences* 1 (2): 112–119.

Weng, K. K. (2010). Lycopene content and antioxidant properties of pink guava industry by-products. Master of science thesis. Universiti Putra Malaysia.

Yang, B., Zhao, M., Shi, J. et al. (2008). Effect of ultrasonic treatment on the recovery and DPPH radical scavenging activity of polysaccharides from longan fruit pericarp. *Food Chemistry* 106 (2): 685–690.

Yao, P., Gao, Y., Simal-Gandara, J. et al. (2021). Litchi (*Litchi chinensis* Sonn.): a comprehensive review of phytochemistry, medicinal properties, and product development. *Food and Function* 12 (20): 9527–9548.

Yi, Y., Zhang, M.W., Liao, S.T. et al. (2012). Structural features and immunomodulatory activities of polysaccharides of longan pulp. *Carbohydrate Polymers* 87 (1): 636–643.

Yuan, Q. and Zhao, L. (2017). The mulberry (*Morus alba* L.) fruit: a review of characteristic components and health benefits. *Journal of Agricultural and Food Chemistry* 65 (48): 10383–10394.

Zeb, A. (2006). Anticarcinogenic potential of lipids from hippophae; evidence from the recent literature. *Asian Pacific Journal of Cancer Prevention* 7 (1): 32.

Zebrowska, J., Gawronski, J., Kaczmarska, E. et al. (2016). Nutraceutical value of the new strawberry cultivars. *Acta Scientiarum Polonorum. Hortorum Cultus* 15 (3): 71–82.

Zhang, H., Ma, Z.F., Luo, X. et al. (2018). Effects of mulberry fruit (*Morus alba* L.) consumption on health outcomes: a mini-review. *Antioxidants* 7: 69.

16

Nutraceutical Potential of Staple Food Crops

Aroma Joshi[1], *Arun Kumar Gupta*[1,2], *Avinash Kumar Jha*[3], *Bindu Naik*[1], *Vijay Kumar*[4], *and Sarvesh Rustagi*[5]

[1] *Department of Food Science & Technology, Graphic Era (Deemed to be University), Dehradun, Uttarakhand, India*
[2] *Graphic Era Hill University, Dehradun, Uttarakhand, India*
[3] *Department of Food Technology and Nutrition, School of Agriculture, Lovely Professional University, Phagwara, Punjab, India*
[4] *Himalayan School of Biosciences, Swami Rama Himalayan University, Dehradun, Uttarakhand, India*
[5] *Department of Food Technology, SALS, Uttaranchal University, Dehradun, Uttarakhand, India*

16.1 Introduction

In general terms, nutraceuticals are defined as food or a portion of food providing health benefits with medicinal properties, and preventing and treating diseases. Nutraceuticals can also be natural constituents, regulated as foods, and utilized as dietary supplements (Das et al. 2023; Nwosu and Ubaoji 2020; Rawat et al. 2023). The field of functional food science has given rise to a new concept called "nutraceutical," which combines "nutrition" and "pharmaceutical." However, DeFelice L. Stephen coined this term in 1989. The Greek philosopher and physician Hippocrates advocated using food-as-medicine from 370 to 460 BC, providing a historical perspective on the utilization of food for health. With the increasing healthcare expenditure and concerns over modern drugs and medicines, people are turning to bioactive molecules of plant origin for their well-being and health, creating a significant opportunity for expansion and growth in the field of functional food science (Das et al. 2023; Ranjan et al. 2023; Rather et al. 2023; Rawat et al. 2023).

The term "nutraceutical" was first used as a marketing language in Japan almost four decades ago, marking a pivotal shift in dietary preferences. These bioactive components operate on the principle that prevention is better than cure, aiming to improve the quality of human life (Jahan et al. 2019; Roy et al. 2018; Samanta et al. 2015). Cereals, as staple food crops, hold great potential for the development of nutraceuticals, with wheat, oats, buckwheat, maize, brown rice, barley, flaxseed, psyllium, and soy being commonly utilized for functional foods. These cereals contain essential nutrients that can alleviate the threat of high blood pressure, coronary heart disease, and cancer while also promoting gastrointestinal health. A diet regularly including cereals can significantly contribute to good health and disease prevention (Saikia and Deka 2011).

Cereals retain 20–80% of their original vitamin and mineral content during milling, with 90% of their fiber content found in the bran. An essential part of a healthy diet, cereals boast high-quality carbohydrate content, along with beneficial components such as fiber, antioxidants, vitamin B, and proteins (Pal and Molnár 2021). In this chapter, we discuss the nutraceutical potential of various

Herbal Nutraceuticals: Products and Processes, First Edition. Edited by Santosh Kumar Upadhyay and Sudhir Pratap Singh.

staple cereals and pseudocereals. We also explore emerging trends in staple food crop research, the integration of traditional and modern knowledge for sustainable nutrition, the importance of educating consumers about nutraceutical potential, and the significance of promoting staple food crops as nutraceutical sources.

16.2 Overview of Major Staple Food Crops and Their Nutritional Composition

For food security, cereal crop production is of fundamental importance. Among more than 50,000 staple food crops in the world, only 15 food crops offer 90% of the energy intake in the world. Rice, maize, and wheat cereal crops dominate two-thirds of the total production with over 90% produced worldwide (Stewart and Lal 2018). Millions of people around the world depend on wheat, rice, maize, and to a minor extent, sorghum, barley, millet, oats, rye, and triticale for their daily survival (Awika 2011; Horvat et al. 2020). According to numerous evaluations, it was found that cereal crops acquire an excellent amount of phenolic acids. This specifies that the intake of whole cereal grains might accord with maintaining human health. Phenolic acids, present in staple food crops, are categorized into derivatives of cinnamic acid (e.g. *p*-coumaric, ferulic, sinapic acids, and caffeic), and benzoic acid (e.g. protocatechuic, *p*-hydroxybenzoic, and vanillic acids). Most of them are bound and esterified to cell wall material in the bran. Ferulic acid, 75% of which is found in the husk of the kernel, 15% in the endosperm of the grain, and the rest in the aleurone layer, is the most abundant phenolic acid in cereals (Hung 2016; Stuper-Szablewska et al. 2019). The nutritional composition of the five highest-produced food crops is given in (Table 16.1). Cereals are majorly consumed foods and essential nutrient sources in both developing and developed countries. Cereals and cereal products are a vital source of protein, fiber, carbohydrate, and energy, and hold a range of micronutrients like vitamin E, magnesium, zinc, and some of the B vitamins. Cereals and their products may have bioactive substances with potential health benefits (Dixit et al. 2021a,b; McKevith 2004; Upadhyay and Singh 2021, 2023).

16.3 Bioactive Compounds Present in Staple Cereal Crops

Wheat is a cereal crop belonging to the family Poaceae (Gramineae) and is a vital food item worldwide (Shewry 2018). The three main components of wheat grains are endosperm, germ, and bran. Bran, the outer layer of wheat grain, contains up to 80% minerals (copper, zinc, magnesium, and iron), anthocyanin, fiber (50%), and vitamin B (Poudel and Bhatta 2017). Amber-colored wheat varieties are commonly cultivated, but due to associated health benefits, wheat rich in phytochemicals and anthocyanins is becoming popular. Cross-breeding between diverse wild wheat species and *Triticum aestivum,* the two varieties of wheat i.e. purple wheat and blue wheat, was done in the nineteenth and first half of the twentieth century respectively (Zeven 1991; Morrison et al. 2004; Buresova et al. 2015). Phytochemicals including phenolic compounds, anthocyanin, flavonoids, and carotenoids, are accountable for diverse colors in wheat (Garg et al. 2016). Anthocyanins, found primarily in the pericarp, give the grain its blue, purple, or combined colors, depending on their concentration. The yellow color comes from carotenoids in the endosperm, while the presence of phlobaphenes in the outer layer causes a red color (Lachman et al. 2017). Purple wheat is chiefly categorized by five pigments, specifically, peonidin-3- (6-malonyl glucoside), cyanidin-3-(6-malonyl glucoside), cyanidin-3-rutinoside, and cyanidin-3-glucoside. Anthocyanin concentration is highest in black wheat and reduces in order of blue-purple-red-amber wheat. Anthocyanins, which are

Table 16.1 Nutritional compositions of major food crops.

Nutritional composition	*Zea mays* (maize)	*Triticum aestivum* (wheat)	*Oryza sativa* (rice)	*Hordeum vulgare* (barley)	*Avena sativa* L. (oats)	*Sorghum bicolor* (broomcorn/ sorghum)	References
Starch	72%	60–75%	34–52%	54–75%	31.55–48.08%	32.1–72.5 g	Shevkani et al. (2017), Kumar et al. (2011), Kahlon (2009), Sampson (2013), Sterna et al. (2015), Gulvady et al. (2013), Jaćimović et al. (2023), Xiong et al. (2019)
Protein	9.5%	14%	10–16%	10–20%	10.58–15.71%	70%	
Lipids	Up to 17%	2.1–3.8%	15–22%	2–3%	5.15–9.66%	3.5%	
Dietary fibers	9.5% β-glucan: 1.5%	11.5–15.5% Arabinoxylan-5.5–7.4% beta-glucan-0.51–0.96% Cellulose-1.67–3.05% Klason-lignin-0.74–2.03%	7–11% Soluble: 1.9%	11–34% Soluble: 3–20% beta-glucan: 2–10% Arabinoxylan: 3.8–6.05%	17.63–22.97% Soluble- 14.32–17.63% β-glucan-3.15–3.29%	Insoluble fibers-75–90% Soluble fibers-10–25%	
Vitamin	B_1, B_3, B_2, B_5, B_6, E, B_9, and K	B_1, B_9, B_2, B_6, B_3, and E	B_1, B_9, B_2, B_5, B_3, B_6, tocotrienols, K, and tocopherols	B_1, B_2, B_3, tocotrienols, and tocopherols	Vitamin E, Thiamin, Riboflavin, Niacin, Vitamin K, Vitamin B_6, and folate	B_1, B_3, B_6	
Minerals	Ca, Mn, Fe, Cu, Mg, P, Se, Na, Zn	Ca, P, Mg, Fe, Zn, Na, Cu, K	Fe, Cu, Se, Mg, Zn	Ca, Fe, P, Mg, Zn, Mn, Se	Ca, Fe, Mg, K, P, Na, Zn, Cu, Mn	Mg, P, K, Ca, Na, Mn, Fe, Cu, Zn	

acylated and contain succinyl and malonyl components, are exclusively found in the purple variety of wheat (Abdel-Aal et al. 2018; Sharma et al. 2019; Garg et al. 2016).

Rice (*Oryza Sativa* L.) is one of the major and regularly consumed staple food crops by humans as well as by animals (Verma and Srivastav 2020). Rice is a rich source of numerous bioactive compounds, including anthocyanin, proanthocyanidin, caffeic acid, ferulic acid, α–, β–carotene, lutein, stigmasterol, lycopene, β-sitosterol, capesterol, coumaric acid, vitamin E isoforms (γ–, α–, δ–tocotrienols, and tocopherols), tricin, phytic acid, and γ-oryzanol (Goufo and Trindade 2014; Hudson et al. 2000; Samyor et al. 2017). These constituents are found in higher levels in the outermost layer of the rice grain, termed bran or germ fraction. These rice nutrients and phytochemicals are akin and expected to be found in other cereals.

Maize, or corn, is widely grown worldwide, with over a billion tons produced on 185 million hectares of land. Top producers include the US, China, India, France, Indonesia, Argentina, and Brazil. Corn comes in various types, such as flour corn, waxy corn, flint corn, sweet corn, dent corn, popcorn, and amylo-maize. It also comes in different colors, ranging from white and yellow to red and purple (Singh et al. 2019). The type of pigments found in corn determines its color. Blue corn contains anthocyanins from cyanidin and malvidin, while red corn contains anthocyanins from cyanidin, malvidin, and pelargonidin. The yellow color in corn comes from carotenoids, which are also known as xanthophylls. The type and genotype of corn determine the carotenoid present in it. Yellow corn has more carotenoids than floury corn. The major carotenoids in corn include lutein and zeaxanthin, along with α-carotene, α-cryptoxanthin, β-carotene, and β-cryptoxanthin. Different types of corn grains contain various phenolic compounds, including vanillic, protocatechuic, sinapic, syringic, *p*-hydroxybenzoic, caffeic, *p*-coumaric, isoferulic and ferulic acids, quercetin, cyanidin-3-O-glucoside, and kaempferol. All these phenolic compounds are primarily concentrated in the embryo/germ and pericarp of cereal grains, with the pericarp containing the highest content of total phenolics (Das and Singh 2016; Kandil et al. 2012; Thakur et al. 2017). In other grains and maize, ferulic acid is an essential phenolic compound that is especially prominent in purple corn varieties. Consumption of maize provides numerous therapeutic properties, including anti-inflammatory, neuroprotective, antidiabetic, hepatoprotective, antiatherogenic, radioprotective, and antiaging effects (Cuevas Montilla et al. 2011; Srinivasan et al. 2007).

Barely grains present bioactive components such as protein, fiber, vitamins, phytochemicals, and minerals. Remarkably, among the many bioactive compounds found in barley, the fiber component, particularly β-glucan fiber, is primarily responsible for the health benefits of barley (Agostini et al. 2015; Baik and Ullrich 2008; Thondre et al. 2011). In barley, dietary fiber is a vital component that serves health benefits. However, no sufficient evidence is available that suggests that phytochemicals can also prevent chronic diseases (Seo et al. 2015). Barley comprises several phytochemicals that are beneficial for health, including phenolic acids, lignans, flavonoids, tocols, folates, and sterols (Malik 2012). In addition, this cereal grain has some exclusive phytochemical properties, like the presence of all eight tocols vitamins, which are usually incomplete in some cereals (Idehen et al. 2017).

Worldwide, oats are one of the vastly consumed cereal grains, as the consumption of oats provides numerous health benefits. They are tremendous sources of bioactive compounds like β-glucans, amino acids, dietary fibers, peptides, resistant starch, protein, avenanthramides, polyphenols, β-sitosterol, and saponins (Tang et al. 2023).

Sorghum also known as *Sorghum moench* is the fifth main cereal worldwide in terms of food crop. Sorghum can be moderately consumed as a substitute for wheat flour, as it does not contain gluten and fulfills the necessity of people having celiac disease (Bin 2007; Cao et al. 2017; Cole et al. 2017; Hermuth and Kosova 2017; Sebayang et al. 2017; Windpassinger et al. 2015; Zhang et al. 2016). This cereal grain contains numerous bioactive compounds including vitamins (B, D, E, and K), polyphenols, and carotenoids (De Morais et al. 2017; Dykes and Rooney 2006; Nogowska et al. 2000). Sorghum is especially rich in phenolic acids including tannins, polycosanols, 3-deoxyanthocyanidins, vitamin E, phytosterols, stilbenoids, and amines (De Morais et al. 2017; Li et al. 2021). Many research studies proved that sorghum has several biological activities including antidiabetic, antiobesity, anti-inflammatory, and anticancer properties (Li et al. 2021). Sorghum is the only source of 3-deoxyanthocyanidins among the food sources (Rizzello et al. 2015).

Millets are cereal food crops with small grains that are known for their ability to withstand drought and adverse weather conditions, requiring minimal use of chemical materials like fertilizers and pesticides. Additionally, millets are gluten-free and have a low glycemic index. They are abundant sources of minerals and dietary fiber such as iron, calcium, and phosphorus. Only a limited number of millets, such as jowar (sorghum), bajra, a whole and decorticated pearl millet grain, ragi (finger millet), and amaranth, have their specific requirements under the Food Safety and Standards (Food Product Standards and Food Additives) Regulations of 2011 (Rawat et al. 2023). Millets are an excellent source of iron, potassium, calcium, zinc, and magnesium. Millet contains vitamin B_6, folic acid, niacin, and phytochemicals like phytate that help to prevent cancer as they have antibacterial and antioxidant properties, and phytic acid that can facilitate lower cholesterol (Amadou et al. 2011; Rizzello et al. 2015). In terms of other cereals, millets play an equivalent role when it comes to their protein content, carbohydrates, and calories. Millets contain a high amount of phytochemicals like flavonoids, phytic acids, phenolic acids, phytosterols, and catechins (Amadou et al. 2011). Millets are an abundant source of phenolic acid, phosphorus, and dietary fiber (Das and Singh 2015). Major millets that are grown and consumed worldwide are proso millet (*Panicum miliaceum*), pearl millet (*Pennisetum glaucum*), finger millet (*Eleusine coracana*), and foxtail millet (*Setaria italica*) (Himanshu et al. 2018).

16.4 Health Benefits of Staple Food Crops

Technology has allowed people to boost land productivity since the Neolithic era, which has helped to fulfill the rising need for food. Most industrialized nations saw food surpluses by the second half of the twentieth century because of the widespread exponential growth in the yields of major food crops that began in the 1950s and was facilitated by advancements in plant breeding, agricultural sciences, and the use of inorganic fertilizers and pesticides. However, most productive soils were exhausted as a result of this lengthy, intense exploitation, which degraded the environment. The fact that the population of the world is foreseen to touch nine billion by 2050 and that a growing percentage of them will demand diets higher in calories and protein, only serves to heighten concerns. The population of the planet is expected to grow unevenly. For example, it is predicted that by 2050, the population of Africa will have doubled to almost two billion people. In addition, it is anticipated that the population of cities will increase more quickly than that of rural areas, with low-income nations experiencing the most urbanization. As a result, fewer people will be working in agriculture. Meeting the demand for food came out as a serious issue when considering the consequences of change in climate while producing food (higher temperatures, more extreme and frequent events of weather, and increasing threat of drought and flooding).

Studies associated with epidemiology stated that consuming barely regularly could decrease the threat of specific diseases including gallstones, chronic heart disease, high blood pressure, and colonic cancer (Bays et al. 2011; Behall et al. 2006; Dongowski et al. 2002; Hoang et al. 2011; Sullivan et al. 2013). Numerous studies have testified to the beneficial effects of barley on human health. Barley has been found to play a key role in maintaining a healthy colon, boosting the immune system, inducing immunostimulation, and much more (Idehen et al. 2017; Kanauchi et al. 1999; Tada et al. 2009). It is worth noting that barley is a rich source of a bioactive soluble fiber called β-glucan. β-glucan has been extensively studied over the past few decades and is believed to have numerous health benefits, including cardioprotective, antioxidative, antiobesity, antidiabetic, anti-inflammatory, and anticancer effects (Li et al. 2022). Phenolic compounds are essential for growth and reproduction, act as defense mechanisms against parasites, predators, and pathogens, and contribute to the color of plants (Dykes and Rooney 2007; Mithöfer and Boland 2012). Sterols and tocols are mainly present in plant oils and offer benefits including defense against toxins, and neurological diseases such as Alzheimer's disease, and diabetes (Bartłomiej et al. 2012).

With the presence of these bioactive compounds, this cereal grain exhibits antiobesity, antioxidant, anticancer, antihyperlipidemic, antidiabetic, anticancer, cardioprotective, and immunomodulatory properties (Tang et al. 2023). Avenanthramides and β-glucans ameliorate the immune system; expel detrimental substances from the body, enhance the breaking down of fat and lipid profile in the body, and lower blood cholesterol. β-glucans helps to regulate insulin secretion which prevents diabetes. Progladins lower cholesterol and blood sugar levels, reduce inflammation, and improve skin health. Saponins and flavone glycosides boost immune function, prevent skin infiltration, and control inflammation. Lignin and phytoestrogen are known to prevent hormone-related cancer and improve the quality of life for postmenopausal women. Sprouted oats detoxify the liver with saponarin (Kim et al. 2021).

Sorghum is rich in polyphenolic compounds due to which it has excellent antioxidant properties to fight diseases associated with antiproliferative properties, and oxidative stress, and enhances glucose metabolism (Espitia-Hernández et al. 2020). Consumption of maize provides numerous therapeutic properties, including antiatherogenic, anti-inflammatory, antidiabetic, neuroprotective, antiaging, hepatoprotective, and radioprotective effects (Cuevas Montilla et al. 2011; Srinivasan et al. 2007). Lutein has antitumor-promoting activity and has been shown to suppress the growth of tumors in mice. Zeaxanthin and lutein have been allied with preventing age-linked macular degeneration, a syndrome that causes early blindness in humans (de la Parra et al. 2007). Anthocyanins present in blue, red, and purple, blue-pigmented corn have been found to have antimutagenic and free radical scavenging activities (Singh et al. 2019). In pigmented rice, the anthocyanins have high scavenging and antioxidant activity. Further research and development studies on rice, both pigmented and nonpigmented varieties, will help elucidate the compound mechanisms present in them. These compounds have diverse roles, including antiallergic, antioxidant, anti-influenza, free radical scavenging, antiatherosclerosis, antiobesity, and antitumor actions against various human degenerative and chronic diseases (Verma and Srivastav 2020). Colored wheat serves various health advantages and has been confirmed to help avert and battle several chronic illnesses like cancer, diabetes, cardiovascular disease, obesity, aging, and inflammation. They have also proven to have a therapeutic influence against hyperglycemia, oxidative liver damage, and capillary fragility (Ficco et al. 2014). Among all the bioactive compounds present in oats, beta-glucan, a dietary fiber has lipid-lowering and hypoglycemic effects that can decrease the threat of obesity, cardiovascular diseases, and diabetes (Nguyen et al. 2020; Salmeron 2017). The latest research work/studies exhibit that polyphenols have antioxidant, antithrombotic, and anti-inflammatory activities, which enhance the health of the intestine, reduce plasma inflammation, and participate in cell transduction signal pathways (Nignpense et al. 2021).

Dietary fiber and phenolic acids are known to aid or prevent several diseases such as cataractogenesis, diabetes, and cardiovascular diseases. Phosphorus is an essential mineral to generate energy in the human body, as it is a vital part of ATP. Due to their nonacid-producing and nonglutinous nature as well as having high nutrition, millets are easily digestible (Das and Singh 2015; Himanshu et al. 2018). Millets are called one of the least allergic and easy-to-digest staple food crops (Rizzello et al. 2015). The major health benefits of consuming millet regularly are that it heals ulcers, lowers cholesterol, and lowers blood sugar (Das et al. 2014). Magnesium present in staple food crops helps to enhance nutrient delivery, enhances blood flow, relaxes blood vessels, and protects cardiovascular system. It also lowers lipids and enhances insulin sensitivity (Himanshu et al. 2018). A great source of magnesium, consuming these cereal grains can reduce the acuteness of asthma and the occurrence of migraine effects. Studies have shown that this can lower blood pressure and decrease the threat of heart attacks, especially in those with diabetes or atherosclerosis (Srivastava et al. 2012). Ferulic acid, found in these grains, has numerous health benefits such as reducing total cholesterol levels, increasing vitamin E absorption, improving sperm vitality, and protecting against UV-induced skin damage. According to Himanshu et al. (2018), ferulic acid has powerful free radical scavenging, antioxidants, and anti-inflammatory properties, which can also have positive effects on cancer and tumors.

16.5 Emerging Trends in Staple Food Crop Research

16.5.1 The Trend of Plant-based Ingredients/Foods

Recent years have seen a significant increase in the popularity of the plant-based diet trend, which has been fueled by a number of factors including innovation, worries about food security, and sustainability. But there are still a number of obstacles in their way of being more widely available, chief among them being consumer acceptance of the flavor, texture, and color of plant-based foods and ingredients.

16.5.2 Beyond Vegan and Vegetarian: Flexitarian Trend

Several consumers are accepting flexitarian diets, which include the consumption of fewer goods produced originally from animals and increasing their intake of plant-based foods, in addition to the developments toward vegetarianism and vegan. Public health guidelines supported and promoted the use of plant-based diets as healthier alternatives to animal-derived goods, citing their lower risk of obesity, cardiovascular disease, and cancer.

16.5.3 New Generation in Plant-based Alternatives

The creation of a new generation of plant-based foods that have characteristics (taste, texture, and flavor) similar to those of traditional animal-based products was primarily fueled by innovation in formulation and processing. For example, a number of businesses have made investments in flavoring and coloring products to be used in plant-based meals in order to replicate the real taste and texture of beef and chicken meat. Masking chemicals have also been useful in reducing the astringency and off-flavors derived from plants. Hydrocolloids, among other binding and thickening agents, are crucial for producing a texture similar to that of conventional products.

16.5.4 Potential for Genetic Modification to Enhance Nutraceutical Content

In the last 10 years, remarkable innovations have been achieved in the use of biotechnology to produce food crops with enhanced nutritional value. Along with new crop varieties that can fight chronic disease, biofortified staple food crops like wheat, maize, and rice, which comprise important micronutrients providing benefit to the world's impoverished, are being developed (Upadhyay 2021, 2024).

16.5.5 Nutritionally Enhanced Food Crops to Improve Global Health

Nearly 842 million people worldwide suffer from undernourishment, which is defined as insufficient food intake. Furthermore, nearly 2 billion people worldwide experience "hidden hunger," a kind of malnutrition brought on by an insufficient intake of vital micronutrients in their regular diet rather than a lack of calories. Malnourished individuals frequently eat meals that revolve around a staple crop, depriving them of access to the vast range of fruits and vegetables that are necessary for a balanced diet.

16.5.5.1 Biofortified Rice

Three million preschool-aged children suffer from eye impairment each year as a result of vitamin A deficiency. Out of them, half a million develop blindness, and two-thirds pass away soon after. Cereals like rice lack β-carotene, the precursor molecule needed for the manufacture of vitamin A. Many people who eat a primarily monotonous diet are therefore susceptible to vitamin A insufficiency. Ingo Potyrus, a research group that created golden rice, which is named for its high content of β-carotene and has a golden color. It is produced through biotechnology and provides a workable option. Two genes from other organisms—the bacteria Erwinia uredovoia and the daffodil—were inserted into the rice genome to create this transgenic crop. These genes rebuild the carotenoid biosynthesis pathway. Subsequently, through marker-assisted backcross breeding, this novel characteristic was incorporated into highly productive local commercial varieties. GR2, the most recent iteration of golden rice technology, substitutes the daffodil gene utilized in GR1 rice with the phytoene synthesis gene from maize, drawing on genes from two different provitamin A pathways.

16.5.5.2 Biofortified Maize and Cassava

Additionally, β-carotene and other vital micronutrients required for sustaining one's health have been added to maize by biofortification. The Li et al. (2010) study measured the triglyceride-rich lipoprotein fraction of blood from female North American volunteers who consumed maize porridge that was biofortified. The researchers discovered that the biofortified corn had a β-carotene vitamin A equivalency value that was 3.1 times more than that of regular white porridge maize. In a related trial with males from Zimbabwe, biofortified yellow maize porridge was found to supply 40–50% of the recommended daily allowance of vitamin A in the US. The goal of the BioCassava Plus initiative is to improve the nutritional status of cassava, which is a staple food for a quarter of a billion people living in sub-Saharan Africa. A porridge made of cassava with a high β-carotene content was given to healthy subjects. These subjects' blood samples showed that biofortified cassava raises the amounts of retinyl palmitate and β-carotene in TRL plasma.

16.5.5.3 Biofortified Wheat

Biotechnology has been used to modify wheat for a variety of health advantages. For instance, RNAi-based technologies have been used to enhance the amount of free lysine, an important amino acid that is typically lacking in wheat, and decrease the quantities of gliadins, the protein that causes celiac disease, in wheat. Experiments on the quality and flavor of dough made from genetically modified wheat have produced promising results. For the fraction of the population that is gluten-sensitive or intolerant, biofortified wheat offers more options. It can also give people in developing nations that rely heavily on wheat as a staple better levels of minerals like iron and zinc. Another crop that has been researched as a model for zinc biofortification is wheat. Millions of people worldwide, whose diet primarily consists of cereal grains low in zinc content, suffer from zinc (Zn) insufficiency, which is the sixth most common cause of disease in low-income nations. A zinc shortage can lead to birth difficulties, impaired brain development, low immunity, and limited growth. Recent advancements in nitrogen (N) management have made it possible to increase zinc concentrations in grain crops like wheat by applying zinc topically as well as through increased soil availability. It has been demonstrated that radiolabeled 65Zn is absorbed by the roots, moved to the shoots, and accumulated in the wheat grain. Increased soil nitrogen availability has been shown by Erenoglu et al. (2011) to promote zinc translocation from roots to shoots and improve zinc accumulation in wheat grains, potentially by raising the quantity of transporter proteins in the presence of nitrogen.

16.5.5.4 Nutritionally Enhanced Feed Crops

Crops that have been nutritionally modified are intended to improve animal and poultry feed. For instance, higher levels of limiting amino acids are produced by animal feed crops, reducing the need for supplementation. Additionally, feed crops have been created to produce manure that is more environmentally friendly. For instance, swine and poultry fed transgenic maize with higher levels of free lysine experienced a similar rise in body weight gain as animals given diets supplemented with lysine. There was also an increase in the quantity of other vital amino acids, like lysine and aspartate. These kinds of tactics can improve the nutritional content of feeding crops that are fed to cattle. Additionally, transgenic maize crops that express cell wall invertase have been created. The transgenic maize's increased grain size and quantity resulted in a significant increase in grain yield (up to 145.3%) when compared to the wild varieties. Biotechnology has been investigated in the elimination of antinutrients from animal feed to improve micronutrient uptake. For instance, transgenic maize that produced a high level of β-glucanase Bgl7AM (779,800 U/kg) was shown to be stable in the pH range of 1.0–8.0, which is typical for the digestive tract.

16.5.5.5 Integrating Traditional and Modern Knowledge for Sustainable Nutrition

The way we manufacture and intake food is not sustainable for our planet, and we are not moving quickly enough toward a world without hunger and a healthy diet for everyone. As per the United Nations' Food and Agriculture Organization (FAO), the number of hungry people has increased by around 2% from 2019 to 828 million people now and three billion people cannot afford a decent diet (FAO, IFAD, UNICEF, WFP and WHO, 2022). Four crops accounted for more than half of world production in 2019: sugar cane (1.9 billion tons), maize (1.1 billion tons), rice (0.8 billion tons), and wheat (0.8 billion tons), although thousands of species have been used as food. The speediness and scope of rising fuel, food, and fertilizer expenses in the wake of recent world events, along with the ensuing shortages, have highlighted the urgent requirement for global effort coordination toward increasing diversity in diets, production systems, and markets, and illustrated the profound effects of our reliance on a small number of crops and commodities.

Approximately, 80% of deforestation and 30% of greenhouse gas emissions are caused by our food systems. 34% of the planet's land is used for agriculture, while 70% of its freshwater is extracted. This puts 86% of at-risk species in danger. With 34% of fish stocks at a biologically unsustainable level, oceans are severely overfished. Examples exist that show how to go on with developing the kinds of food systems we require. Several studies have confirmed that food biodiversity, or the diversity of plants, animals, and other organisms used as food, including the genetic resources found within species, between species, and provided by ecosystems, is one way to address the global issues of food insecurity, malnutrition, and unsustainable food production. Most people think that to include more food species in sustainable food system strategies, food system transformation must place a larger emphasis on food biodiversity.

16.6 Importance of Educating Consumers About the Nutraceutical Potential

Nutraceuticals, food or food components that provide health benefits beyond their basic nutritional value, have gained significant popularity in recent years. As consumers seek natural ways to enhance their health and well-being, the demand for nutraceuticals continues to grow. However, this growing interest also highlights the need for robust consumer education to ensure informed decision-making and responsible use of these products.

16.6.1 Significance of Consumer Education

1) ***Informed decision-making***: Educating consumers about nutraceuticals empowers them to make informed choices about their health. Understanding the potential benefits, risks, and interactions of nutraceuticals allows individuals to assess whether these products align with their health goals and preferences (Boccia and Punzo 2020).
2) ***Responsible use***: Consumer education promotes responsible use of nutraceuticals, minimizing potential misuse or overuse. By understanding appropriate dosages, potential side effects, and interactions with medications, consumers can avoid adverse health outcomes (Teoh et al. 2021).
3) ***Countering misinformation***: The rise of nutraceuticals has also led to the spread of misinformation and misleading claims. Consumer education helps individuals identify credible sources of information and critically evaluate health claims made about nutraceuticals (Teoh et al. 2019).

16.6.2 Strategies for Effective Consumer Education

1) ***Transparent labeling***: Clear and transparent labeling of nutraceutical products is crucial for consumer understanding. Labels should provide accurate and easily accessible information about ingredients, dosages, potential side effects, and interactions with medications.
2) ***Health claim regulation***: Regulating health claims on nutraceutical products ensures that these claims are based on sound scientific evidence and clearly distinguish between substantiated and unsubstantiated claims.
3) ***Healthcare professional involvement***: Healthcare professionals play a key role in educating consumers about nutraceuticals and providing individualized advice based on individual needs and health conditions.
4) ***Public awareness campaigns***: Public awareness campaigns can be effective in disseminating accurate information about nutraceuticals and promoting responsible use. These campaigns should employ various communication channels, including traditional media, social media, and community outreach programs.
5) ***Educational resources***: Developing and disseminating educational resources, such as brochures, websites, and online courses, can provide consumers with in-depth knowledge about

nutraceuticals. These resources should be accessible, engaging, and tailored to different audiences (Hilton 2017; Losasso et al. 2012; Turvey et al. 2021).

16.7 Importance of Promoting Staple Food Crops as Nutraceutical Sources

Owing to the constantly evolving human lifestyle, oxidative stress is frequently caused by an overloading of the antioxidant defense systems. Additionally, as people age, their levels of antioxidant defense mechanisms decline noticeably. Numerous diseases could arise as a result of them. As a result, research during the previous few decades has mostly concentrated on various nutraceuticals. Products containing antioxidants can either naturally scavenge free radicals (such as vitamins and polyunsaturated fats) or they can specifically boost the body's defenses (Maurya et al. 2021; Sachdeva et al. 2020). The possible benefits and drawbacks of nutraceuticals for healthy people are discussed in this chapter. However, the main factors influencing a person's vulnerability to any given disease are their genetic makeup and unhealthy lifestyle choices, such as heavy drinking and smoking. Therefore, each person's reaction to nutraceuticals may be different. When used in moderation within the appropriate recommended dietary intakes, nutraceuticals have been shown to provide numerous health advantages and can prevent disease, and promote general well-being in people.

The relationship between nutrition and disease is being hailed by scientists and nutritionists as a recent finding, after the field has mostly ignored nutrition for the past century, even going so far as to call anyone who claims to have health advantages from vitamins or other foods as a quack or food faddist. For scientists who have been dismissing the first-hand accounts and positive anecdotal reports of millions of consumers worldwide for the past 50 years, this may be a groundbreaking discovery, but for food faddists, whose views on nutrition have historically been far more stable than those of scientists, not much has changed. The best foods are natural ones. Nutrients such as vitamins guard against illnesses. This is nothing new for food fadists. But science deserves credit for at least paying attention to the food faddists. Food marketers' perception of functional foods as a means of promoting junk food is a concerning development that casts serious doubt on the amount of true nutritional research that has been done in recent decades, even though current trends in nutrition are undoubtedly a welcome departure from the pessimistic, oversimplified, and biased approach of most nutrition research conducted a century ago. It appears that there are too many functional foods that are merely gimmicks.

The task at hand is for public health officials to unequivocally show that the public's health comes before the financial interests of massive food, chemical, and pharmaceutical corporations. In this regard, it is interesting to observe that the government is pushing for regulation reforms in the food business that increasingly favor political meddling and are not optimal for customers or public health.

16.8 Future Scope and Conclusion

The diverse range of staple food crops such as wheat, maize, oats, rice, millet, sorghum, and barley covers most of the essential nutrients, which are important for maintaining human health. These nutrients and bioactive compounds are majorly found in the endosperm, bran, and germ of food crops. The presence of bioactive compounds such as phytochemicals, polyphenols (phenolic acids, flavonoids, and phytic acids), vitamins, minerals, amino acids, dietary fibers, and proteins contribute to various health benefits. There are some health benefits that are supported by many research studies. Although finite literature and research data are available on epidemiology and experimental studies produced from human studies. Emerging research suggests that a whole-grain cereal diet can prevent cancer, heart disease, diabetes, and other illnesses. Epidemiological studies also specify that the intake of food crops regularly offers benefits beyond basic nutrition. The fallouts of these studies show that the more polyphenols (phenolic acids and flavonoids) and fibers a cereal grains diet contain, the lower the risk of chronic diseases. Although there is evidence suggesting the health benefits of consuming whole grains, there are still some inconsistencies in human intervention studies and mechanisms of action that require further research and identification. Future studies that can clarify the complex links between whole grains and health may also help to identify the nutraceutical potential of food staple crops.

References

Abdel-Aal, E.S.M., Hucl, P., and Rabalski, I. (2018). Compositional and antioxidant properties of anthocyanin-rich products prepared from purple wheat. *Food Chemistry* 254: 13–19.

Agostini, S., Chiavacci, E., Matteucci, M. et al. (2015). Barley beta-glucan promotes MnSOD expression and enhances angiogenesis under oxidative microenvironment. *Journal of Cellular and Molecular Medicine* 19: 227e38.

Amadou, I., Gbadamosi, O.S., and Le, G.W. (2011). Millet-based traditional processed foods and beverages—a review. *Cereal Foods World* 56 (3): 115.

Awika, J.M. (2011). Major cereal grains production and use around the world. In: *Advances in Cereal Science: Implications to Food Processing and Health Promotion* (ed. J.M. Awika, V. Piironen, and S. Bean), 1–13. American Chemical Society.

Baik, B.K. and Ullrich, S.E. (2008). Barley for food: characteristics, improvement, and renewed interest. *Journal of Cereal Science* 48: 233.e42.

Bartłomiej, S., Justyna, R.K., and Ewa, N. (2012). Bioactive compounds in cereal grains–occurrence, structure, technological significance and nutritional benefits–a review. *Food Science and Technology International* 18 (6): 559–568.

Bays, H., Frestedt, J.L., Bell, M. et al. (2011). Reduced viscosity barley β-Glucan versus placebo: a randomized controlled trial of the effects on insulin sensitivity for individuals at risk for diabetes mellitus. *Nutrition & Metabolism (London)* 8: 1.

Behall, K.M., Scholfield, D.J., and Hallfrisch, J. (2006). Whole-grain diets reduce blood pressure in mildly hypercholesterolemic men and women. *Journal of the American Dietetic Association* 106: 1445.e9.

Bin, T. (2007). Properties of grain sorghum and its prospects of utilization in food industry. *Cereal Feed India* 7: 9.

Boccia, F. and Punzo, G. (2020). Nutraceuticals: some remarks by a choice experiment on food, health and new technologies. *Food Research International* 130: 108888.

Burešová, V., Kopecký, D., Bartoš, J. et al. (2015). Variation in genome composition of blue-aleurone wheat. *Theoretical and Applied Genetics* 128: 273–282. https://doi.org/10.1007/s00122-014-2427-3.

Cao, W., Sun, C., Li, X. et al. (2017). Methane production enhancement from products of alkaline hydrogen peroxide pretreated sweet sorghum bagasse. *RSC Advances* 10: 5701–5707.

Cole, M.R., Eggleston, G., Petrie, E. et al. (2017). Cultivar and maturity effects on the quality attributes and ethanol potential of sweet sorghum. *Biomass and Bioenergy* 96: 183–192.

Cuevas Montilla, E., Hillebrand, S., Antezana, A., and Winterhalter, P. (2011). Soluble and bound phenolic compounds in different Bolivian purple corn (Zeamays L.) cultivars. *Journal of Agricultural and Food Chemistry* 59: 7068–7074.

Das, A.K. and Singh, V. (2015). Antioxidative free and bound phenolic constituents in pericarp, germ and endosperm of Indian dent (*Zea mays* var. indentata) and flint (*Zea mays* var. indurata) maize. *Journal of Functional Foods* 13: 363–374.

Das, A.K. and Singh, V. (2016). Antioxidative free and bound phenolic constituents in botanical fractions of Indian specialty maize (*Zea mays* L.) genotypes. *Food Chemistry* 201: 298–306.

Das, A.K., Sreerama, Y.N., and Singh, V. (2014). Diversity in phytochemical composition and antioxidant capacity of dent, flint, and specialty corns. *Cereal Chemistry* 91 (6): 639–645.

Das, T., Thakur, R., Dhua, S. et al. (2023). Processing of cereals. In: *Cereal Grains* (ed. G.A. Nayik, T. Tufail, F.M. Anjum, and M.J. Ansari), 195–223. CRC Press.

De Morais, C.L., Pinheiro, S.S., and Martino, H.H.D. (2017). Sorghum (*Sorghum bicolor* L.): nutrients, bioactive compounds, and potential impact on human health. *Critical Reviews in Food Science and Nutrition* 57: 372–390.

Dixit, S., Shukla, A., Singh, V., and Upadhyay, S.K. (2021a). Bioprospecting of natural compounds for industrial and medical applications: current scenario and bottleneck. In: *Bioprospecting of Plant Biodiversity for Industrial Molecules*. Wiley; Portico. https://doi.org/10.1002/9781119718017.ch3.

Dixit, S., Shukla, A., Singh, V., and Upadhyay, S.K. (2021b). Engineering of plant metabolic pathway for nutritional improvement: recent advances and challenges. In: *Genome Engineering for Crop Improvement*. Wiley; Portico. https://doi.org/10.1002/9781119672425.ch20.

Dongowski, G., Huth, M., Gebhardt, E., and Flamme, W. (2002). Dietary fiber-rich barley products abeneficially affect the intestinal tract of rats. *The Journal of Nutrition* 132: 3704.e14.

Dykes, L. and Rooney, L.W. (2006). Sorghum and millet phenols and antioxidants. *Journal of Cereal Science* 44: 236–251.

Dykes, L. and Rooney, L.W. (2007). Phenolic compounds in cereal grains and their health benefits. *Cereal foods World* 52 (3): 105–111.

Erenoglu, E.B., Kutman, U.B., Ceylan, Y. et al. (2011). Improved nitrogen nutrition enhances root uptake, root-to-shoot translocation and remobilization of zinc (65Zn) in wheat. *New Phytologist* 189 (2): 438–448.

Espitia-Hernández, P., González-Chávez, M.L., AscacioValdés, J.A. et al. (2020). Sorghum (*Sorghum bicolor* L.) as a potential source of bioactive substances and their biological properties. *Critical Reviews in Food Science and Nutrition* 62 (8): 2269–2280.

Ficco, D.B., De Simone, V., Colecchia, S.A. et al. (2014). Genetic variability in anthocyanin composition and nutritional properties of blue, purple, and red bread (Triticum aestivum L.) and durum (Triticum turgidum L. ssp. turgidum convar. durum) wheats. *Journal of Agricultural and Food Chemistry* 62 (34): 8686–8695.

Garg, M., Chawla, M., Chunduri, V. et al. (2016). Transfer of grain colors to elite wheat cultivars and their characterization. *Journal of Cereal Science* 71: 138–144.

Goufo, P. and Trindade, H. (2014). Rice antioxidants: phenolic acids, flavonoids, anthocyanins, proanthocyanidins, tocopherols, tocotrienols, γ-oryzanol, and phytic acid. *Food Science and Nutrition* 2 (2): 75–104.

Gulvady, A.A., Brown, R.C., and Bell, J.A. (2013). Nutritional comparison of oats and other commonly consumed whole grains. In: *Oats Nutrition and Technology* (ed. Y. Chu), 71–93. Wiley.

Hermuth, J. and Kosova, K. (2017). Characterization of the first Czech sorghum variety Ruzrok tested in the Czech Republic. *Czech Journal of Genetics and Plant Breeding* 53: 37–44.

Hilton, J. (2017). Growth patterns and emerging opportunities in nutraceutical and functional food categories: Market overview. In: *Developing New Functional Food and Nutraceutical Products* (ed. D. Bagchi and S. Nair), 1–28. Academic Press.

Himanshu, M., Chauhan, S.K., and Sonawane, S.S.A. (2018). Nutritional and nutraceutical properties of millets: a review. *Clinical Journal of Nutrition and Dietetics* 1 (1): 1–10.

Hoang, M.H., Houng, S.J., Jun, H.J. et al. (2011). Barley intake induces bile acid excretion by reduced expression of intestinal ASBT and NPC1L1 in C57BL/6J mice. *Journal of Agricultural and Food Chemistry* 59: 6798.e805.

Horvat, D., Šimić, G., Drezner, G. et al. (2020). Phenolic acid profiles and antioxidant activity of major cereal crops. *Antioxidants* 9 (6): 527.

Hudson, E.A., Dinh, P.A., Kokubun, T. et al. (2000). Characterization of potentially chemopreventive phenols in extracts of brown rice that inhibit the growth of human breast and colon cancer cells. *Cancer Epidemiology, Biomarkers & Prevention* 9 (11): 1163–1170.

Hung, P.V. (2016). Phenolic compounds of cereals and their antioxidant capacity. *Critical Reviews in Food Science and Nutrition* 56: 25–35.

Idehen, E., Tang, Y., and Sang, S. (2017). Bioactive phytochemicals in barley. *Journal of Food and Drug Analysis* 25 (1): 148–161.

Jaćimović, S., Kiprovski, B., Ristivojević, P. et al. (2023). Chemical composition, antioxidant potential, and nutritional evaluation of cultivated sorghum grains: a combined experimental, theoretical, and multivariate analysis. *Antioxidants* 12 (8): 1485. https://doi.org/10.3390/antiox12081485.

Jahan, F.N., Rahim, M.A., Bokhtiar, S.M., and Samanta, A.K. (2019). Potentiality of underutilized crop Dioscorea spp.: a source of nutraceutical. *SAARC Journal of Agriculture* 17 (2): 1–13.

Kahlon, T.S. (2009). *Rice Bran: Production, Composition, Functionality and Food Applications, Physiological Benefits*. Boca Raton: CRC Press, Taylor and Francis Group.

Kanauchi, O., Fujiyama, Y., Mitsuyama, K. et al. (1999). Increased growth of Bifidobacterium and Eubacterium by germinated barley foodstuff, accompanied by enhanced butyrate production in healthy volunteers. *International Journal of Molecular Medicine* 3: 175.e84.

Kandil, A., Li, J., Vasanthan, T., and Bressler, D.C. (2012). Phenolic acids in some cereal grains and their inhibitory effect on starch liquefaction and saccharification. *Journal of Agricultural and Food Chemistry* 60: 8444–8449.

Kim, I., Hwang, C., Yang, W., and Kim, C. (2021). Multiple antioxidative and bioactive molecules of oats (*Avena sativa* L.) in human health. *Antioxidants* 10 (9): 1454.

Kumar, P., Yadava, R., Gollen, B. et al. (2011). Nutritional contents and medicinal properties of wheat: a review. *Life Sciences and Medicine Research* 22: 1–10.

Lachman, J., Martinek, P., Kotíková, Z. et al. (2017). Genetics and chemistry of pigments in wheat grain–a review. *Journal of Cereal Science* 74: 145–154.

Li, S., Nugroho, A., Rocheford, T., and White, W.S. (2010). Vitamin A equivalence of the β-carotene in β-carotene–biofortified maize porridge consumed by women. *The American Journal of Clinical Nutrition* 92 (5): 1105–1112.

Li, Z., Zhao, X., Zhang, X., and Liu, H. (2021). Bioactive compounds and biological activities of sorghum grains. *Foods* 10 (11): 2868.

Li, Y., Li, T., and Liu, R.H. (2022). Bioactive compounds of highland barley and their health benefits. *Journal of Cereal Science* 103: 103366.

Losasso, C., Cibin, V., Cappa, V. et al. (2012). Food safety and nutrition: improving consumer behavior. *Food Control* 26 (2): 252–258.

Malik, A. H. (2012). Governing grain protein concentration and composition in wheat and barley: use of genetic and environmental factors. Doctoral Dissertation, Swedish University of Agricultural Sciences.

Maurya, A.P., Chauhan, J., Yadav, D.K. et al. (2021). Nutraceuticals and their impact on human health. In: *Preparation of Phytopharmaceuticals for the Management of Disorders*, 229–254. Academic Press.

McKevith, B. (2004). Nutritional aspects of cereals. *Nutrition Bulletin* 29 (2): 111–142. https://doi.org/10.1111/j.1467-3010.2004.00418.x.

Mithofer, A. and Boland, W. (2012). Plant defense againts herbivores: chemical aspects. *Annual Review of Plant Biology* 63: 25–20.

Morrison, L.A., Metzger, R.J., and Lukaszewski, A.J. (2004). Origin of the blue-aleurone gene in Sebesta Blue wheat genetic stocks and a protocol for its use in apomixis screening. *Crop Science* 44 (6): 2063–2067.

Nguyen, T.T.L., Gilbert, R.G., Gidley, M.J., and Fox, G.P. (2020). The contribution of beta-glucan and starch fine structure to texture of oat-fortified wheat noodles. *Food Chemistry* 324: 8.

Nignpense, B.E., Francis, N., Blanchard, C., and Santhakumar, A.B. (2021). Bioaccessibility and bioactivity of cereal polyphenols: a review. *Foods* 10: 1595.

Nogowska, M., Jelińska, A., Muszalska, I., and Stanisz, B. (2000). Biological functions of macro- and micronutrients. *Farmacja Polska* 56: 995–1003.

Nwosu, O.K. and Ubaoji, K.I. (2020). Nutraceuticals: history, classification and market demand. In: *Functional Foods and Nutraceuticals* (ed. C. Egbuna and G. Dable Tupas). Cham: Springer.

Pal, M. and Molnár, J. (2021). Growing importance of cereals in nutrition and healthy life. *International Journal of Food Science and Agriculture* 5: 275–277.

de la Parra, C., Serna, S.O., and Liu, R.H. (2007). Effect of processing on the phytochemical profiles and antioxidant activity of corn for production of masa, tortilla and tortilla chips. *Journal of Agricultural and Food Chemistry* 55: 4177–4183.

Poudel, R. and Bhatta, M. (2017). *Review of Nutraceuticals and Functional Properties of Whole Wheat*, 264. Faculty Publications in Food Science and Technology.

Ranjan, R., Singh, S., Dhua, S. et al. (2023). Kodo millet (*Paspalum scrobiculatum*). In: *Nutri-Cereals: Nutraceutical and Techno-Functional Potential* (ed. R. Sharma, V. Nanda, and S. Sharma). Routledge.

Rather, M.A., Thakur, R., Hoque, M. et al. (2023). Sorghum (*Sorghum bicolor*): phytochemical composition, bio-functional, and technological characteristics. In: *Nutri-Cereals*, 57–89. CRC Press.

Rawat, M., Varshney, A., Rai, M. et al. (2023). A comprehensive review on nutraceutical potential of underutilized cereals and cereal-based products. *Journal of Agriculture and Food Research* 12: 100619.

Rizzello, C.G., Cavoski, I., Turk, J. et al. (2015). Organic cultivation of *Triticum turgidum* subsp. durum is reflected in the floursourdough fermentation-bread axis. *Applied and Environmental Microbiology* 81 (9): 3192–3204.

Roy, S., Chikkerur, J., Roy, S.C. et al. (2018). Tagatose as a potential nutraceutical: production, properties, biological roles, and applications. *Journal of Food Science* 83 (11): 2699–2709.

Sachdeva, V., Roy, A., and Bharadvaja, N. (2020). Current prospects of nutraceuticals: a review. *Current Pharmaceutical Biotechnology* 21 (10): 884–896.

Saikia, D. and Deka, S.C. (2011). Cereals: from staple food to nutraceuticals. *International Food Research Journal* 18 (1): 21–30.

Salmeron, I. (2017). Fermented cereal beverages: from probiotic, prebiotic and synbiotic towards nanoscience designed healthy drinks. *Letters in Applied Microbiology* 65: 114–124.

Samanta, A.K., Jayapal, N., Jayaram, C. et al. (2015). Xylooligosaccharides as prebiotics from agricultural by-products: production and applications. *Bioactive Carbohydrates and Dietary Fibre* 5 (1): 62–71.

Sampson, G. O. 2013. Beta-glucan content in tropical maize genotypes: extractability, structural analysis, rheology and function. PhD dissertation. Kwame Nkrumah University of Science and Technology.

Samyor, D., Das, A.B., and Deka, S.C. (2017). Pigmented rice a potential source of bioactive compounds: a review. *International Journal of Food Science and Technology* 52: 1073–1081.

Sebayang, A.H., Masjuki, H.H., Ong, H.C. et al. (2017). Optimization of bioethanol production from sorghum grains using artificial neural networks integrated with ant colony. *Industrial Crops and Products* 97: 146–155.

Seo, C.R., Yi, B., Oh, S. et al. (2015). Aqueous extracts of hulled barley containing coumaric acid and ferulic acid inhibit adipogenesis in vitro and obesity in vivo. *Journal of Functional Foods* 12: 208–218.

Sharma, R., Adhikari, P., Shrestha, J., and Acharya, B.P. (2019). Response of maize (Zea mays L.) hybrids to different levels of nitrogen. *Archives of Agriculture and Environmental Science* 4 (3): 295–299.

Shevkani, K., Singh, N., Bajaj, R., and Kaur, A. (2017). Wheat starch production, structure, functionality and applications—a review. *International Journal of Food Science and Technology* 52 (1): 38–58.

Shewry, P.R. (2018). Do ancient types of wheat have health benefits compared with modern bread wheat? *Journal of Cereal Science* 79: 469–476.

Singh, N., Singh, S., and Shevkani, K. (2019). Maize: composition, bioactive constituents, and unleavened bread. In: *Flour and Breads and their Fortification in Health and Disease Prevention*, 111–121. Academic Press.

Srinivasan, M., Sudheer, A.R., and Menon, V.P. (2007). Ferulic acid: therapeutic potential through its antioxidant properties. *Journal of Clinical Biochemistry and Nutrition* 40: 92–100.

Srivastava, S.C., Sharma, U.C., Singh, B.K., and Yadava, H.S. (2012). A profile of garlic production in India: facts, trends and opportunities. *International Journal of Agriculture Environment and Biotechnology* 5 (4): 477–482.

Sterna, V., Zute, S., and Brunava, L. (2015). Oat grain composition and its nutrition benefice. *Agriculture and Agricultural Science Procedia* 8: 252–256.

Stewart, B. and Lal, R. (2018). Increasing world average yields of cereal crops: it's all about water. *Advances in Agronomy* 151: 1–44.

Stuper-Szablewska, K., Kurasiak-Popowska, D., Nawracała, J., and Perkowski, J. (2019). Quantitative profile of phenolic acids and antioxidant activity of wheat grain exposed to stress. *European Food Research and Technology* 245: 1595–1603.

Sullivan, P., Arendt, E., and Gallagher, E. (2013). The increasing use of barley and barley by-products in the production of healthier baked goods. *Trends in Food Science and Technology* 29: 124.e34.

Tada, R., Ikeda, F., Aoki, K. et al. (2009). Barley-derived β-d-glucan induces immunostimulation via a dectin-1-mediated pathway. *Immunology Letters* 123 (2): 144–148.

Tang, Y., Li, S., Yan, J. et al. (2023). Bioactive components and health functions of oat. *Food Reviews International* 39 (7): 4545–4564.

Teoh, S.L., Ngorsuraches, S., Lai, N.M. et al. (2019). Factors affecting consumers' decisions on the use of nutraceuticals: a systematic review. *International Journal of Food Sciences and Nutrition* 70 (4): 491–512.

Teoh, S.L., Ngorsuraches, S., Lai, N.M., and Chaiyakunapruk, N. (2021). Consumer preferences and willingness to pay for nutraceuticals: a discrete choice experiment. *Value in Health Regional Issues* 24: 167–172.

Thakur, S., Singh, N., Kaur, A., and Singh, B. (2017). Effect of extrusion on physicochemical properties, digestibility, and phenolic profiles of grit fractions obtained from dry milling of normal and waxy corn. *Journal of Food Science* 82: 1101–1109.

Thondre, P., Ryan, L., and Henry, C. (2011). Barley b-glucan extracts as rich sources of polyphenols and antioxidants. *Food Chemistry* 126: 72.e7.

Turvey, C., Moran, M., Sacheck, J. et al. (2021). Impact of messaging strategy on consumer understanding of food date labels. *Journal of Nutrition Education and Behavior* 53 (5): 389–400.

Upadhyay, S.K. (Éd.) (2021). *Genome Engineering for Crop Improvement*. Wiley. https://doi.org/10.1002/9781119672425.

Upadhyay, S.K. (Éd.) (2024). *Applications of Genome Engineering in Plants*. Wiley. https://doi.org/10.1002/9781394183913.

Upadhyay, S.K. and Singh, S.P. (Eds.) (2021). *Bioprospecting of Plant Biodiversity for Industrial Molecules*. Wiley. https://doi.org/10.1002/9781119718017.

Upadhyay, S.K. and Singh, S.P. (Éds.) (2023). *Plants as Bioreactors for Industrial Molecules*. Wiley. https://doi.org/10.1002/9781119875116.

Verma, D.K. and Srivastav, P.P. (2020). Bioactive compounds of rice (*Oryza sativa* L.): review on paradigm and its potential benefit in human health. *Trends in Food Science and Technology* 97: 355–365.

Windpassinger, S., Friedt, W., Frauen, M. et al. (2015). Designing adapted sorghum silage types with an enhanced energy density for biogas generation in temperate Europe. *Biomass and Bioenergy* 81 (496–504): 18.

Xiong, Y., Zhang, P., Warner, R.D., and Fang, Z. (2019). Sorghum grain: from genotype, nutrition, and phenolic profile to its health benefits and food applications. *Comprehensive Reviews in Food Science and Food Safety* 18 (6): 2025–2046.

Zeven, A.C. (1991). Wheats with purple and blue grains: a review. *Euphytica* 56: 243–258.

Zhang, Z., Zhang, G., Li, W. et al. (2016). Enhanced biogas production from sorghum stem by co-digestion with cow manure. *International Journal of Hydrogen Energy* 41: 9153–9158.

17

Pseudocereals Nutraceuticals

Sonia Morya[1], Aniket More[1], Arno Neumann[2], and Shikha Chauhan[1]

[1] *Department of Food Technology and Nutrition, School of Agriculture, Lovely Professional University, Phagwara, Punjab, India*
[2] *BET Bioscience Extraction Technologies Inc., Abbotsford, British Columbia, Canada*

17.1 Introduction

In the present era, characterized by a heightened focus on health and an increasing preference for nutritious foods, pseudocereals have gained substantial attention over the past two decades (Ram et al. 2023). In 2011, the Food and Agriculture Organization designated pseudocereals as "the grains of the twenty-first century," these grains are rich in fiber, proteins, and starch, offering a balanced essential amino acid composition, particularly abundant in sulfur-rich amino acids (Wanniarachchi et al. 2023). The adaptability of pseudocereals to diverse environments, ranging from tropical to temperate climates, positions them as promising crops (Zenda et al. 2021). Their adaptability is credited to a substantial level of genetic variability. Key species in this category include amaranth (*Amaranthus* sp.), quinoa (*Chenopodium quinoa* Willd), and buckwheat (*Fagopyrum* sp.) (Gupta and Morya 2022; Singh et al. 2021; Upadhyay and Singh 2021, 2023).

Beyond their nutritional excellence, pseudocereals serve as valuable suppliers of essential minerals like calcium, iron, and zinc, along with a variety of vitamins (Rawat et al. 2023). Additionally, they contain health-promoting phytochemicals such as betalains, saponins, polyphenols, phytosterols, and phytosteroids (Morya et al. 2022a). Collectively, these qualities make pseudocereals noteworthy for their nutritional benefits and hold promise for future agricultural practices and human health (Sánchez-Velázquez et al. 2022).

Even though pseudocereals are both gluten-free (GF) and protein-rich, their underutilization is unexpected, especially considering their wealth of essential nutrients and valuable saponins, which hold significant applications in agro-pharma and agro-industry (Thakur et al. 2021). The rising global incidence of celiac disease has heightened the need for GF alternatives, emphasizing the untapped potential of pseudocereals (Breslauer et al. 2023).

Pseudocereals have gained worldwide acknowledgment in the nutraceutical industry, witnessing a surge in popularity owing to their outstanding nutritional profile that surpasses that of traditional cereals (Rao and Poonia 2023). Rich in active principles such as amino acids, antioxidants, dietary fiber, flavonoids, lignans, minerals, polyphenols, unsaturated fatty acids, vitamins, and notable fagopyritols, they stand out in the functional food sector, offering superior protein content and

Herbal Nutraceuticals: Products and Processes, First Edition. Edited by Santosh Kumar Upadhyay and Sudhir Pratap Singh.

quality compared to cereals. This positions them as excellent candidates for producing functional foods (Awuchi and Morya 2023; Peñas et al. 2023).

The amino acid composition of pseudocereal proteins, featuring crucial elements like arginine, tryptophan, lysine, and histidine, highlights their potential as valuable supplements for child nutrition (Malik and Singh 2022). Parameters like net protein use (NPU), protein efficiency ratio (PER), protein digestibility, and available lysine underscore the superior nutritional quality of pseudocereal proteins, bringing them close to the levels found in casein (Das 2016). Pseudocereal proteins, with compositions resembling those in legumes and low prolamin content, are favorable for individuals with celiac disease. These grains exhibit various health benefits, including anticancer effects, anti-inflammatory actions, antioxidant capabilities, prevention of obesity, protection for the cardiovascular system, and reduction of cholesterol levels (Rao and Poonia 2023).

Rich in fatty acids, especially unsaturated ones like linolenic acid, pseudocereals distinguish themselves from conventional cereals. Their mineral content, nearly double that of other cereals, holds promise in addressing hidden hunger (Karaca et al. 2023). Cultivating pseudocereals on marginal lands, often afflicted by abiotic stresses, showcases their resilience and positions them as potential solutions for underdeveloped and developing countries striving to maintain national food security (Dubey et al. 2020).

Pseudocereals like amaranth, buckwheat, chia seeds, kaniwa, and quinoa offer a multitude of health benefits, promoting heart health, aiding in diabetes management, possessing anti-inflammatory properties, and catering to individuals with celiac disease due to their low gluten content (Ashik and Morya 2022; Raungrusmee 2023). Their fiber richness contributes to weight management and supports gastrointestinal health. With versatile applications, these pseudocereals have evolved into essential components of functional foods, contributing not only to taste preferences but also enhancing the health benefits of these products (Schmidt et al. 2023).

Recognizing the potential of pseudocereals as nutritional powerhouses and climate-resilient crops, recent biotechnological interventions are urgently needed to assist in enhancing the cultivation and domestication of crops (Dixit et al. 2021; Kumar et al. 2023; Upadhyay 2021, 2024). Exploring the nutritional intricacies of emerging pseudocereals while addressing challenges hindering their integration into the broader nutraceutical landscape is crucial. The ecological dimension of pseudocereals adds to their appeal, presenting them as an eco-friendly, functional, cost-effective, and economically viable choice that contributes to environmental sustainability (Morya et al. 2022b). This dual advantage positions pseudocereals as promising contributors to future agricultural practices and the food industry (Lazou 2022).

17.2 Types and Classification of Pseudocereals

As we embark on the exploration of pseudocereals, a group of nongrass grains that defy conventional cereal classifications, it becomes essential to delve into their types and classifications. In an era where maintaining a diverse diet and nutritional resilience is crucial, pseudocereals have garnered recognition as versatile and sustainable choices. This section reveals the diverse realm of key pseudocereals, each offering unique nutritional qualities and adaptability to various ecological environments.

17.2.1 Amaranth

Amaranth, with a cultivation history of more than 7000 years, stands as an ancient grain. Its roots trace back to the Andean region of South America, and over time, it has evolved into a globally recognized culinary staple. Grouped into four classifications—grain-producing, vegetables, weed, and ornamental—amaranth exhibits adaptability to diverse climates. Its global reverence in religious traditions, culinary versatility, and rich nutritional profile make it a highly promising crop.

Amaranth is revered globally in religious traditions. In Hindu culture, known as "rajgira" or "ramdana," and Christian rituals and in indigenous cultures like the Hopi and Navajo, amaranth, known as "wuhti" or "tumbleweed" is crucial for ceremonial offerings, symbolizing spiritual nourishment and connection. Its widespread use across cultures emphasizes its enduring significance in fostering spiritual connections and honoring traditions worldwide. In Mesoamerican civilizations like the Aztecs, amaranth, or "huautli," was a revered staple and symbol of immortality, and featured prominently in ceremonies and rituals, including the creation of elaborate sculptures as offerings to gods.

Globally recognized for its health-oriented benefits, grain amaranth supports muscle development, immunity, and overall health, with abundant proteins and micronutrients. Its carbohydrate-rich composition, coupled with antioxidants and lipids like squalene, enhances dietary significance. Notably, *Amaranthus caudatus* is among the top five vegetable plants known for its impressive antioxidant capacity. The remarkable antioxidant properties of amaranth are attributed to the presence of flavonoids, phenolic acids, phytosterols, and squalene (Figure 17.1). Beyond the grain, amaranth's leaves offer high nutritional value, and its economic appeal with lower production cost adds to its significance.

Amaranth features a dual fiber content, comprising both soluble and insoluble fibers, contributing significantly to digestion promotion and the regulation of blood sugar levels. Its applications span a broad spectrum, from demonstrating antimicrobial and anticancer properties to finding utility in traditional medicine. Amaranthus, as a genus, is underscored by an abundance of in vitro and in vivo research findings, highlighting its robust antioxidant potential and, consequently, its health-preserving and healing attributes (Park et al. 2020).

The culinary landscape of amaranth is marked by versatility, offering accessible and affordable nutrition with extended shelf life. From preventing anemia to showcasing antioxidant and antimicrobial properties, amaranth's medicinal benefits are acknowledged. Its dual role as a vegetable and grain contributes to diverse culinary traditions, finding use in salads, soups, and various dishes worldwide (Baraniak and Kania-Dobrowolska 2022).

In the realm of global cuisine, amaranth takes center stage with enticing dishes like the amaranth salad with roast cauliflower and amaranth cakes featuring lentils, kale, and chipotle aioli. Its rich history in ancient Mexico, where it was a dietary staple, and its global gastronomic significance from Ethiopia (unleavened bread "kita") to Vietnam (potherbs) and India (the seeds are consumed with boiled rice, used in the preparation of "laddoos," and popped grains when mixed with molasses and honey, create a confectionery product) underscore its culinary appeal. The fusion of flavors across the globe showcases amaranth's ability to seamlessly integrate into various culinary traditions (Salvador-Reyes et al. 2023).

The journey continues with the global resonance of amaranth in dishes like amaranth tikkis in Mexico, vegetable coconut amaranth pilaf in Southeast Asia, carrot and amaranth polenta in India, cauliflower schnitzel with a puffed amaranth crust in Germany, and grilled peach and papaya salad with amaranth granola in the Caribbean. These examples highlight amaranth's versatility and its ability to create a global tapestry of taste and nutrition.

While generally safe, gradual introduction is advised for individuals with digestive concerns, and its GF nature suits those with celiac disease (Dhondiram et al. 2023). Amaranth's incorporation into various dishes enhances diets and contributes to a wholesome culinary experience. Culinary innovation shines through diverse dishes, utilizing both grains and the entire plant. Agronomic resilience makes it suitable for intercropping and crop rotation, showcasing the adaptability that defines the culinary landscape of amaranth.

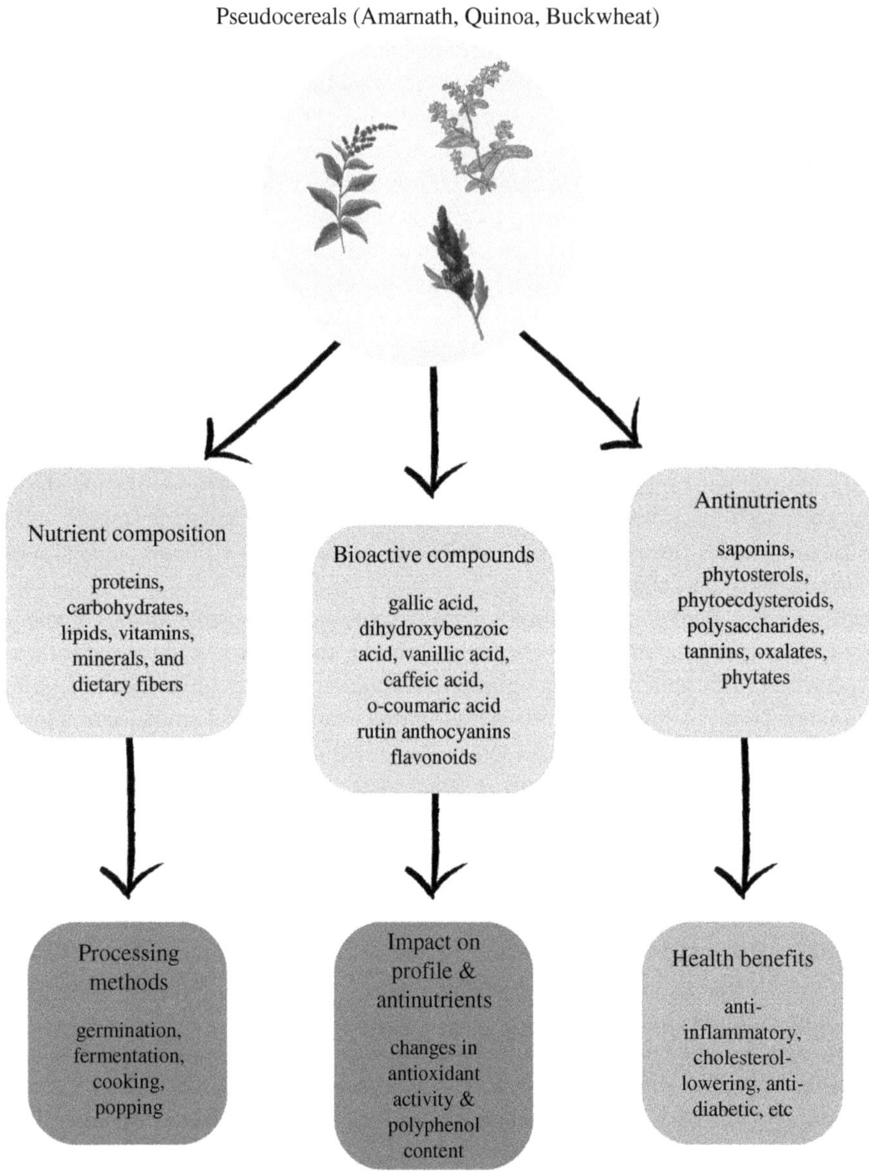

Figure 17.1 Bioactive components in amaranth.

17.2.2 Quinoa

Quinoa, originating in Peru and belonging to the Chenopodiaceae plant family, has evolved into a global sensation celebrated for its adaptability, nutritional richness, and culinary versatility (Rosentrater and Evers 2018). Once regarded as humble, quinoa has transcended its origins, spreading to Europe, North America, Australia, China, and Japan. Its remarkable adaptability to challenging environments sets it apart, thriving where other crops face difficulties (Bazile et al. 2016). The pivotal moment arrived in 2013 with the International Year of Quinoa, catapulting it into global recognition beyond its modest beginnings (Alandia et al. 2021).

Quinoa's resilience is evident in its ability to thrive in diverse climates and challenging conditions, showcasing its robustness as a versatile crop (Ain et al. 2023). Both quinoa leaves and seeds are fit for consumption, with the seeds being the main form consumed by humans. Boasting over 250 distinct varieties, quinoa demonstrates exceptional diversity, featuring notable types like bear, cherry vanilla, Cochabamba, and temuco (Ranjan et al. 2023).

The scientific community has shown interest in the nutritional richness of quinoa, which includes fibers, lipids, proteins, vitamins, and minerals, along with a unique balance of essential amino acids (Angeli et al. 2020). Acknowledged as a "functional food," quinoa unfolds potential health benefits, contributing to the mitigation of various diseases (Landi et al. 2021). Its components, encompassing antioxidants, fatty acids, fiber, minerals, vitamins, and plant hormones, play a crucial role in human nutrition. They emphasize the functional aspects of quinoa in safeguarding cell membranes and enhancing brain function. (Kumar et al. 2023). Studies conducted both in vitro and in vivo have provided evidence that quinoa and its bioactive components offer a range of health-promoting effects (Hernández-Ledesma 2019).

Quinoa is rich in bioactive compounds (Figure 17.2), including betacyanins, betalains, kaempferol, quercetin, and tocopherol. These compounds play a variety of roles in the body, showcasing diverse biological functions. Among their notable effects are antihypertensive properties, antiviral activity, antioxidant capabilities, anticancerous effects, antidiabetic potential, antiallergic attributes, as well as antiatherosclerosis and antithrombosis effects. Polyphenols, known for diverse pharmacological effects including antiallergic, antiviral, anti-inflammatory, cardiovascular protection, and anticarcinogenic properties, add another layer to quinoa's nutritional prowess (Mudau et al. 2022). In a comprehensive review by (Ng and Wang 2021), the exploration of quinoa's functional role unveils its potential to mitigate chronic diseases. The study illustrates how quinoa's fiber content, along with its array of vitamins, minerals, and antioxidants, plays a pivotal role in enhancing cardiovascular health and managing diabetes.

Furthermore, evidence suggests that antioxidants like flavanols, found in quinoa, may help reduce elevated oxidative stress associated with conditions such as Alzheimer's disease, CSD, depression, and cancer (Sharma et al. 2023). Quinoa's robust antioxidant activity positions it as a promising contender for consideration as a functional food in disease prevention. This comprehensive exploration underscores the interconnected roles of quinoa's nutritional richness, polyphenols, and antioxidant properties in promoting health and preventing chronic diseases (Thambiliyagodage et al. 2023).

In 2023, quinoa maintained its status as a highly sought-after and popular ingredient, showcasing enduring appeal across diverse global cuisines. Its adaptability is evident in versatile applications, featuring prominently in dishes ranging from salads and soups to tacos and burgers (Angeli et al. 2020). The trajectory of quinoa's future in the culinary landscape is shaped by nutritional advancements and evolving health trends, with chefs valuing its remarkable versatility in various dishes (Reisman 2020). Recognized for its distinctive texture and nutty flavor, quinoa is a favored ingredient for those seeking innovation and health-conscious menu options, providing a GF choice for individuals with sensitivities, including those with celiac disease (Sae-Eaw et al. 2022). The association with the gourmet French brand "Picard" emphasizes quinoa's premium status, offered alongside other delectable options. Quinoa's widespread popularity stems from its ability to complement a diverse range of dishes, establishing itself as a superfood with broad appeal. This ascent to superfood status is fueled not only by culinary prowess but also by inherent sustainability, contributing an eco-friendly dimension to its culinary charm (Afzal et al. 2023).

Quinoa's global journey is exemplified by its integration into various global cuisines. In South America, its place of origin, quinoa remains a fundamental component of traditional dishes,

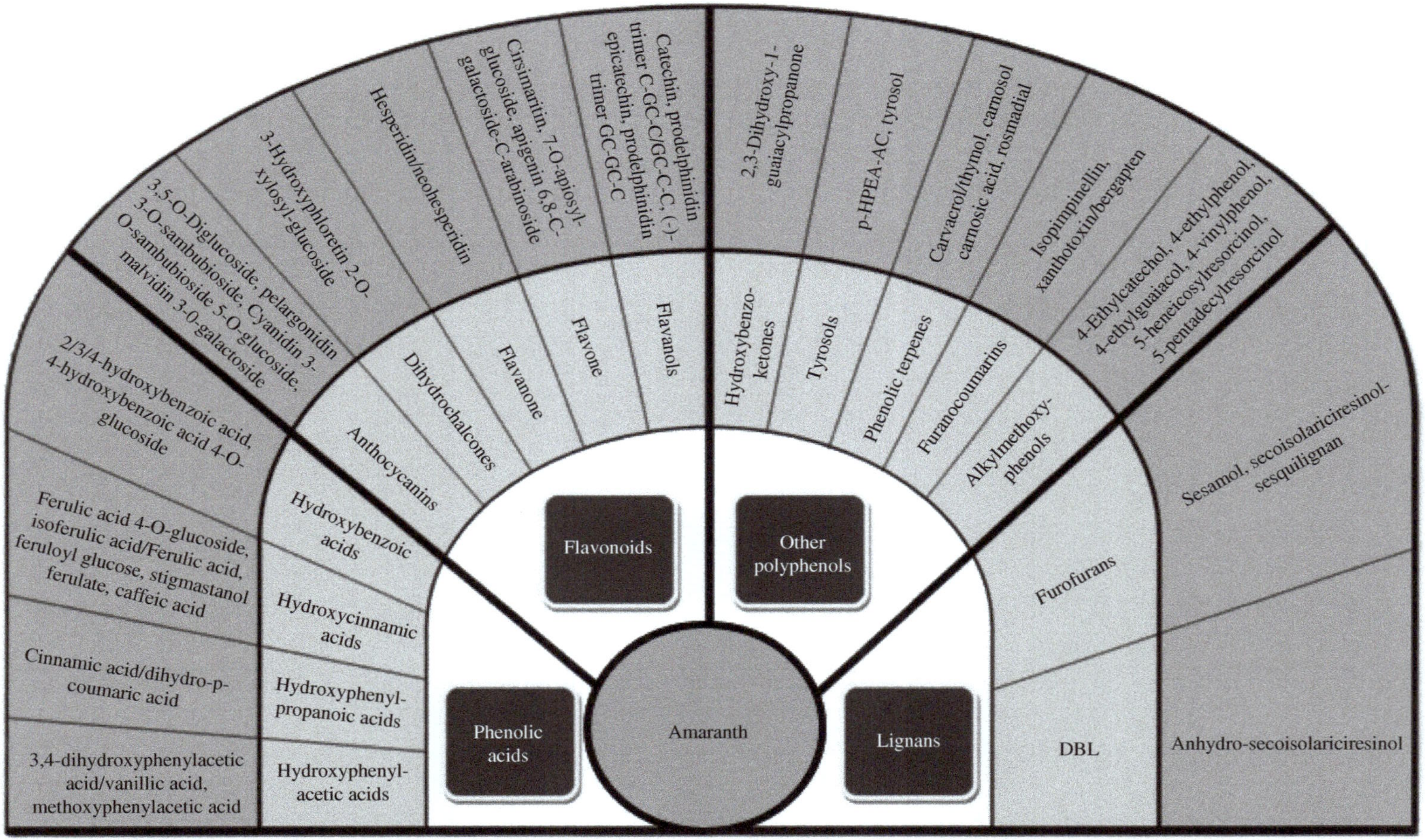

Figure 17.2 Bioactive components in quinoa.

providing a rich source of nutrients (Schmidt et al. 2023). Moving eastwards, Asian cuisines are embracing quinoa in innovative sushi rolls, stir-fries, and desserts. In the Middle East, quinoa seamlessly blends into hearty salads and grain-based dishes, celebrating its compatibility with the region's flavorful cuisine. North American chefs, across the Atlantic, showcase quinoa's versatility in tex-mex bowls and protein-packed veggie burgers. Extending to Africa, quinoa contributes to nutritional diversity and sustainable agriculture practices. The international culinary community appreciates quinoa not just for its health benefits but also for its ability to enhance the taste and texture of a wide array of dishes, reflecting a truly global appreciation for this pseudocereal (Montemurro et al. 2019).

The FAO has played a significant role in promoting the cultivation and spread of quinoa, contributing to its enhancement in the food sector. As a pseudocereal, quinoa's journey from a traditional staple to a global phenomenon underscores its adaptability, nutritional excellence, and culinary appeal. This evolution positions quinoa as a valuable pseudocereal in the dynamic landscape of functional foods.

17.2.3 Buckwheat

Buckwheat, originating in Southern China, has evolved into a globally cultivated pseudocereal, finding prominence in countries such as China, the United States, Russia, Kazakhstan, Ukraine, France, and Poland (Pipan et al. 2023). Recognized for its ecological adaptability, buckwheat is gaining traction in Canada, Europe, and the USA, particularly in the form of noodles and pancakes. The two main species, tartary buckwheat (*Fagopyrum tataricum* (L.) Gaertn) and common buckwheat (*Fagopyrum esculentum* Moench) are widely grown and consumed worldwide. Interestingly, buckwheat was a major crop in certain Asian and European countries before wheat production took precedence (Li et al. 2022).

Recent years have witnessed a renewed interest in buckwheat for food production, and its global production quantity has reached about 4.83 million tons in 2023 (EMR 2023). This resurgence in popularity is not only due to its versatile culinary applications but also stems from its potential health benefits, making it a subject of interest among food scientists. Buckwheat is increasingly recognized for its similar chemical compounds (Figure 17.3) and functional properties that promote health (Jin et al. 2022).

Beyond its culinary appeal, buckwheat serves as a resilient crop, capable of thriving on marginal lands and under harsh conditions. Its fast growth cycle, producing grains within approximately 60–70 days, adds to its significance in ensuring food security in different regions (Joshi et al. 2019). Breeding programs are actively underway to enhance the agronomic traits of buckwheat, reflecting the commitment to its sustainable cultivation (Solanki and Singh 2023).

The rising popularity of buckwheat is notably influenced by its attractive nutritional profile, characterized by its GF property (Skřivan et al. 2023). Rich in B group vitamins, essential macro, and microelements, including thiamine, riboflavin, pyridoxine, sodium, potassium, copper, zinc, magnesium, iron, calcium, and manganese, buckwheat becomes a versatile and nutritious choice for various food products, ranging from bakeries to confectionery items (Kaur et al. 2023). The consumption or supplementation of buckwheat has been found to impact the gastrointestinal (GI) microbiome, resulting in either increased or decreased alpha diversity. Alpha diversity serves as a measure of both the richness and evenness of microorganisms within samples (Valido et al. 2023).

In seven in vitro studies using human gastrointestinal-derived cancer cell lines and one animal study, buckwheat-derived compounds were explored for their potential to inhibit tumor growth. Notably, a study by Dziedzic et al. (2018) found that buckwheat bran demonstrated the highest cytotoxicity against human colon adenocarcinoma cell line HT-29, while raw and roasted grains

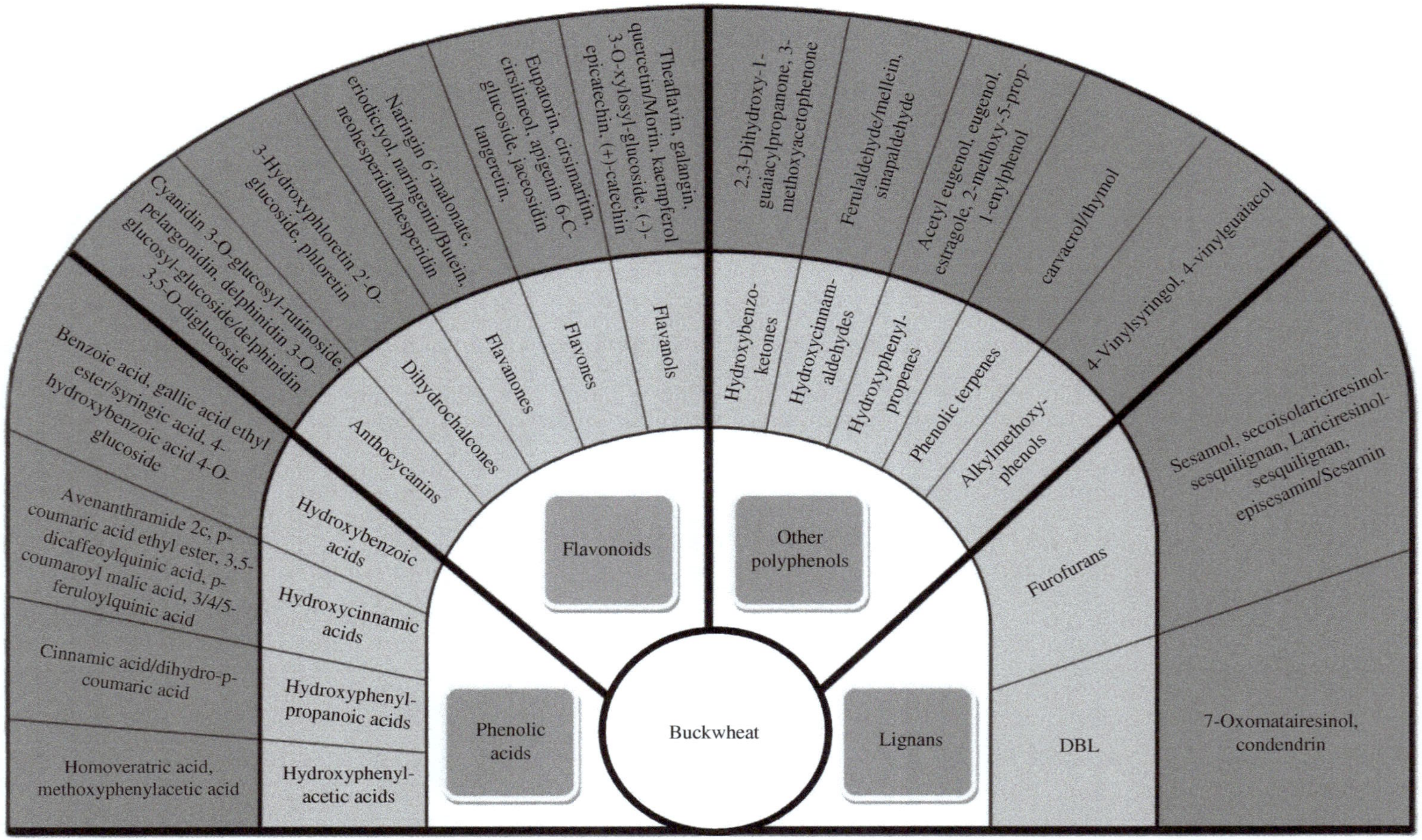

Figure 17.3 Bioactive components in buckwheat.

exhibited lower activity. Specific components, including flavonoids like catechin and quercetin, along with amino acids such as serine, proline, glycine, histidine, and arginine, were identified as the most cytotoxic to cancer cells. Interestingly, the most abundant flavonoid, rutin, did not show significant cytotoxicity. These findings provide valuable insights into the potential anticancer properties of certain components in buckwheat (Valido et al. 2023).

The multifaceted appeal of buckwheat, spanning culinary versatility, health benefits, and agronomic resilience, positions it as a valuable and increasingly sought-after pseudocereal on the global stage.

As we conclude our exploration into the diverse world of pseudocereals, each with its unique cultural, historical, and nutritional significance, the subsequent section will delve into the nutritional composition of pseudocereals. Unveiling the essential components that contribute to their status as nutritional powerhouses, this next segment aims to provide a detailed analysis of the specific elements that make pseudocereals not only a culinary delight but also a key player in fostering human health.

17.3 Nutritional Composition of Pseudocereals

Pseudocereal grains, such as amaranth, buckwheat, and quinoa, are esteemed for their exceptional nutritional value. They feature energy-rich carbohydrates, encompassing dietary fiber, starch, and sugar. This nutritional profile is further enriched with essential substances like fatty acids, flavonoids, minerals, phenolic acids, and vitamins. The inclusion of bioactive compounds such as betalains, phenols, phytosterols, and phytoecdysteroids contributes to the overall nutritional richness of these grains (Kaur et al. 2023).

The abundant presence of fiber and essential fats in pseudocereals translates to significant health benefits. These grains not only provide essential macronutrients like proteins, carbohydrates, and fatty acids but also serve as plentiful sources of vital vitamins and minerals. Calcium, folate, lutein, potassium, vitamin A, and B6 are among the essential nutrients found in noteworthy quantities in pseudocereals.

Before delving into the detailed breakdown in Table 17.1, let us take a glance at the key nutritional highlights of pseudocereals. This table encapsulates the essential components of amaranth, quinoa, and buckwheat, offering a quick reference for the rich nutritional content we are about to explore.

17.4 Health Benefits of Pseudocereals

Pseudocereals, such as amaranth, buckwheat, chia, and quinoa, offer a wealth of nutrients like fiber, healthy fats, and plant-based protein. Rich in antioxidants, minerals (iron, magnesium, potassium, and zinc), and vitamins (E and B complex), they are GF (Rawat et al. 2023). Their soluble fiber aids in regulating blood sugar, supporting digestion, and reducing cholesterol due to a lower glycemic index compared to refined grains (Malleshi et al. 2020). This makes them a nutritious and versatile dietary option, particularly beneficial for those with celiac disease (Bangar et al. 2022).

Delving deeper into their health implications, replacing common cereals with these pseudocereals provides essential daily nutrients while lowering the risk of various diseases, thanks to their rich bioactive compounds (Shahbaz et al. 2023). Often overlooked, these plant seeds should become a regular part of our diet, functioning as nutraceuticals to enhance overall health and combat diseases. Compounds like angiotensin I, peptides, quercetin, rutin, and antioxidants found in these seeds play a crucial role in reducing oxidative stress, contributing to better health (Rahman et al. 2022).

Table 17.1 Nutritional components of pseudocereals.

Nutritional profile	Amaranth	Quinoa	Buckwheat
Moisture (% wet basis)	8.9–9.4	8.2–13.1	11.0
Carbohydrates (% dry basis)	63.1–70.0	48.5–77.0	63.1–82.1
Starch (% dry basis)	65.0–75.0	58.1–64.2	54.5–57.4
Total dietary fiber (% dry basis)	2.7–17.3	7.0–26.5	17.8
Insoluble (% total fiber)	78.0–86.0	78.0	70.3
Soluble (% total fiber)	14.0–22.0	22.0	16.0
Crude protein (%)	13.1–21.5	9.1–16.7	5.7–14.2
Essential amino acid (g/100 g protein)			
Threonine	3.3–5.0	2.1–8.9	3.9–4.0
Valine	3.9–5.0	0.8–6.1	2.3–6.1
Phenylalanine	3.7–4.7	3.0–4.7	1.3–7.2
Isoleucine	2.7–4.2	0.8–7.4	1.1–4.1
Leucine	4.2–6.9	2.3–9.4	2.2–7.6
Methionine	1.6–4.6	0.3–9.1	0.5–2.5
Tryptophan	0.9–1.8	0.6–1.9	1.83
Lysine	4.8–8.0	2.4–7.8	4.2–8.6
Non-essential amino acids (g/100 g protein)			
Aspartate	7.3–10.7	8.0	7.6–16.6
Glutamate	14.4–17.7	13.2	23.2–24.4
Serine	4.9–9.3	3.4–5.7	3.2–8.6
Histidine	1.9–3.8	1.4–5.4	1.8–4.9
Glycine	6.7–15.2	2.2–6.1	6.2–13.2
Arginine	8.7–15.6	6.9–13.6	10.5–11.3
Alanine	3.5–6.2	3.2–5.7	4.6–11.3
Tyrosine	3.3–3.7	2.5–3.7	0.6–4.9
Cysteine	2.1–3.6	0.1–2.7	0.8–3.5
Proline	2.82–4.6	2.3–5.5	2.6–8.8
Lipids (% dry basis)	5.6–10.9	4.0–7.6	0.7–7.4
Saturated fatty acid (% lipids)	20.1–30.9	15.5–29.0	18.8–19.5
Lauric (C12:0)	nd	nd	0.02–0.04
Myristic (C14:0)	nd	nd	0.07–0.1
Pentadecylic (C15:0)	nd	nd	0.05–0.06
Palmitic (16:0)	18.8–20.2	9.3–10.7	13.2–18.5
Margaric (17:0)	nd	nd	0.05–0.06
Stearic (C18:0)	3.7–4.2	0.7–1.1	1.4–6.3
Arachidic (C20:0)	nd	nd	1.1–1.2
Behemic (C22:0)	nd	nd	1.1–1.3
Lignoceric (C24:0)	nd	nd	0.7–0.8
Unsaturated fatty acid (% lipids)	61.0–87.3	71.0–84.5	80.1–80.9

Table 17.1 (Continued)

Nutritional profile	Amaranth	Quinoa	Buckwheat
Palmitoleic (16:1)	nd	nd	0.15–0.20
Oleic (18:1-9c)	22.7–31.8	15.7–31.1	35.7–47.9
Vaccenic (18:1-11c)	1.4–2.0	1.3–1.7	—
Linoleic (18:2 ω-6)	37.1–45.9	44.9–58.6	31.4–44.6
Linolenic (18:3 ω-3)	0.6–1.4	3.0–11.1	0.0–5.3
Gondoic (C20:1)	0.2–0.3	0.6–1.6	1.8–3.1
Erucic (C22:1)	nd–0.1	nd–1.5	0.2–0.5
ω-6/ω-3	33.0–68.9	4.7–19.6	—
Mineral composition (mg/100 g, dry basis)			
Potassium	290–434	656–1475	450
Phosphorous	441–455	140–530	330–395.3
Calcium	175–206	27.5–148.7	46.5–50.4
Magnesium	254–266	207–502	390
Sodium	0.6	11–31	—
Iron	12–17.4	1.1–16.7	11.8–14.9
Zinc	3.7–5.2	0.8–4.8	2.1–2.4
Manganese	4	—	1.2–1.8
Copper	0.77	1–9.5	0.9–1.6
Vitamins (mg/100 g, dry basis)			
Vitamin B1 (Thiamine)	0.01–0.1	0.3–0.4	0.1–3.3
Vitamin B2 (Riboflavin)	0.04–0.41	0.3–0.4	0.06–10.6
Vitamin B3 (Niacin)	<0.01–8.04	1.1–1.5	2.1–18.0
Vitamin B6 (Pyridoxine)	0.04–0.6	0.5	0.27–0.33
Folate	0.05–0.07	0.18	—
Vitamin E (mg/kg, dry basis)	15.4	24.7	9.5–16.4
Total carotenoids (mg/kg, dry basis)	3.7–4.7	4.6–4.8	—
Lutein	3.6–4.4	5.8–12	3.71
Zeaxanthin	trace–0.3	0.3–5.4	nd
β-carotene	nd	nd–1.1	1.05

Source: Dakhili et al. (2019), Joshi et al. (2018), Martínez-Villaluenga et al. (2020), and Tien et al. (2018).

Furthermore, pseudocereals, being high in soluble fiber, play a key role in regulating bowel movement and controlling issues such as blood pressure, high cholesterol, and heart diseases (Garutti et al. 2022). Whether consumed alone or alongside traditional cereals, incorporating pseudocereals into our diet may contribute to maintaining overall well-being, potentially extending life expectancy by slowing down the aging process (Skendi et al. 2020). This comprehensive narrative underscores the significant therapeutic potential of pseudocereals for health, weaving a story of diverse benefits for individuals seeking a holistic approach to nutrition.

17.4.1 Nutraceutical Properties of Amaranth

Amaranth, categorized as a pseudocereal, was studied in the context of hypertension using spontaneously hypertensive rat biotechnology as the animal model. The study, carried out by Fritz et al. (2011), employed the design of intragastric administration of amaranth protein hydrolyzate. The observed outcomes revealed a significant reduction in blood pressure. Also, the context of hypertension using Wistar rats as the animal model. In the research carried out by Lado et al. (2015), the research design involved comparing a basal diet with an amaranth protein diet in cholesterol-enriched diets for 42 animals over a period of fourweeks. The observed outcomes included a reduction in plasma total cholesterol and triglycerides, an increase in fecal cholesterol excretion, and antioxidant effects demonstrated by elevated ferric-reducing antioxidant power (FRAP) values. Additionally, there was a decrease in TBA values in plasma and liver, along with enhanced superoxide dismutase (SOD) activity.

Amaranth was examined in the context of diabetes using streptozotocin-induced CD1 mice as the animal model. The research design, as outlined by Jorge et al. (2015), involved administering amaranth protein hydrolyzates to 25 animals, with a 30-minute duration for the acute study and a four-week duration for the chronic study. The observed outcomes revealed a significant improvement in glucose tolerance and increased plasma insulin levels in both acute and chronic studies. They were also investigated for their impact on diabetes using Wistar albino rats as the animal model. In the study designed by Mithila and Khanum (2015), 24 animals were divided into groups receiving either a basal diet or an amaranth-based diet for 15 days. The outcomes revealed a notable decrease in food intake and body weight gain, along with changes in plasma ghrelin, leptin, and cholecystokinin levels. Additionally, there was an improvement in the overall blood glucose profile.

Investigating the impact of amaranth on iron deficiency in Wistar albino rats, Laparra and Haros (2016) designed a study comparing a control group with rats fed amaranth bread. The outcomes revealed an increase in iron content, a significant elevation in hemoglobin concentrations, down-regulation of transferrin receptor 2 (TfR2) and IL-6 expression levels, and an influence on iron bioavailability.

In a study examining the impact of amaranth on cardiovascular health in Wistar rats, Sabbione et al. (2016) conducted a two-week investigation comparing casein with amaranth protein. The observed outcomes revealed a significant induction of clotting tests, thrombin time, and activated partial thromboplastin time. The presence of magnesium and phenolic acid in amaranth contributes to its ability to counteract heart diseases and enhance the strength of heart muscles. A study conducted by Berger et al. (2003), published in the International Journal for Vitamin and Nutrition Needs, highlights that amaranth also plays a role in managing cholesterol levels. The study reports a significant reduction in very low-density lipoprotein (VLDL) cholesterol ranging from 21% to 50%.

In an eight-week study focusing on obesity in C57BL/6 mice, Olguín-Calderón et al. (2019), compared a basal diet with an amaranth protein diet in diet-induced obesity. The observed outcomes included an increase in cecal crypt depth and calceiform cells, a reduction in the abundance of *Ruminococcaceae* and *Lachnospiracea* induced by a high-fat diet, and an induction of the abundance of *Prevotellaceae*.

17.4.2 Nutraceutical Properties of Quinoa

Exploring prediabetes in individuals, Abellán Ruiz et al. (2017) conducted a 28-day study with 30 participants, comparing a control group to a processed quinoa group. The observed outcomes included a significant decrease in body mass index and glycated hemoglobin, along with an increase in satiation and fullness degree. However, no significant differences were noted in fasting plasma glucose levels. The glycemic index (GI) of quinoa varies from 35 to 53, depending on the cooking time. Even after being subjected to overcooking, if 150 g of quinoa grains undergo cooking, refrigeration, and reheating in the microwave for 1.5 minutes, the reported GI is 53, which still falls within the low range (Mohamed Ahmed et al. 2021). According to Berti et al. (2004), consuming quinoa resulted in a significant reduction in triglycerides and free fatty acids (FFA). Elevated FFA levels can disrupt glucose regulation through insulin, leading to increased triglyceride levels. Therefore, regular quinoa consumption may contribute to suppressing FFA, enhancing insulin sensitivity, and reducing blood glucose and triglyceride levels.

In a four-week study involving 35 postmenopausal and overweight women, De Carvalho et al. (2014) compared the effects of quinoa to corn flakes. The observed outcomes revealed positive health effects by significant reduction in serum triglycerides (TG), thiobarbituric acid reactive substances (TBARS), and vitamin E concentrations. Additionally, there was an increase in urinary excretion of enterolignans, a reduction in total and LDL-cholesterol, and an increase in glutathione (GSH) levels.

In a 30-day study involving 22 students, Farinazzi-Machado et al. (2012) investigated the impact of consuming quinoa bars (19.5 g/day) on cardiovascular health. The observed outcomes indicated a reduction in total and LDL-cholesterol and triglycerides (TG). However, there was no significant reduction in blood glucose levels, body weight, or blood pressure.

In an eight-week study involving 37 overweight male subjects, (Li et al. 2018) investigated the impact of consuming quinoa-enriched bread compared to refined wheat bread on cardiovascular health. The observed outcomes revealed alterations in glucose response, with minimal effects on other cardiovascular risk biomarkers such as LDL-cholesterol.

In a study with 40 C57BL/6 mice, (Liu et al. 2018a) investigated the impact of a quinoa diet compared to a control diet on colonic colitis induced by DSS exposure (five days, 2.5%). The observed outcomes included a reduction in clinical symptoms, as evidenced by a decrease in the disease activity index and the degree of histological damage. Additionally, the quinoa diet alleviated DSS-induced dysbiosis, characterized by a decrease in the abnormal expansion of proteobacteria, and reduced overgrowth of *Escherichia/Shigella* and *Peptoclostridium*.

In an eight-week study involving 30 obese diabetic db/db mice, (Noratto et al. 2019) explored the effects of a quinoa-based diet compared to a basal diet on obesity-associated inflammation. The study outcomes revealed several positive changes, including reductions in total and LDL-cholesterol, oxidized LDL, and protein carbonyls. Additionally, there was a decrease in the inflammatory marker IL-6, as well as a reduction in hepatic steatosis and total cholesterol accumulation in the liver.

17.4.3 Nutraceutical Properties of Buckwheat

Consuming pseudocereals, particularly buckwheat, and products containing buckwheat, has shown promising health benefits. This includes notable reductions in blood glucose (14.5%), low-density lipoprotein (9.05%), total cholesterol (4–13.56%), and triglycerides (4.85%) (Thakur et al. 2021). The lipid-lowering properties of buckwheat proteins are linked to their ability to stimulate bile acid secretions. Introducing buckwheat protein into diets has been observed to significantly lower cholesterol levels in the serum, gall bladder, and liver of hamsters (Verma et al. 2020). This underscores the potential of incorporating buckwheat-based products for promoting cardiovascular health (Solanki and Singh 2023).

In a six-week study focusing on a high-fat diet in C57BL/6 mice, Zhou et al. (2018) examined the effects of tartary buckwheat protein compared to casein in control and high-fat diet conditions.

The study involved 27 animals, and the observed outcomes included a reduction in plasma total cholesterol and triglycerides, promotion of the growth of beneficial bacteria such as *Lactobacillus*, *Bifidobacterium*, and *Enterococcus*, inhibition of the growth of *E. coli*, decrease in levels of plasma inflammation factors (LPS, TNF-α, and IL-6), and an increase in the excretion of total bile acids and short-chain fatty acids in feces.

In a study examining type-2 diabetes in diabetic subjects, (Stringer et al. 2013) investigated the effects of acute or chronic (1 week) consumption of buckwheat compared to rice crackers. The observed outcomes included alterations in post-prandial plasma GLP-1, GIP, and pancreatic polypeptide after acute consumption of buckwheat. However, no changes were noted in post-prandial glucose, insulin, or C-peptide concentrations. Additionally, the consumption of buckwheat showed no effects on apolipoproteins, fasting glucose, or lipids in both diabetic and healthy subjects. In another study, Liu et al. (2018b) conducted a four-week investigation with 165 individuals. The study involved comparing a control group with a tartary buckwheat group consuming 150 g/day. The observed outcomes included a decrease in fast insulin, LDL-cholesterol, and total cholesterol. However, there were no observed changes in blood glucose or glycated hemoglobin levels.

Amaranth, quinoa, and buckwheat, studied in various health contexts, show promising outcomes, making them valuable additions to a balanced diet. Incorporating these pseudocereals is not just a culinary choice; it is a holistic approach to nutrition, unlocking a world of health benefits for a vibrant life.

17.5 Functional Properties and Nutraceutical Potential

Pseudocereals like amaranth, buckwheat, and quinoa, have evolved from being overlooked grains to nutritional powerhouses, celebrated for their rich nutrient profiles and associated health benefits (Figure 17.4). These grains are abundant in high-quality carbohydrates, dietary fibers, lipids, proteins, minerals, and vitamins. They are further enriched by a diverse array of bioactive compounds such as flavonoids, phenolic acids, and saponins, contributing to their nutritional appeal (Fernández-López et al. 2020).

Despite their nutritional richness, pseudocereals harbor various antinutrient components, with polyphenols found at higher levels compared to traditional cereals. Notably, buckwheat, quinoa, and amaranth display varying phenolic acid content, with buckwheat taking the lead. Identified phenolic acids encompass caffeic acid, dihydroxybenzoic acid, gallic acid, *o*-coumaric acid, rutin, and vanillic acid. These grains also house anthocyanins and flavonoids, with buckwheat exhibiting the highest content. Antinutrient components include oxalates, phytoecdysteroids, phytosterols, phytates, polysaccharides, saponins, and tannins.

While pseudocereals offer myriad health benefits, certain compounds, like phenolic compounds in *Amaranthus cruentus*, can have adverse effects on protein metabolism. Similarly, trypsin inhibitors and phytate in buckwheat seeds can impact protein digestion and mineral bioavailability (Agregán et al. 2023). It is crucial to address saponins in quinoa seeds, as their elimination or reduction before food processing is essential due to their adverse effects on human and animal health, with buckwheat maintaining a higher range of saponin content. Saponins form complexes with iron and zinc, leading to diminished bioavailability (Mohamed Ahmed et al. 2021). Moreover, the presence of phytic acid in amaranth and quinoa contributes to the diminished bioavailability of essential elements (Miranda-Ramos and Haros 2020). For instance, amaranth oil, abundant in tocotrienols and squalene, shows potential in managing cholesterol levels. Amaranth's tannin concentrations exhibit antidiabetic properties, while darker quinoa varieties display antioxidant activity (Martinez-Lopez et al. 2020).

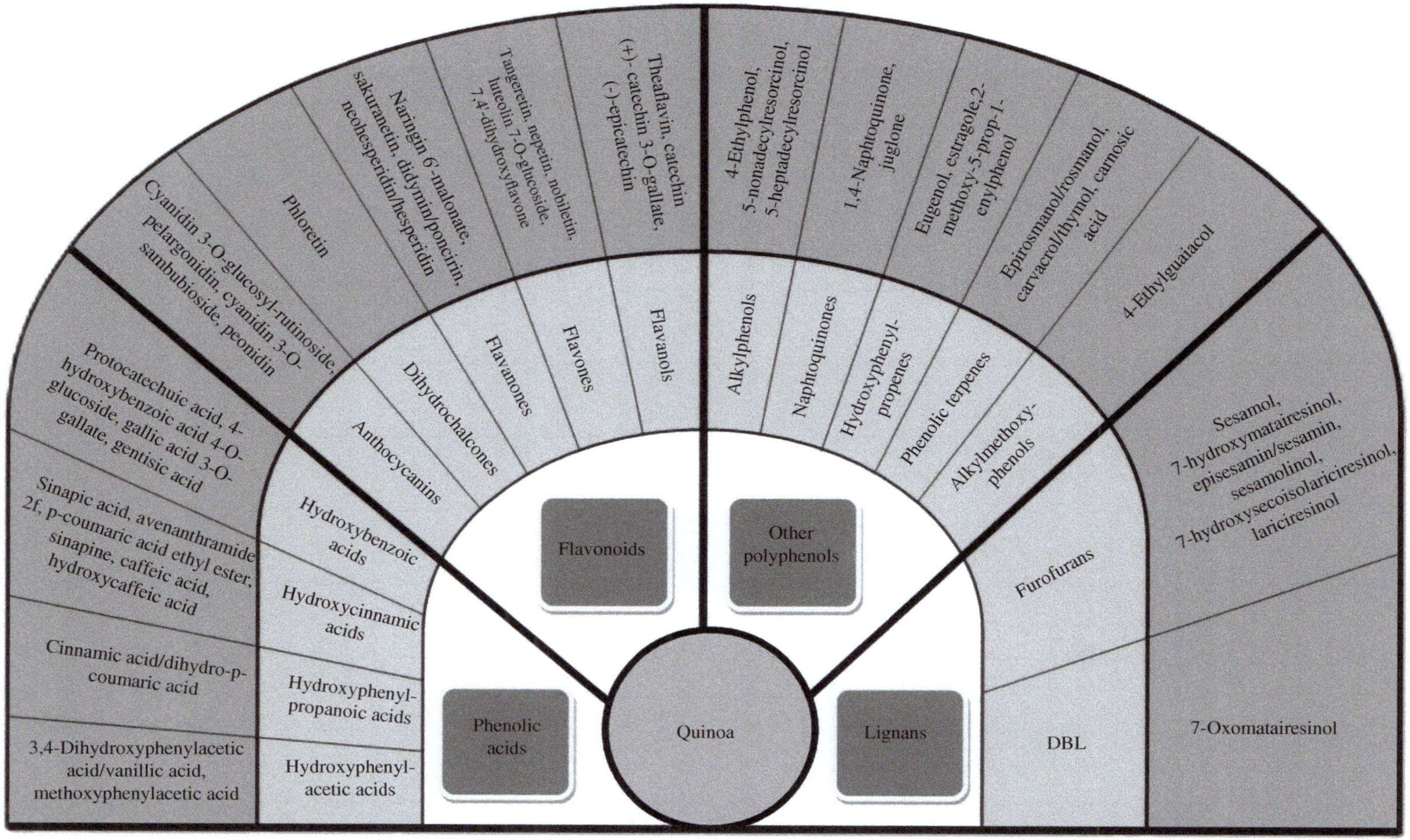

Figure 17.4 Functional properties of pseudocereals.

Different processing methods exert a notable impact on the nutritional and bioactive constituents of pseudocereals. The bioavailability of calcium, iron, and zinc in amaranth and quinoa seeds is notably enhanced through cooking methods like boiling and roasting (Cilla et al. 2019). However, subjecting tartary buckwheat flour to thermal processing leads to a significant decrease in antioxidant activity, total flavonoids, and total phenolics (Ge and Wang 2020). Popping, especially in amaranth, results in an increase in acid detergent fiber (ADF), ash, fat, and neutral detergent fiber (NDF) content, along with a decrease in protein digestibility (Nazeer and Firincioglu 2022). Both red and yellow varieties of quinoa exhibit increased antioxidant activity and total phenolic content after baking (Ballester-Sánchez et al. 2019).

Germination proves to be a beneficial processing method, enhancing antioxidant activity and polyphenols in amaranth seeds (Popoola 2022). Buckwheat, when germinated, shows a significant rise in polyphenol content, antioxidant activity, and total dietary fiber. Moreover, there is a significant reduction in crude fat, phytic acid, and trypsin inhibitor activity (Sofi et al. 2022). Tannin content increases in buckwheat after 168 hours of germination, and quinoa experiences elevated levels of vitamin E, vitamin C, and antioxidant activity (Bhinder et al. 2022).

In contrast, the fermentation of cereals, facilitated by microorganisms such as lactic acid bacteria (LAB), improves the sensory attributes and nutritional qualities of the end product. (Dong et al. 2023). Pseudocereals, rich in polysaccharides, serve as favorable substrates for microbial fermentations. LABs, known for their safety and fermentative capabilities, significantly contribute to the flavor, aroma, and nutritional enhancement of fermented cereals and pseudocereals (Petrova and Petrov 2020). The process of fermentation results in a decline in in vitro protein digestibility in amaranth, coupled with an elevation in ash content, fat, and protein in amaranth grains (Chiofalo et al. 2023). During the fermentation process, the chemical composition experiences a transformation, marked by a 3% reduction in vitamin E and a substantial 78% rise in antioxidant activity in quinoa. An intriguing aspect of LAB fermentation is its impact on the release and modification of phenolic compounds (Li et al. 2021). The metabolic pathways involved, including the production of bioactive peptides, highlight LAB's potential in enhancing the functional value of cereals and pseudocereals. Particularly, the modulation of phenolic profiles during fermentation contributes to increased antioxidant activity (Rollán et al. 2019).

In summary, these diverse processing methods significantly impact the nutritional and bioactive profiles of pseudocereals, emphasizing the need for careful consideration in food preparation to maximize their health benefits. Pseudocereals, encompassing a wealth of nutritional and bioactive components such as flavonoids, phenolic compounds, carotenoids, and betalains, offer a spectrum of health benefits. Recognized for their GF nature, these grains play a pivotal role in diverse physiological aspects, showcasing anti-inflammatory, antiaging, and anticarcinogenic properties for overall well-being. Rich in antioxidants, pseudocereals exhibit potent antioxidative effects, combating oxidative stress in the body. Additionally, their antimicrobial properties fortify the immune system, promoting gastrointestinal health and modulating immune responses.

The cholesterol-lowering and antidiabetic attributes of pseudocereals make them valuable in managing cardiovascular health and diabetes (Agregán et al. 2023). Furthermore, their prebiotic nature supports gut health, and their antihypertensive qualities contribute to overall cardiovascular well-being (Tsafrakidou et al. 2020). Embracing pseudocereals in one's diet can serve as a preventive measure against osteoporosis and obesity (Patil et al. 2022).

Celiac syndrome, associated with gluten intolerance, is a condition that can affect individuals of all age groups. Yet, ensuring the successful adherence to a GF diet faces obstacles associated with the cost, cost-effectiveness, and precise labeling of GF foods. In exploring alternative food options, specific products exhibit distinctive characteristics. Buckwheat noodles, featuring varying buckwheat flour percentages (40%, 50%, 60%), showcase reduced cooking time and achieve the highest swelling volume. Amaranth spaghetti, composed of 10% amaranth, 50% rice, and 40% quinoa, presents a texture that is sticky with insufficient firmness and a taste profile that is soft with a musty flavor. Quinoa lady finger biscuits, incorporating quinoa flour (QF) at different proportions (25%, 50%, 75%, 100%) with rice flour (RF), stand out for their high protein and lipid content, accompanied by improved color and texture. This detailed analysis provides valuable insights into the characteristics of these products, offering considerations for individuals navigating gluten intolerance and seeking suitable dietary options (Kumari and Morya 2021). In summary, pseudocereals emerge as multifaceted functional foods, providing an array of health-promoting effects for holistic well-being.

17.6 Challenges and Opportunities

Pseudocereals, though rich in nutritional content, face a notable challenge in terms of limited awareness and acceptance among consumers compared to traditional grains. Encouraging educational campaigns and strategic marketing becomes a crucial pathway to familiarize the public with the nutritional content and wellness advantages linked to pseudocereals.

Nonetheless, despite their considerable potential, challenges such as the high expenses associated with imported grains and a lack of awareness among consumers have impeded their broad adoption (Taylor et al. 2023).

The comprehensive exploration of the functional and nutritional advantages of pseudocereals, along with a careful examination of potential adverse effects, necessitates further research. While pseudocereals exhibit elevated levels of protein, minerals, amino acids, and other advantageous components, challenges arise due to insufficient research on their nutritional composition and a lack of advanced processing technologies. These factors hinder the widespread consumption and commercialization of these nutrient-rich grains. Challenges are compounded by an underdeveloped supply chain, limiting access to these essential foods (Ali and Bhattacharjee 2023).

Resistance to incorporating pseudocereals into daily diets is prevalent due to unfamiliar tastes and textures. To overcome this hurdle, culinary innovations, recipe development, and highlighting the versatility of pseudocereals become essential in enhancing their palatability and broadening their appeal. Moreover, the limited cultivation and accessibility of pseudocereal varieties compared to major crops underscore the need for investment in research, sustainable farming practices, and supportive agricultural policies (Bakshi et al. 2023).

Despite these challenges, pseudocereals present an opportunity for the development of functional foods targeting specific health concerns, including cardiovascular health, diabetes, and cancer prevention. However, research and development efforts must concentrate on creating innovative and appealing functional food products to drive adoption (Kasote et al. 2021). Global culinary integration offers a pathway to open new markets and diversify consumption patterns, but it requires overcoming cultural barriers and preconceptions associated with traditional dietary habits.

Expanding the range of pseudocereal-based products, such as snacks, cereals, and beverages, can cater to diverse consumer preferences (Tiwari et al. 2021). Nevertheless, addressing challenges related to processing, shelf stability, and scalability in product development remains a significant hurdle that needs attention.

In the realm of research and innovation, continuous exploration of bioactive compounds, health benefits, and technological applications of pseudocereals is essential. However, securing funding and fostering collaboration between academia, industry, and policymakers is crucial to drive these research initiatives. Notably, while pseudocereals are GF, concerns about cross-contamination and allergen labeling persist. Stringent quality control measures, clear labeling, and certification processes present an opportunity to address these concerns, build consumer trust, and pave the way for wider acceptance (Vriesekoop 2021).

The current landscape reflects a juxtaposition of challenges and opportunities for pseudocereals. Rising health consciousness provides a favorable environment for adoption, yet effective communication of these benefits to consumers remains a challenge. Navigating this intricate balance requires strategic efforts in marketing and educational initiatives to promote the recognition and integration of pseudocereals into mainstream dietary practices (April-Lalonde et al. 2023).

17.7 Conclusion

The exploration of pseudocereals unveils a compelling narrative marked by resilience, nutritional richness, and vast untapped potential. These grains, acclaimed as nutritional powerhouses and crops resilient to climate variations, showcase exceptional adaptability and a diverse array of health benefits. Despite being GF and rich in proteins, currently underutilized pseudocereals present an opportune moment for wider recognition and application, especially given the global demand for GF alternatives driven by the escalating cases of celiac disease. Positioned as excellent candidates for functional food production in the nutraceutical industry, pseudocereals offer a multitude of active principles that hold immense promise for human health. The amino acid composition of pseudocereal proteins, particularly their suitability as supplements for child nutrition, further underscores their nutritional value. Their low prolamin content enhances versatility and inclusivity, catering to individuals with celiac disease. Pseudocereals, boasting diverse health benefits encompassing cardioprotection, anticancerous, and anti-inflammatory properties, exhibit a unique fatty acid composition and heightened mineral content, effectively addressing hidden hunger. Cultivating these grains on marginal lands, resilient to abiotic stresses, emerges as a promising solution for underdeveloped and developing countries striving for food security. Beyond their nutritional contributions, pseudocereals play a pivotal role as essential components of functional foods, elevating taste preferences and significantly contributing to the health benefits of these products. Recognizing them as nutritional powerhouses and resilient crops, recent biotechnological interventions have become imperative for sustained improvement and domestication. Looking ahead, acknowledging, and harnessing the dual advantage of pseudocereals as eco-friendly and economically viable choices positions them as valuable contributors to future agricultural practices and the food industry. Their potential as nutraceuticals is particularly promising, with the promotion of educational campaigns and strategic marketing serving as crucial avenues to acquaint the public with the nutritional value and health benefits associated with pseudocereals. Pseudocereals emerge not merely as grains of the twenty-first century but as potential transformative agents in advancing holistic well-being and sustainable nutrition, poised to carve a significant niche in the nutraceutical realm.

References

Abellán Ruiz, M.S., Barnuevo Espinosa, M.D., García Santamaría, C. et al. (2017). Efecto del consumo de quinua (*Chenopodium quinoa*) como coadyuvante en la intervención nutricional en sujetos prediabéticos. *Nutricion Hospitalaria* 34 (5): 1163–1169. https://doi.org/10.20960/nh.843.

Afzal, I., Haq, M.Z.U., Ahmed, S. et al. (2023). Challenges and perspectives for integrating quinoa into the agri-food system. *Plants* 12 (19): 3361.

Agregán, R., Guzel, N., Guzel, M. et al. (2023). The effects of processing technologies on nutritional and anti-nutritional properties of pseudocereals and minor cereal. In: *Food and Bioprocess*

Technology, vol. 16, Issue 5 (ed. D.W. Sun), 961–986. Springer. https://doi.org/10.1007/s11947-022-02936-8.

Ain, Q.T., Siddique, K., Bawazeer, S. et al. (2023). Adaptive mechanisms in quinoa for coping in stressful environments: an update. *PeerJ* 11: e14832. https://doi.org/10.7717/peerj.14832.

Alandia, G., Odone, A., Rodriguez, J.P. et al. (2021). Quinoa—evolution and future perspectives. In: *The Quinoa Genome* (ed. K. Chittaranjan), 179–195. Springer.

Ali, A. and Bhattacharjee, B. (2023). Nutrition security, constraints, and agro-diversification strategies of neglected and underutilized crops to fight global hidden hunger. In: *Frontiers in Nutrition*, vol. 10 (ed. R. Singh). Frontiers Media SA. https://doi.org/10.3389/fnut.2023.1144439.

Angeli, V., Silva, P.M., Massuela, D.C. et al. (2020). Quinoa (*Chenopodium quinoa* Willd.): an overview of the potentials of the "golden grain" and socio-economic and environmental aspects of its cultivation and marketization. *Foods* 9 (2). https://doi.org/10.3390/foods9020216.

April-Lalonde, G., Deaconu, A., Cole, D.C., and Batal, M. (2023). Traditional food consumption in Andean Ecuador and associated consumer characteristics, shopping and eating habits. *Ecology of Food and Nutrition* 62 (5-6): 308–333. https://doi.org/10.1080/03670244.2023.2264197.

Ashik, S.K. and Morya, S. (2022). A review on finger millet properties, processing, health benefits, and applications. *The Pharma Innovation Journal* 11 (7): 3388–3395.

Awuchi, C.G. and Morya, S. (2023). Herbs of asteraceae family: nutritional profile, bioactive compounds, and potentials in therapeutics. In: *Harvesting Food from Weeds* (ed. P. Gupta, N. Chhikara, and A. Panghal), 21–78. Wiley.

Bakshi, S., Kamble, S., Jambhulkar, S.J. et al. (2023). Transgenic approaches for nutritional quality improvement of cereals and pseudocereals. In: *Developing Sustainable and Health Promoting Cereals and Pseudocereals* (ed. M. Rakszegi, M. Papageorgiou, and J.M. Rocha), 285–313. Academic Press. https://doi.org/10.1016/B978-0-323-90566-4.00004-7.

Ballester-Sánchez, J., Gil, J.V., Haros, C.M., and Fernández-Espinar, M.T. (2019). Effect of incorporating white, red or black quinoa flours on free and bound polyphenol content, antioxidant activity and colour of bread. *Plant Foods for Human Nutrition* 74 (2): 185–191. https://doi.org/10.1007/s11130-019-00718-w.

Bangar, S.P., Sharma, N., Singh, A. et al. (2022). Glycaemic response of pseudocereal-based gluten-free food products: a review. *International Journal of Food Science & Technology* 57 (8): 4936–4944.

Baraniak, J. and Kania-Dobrowolska, M. (2022). The dual nature of amaranth—functional food and potential medicine. *Foods* 11 (4): 618. https://doi.org/10.3390/foods11040618.

Bazile, D., Jacobsen, S.E., and Verniau, A. (2016). The global expansion of quinoa: trends and limits. In: *Frontiers in Plant Science*, vol. 7 (ed. P.C. Struik). Switzerland: Frontiers Media. https://doi.org/10.3389/fpls.2016.00622.

Berger, A., Gremaud, G., Baumgartner, M. et al. (2003). Cholesterol-lowering properties of amaranth grain and oil in hamsters 1. *International Journal for Vitamin and Nutrition Research* 73 (1): 39–47.

Berti, C., Ballabio, C., Restani, P. et al. (2004). Immunochemical and molecular properties of proteins in *Chenopodium quinoa*. *Cereal Chemistry* 81 (2): 275–277.

Bhinder, S., Singh, N., and Kaur, A. (2022). Impact of germination on nutraceutical, functional and gluten free muffin making properties of Tartary buckwheat (*Fagopyrum tataricum*). *Food Hydrocolloids* 124: 107268.

Breslauer, R., Nalbandian, E., Reinman, T. et al. (2023). Buckwheat production and value-added processing: a review of potential western Washington cropping and food system applications. *Sustainability* 15 (20): 14758. https://doi.org/10.3390/su152014758.

Chiofalo, B., Vastolo, A., Oteri, M. et al. (2023). Productive, qualitative, and in vitro fermentation traits of amaranthus grains as potential ingredients for pig diet. *Agriculture (Switzerland)* 13 (7): 1445. https://doi.org/10.3390/agriculture13071445.

Cilla, A., Barberá, R., López-García, G. et al. (2019). Impact of processing on mineral bioaccessibility/bioavailability. In: *Innovative Thermal and Non-Thermal Processing, Bioaccessibility and Bioavailability of Nutrients and Bioactive Compounds* (ed. F.J. Barba, J.M.A. Saraiva, G. Cravotto, and J.M. Lorenzo), 209–239. Elsevier. https://doi.org/10.1016/B978-0-12-814174-8.00007-X.

Dakhili, S., Abdolalizadeh, L., Hosseini, S.M. et al. (2019). Quinoa protein: composition, structure and functional properties. In: *Food Chemistry*, vol. 299 (ed. C. Lavelle). Elsevier Ltd. https://doi.org/10.1016/j.foodchem.2019.125161.

Das, S. (2016). Pseudocereals: an efficient food supplement. In: *Amaranthus: A Promising Crop of Future* (ed. S. Das). Singapore: Springer. https://doi.org/10.1007/978-981-10-1469-7_2.

De Carvalho, F.G., Ovídio, P.P., Padovan, G.J. et al. (2014). Metabolic parameters of postmenopausal women after quinoa or corn flakes intake-a prospective and double-blind study. *International Journal of Food Sciences and Nutrition* 65 (3): 380–385. https://doi.org/10.3109/09637486.2013.866637.

Dhondiram, D.S., Sawant, A.A., Mamoon, F. et al. (2023). Nutritional characterization of pseudocereals: a review. *The Pharma Innovation 12* (3): 109–118. https://doi.org/10.22271/tpi.2023.v12.i3b.19413.

Dixit, S., Shukla, A., Singh, V., and Upadhyay, S.K. (2021). Engineering of plant metabolic pathway for nutritional improvement: recent advances and challenges. In: *Genome Engineering for Crop Improvement*. Wiley; Portico. https://doi.org/10.1002/9781119672425.ch20.

Dong, Y., Wang, N., Wang, S. et al. (2023). A review: the nutrition components, active substances and flavonoid accumulation of Tartary buckwheat sprouts and innovative physical technology for seeds germinating. In: *Frontiers in Nutrition*, vol. 10 (ed. M. Schmiele). Frontiers Media SA. https://doi.org/10.3389/fnut.2023.1168361.

Dubey, P.K., Singh, A., Chaurasia, R., and Abhilash, P.C. (2020). Climate adaptive agricultural interventions for food, nutritional, health and livelihood security. In: *Nature-based Solutions for Resilient Ecosystems and Societies* (ed. S. Dhyani, A. Gupta, and M. Karki), 267–288. Singapore: Springer.

Dziedzic, K., Górecka, D., Szwengiel, A. et al. (2018). The cytotoxic effect of artificially digested buckwheat products on HT-29 colon cancer cells. *Journal of Cereal Science* 83: 68–73. https://doi.org/10.1016/j.jcs.2018.07.020.

EMR (2023). EMR. https://www.expertmarketresearch.com/reports/buckwheat-market (accessed 5 July 2024)

Farinazzi-Machado, F.M.V., Barbalho, S.M., Oshiiwa, M. et al. (2012). Use of cereal bars with quinoa (*Chenopodium quinoa* W.) to reduce risk factors related to cardiovascular diseases. *Food Science and Technology* 32 (2): 239–244. https://doi.org/10.1590/s0101-20612012005000040.

Fernández-López, J., Botella-Martínez, C., de Vera, C.N.R. et al. (2020). Vegetable soups and creams: raw materials, processing, health benefits, and innovation trends. *Plants* 9 (12): 1–33. https://doi.org/10.3390/plants9121769.

Fritz, M., Vecchi, B., Rinaldi, G., and Añón, M.C. (2011). Amaranth seed protein hydrolysates have in vivo and in vitro antihypertensive activity. *Food Chemistry* 126 (3): 878–884. https://doi.org/10.1016/j.foodchem.2010.11.065.

Garutti, M., Nevola, G., Mazzeo, R. et al. (2022). The impact of cereal grain composition on the health and disease outcomes. In: *Frontiers in Nutrition*, vol. 9 (ed. Q. Shen). Switzerland: Frontiers Media. https://doi.org/10.3389/fnut.2022.888974.

Ge, R.H. and Wang, H. (2020). Nutrient components and bioactive compounds in tartary buckwheat bran and flour as affected by thermal processing. *International Journal of Food Properties* 23 (1): 127–137. https://doi.org/10.1080/10942912.2020.1713151.

Gupta, N. and Morya, S. (2022). Bioactive and pharmacological characterization of Chenopodium quinoa, Sorghum bicolor and Linum usitassimum: a review. *Journal of Applied and Natural Science* 14 (3): 1067–1084. Applied and Natural Science Foundation. https://doi.org/10.31018/jans.v14i3.3796.

Hernández-Ledesma, B. (2019). Quinoa (*Chenopodium quinoa* Willd.) as source of bioactive compounds: a review. *Bioactive Compounds in Health and Disease* 2 (3): 27–47.

Jin, J., Ohanenye, I.C., and Udenigwe, C.C. (2022). Buckwheat proteins: functionality, safety, bioactivity, and prospects as alternative plant-based proteins in the food industry. *Critical Reviews in Food Science and Nutrition* 62 (7): 1752–1764.

Jorge, S.S., Raúl, R.B., Isabel, G.L. et al. (2015). Dipeptidyl peptidase iv inhibitory activity of protein hydrolyzates from Amaranthus hypochondriacus L. grain and their influence on postprandial glycemia in streptozotocin-induced diabetic mice. *African Journal of Traditional, Complementary and Alternative Medicines* 12 (1): 90–98. https://doi.org/10.4314/ajtcam.v12i1.13.

Joshi, D. C., Chaudhari, G. V., Sood, S., Kant, L., Pattanayak, A., Zhang, K., Fan, Y., Janovská, D., Meglič, V., & Zhou, M. (2019). Revisiting the versatile buckwheat: reinvigorating genetic gains through integrated breeding and genomics approach. In Joshi D. C. (Ed), *Planta* (Vol. 250, Issue 3, pp. 783–801). Springer Verlag. https://doi.org/10.1007/s00425-018-03080-4

Joshi, D. C., Sood, S., Hosahatti, R., Kant, L., Pattanayak, A., Kumar, A., Yadav, D., & Stetter, M. G. (2018). From zero to hero: the past, present and future of grain amaranth breeding. In *Theoretical and Applied Genetics* (Vol. 131, Issue 9, pp. 1807–1823). Springer Verlag. https://doi.org/10.1007/s00122-018-3138-y

Karaca, A.C., Nickerson, M., Caggia, C. et al. (2023). Nutritional and functional properties of novel protein sources. *Food Reviews International* 39 (9): 6045–6077. https://doi.org/10.1080/87559129.2022.2067174.

Kasote, D., Tiozon, R.N., Sartagoda, K.J.D. et al. (2021). Food processing technologies to develop functional foods with enriched bioactive phenolic compounds in cereals. In: *Frontiers in Plant Science*, vol. 12. Switzerland: Frontiers Media. https://doi.org/10.3389/fpls.2021.771276.

Kaur, H., Shams, R., Dash, K.K., and Dar, A.H. (2023). A comprehensive reiew of pseudo-cereals: nutritional profile, phytochemicals constituents and potential health promoting benefits. *Applied Food Research* 3 (24): 100351. https://doi.org/10.1016/j.afres.2023.100351.

Kumar, S., Malviya, R., and Sundram, S. (2023). Nutritional neurology: unraveling cellular mechanisms of natural supplements in brain health. *Human Nutrition & Metabolism* 35 (3): 200232.

Kumar, A., Shumayla, and Upadhyay, S.K. (2023). CRISPR/Cas-mediated genome editing in plants: a historical perspective. In: *Applications of Genome Engineering in Plants*. Wiley; Portico. https://doi.org/10.1002/9781394183913.ch1.

Kumari, J. and Morya, S. (2021). Celiac disease: an epidemiological condition: Insight on gluten free diet, significance and regulatory recommendations. *The Pharma Innovation* 6 (7): 641–654. https://doi.org/10.22271/tpi.2021.v10.i5i.6280.

Lado, M.B., Burini, J., Rinaldi, G. et al. (2015). Effects of the dietary addition of amaranth (*Amaranthus mantegazzianus*) protein isolate on antioxidant status, lipid profiles and blood pressure of rats. *Plant Foods for Human Nutrition* 70 (4): 371–379. https://doi.org/10.1007/s11130-015-0516-3.

Landi, N., Ruocco, M.R., Ragucci, S. et al. (2021). Quinoa as source of type 1 ribosome inactivating proteins: a novel knowledge for a revision of its consumption. *Food Chemistry* 342: 128337. https://doi.org/10.1016/J.FOODCHEM.2020.128337.

Laparra, J.M. and Haros, M. (2016). Inclusion of ancient Latin-American crops in bread formulation improves intestinal iron absorption and modulates inflammatory markers. *Food and Function* 7 (2): 1096–1102. https://doi.org/10.1039/c5fo01197c.

Lazou, A.E. (2022). Food extrusion: an advanced process for innovation and novel product development. *Critical Reviews in Food Science and Nutrition* 64 (14): 4532–4560.

Li, L., Lietz, G., Bal, W. et al. (2018). Effects of quinoa (*Chenopodium quinoa* Willd.) consumption on markers of CVD risk. *Nutrients* 10 (6): 777. https://doi.org/10.3390/nu10060777.

Li, L., Lietz, G., and Seal, C.J. (2021). Phenolic, apparent antioxidant and nutritional composition of quinoa (*Chenopodium quinoa* Willd.) seeds. *International Journal of Food Science and Technology* 56 (7): 3245–3254. https://doi.org/10.1111/ijfs.14962.

Li, H., Lv, Q., Liu, A. et al. (2022). Comparative metabolomics study of Tartary (*Fagopyrum tataricum* (L.) Gaertn) and common (*Fagopyrum esculentum* Moench) buckwheat seeds. *Food Chemistry* 371: 131125.

Liu, Y., Qiu, J., Yue, Y. et al. (2018b). Dietary black-grained wheat intake improves glycemic control and inflammatory profile in patients with type 2 diabetes: a randomized controlled trial. *Therapeutics and Clinical Risk Management* 14: 247–256. https://doi.org/10.2147/TCRM.S151424.

Liu, W., Zhang, Y., Qiu, B. et al. (2018a). Quinoa whole grain diet compromises the changes of gut microbiota and colonic colitis induced by dextran Sulfate sodium in C57BL/6 mice. *Scientific Reports* 8 (1): 14916. https://doi.org/10.1038/s41598-018-33092-9.

Malik, A.M. and Singh, A. (2022). Pseudocereals proteins-a comprehensive review on its isolation, composition and quality evaluation techniques. *Food Chemistry Advances* 1: 100001.

Malleshi, N.G., Agarwal, A., Tiwari, A., and Sood, S. (2020). Nutritional quality and health benefits. In: *Millets and Pseudo Cereals: Genetic Resources and Breeding Advancements* (ed. M. Singh and S. Sood), 159–168. Elsevier. https://doi.org/10.1016/B978-0-12-820089-6.00009-4.

Martinez-Lopez, A., Millan-Linares, M.C., Rodriguez-Martin, N.M. et al. (2020). Nutraceutical value of kiwicha (*Amaranthus caudatus* L.). *Journal of Functional Foods* 65 (2): 103735. Elsevier Ltd. https://doi.org/10.1016/j.jff.2019.103735.

Martínez-Villaluenga, C., Peñas, E., and Hernández-Ledesma, B. (2020). Pseudocereal grains: nutritional value, health benefits and current applications for the development of gluten-free foods. *Food and Chemical Toxicology* 137: 111178. Elsevier Ltd. https://doi.org/10.1016/j.fct.2020.111178.

Miranda-Ramos, K.C. and Haros, C.M. (2020). Combined effect of chia, quinoa and amaranth incorporation on the physico-chemical, nutritional and functional quality of fresh bread. *Foods* 9 (12): 1859. https://doi.org/10.3390/foods9121859.

Mithila, M.V. and Khanum, F. (2015). Effectual comparison of quinoa and amaranth supplemented diets in controlling appetite; a biochemical study in rats. *Journal of Food Science and Technology* 52 (10): 6735–6741. https://doi.org/10.1007/s13197-014-1691-1.

Mohamed Ahmed, I.A., Al Juhaimi, F., and Özcan, M.M. (2021). Insights into the nutritional value and bioactive properties of quinoa (*Chenopodium quinoa*): past, present and future prospective. *International Journal of Food Science and Technology* 56 (8): 3726–3741. John Wiley and Sons Inc. https://doi.org/10.1111/ijfs.15011.

Montemurro, M., Pontonio, E., and Rizzello, C.G. (2019). Quinoa flour as an ingredient to enhance the nutritional and functional features of cereal-based foods. In: *Flour and breads and their fortification in health and disease prevention* (ed. V.R. Preedy and R.R. Watson), 453–464. Academic Press.

Morya, S., Awuchi, C.G., and Menaa, F. (2022b). Advanced functional approaches of nanotechnology in food and nutrition. In: *Environmental Management Technologies: Challenges and Opportunities* (ed. P. Chowdhary, V. Kumar, S. Kumar, and V. Hare). CRC press.

Morya, S., Menaa, F., Jiménez-López, C. et al. (2022a). Nutraceutical and pharmaceutical behavior of bioactive compounds of miracle oilseeds: an overview. *Foods* 11 (13): 1824.

Mudau, F.N., Chimonyo, V.G.P., Modi, A.T., and Mabhaudhi, T. (2022). Neglected and underutilised crops: a systematic review of their potential as food and herbal medicinal crops in South Africa. *Frontiers in Pharmacology* 12: 809866.

Nazeer, S. and Firincioglu, S.Y. (2022). Amaranth in animal nutrition. *Journal of Agriculture, Food, Environment and Animal Sciences* 3 (2): 195–211.

Ng, C.Y. and Wang, M. (2021). The functional ingredients of quinoa (*Chenopodium quinoa*) and physiological effects of consuming quinoa: a review. In: *Food Frontiers*, vol. 2, Issue 3 (ed. F. Xie), 329–356. John Wiley and Sons Inc. https://doi.org/10.1002/fft2.109.

Noratto, G.D., Murphy, K., and Chew, B.P. (2019). Quinoa intake reduces plasma and liver cholesterol, lessens obesity-associated inflammation, and helps to prevent hepatic steatosis in obese db/db mouse. *Food Chemistry 287*: 107–114. https://doi.org/10.1016/j.foodchem.2019.02.061.

Olguín-Calderón, D., González-Escobar, J.L., Ríos-Villa, R. et al. (2019). Modulation of caecal microbiome in obese mice associated with administration of amaranth or soybean protein isolates. *Polish Journal of Food and Nutrition Sciences* 69 (1): 35–44. https://doi.org/10.31883/pjfns-2019-0002.

Park, S.J., Sharma, A., and Lee, H.J. (2020). A review of recent studies on the antioxidant activities of a third-millennium food: Amaranthus spp. *Antioxidants* 9 (12): 1–22. https://doi.org/10.3390/antiox9121236.

Patil, P.J., Usman, M., Zhang, C. et al. (2022). An updated review on food-derived bioactive peptides: focus on the regulatory requirements, safety, and bioavailability. *Comprehensive Reviews in Food Science and Food Safety* 21 (2): 1732–1776.

Peñas, E., Tomé-Sánchez, I., and Martínez-Villaluenga, C. (2023). Comparison of the nutritional value of minor and pseudocereal crops to major crops and the barriers in their breeding for developing healthy grains. In: *Developing Sustainable and Health Promoting Cereals and Pseudocereals* (ed. M. Marianna Rakszegi, M. Papageorgiou, and J.M. Rocha), 99–133. Academic Press.

Petrova, P. and Petrov, K. (2020). Lactic acid fermentation of cereals and pseudocereals: ancient nutritional biotechnologies with modern applications. In: *Nutrients*, vol. 12), Issue 4 (ed. M.L. Fernandez and L.S. Majem). MDPI AG. https://doi.org/10.3390/nu12041118.

Pipan, B., Sinkovič, L., Neji, M. et al. (2023). Agro-morphological and molecular characterization reveal deep insights in promising genetic diversity and marker-trait associations in *Fagopyrum esculentum* and *Fagopyrum tataricum*. *Plants* 12 (18): 3321.

Popoola, O.O. (2022). Phenolic compounds composition and in vitro antioxidant activity of Nigerian *Amaranthus viridis* seed as affected by autoclaving and germination. *Measurement: Food* 6: 100028.

Rahman, M.M., Rahaman, M.S., Islam, M.R. et al. (2022). Role of phenolic compounds in human disease: current knowledge and future prospects. In: *Molecules*, vol. 27, Issue 1. MDPI. https://doi.org/10.3390/molecules27010233.

Ram, A., Thattantavide, A., and Kumar, A. (2023). Re-emergence of Pseudocereals as Superfoods for Food Security and Human Health: Current Progress and Future Prospects. In: *Wild Food Plants for Zero Hunger and Resilient Agriculture*, 207–236. Springer.

Ranjan, S., Sow, S., Ghosh, M. et al. (2023). Nutraceutical properties and secondary metabolites of quinoa (*Chenopodium quinoa* Willd.): a review. *International Journal of Food Properties* 26 (2): 3477–3491. https://doi.org/10.1080/10942912.2023.2286895.

Rao, V. and Poonia, A. (2023). Protein characteristics, amino acid profile, health benefits and methods of extraction and isolation of proteins from some pseudocereals—a review. In: *Food Production, Processing and Nutrition*, vol. 5), Issue 1 (ed. A.K. Anal). BioMed Central Ltd. https://doi.org/10.1186/s43014-023-00154-z.

Raungrusmee, S. (2023). Underutilized cereals and pseudocereals' nutritional potential and health implications. In: *Pandemics and Innovative Food Systems* (ed. A.K. Anal), 163–193. CRC press.

Rawat, M., Varshney, A., Rai, M. et al. (2023). A comprehensive review on nutraceutical potential of underutilized cereals and cereal-based products. *Journal of Agriculture and Food Research* 12: 100619. https://doi.org/10.1016/J.JAFR.2023.100619.

Reisman, E. (2020). Superfood as spatial fix: the ascent of the almond. *Agriculture and Human Values* 37 (2): 337–351.

Rollán, G.C., Gerez, C.L., and Leblanc, J.G. (2019). Lactic fermentation as a strategy to improve the nutritional and functional values of pseudocereals. In: *Frontiers in Nutrition*, vol. 6

(ed. J.L. Coutre, J.M. Riera, E. Berry, et al.). Switzerland: Frontiers Media. https://doi.org/10.3389/fnut.2019.00098.

Rosentrater, K.A. and Evers, A.D. (2018). Introduction to cereals and pseudocereals and their production. In: *Kent's Technology of Cereals* (ed. K. Rosentrater), 1–76. Elsevier. https://doi.org/10.1016/b978-0-08-100529-3.00001-3.

Sabbione, A.C., Rinaldi, G., Añón, M.C., and Scilingo, A.A. (2016). Antithrombotic effects of amaranthus hypochondriacus proteins in rats. *Plant Foods for Human Nutrition* 71 (1): 19–27. https://doi.org/10.1007/s11130-015-0517-2.

Sae-Eaw, A., Wongsaichia, S., Giacalone, D. et al. (2022). Conceptualizing a gluten-free instant noodle prototype using environmental sustainability aspects: a cross-national qualitative study on Thai and Danish consumers. *Foods* 11 (16): 2437. https://doi.org/10.3390/foods11162437.

Salvador-Reyes, R., Furlan, L., Martínez-Villaluenga, C. et al. (2023). From ancient crop to modern superfood: exploring the history, diversity, characteristics, technological applications, and culinary uses of Peruvian fava beans. *Food Research International* 173: 113394.

Sánchez-Velázquez, O.A., Manzanilla-Valdez, M.L., Mondor, M. et al. (2022). Pseudocereals: ancient grains as sources of bioactive compounds for functional foods and modulation of molecular markers. In: *Molecular Mechanisms of Functional Food* (ed. R.C. Vega and B.D. Oomah), 520–595. Wiley.

Schmidt, D., Verruma-Bernardi, M.R., Forti, V.A., and Borges, M.T.M.R. (2023). Quinoa and amaranth as functional foods: a review. *Food Reviews International* 39 (4): 2277–2296. https://doi.org/10.1080/87559129.2021.1950175.

Shahbaz, M., Raza, N., Islam, M. et al. (2023). The nutraceutical properties and health benefits of pseudocereals: a comprehensive treatise. *Critical Reviews in Food Science and Nutrition* 63 (29): 10217–10229. https://doi.org/10.1080/10408398.2022.2071205.

Sharma, A., Alam, M., Pant, K., and Nanda, V. (2023). Amaranth & quinoa sprouts. In: *Advances in Plant Sprouts: Phytochemistry and Biofunctionalities* (ed. I. Majid, B.A. Kehinde, B. Dar, and V. Nanda), 127–151. Cham: Springer International Publishing.

Singh, A., Kumari, A., and Chaudhary, H.K. (2021). Amaranth, buckwheat, and chenopodium: the "ABC" nutraceuticals of northwestern himalayas. In: *Agricultural Biotechnology: Latest Research and Trends* (ed. D.K. Srivastava, A.K. Thakur, and P. Kumar), 587–634. Springer.

Skendi, A., Zinoviadou, K. G., Papageorgiou, M., & Rocha, J. M. (2020). Advances on the valorization and functionalization of by-products and wastes from cereal-based processing industry. In Vinson, J (Ed), *Foods* (Vol. 9, Issue 9). MDPI Multidisciplinary Digital Publishing Institute. https://doi.org/10.3390/foods9091243

Skřivan, P., Chrpová, D., Klitschová, B. et al. (2023). Buckwheat flour (*Fagopyrum esculentum* Moench)—a contemporary view on the problems of its production for human nutrition. In: *Foods*, vol. 12, Issue 16 (ed. J. Vinson), 3055. Multidisciplinary Digital Publishing Institute (MDPI). https://doi.org/10.3390/foods12163055.

Sofi, S.A., Ahmed, N., Farooq, A. et al. (2022). Nutritional and bioactive characteristics of buckwheat, and its potential for developing gluten-free products: an updated overview. In: *Food Science and Nutrition* (ed. S. Safren). John Wiley and Sons Inc. https://doi.org/10.1002/fsn3.3166.

Solanki, C. and Singh, R. (2023). Buckwheat processing and its utilization in value-added products: a comprehensive review. ~ *2746* ~. *The Pharma Innovation Journal* 12 (8): 2746–2752. www.thepharmajournal.com.

Stringer, D.M., Taylor, C.G., Appah, P. et al. (2013). Consumption of buckwheat modulates the post-prandial response of selected gastrointestinal satiety hormones in individuals with type 2 diabetes mellitus. *Metabolism: Clinical and Experimental* 62 (7): 1021–1031. https://doi.org/10.1016/j.metabol.2013.01.021.

Taylor, J.R.N., Rehm, C.D., de Kock, H.L. et al. (2023). South African consumers' knowledge, opinions and awareness of whole grains and their health benefits: a cross-sectional online survey. *Nutrients* 15 (16): 3522. https://doi.org/10.3390/nu15163522.

Thakur, P., Kumar, K., and Dhaliwal, H.S. (2021). Nutritional facts, bio-active components and processing aspects of pseudocereals: a comprehensive review. *Food Bioscience* 42: 101170. https://doi.org/10.1016/J.FBIO.2021.101170.

Thambiliyagodage, C., Jayanetti, M., Mendis, A. et al. (2023). Recent advances in chitosan-based applications—a review. *Materials* 16 (5): 2073.

Tien, N.N.T., Trinh, L.N.D., Inoue, N. et al. (2018). Nutritional composition, bioactive compounds, and diabetic enzyme inhibition capacity of three varieties of buckwheat in Japan. *Cereal Chemistry* 95 (5): 615–624. https://doi.org/10.1002/cche.10069.

Tiwari, U., Muthugopal Sasthri, V., Krishnakumar, N., and Prabhasankar, P. (2021). *Innovative processing technologies for healthy grains.* In: *Advances in Conventional Cereal and Pseudocereal Processing, 1e*e (ed. M. Pojicánd). John Wiley & Sons.

Tsafrakidou, P., Michaelidou, A.M., and Biliaderis, C.G. (2020). Fermented cereal-based products: nutritional aspects, possible impact on gut microbiota and health implications. In: *Foods*, vol. 9, Issue 6 (ed. J. Vinson). MDPI Multidisciplinary Digital Publishing Institute. https://doi.org/10.3390/FOODS9060734.

Upadhyay, S.K. (Éd.) (2021). *Genome Engineering for Crop Improvement.* Wiley. https://doi.org/10.1002/9781119672425.

Upadhyay, S.K. (Éd.) (2024). *Applications of Genome Engineering in Plants.* Wiley. https://doi.org/10.1002/9781394183913.

Upadhyay, S.K. and Singh, S.P. (Éds.) (2021). *Bioprospecting of Plant Biodiversity for Industrial Molecules.* Wiley. https://doi.org/10.1002/9781119718017.

Upadhyay, S.K. and Singh, S.P. (Éds.) (2023). *Plants as Bioreactors for Industrial Molecules.* Wiley. https://doi.org/10.1002/9781119875116.

Valido, E., Stoyanov, J., Gorreja, F. et al. (2023). Systematic review of human and animal evidence on the role of buckwheat consumption on gastrointestinal health. In: *Nutrients*, vol. 15, Issue 1 (ed. M.L. Fernandez and L.S. Majem). MDPI. https://doi.org/10.3390/nu15010001.

Verma, K.C., Rana, A.S., Joshi, N., and Bhatt, D. (2020). Review on common buckwheat (*Fagopyrum esculentum* Moench): a potent Himalayan crop. *Annals of Phytomedicine: An International Journal* 9 (2): 125–133. https://doi.org/10.21276/ap.2020.9.2.10.

Vriesekoop, F. (2021). Beer and allergens. In: *Beverages*, vol. 7, Issue 4 (ed. E. Chambers). MDPI. https://doi.org/10.3390/beverages7040079.

Wanniarachchi, P.C., Pavalakumar, D., and Jayasinghe, M.A. (2023). Pseudocereals: nutrition, health benefits, and potential applications in gluten-free food product developments. *Current Nutrition & Food Science* 19 (4): 377–385.

Zenda, T., Liu, S., Dong, A., and Duan, H. (2021). Advances in cereal crop genomics for resilience under climate change. In: *Life*, vol. 11), Issue 6 (ed. M. Burnett). MDPI AG. https://doi.org/10.3390/life11060502.

Zhou, X.L., Yan, B.B., Xiao, Y. et al. (2018). Tartary buckwheat protein prevented dyslipidemia in high-fat diet-fed mice associated with gut microbiota changes. *Food and Chemical Toxicology* 119: 296–301. https://doi.org/10.1016/j.fct.2018.02.052.

18

Safety Concerns Associated with Microalgal Nutraceuticals

Freny Shah[1]*, Niranjan Mishra*[1]*, Kanti Bhooshan Pandey*[2]*, and Sandhya Mishra*[1]

[1] *Applied Phycology and Biotechnology Division, CSIR-Central Salt and Marine Chemicals Research Institute, Bhavnagar, Gujarat, India*
[2] *BDIM Division, CSIR-Central Salt and Marine Chemicals Research Institute, Bhavnagar, Gujarat, India*

Abbreviations

ADHD	Attention-deficit/hyperactivity disorder
APX	Ascorbate peroxidase
CAGR	Compound Annual Growth Rate
CAT	Catalase
DHA	Docosa Hexaneoic Acid
DPPH	2,2-diphenyl-1-picrylhydrazyl
H_2O_2	Hydrogen peroxide
FDA	Food and Drug Administration
FSANZ	Food Standard Australia and New Zealand
GMO	Genetically Modified Organism
GRAS	Generally Recognized as Safe
MDA	Malondialdehyde
NaClO	Sodium hypochlorite
PUFA	Poly Unsaturated Fatty Acid
ROS	Reactive oxygen species
SOD	Superoxide dismutase

18.1 Introduction

Cyanophyceae and Chlorophyceae are the best microbes for several bioactive compounds having pharmaceutical, nutraceutical, and cosmeceutical applications. This chapter provides an insight into some precautionary measures one needs to follow while consuming microalgae as nutrient supplement. It has been a well-established fact that safety is the most significant aspect to be taken into the consideration in microalgae-based food technology, more specifically during open pond biomass production (Enzing et al. 2014). Some principles of the production/utilization of microalgae-based products i.e. functional foods are gaining so much focus in some countries by

Herbal Nutraceuticals: Products and Processes, First Edition. Edited by Santosh Kumar Upadhyay and Sudhir Pratap Singh.

their respective concerning organization, but in developing countries like India, one has to be very alert, and aware of its source/origin and quality control. EU regulations mostly affect microalgal products for feed and food along with their ingredients at large scale (Vigani et al. 2015). Restriction GMO-based edible nutritional products are excluded from the list at present, as no recombinant microalgae are available at a marketable scale. The European Parliament, along with the Council, implemented common law for food regulation (EC No 178/2002). The food law was based on the general principles and requirements, which can be a basis for the feed and food development legislation. Food that was not taken by humans previously was named as novel food during the year 1975. During this period, the food regulation law was adopted by the EU. This law significantly points toward the enhancement of authorization process to ensure fast food delivery to the needy.

The other renowned food agency that regulates food-based law in the US is FDA, which controls the quality of microalgal byproducts. Based on the classification by Center for Food Safety and Applied Nutrition, microalgal biomass is considered as "other dietary supplement." *Dunaliella, Haematococcus, Chlorella, Schizochytrium, Porphyridium cruentum, Spirulina,* and *Crypthycodinium cohnii* are considered in the GRAS category. The USA and EU, along with other countries, have their respective regulations, for e.g. FSANZ regulates the consumption of new food supplements. The Food Standard Code department has permitted the strain of *Schizochytrium*, rich in omega-3 long-chain PUFA and DHA. Canada guides consumers to take precautions while using anything other than *Spirulina*-based products, besides not allowing therapeutic claims for things such as functional food. Considering such scenarios, the aim of the chapter is to signify the potential of microalgal biomass as nutrient supplement-based food products with certain parameters affecting its food/feed with respect to its consumption. Since they are at the base of the food chain in the aquatic ecosystem, one of the major issues in their direct/indirect intake is biomass contamination through pollutants e.g. herbicides, heavy metals, insecticides, hexachlorobenzene, and phenols (Baldiris-Navarro et al. 2018). The human world population is expected to reach more than nine billion by 2075, which has created the need for an alternate source of healthy food. The knowledge gained on the correlation between human health and food biochemistry indicates the importance of utilizing novel sources that are eco-friendly and also impact overall health besides providing essential nutrients. The exploitation of biomass commercially is facing challenges because of the diversity in species with deviations in biomass and several biotic and abiotic factors. Some more challenges are related to alteration during downstream processing, less safety information on the bioactive compounds, besides its fishy odor and unique taste in food formulations, so people have to develop their taste. Even then, these critical challenges provide great opportunities for exploiting them for their development as functional foods. Thorough research is required to overcome these bottlenecks to lead forward during the commercialization of microalgal functional foods.

18.2 Market Trends

In the last few years, people have had great awareness of healthy food, especially during the COVID-19 pandemic. Generally, healthy food contains such ingredients or nutrient supplements having prophylactic measures that may be beneficial for overall human health. Health conscious consumers are increasing day by day, which leads to increased demand for functional food products that aim for better health, preventing some chronic diseases, and increasing longevity (Pandey and Suttajit 2022; Price Waterhouse and Coopers 2009). Hence, microalgae-based metabolites have become the pioneering approach for the development of healthy edible products

(Gouveia et al. 2008). High nutritional value products from *Spirulina* is approximately 12,000 tons/year, with 70% being manufactured in developing countries of Asia. Worldwide cultivation of *Chlorella* is very high and estimated as 5000 tons/year. The market volume of other microalgae is also high such as *D. salina* (~3000 tons for carotene), *A. flosaquae* (~1500 tons for food), *H. pluvialis* (~700 tons for astaxanthin), *C. cohnii* (~500 tons for DHA), and *Shizochytrium* (~20 tons for DHA). The *Spirulina* market globally is estimated to surpass a CAGR of more than 10%, reaching around US$ 2000 by 2026. The functional food industry gradually continues to see relevant sales increase, driven by novel foods and soft drinks (Bagchi and Nair 2016). Promising niches in the market exist for such nutraceuticals from *Chlorella* spp. with a global market for one of the carotenoids in food, feed, and beverages. In addition, it is complementary to polyunsaturated fatty acids, which were conventionally obtained from fish oil.

18.3 Microalgal Source and Nutrient Profile

Several microalgal species have been used since long for food and feed, but presently they are more widely used at industrial scale for several products. Nutraceutical/pharmaceutical/cosmeceutical products derived from microalgae are being tested; however, there has been a great scope for the discovery of other health benefits along with their enhanced consumption. It will also improve interest in considering it for several treatments/prophylaxis. There is a need for intense investigation on characterizing novel bioactives from microalgal strains for better human welfare. It is a sustainable and alternative source of PUFAs; however, fewer species have been studied for PUFAs and more critical research is required for the isolation of new species. The most significant areas for microalgal research that need attention are system biology and omics technologies (Gimpel et al. 2015). To date, >40 microalgal species have been genetically sequenced, still, the new omics data are required for the characterization of several metabolic cycles associated with nutraceuticals (Guarnieri and Pienkos 2015), along with the controlling pathways responsible for the metabolic processes (Paliwal et al. 2016).

Conventional drugs to control postprandial hyperglycemia contain extracts of plants/herbs and some synthetic medicine e.g. acarbose. Artificial/synthetic medicines have several side effects like diarrhea and flatulence. It has already been established that pigments of cyanobacteria have magnificent free radical scavenging activity and antioxidant properties, thus, being a good source of naturally derived nutraceuticals with significant metabolite. Majorly, three cyanobacterial species, *Microcoleus, Lyngbya*, and *Synechocystis* spp., were studied extensively for their use as food and feed. The EPA-rich *Nannochloropsis oceanica* was studied for its use as a potential organic source of nutraceuticals grown in wastewaters i.e. pharmaceutical industry, pesticides industry, petroleum (oil) industry, and activated sludge treatment plant of municipality sewage (Mitra et al. 2015).

Arthrospira spp. is considered one of the most significant microalgal species exploited for the high- and low-value microalgal substances and byproducts. However, in recent times, the biomass of this microalgae is being exploited for a specific product resulting in the rise of the overall cost of the production process. Therefore, to lift the economic feasibility of the biorefinery process using *Arthrospira* spp., every single high- as well as low-value product derived from it should be valorized (Mitra and Mishra 2019).

Other microalgal species such as *Lyngbya* spp., *Pseudoanabaena* spp., and *Spirulina* spp. have been researched for their iso-bolographic studies, DPPH radical scavenging, and antinephrolithe activities. *Spirulina* is used as one of the greatest nutraceuticals for the improvement of overall human health having an immunity-boosting capacity. It is also beneficial for the heart, eyes, gut, hair growth, brain, glucose level, and cholesterol. *Spirulina* is also prescribed for the aid of several emotional and various mental disorders and ADHD.

18.4 Functional Foods with Microalgae

18.4.1 Vitamins

Some of the microalgae are considered significant sources of vitamins, especially vitamin B12 and vitamin E (Sharma et al. 2011). The peroxide stress induced in *Arthrospira platensis* leads to an increase of 4-fold in vitamin C and 8-fold α-tocopherol. Some strains are reported to produce vitamins that plants lack e.g. cyanocobalamin (B12) (Gupta and Gupta 2015). *Arthrospira was* also found rich in pro-vitamins A, B1, B2, B6, D, and E (Abidizadegan et al. 2021). Vitamin A equivalence of *Arthrospira* β-carotene improves the vitamin A content in the human body, besides other minerals and vitamins vital for natural immunity, like, **B6, E, and C** (Wang et al. 2020). Thus, making solid proof for its future possible exploitation in various deficiency diseases. They also regulate the production of white blood cells and antibodies that fight bacteria and viruses in our body. *Spirulina* contains a substantial amount of vitamin E (Falquet and Hurni 1997), which is comparable to other known and used sources.

18.4.2 Minerals

Mineral contents in the microalgal biomass are greatly attuned to the cultivation conditions. Most elements like Ca^{2+}, P^{3+}, and Mg^{2+} are present in amounts comparable to those in milk. *A. platensis* fits well into the human diet chart since the iron content meets the recommendation for infants, children, and adults, which is reported to contain 1413 mg potassium (Wells et al. 2017). According to the guidelines provided by the European Food Safety Authority (EFSA) and Scientific Committee on Food (SCF), the upper safe level for total iodine uptake/day is 600 μg for an adult with 60kg body weight. The iodine content of *Arthrospira* is 3 μg/10 g of dried biomass, resulting in no risk of excessive iodine upon consumption. Though very little information on the level of iodine in *Spirulina* is available, some data that shows its possibility through adaptation/selection (Sharma et al. 2005) could be beneficial in controlling goiter and thyroid-related hormones too.

18.5 Safety Concerns and Measures to be Followed

Chlorella spp. and *Arthrospira* spp. are valuable as nutritional supplements due to their bioactive properties, which are cultivated for distribution globally. The standard along with its safety mainly relies on the upstream and downstream processes. The commonly recommended daily dosage for any product is 3.0, with any of the studied products contributing to mineral intake, with the exception of Fe, found at higher but acceptable levels in *Chlorella and Spirulina* spp.

Chlorella and *Spirulina* spp. are rich sources of proteins and other essential trace elements and enzymes, are not known to produce any toxic compounds, and their safety is well-established (Buono et al. 2014; Wells et al. 2017; García et al. 2017). In clinical trials, it showed promising bioactive potential with antihypertensive, immunomodulatory hypoglycemic, and antilipidemic effects (Nielsen et al. 2010; Kim et al. 2016; Juszkiewicz et al. 2018. It also decreased the viral load and increased CD4 cells in HIV-infected patients, modulating lipid metabolism (Zeinalian et al. 2017), and increasing the ability to resist mental and physical fatigue in men (Johnson et al. 2016). *Chlorella* supplements have an antioxidant effect in smokers and may be used in therapy for depressive disorders (Panahi et al. 2013, 2015), and for improving glycemic status in patients with nonalcoholic fatty liver disease (Ebrahimi-Mameghani et al. 2017). The increased levels of arsenic, aluminum, and lead are a consequence

Table 18.1 Oxidative stress-induced carotenoids in microalgae.

Astaxanthin, lutein, β-carotene	*Haematococcus MT 2877*	High light (250 μmol/m^2/s) induction	—
Lutein	*Chlorella protothecoides*	Addition of 0.1 mM H_2O_2 and 0.01 mM NaClO plus 0.5 mM Fe^{2+}	—
β-Carotene	*Dunaliella salina*	1–3 M NaCl coupled with addition of 1 mM n-propyl gallate	Ascorbate, H_2O_2, superoxide dismutase (SOD), catalase (CAT), ascorbate peroxidase (APX), Malondialdehyde (MDA)
β-Carotene	*Dunaliella bardawil*	High light (500 W/m^2); addition of ROS generating chemicals	SOD, CAT
Astaxanthin	*Chlorococcum sp.*	Addition of 0.1 mM H_2O_2	—
Astaxanthin	*Chlorella zofingiensis*	Addition of 0.001–10 mM H_2O_2 and NaClO	—
Astaxanthin	*Haematococcus pluvialis*	Addition of 450 μM iron coupled with high temperature (30–36°C)	H_2O_2, $O{\bullet}_2-$, $^1O_2$, OH

Source: Adapted from Chokshi et al. (2017).

of unsuitable cultivation. (Hedegaard et al. 2013; Papazi et al. 2010; Rzymski et al. 2015). However, individuals suffering from autoimmune disorders can be specifically susceptible to microalgal supplements (Lee and Werth 2004), with adverse effects e.g. diarrhea, nausea, abdominal pain, and skin rashes Rzymski and Jaśkiewicz 2017). It is essential to determine the content of elements with toxicity in such products (Al-Dhabi 2013; Hedegaard et al. 2013; Rzymski et al. 2015).

Recently, some food supplements with multi-ingredients revealed high discrepancies between the actual nutritional value and that declared on the product label (Brandon et al. 2014; Niedzielski et al. 2016; Poniedziałek et al. 2018). Commercial scale cultivation of microalgae for foods employs open raceway pond which is prone to various environmental biotic and abiotic factors affecting the quality. So, here the oxidative stress induced carotenoids can be produced (Table 18.1).

Another concern associated with safety and application is nucleic acid in the biomass, which affects gout and kidney stones. The presence of microcystin in *Microcystis aeruginosa*, anatoxin by *Anabaena flos-aquae*, and saxitoxin by *Aphanizomenon flos-aquae* are associated with gastrointestinal tract, liver, synapse, and functions of nerve axon.

18.6 Applications of Microalgal Nutraceuticals

18.6.1 Human Nutrition

Some of the microalgae are widely sold in the market as a nutrient supplement or additive, colorant, and food emulsion. They are well-known protein sources with high nutritional aspects, vitamins, mainly vitamin B12 and pro-vitamin A, minerals, especially iron, γ-tocopherol, and γ-linolenic acid.

Arthrospira has highly valuable and sensitive pigments like phycocyanin, β-carotene, and xanthophyll, besides phenolic compounds with bioactive potential. It has been also used by astronauts of NASA as a dietary supplement. The therapeutic and nutritional aspects have been critically reviewed to prove the beneficial aspects of microalgae which can be attributed to its phycobiliproteins content among other ingredients (Ghosh et al. 2015; Mitra and Mishra 2019). According to the USDA, one tablespoon or 7 grams (g) of dried *Spirulina* is considered safe for intake. It also contains thiamine, riboflavin, niacin, folic acid, and vitamins A, B6, and K. The products are being commercialized worldwide directly for human beings and animals as food or feed. The concentration of inorganic elements was not found to exceed the present regulation levels, and hence, it can be considered safe for consumption (Becker 1988; Borowitzka 1988). The effect of *A. platensis* and *C. vulgaris* on kefir microbiota was studied with better probiotic activity than *C. vulgaris*. Since time immemorial, this nontoxic organism has offered potential benefits concerning its lucrative nutrient profile, besides converting CO_2 into organic elements while producing O_2 during their growth in brine. Such an organism with a plethora of better health for humans, animals, and the environment is limitless and has probably more to offer for the future. An energy bar or a drink with *Spirulina* in the diet of athletes is one of the most important factors in increasing their endurance and performance too. A study by Piotr Machowiec et al. 2021 reported that 1–8 g of *Spirulina* can help lower blood pressure after 2–12 weeks of intake. One can add it to routine diet e.g. in snacks, smoothies, vegetable/fruit salads, soups, bread, desserts, and beverages. It requires no extra procedure or special handling for the enhancement and availability of its protein.

18.6.2 Biomedical Applications

Arthrospira spp. combats several life-threatening diseases, as it has a good nutraceutical potential. Although the beneficial effects of *Arthrospira* have been exploited in numerous clinical trials after *in vitro* assay, there is an utmost requirement for *in vivo* trials. It is also known to possess cholesterol-lowering effects, antiviral effects, anticancer activity, hypo-lipidemic effect, antioxidant activity, mental protective effect, immune-stimulant effect, and protective effect against diabetes, and obesity (Hoseini et al. 2013; Mishra 2023). C-phycocyanin extracted from *Arthrospira* sp. also has potent anticancer activity as it exerts stronger photodynamic action on tumor cells than clinically approved hematoporphyrin. Recently, a good amount of immunoactive insulin was detected in different species of *Arthrospira*, which made them a prospective candidate for antidiabetic research. *Arthrospira* has gained attention in the field of medical applications apropos of its nutraceutical and pharmaceutical contents. However, recent developments in the process of *Arthrospira*-mediated synthesis of gold and silver nanoparticles (Govindaraju et al. 2008) can lead to some novel prospects in medical science. Thus, poly-phenolic compounds present in *Arthrospira* are recommended for the defensive effect against neurotoxicity and for the protection against neurodegenerative disorders of the brain (Paliwal et al. 2015; Patel et al. 2006). C-PC was analyzed for anti-inflammation in humans through its potential to act on cyclooxygenase-2 activity, as a catalyst responsible for inflammation. Further, it was studied *in vivo* for its proper dosage effect in mice. No harmful effect of C-PE on the food intake, body weight, or age was observed. It is a noteworthy to adopt C-PE based on the results of the treatment (Ghosh et al. 2015), as it indicated no negative health issues even through overdose (Soni et al. 2017). Microalgal pigments possess free radical scavenging activity along with the antioxidant properties. Pure C-PE, C-PC, lycopene, and myxoxanthophyll were found to be responsible for reducing α-glucosidase activity. Looking at the potential antagonistic effects of pure pigments toward glucosidase and amylase, these pigments may be used as supplements with dual benefits, i.e. antihyperglycemic and antioxidant activities (Ghosh et al. 2016). Influenza, being the most common respiratory disease, poses several health hazards. Frequent mutations in the viral genome also hamper the development of vaccines.

18.6.3 Nucleic Acids

The content of RNA or DNA is an essential nutritive measure, as the biochemical degradation of the adenine, purines, and guanine ends with the production of uric acid. An escalation in uric acid can cause gout and kidney stones. The maximum acceptable daily dose of it should be ~4 g/day. *Spirulina maxima* and *Spirulina platensis* contain ~4.2–6% of nucleic acid of total dry weight. It has been noted that the total nucleic acid of *Spirulina* is much lower than that of unicellular microbes.

18.6.4 As Anticancer Agents

C-PC was studied for its anticancer activity. Only a small dose of 50 μM was able to decrease the cell proliferation >40% in less than 50 hours (Ghosh et al. 2015). Besides, it is also used to induce apoptosis in the prostate cancer by decreasing the topotecan dosage that generally poses health hazards in the affected individuals (Gantar et al. 2012). A report suggested that phycocyanin results in reduced proliferation of cells leading at a lower dosage of 7 μg/ml, which also results in the loss of nuclear entities. Another study has established the therapeutic efficacy of phycocyanin on the individuals who are working in nuclear power plants and are greatly exposed to nuclear radiation. It has been recorded that C-PC can effectively influence damaged DNA repair, which is important to preserve the genomic integrity. These findings definitely proved that C-PC has anticancer properties that might open up exciting doors for cancer treatment.

18.6.5 Animal/Aquafeed

Some countries use *Spirulina* as a partial nutrient substitute for imported fodder products to promote the development, viability, and immunity of fish, poultry, and shrimps. The demand for animal protein is increasing as food for humans due to the drastic increase in population. This puts pressure on the search for cheap yet nutritional feed for these animals. Smaller doses were investigated, particularly in Eastern Europe, following decades of studies where animals were fed significant amounts of microalgae, up to 50% of the standard meal, to utilize their protein content (Musafarov and Taubayev 1974; Richmond 2004). Small amounts of microalgal biomass from the strains of *Scenedesmus*, *Chlorella*, and *Spirulina* have been shown to have a positive impact on animal physiology when added to livestock feed as a single cell protein (SCP) supplement instead of traditional protein sources like soybean or fish meal. Specifically, the inclusion of microalgal feeds in the animals' regular meals was shown to elicit a nonspecific immune response and strengthen their immune systems (Belay et al. 1993). Studies conducted on pigs, ruminants, cattle, and poultry have shown that up to 10% concentrations of spirulina can satisfactorily replace traditional protein sources, whereas higher concentrations stunt growth (Petterson 1996). To cope with this situation, the identification of new food resources having high nutritive value needs to be optimized using land and water efficiently. Currently, cyanobacteria, especially *Arthrospira*, is emerging as a potential candidate to meet the criteria and is being explored in animal feed trials. However, to ensure the safety measures related to *Arthrospira* consumption, several types of research have been done to check any kind of toxicity upon feeding and has been further confirmed as a nontoxic organism after few trials (Hutadilok-Towatana et al. 2010). To date, enormous research studies have been done to find out the impact of dietary additives of *Arthrospira* on the productivity and health of ruminants.

18.6.6 Aquaculture

Regarding the use of microalgae in aquaculture, at least two trends are emerging globally: (i) the development of increasingly complex species of microalgae to feed both vertebrate young and invertebrate larvae, and (ii) the addition of commonly produced microalgae to fish feed to produce benefits akin to those observed in animal feeding.

For bivalves in all growth stages as well as the larvae of some fish and crustacean species raised in aquaculture, microalgae provide an essential source of food. Additionally, zooplanktons raised to feed the larvae and juveniles of some fish and crab species consume them. (Brown et al. 1997). In aquaculture, microalga is not only important as a food source but, together with bacteria, regulates the balance between oxygen and CO_2 in the aquaculture system. Many species of microalgae are being utilized in aquaculture.

18.7 Problems of Contamination

In outdoor conditions, careful monitoring is required to keep contaminants away or at least under control. If proper culture conditions are maintained for the growth of *Spirulina,* with timely intervention during adverse climatic conditions, other algal contaminants can be kept at bay to a large extent. However, contamination of *Chlorella* may occur sometimes. Vonshak 1993 suggests high levels (0.02 M) of bicarbonate and keeping dissolved organics as low as possible are the necessary conditions to prevent *Chlorella* contamination. Even a 1 mM concentration of ammonia is said to control *Chlorella* (Vonshak 1986). *Spirulina subsalsa* also has been reported to occur frequently as a contaminant (Gupta and Changwal 1992) in the *Chlorella* culture. Amoebas have been reported but can be controlled by the chemical NH_4Cl (Vonshak 1993). Bacterial contamination is hardly seen because of the high pH (alkalinity) of the medium except in sediment and precipitate, which are periodically removed. Further, it can be protected from mosquitoes/insects, especially during monsoon, by covering the top with a net and agitation/aeration through a perforated pipe (metal/plastic) inside the culture (which serves a dual purpose of mixing CO_2 + agitation + pH maintenance). Our strategy was to maintain alkaline pH of around 10 by adding $NaHCO_3$, which was very efficient. It served the purpose of maintaining alkalinity as a source of carbon increasing the growth rate, and protecting the culture inside, and mosquito net on the top for airborne contaminants.

18.8 Future Scope

It has been a well-established fact that microalgae are capable of producing a broad range of primary as well as secondary metabolites that have several antiviral, antibacterial, and antifungal properties that can potentially be exploited by nutraceutical and pharmaceutical sectors on a commercial scale. The rate of acceptance for the microalgal nutraceuticals shows that benefits are too slow because of the general unawareness among the people. The applicability of microalgal nutraceuticals, especially derived from *Spirulina*, as an immunity booster has already been clinically reported. Algal nutraceuticals enhance both innate and adaptive immunity through the use of bioactive substances such as sulfated polysaccharides, phycobiliproteins, and peptides that block the ACE enzyme (Rath et al 2021). Awareness of this kind of microalgal-based nutraceutical may lead to the improvement of overall human health without causing any side reactions if taken as prescribed.

With the projected upsurge in global populations and climatic changes, we need to find an alternative source of sustainable food that is beneficial for the improvement of human health. Microalgae are one of the best natural resources, which helps to meet the lacuna of nutrition with its renewability and sustainability at an affordable rate. Some of the bottlenecks in accepting microalgae as a functional food are its accessibility, availability, and sustainable methods for removing and purifying the metabolites; preservation of its moiety during the upstream and downstream processing of the products is equally important for human consumption. Furthermore, for the success of microalgal commercialization in food or feed, it is necessary to develop eco-sustainable and -innovative processes efficient in transforming the raw biomass into high-value products or isolated nutraceuticals without compromising on their nutrients and environment. Advanced techniques like nonthermal processes e.g. enzyme-assisted processing, fermentation, pulsed electric field, ultrasonication, and microwave are very efficient in protecting their nutritional value; besides ensuring food safety. In the end, there is a necessity for good cooperation between basic R&D and applied aspects within industries to better understand the gaps, driven by gradual increase in consumer demands for such foods. It may guide the researchers to deliver details on the gaps, particularly *in vivo* bioavailability, bioaccessibility, and health benefits of microalgae-based bioactives. In any circumstances, food industries must control the quality, safety, and proper labeling of the functional foods before their commercialization.

18.9 Conclusion

Functional foods are mainly consumed to sustain the health of human body after physical exertion, or as a prophylaxis for ailments/diseases/outbreaks, besides providing a necessary source of nutrients for malnourished or heavily nourished individuals. Success not only depends on the concerned scientist/researchers but also on financial and moral support with encouragement and motivation from the authorities and ministry. It is a reality that too many optimistic outcomes could not be achieved, which were estimated and planned for the translation of the work done so far, but definitely, the deep fundamental studies that were published have contributed significantly to the science to be carried forward at global level for humanity and environment. The future of microalgae is expected to be very impressive on the planet, saving the biosphere through its unique potential benefits.

Acknowledgments

FS and SM would like to acknowledge the CSIR Emeritus Scientist Scheme for the financial support during the preparation of this book chapter. The director, CSIR-CSMCRI, and present DC, Applied Phycology and Biotechnology Division are deeply acknowledged for their encouragement. All the authors would like to express our sincere gratitude to all the individuals and organizations that have contributed their knowledge, time, and hard work to the publication of this book chapter. We would also like to express our appreciation to Wiley for considering our work and providing the opportunity to publish our chapter.

References

Abidizadegan, M., Peltomaa, E., and Blomster, J. (2021). The potential of cryptophyte algae in biomedical and pharmaceutical applications. *Frontiers in Pharmacology* 11: 618836.

Al-Dhabi, N.A. (2013). Heavy metal analysis in commercial *Spirulina* products for human consumption. *Saudi Journal of Biological Sciences* 20 (4): 383–388.

Bagchi, D. and Nair, S. (ed.) (2016). *Developing New Functional Food and Nutraceutical Products*. Academic Press.

Baldiris-Navarro, I., Sanchez-Aponte, J., González-Delgado, A. et al. (2018). Removal and biodegradation of phenol by the freshwater microalgae *Chlorella vulgaris*. *Methods* 6: 7.

Becker, E.W. (1988). Micro-algae for human and animal consumption. In: *Micro-algal Biotechnology* (ed. M.A. Borowitzka and L. Borowitzka), 222–256. Cambridge University Press: Cambridge.

Belay, A., Ota, Y., Miyakawa, K., and Shimamatsu, H. (1993). Current knowledge on potential health benefits of Spirulina. *Journal of Applied Phycology* 5: 235–241.

Borowitzka, M.A. (1988). Vitamins and fine chemicals from micro-algae. In: *Micro-Algal Biotechnology* (ed. M.A. Borowitzka and L.J. Borowitzka), 153–196. Cambridge University Press.

Brandon, E.F.A., Bakker, M.I., Kramer, E. et al. (2014). Bioaccessibility of vitamin A, vitamin C and folic acid from dietary supplements, fortified food and infant formula. *International Journal Offood Sciences and Nutrition* 65 (4): 426–435.

Brown, M.R., Jeffrey, S.W., Volkman, J.K., and Dunstan, G.A. (1997). Nutritional properties of microalgae for mariculture. *Aquaculture* 151 (1-4): 315–331.

Buono, S., Langellotti, A.L., Martello, A. et al. (2014). Functional ingredients from microalgae. *Food & Function* 5 (8): 1669–1685.

Chokshi, K., Pancha, I., Ghosh, A., and Mishra, S. (2017). Oxidative stress-induced bioprospecting of microalgae. In: *Systems Biology of Marine Ecosystems* (ed. M. Kumar and P. Ralph), 251–276. Springer.

Ebrahimi-Mameghani, M., Sadeghi, Z., Farhangi, M.A. et al. (2017). Glucose homeostasis, insulin resistance and inflammatory biomarkers in patients with non-alcoholic fatty liver disease: Beneficial effects of supplementation with microalgae *Chlorella vulgaris*: a double-blind placebo-controlled randomized clinical trial. *Clinical Nutrition* 36 (4): 1001–1006.

Enzing, C., Ploeg, M., Barbosa, M., & Sijtsma, L. (2014). Microalgae-based products for the food and feed sector: an outlook for Europe. *JRC Scientific and Policy Reports*. European Commission; 19–37.

Falquet, J. and Hurni, J.P. (1997). *The Nutritional Aspects of Spirulina. Antenna Foundation.*

Gantar, M., Simović, D., Djilas, S. et al. (2012). Isolation, characterization and antioxidative activity of C-phycocyanin from *Limnothrix* sp. strain 37-2-1. *Journal of Biotechnology* 159 (1-2): 21–26.

García, J.L., De Vicente, M., and Galán, B. (2017). Microalgae, old sustainable food and fashion nutraceuticals. *Microbial Biotechnology* 10 (5): 1017.

Ghosh, T., Bhayani, K., Paliwal, C. et al. (2016). Cyanobacterial pigments as natural anti-hyperglycemic agents: an in vitro study. *Frontiers in Marine Science* 3: 146.

Ghosh, T., Paliwal, C., Maurya, R., and Mishra, S. (2015). Microalgal rainbow colours for nutraceutical and pharmaceutical applications. In: *Plant Biology and Biotechnology: Volume I: Plant Diversity, Organization, Function and Improvement* (ed. B. Bahadur, M.V. Rajam, L. Sahijram, and K.V. Krishnamurthy), 777–791.

Gimpel, J.A., Henríquez, V., and Mayfield, S.P. (2015). In metabolic engineering of eukaryotic microalgae: potential and challenges come with great diversity. *Frontiers in Microbiology* 6: 1376.

Gouveia, L., Batista, A.P., Sousa, I. et al. (2008). Microalgae in novel food products. In: *Food Chemistry Research Developments* (ed. K.N. Papadopoulos) Chapter 2. Nova Science Publishers, Inc.

Govindaraju, K., Basha, S.K., Kumar, V.G., and Singaravelu, G. (2008). Silver, gold and bimetallic nanoparticles production using single-cell protein (*Spirulina platensis*) Geitler. *Journal of Materials Science* 43: 5115–5122.

Guarnieri, M.T. and Pienkos, P.T. (2015). Algal omics: unlocking bioproduct diversity in algae cell factories. *Photosynthesis Research* 123: 255–263.

Gupta, R. S., & Changwal, M. L. (1992). Biotechnology of mass production of *Spirulina* and *Arthrospira* in fresh water. *Proceedings: Spirulina ETTA National Symposium, MCRC*, Madras. p 125–128.

Gupta, U.C. and Gupta, S.C. (2015). Role of vitamins in human health and nutrition: sources and morbidity. *Current Nutrition & Food Science* 11 (2): 105–115.

Hedegaard, R.V., Rokkjær, I., and Sloth, J.J. (2013). Total and inorganic arsenic in dietary supplements based on herbs, other botanicals and algae—a possible contributor to inorganic arsenic exposure. *Analytical and Bioanalytical Chemistry* 405: 4429–4435.

Hoseini, S.M., Khosravi-Darani, K., and Mozafari, M.R. (2013). Nutritional and medical applications of *spirulina* microalgae. *Mini Reviews in Medicinal Chemistry* 13 (8): 1231–1237.

Hutadilok-Towatana, N., Reanmongkol, W., and Panichayupakaranant, P. (2010). Evaluation of the toxicity of *Arthrospira (Spirulina) platensis* extract. *Journal of Applied Phycology* 22: 599–605.

Johnson, M., Hassinger, L., Davis, J. et al. (2016). A randomized, double blind, placebo controlled study of *Spirulina* supplementation on indices of mental and physical fatigue in men. *International Journal of Food Sciences and Nutrition* 67 (2): 203–206.

Juszkiewicz, A., Basta, P., Petriczko, E. et al. (2018). An attempt to induce an immunomodulatory effect in rowers with spirulina extract. *Journal of the International Society of Sports Nutrition* 15 (1): 9.

Kim, S., Kim, J., Lim, Y. et al. (2016). A dietary cholesterol challenge study to assess *Chlorella* supplementation in maintaining healthy lipid levels in adults: a double-blinded, randomized, placebo-controlled study. *Nutrition Journal* 15 (1): 1–7.

Lee, A.N. and Werth, V.P. (2004). Activation of autoimmunity following use of immune stimulatory herbal supplements. *Archives of Dermatology* 140 (6): 723–727.

Machowiec, P., Ręka, G., Maksymowicz, M. et al. (2021). Effect of spirulina supplementation on systolic and diastolic blood pressure: systematic review and meta-analysis of randomized controlled trials. *Nutrients* 13 (9): 3054.

Mishra, S. (2023). *The Dietary Benefits and Potential of Spirulina* (ed. S. Mishra). New Delhi: CSIR-NIScPR.

Mitra, M. and Mishra, S. (2019). Multiproduct biorefinery from *Arthrospira* spp. towards zero waste: current status and future trends. *Bioresource Technology* 291: 121928.

Mitra, M., Patidar, S.K., George, B. et al. (2015). A euryhaline *Nannochloropsis gaditana* with potential for nutraceutical (EPA) and biodiesel production. *Algal Research* 8: 161–167.

Musafarov, A.M. and Taubayev, T.T. (1974). *Chlorella* (in Russian). Tashkent: FAN.

Niedzielski, P., Rudnicka, M., Wachelka, M. et al. (2016). Selenium species in selenium fortified dietary supplements. *Food Chemistry* 190: 454–459.

Nielsen, C.H., Balachandran, P., Christensen, O. et al. (2010). Enhancement of natural killer cell activity in healthy subjects by Immulina®, a *Spirulina* extract enriched for Braun-type lipoproteins. *Planta Medica* 76 (16): 1802–1808.

Paliwal, C., Ghosh, T., George, B. et al. (2016). Microalgal carotenoids: potential nutraceutical compounds with chemotaxonomic importance. *Algalr Research* 15: 24–31.

Paliwal, C., Pancha, I., Ghosh, T. et al. (2015). Selective carotenoid accumulation by varying nutrient media and salinity in *Synechocystis* sp. CCNM 2501. *Bioresource Technology* 197: 363–368.

Panahi, Y., Badeli, R., Karami, G.R. et al. (2015). A randomized controlled trial of 6-week *Chlorella vulgaris* supplementation in patients with major depressive disorder. *Complementary Therapies in Medicine* 23 (4): 598–602.

Panahi, Y., Mostafazadeh, B., Abrishami, A. et al. (2013). Investigation of the effects of *Chlorella vulgaris* supplementation on the modulation of oxidative stress in apparently healthy smokers. *Clinical Laboratory* 59 (5-6): 579–587.

Pandey, K.B. and Suttajit, M. (ed.) (2022). *Plant Bioactives as Natural Panacea Against Age-induced Diseases: Nutraceuticals and Functional Lead Compounds for Drug Development.* Elsevier.

Papazi, A., Makridis, P., and Divanach, P. (2010). Harvesting *Chlorella minutissima* using cell coagulants. *Journal of Applied Phycology* 22 (3): 349–355.

Patel, A., Mishra, S., Pawar, R., and Ghosh, P.K. (2006). Purification and characterization of C-Phycocyanin from cyanobacterial species of marine and freshwater habitat. *Protein Expression and Purification* 40 (2): 248–255.

Petterson, G. (1996). Biotechnological applications of cyanobacteria. *Journal of Scientific and Industrial Research* 55: 669–684.

Poniedziałek, B., Niedzielski, P., Kozak, L. et al. (2018). Monitoring of essential and toxic elements in multi-ingredient food supplements produced in European Union. *Journal of Consumer Protection and Food Safety* 13: 41–48.

Price Waterhouse & Coopers (2009). Leveraging growth in the emerging functional foods industry: trends and market opportunities. In: *Functional Foods Reports*, 1–22. Price Waterhouse & Coopers.

Richmond, A. (ed.) (2004). *Handbook of Microalgal Culture: Biotechnology and Applied Phycology*, vol. 577. Oxford: Blackwell Science.

Rzymski, P. and Jaśkiewicz, M. (2017). Microalgal food supplements from the perspective of Polish consumers: patterns of use, adverse events, and beneficial effects. *Journal of Applied Phycology* 29: 1841–1850.

Rzymski, P., Niedzielski, P., Kaczmarek, N. et al. (2015). The multidisciplinary approach to safety and toxicity assessment of microalgae-based food supplements following clinical cases of poisoning. *Harmful Algae* 46: 34–42.

Sharma, M.K., Patni, R., Kumar, M., and Kumar, A. (2005). Modification of mercury-induced biochemical alterations in blood of Swiss albino mice by *Spirulina fusiformis*. *Environmental Toxicology and Pharmacology* 20 (2): 289–296.

Sharma, N.K., Tiwari, S.P., Tripathi, K., and Rai, A.K. (2011). Sustainability and cyanobacteria (blue-green algae): facts and challenges. *Journal of Applied Phycology* 23: 1059–1081.

Soni, R.A., Sudhakar, K., and Rana, R.S. (2017). *Spirulina*–from growth to nutritional product: a review. *Trends in Food Science & Technology* 69: 157–171.

Vigani, M., Parisi, C., Rodríguez-Cerezo, E. et al. (2015). Food and feed products from micro-algae: market opportunities and challenges for the EU. *Trends in Food Science & Technology* 42 (1): 81–92.

Vonshak, A. (1986). Laboratory techniques for the culturing of microalgae. In: *Handbook for Algal Mass Culture* (ed. A. Richmond), 117–145. Boca Raton: CRC Press.

Vonshak, A. (1993). Microalgae: laboratory growth techniques and the biotechnology of biomass production. In: *Photosynthesis and Production in a Changing Environment: A Field and Laboratory Manual* (ed. D.O. Hall, J.M.O. Scurlock, H.R. Bolhar-Nordenkampf, et al.), 337–355. Dordrecht: Springer Netherlands.

Wang, A., Yan, K., Chu, D. et al. (2020). Microalgae as a mainstream food ingredient: demand and supply perspective. In: *Microalgae Biotechnology for Food, Health and High Value Products* (ed. A. Alam, J.-L. Xu, and Z. Wang), 29–79.

Wells, M.L., Potin, P., Craigie, J.S. et al. (2017). Algae as nutritional and functional food sources: revisiting our understanding. *Journal of Applied Phycology* 29: 949–982.

Zeinalian, R., Farhangi, M.A., Shariat, A., and Saghafi-Asl, M. (2017). The effects of *Spirulina platensis* on anthropometric indices, appetite, lipid profile and serum vascular endothelial growth factor (VEGF) in obese individuals: a randomized double blinded placebo controlled trial. *BMC Complementary and Alternative Medicine* 17: 1–8.

19

Toxicity Evaluation of Nutraceuticals

Sonia Morya[1], Farid Menaa[2], and Sharvary Arun Vichare[1]

[1] *Department of Food Technology and Nutrition, School of Agriculture, Lovely Professional University, Phagwara, Punjab, India*
[2] *Department of Biomedical and Environmental Engineering, California Innovations Corporation, San Diego, CA, USA*

19.1 Introduction

Nutraceuticals, a term combining "nutrition" and "pharmaceutical," describe food-based goods that have both fundamental nutritional value and additional health advantages (Aronson 2017; Morya et al. 2022a). These products, ranging from isolated nutrients, dietary supplements, and diets to genetically engineered foods and herbal products, play a significant role in health promotion and disease prevention (AlAli et al. 2021; Gautam et al. 2022; McGonigle and Ruggeri 2014). Toxicity evaluation of nutraceuticals is crucial for ensuring the safety and efficacy of bioactive compounds, such as vitamins, minerals, and herbal extracts, which have potential health benefits beyond basic nutritional functions (Bhavani et al. 2023; Vilas-Boas et al. 2021). This process involves assessing potential adverse effects on human health, including acute, subchronic, chronic, genotoxic, carcinogenic, and reproductive toxicity (Astashkina et al. 2012; Kokova 2023). The objective is to establish a comprehensive understanding of the safety profile of nutraceuticals, ensuring they do not pose harm to consumers when used as recommended (Coppens et al. 2006). Research in this field uses methodologies such as animal studies, in vitro experiments, and clinical trials to evaluate toxicity, understand the dose-response relationship, identify potential adverse effects, and determine safe consumption levels (Goncharov et al. 2019; Morya et al. 2022b; Raies and Bajic 2016). The evaluation of their toxicity is crucial (Raies and Bajic 2016; Shukla et al. 2010). This process ensures that these substances, often perceived as natural and safe, do not pose unforeseen health risks to consumers (Schilter et al. 2003). This chapter aims to provide the methods and importance of toxicity evaluation in nutraceuticals, highlighting their impact on public health and industry (Gupta et al. 2021). Nutraceuticals encompass a broad range of products, including dietary supplements like vitamins, minerals, herbs, and substances like bioactive food components (Dixit et al. 2021; Rajasekaran et al. 2008; Singh and Upadhyay 2021, 2023; Tripathi et al. 2020; Upadhyay and Singh 2021, 2023). They offer various health benefits, such as prevention and treatment of diseases, improvement of physical and mental health, and enhancement of life expectancy (Bhatia et al. 2020; Keservani et al. 2020). Examining the bioavailability and metabolism of nutraceuticals in the human body is an important part of toxicity assessment (McClements

Herbal Nutraceuticals: Products and Processes, First Edition. Edited by Santosh Kumar Upadhyay and Sudhir Pratap Singh.

et al. 2015). It is possible to forecast these substances' potential toxicity and propose the right dosage by having a thorough understanding of how they are absorbed, distributed, metabolized, and eliminated from the body (Lin et al. 2003). To evaluate the toxicity of nutraceuticals, valid research frequently uses rigorous study methods, including meta-analyses, systematic reviews, and randomized controlled trials (RCTs) (Pagano et al. 2018; Santini et al. 2018). These studies take into account any potential drawbacks along with any advantages of using nutraceuticals. The goal is to offer evidence-based insights that can direct policy choices and suggestions for public health (Grumezescu 2016). The market for nutraceuticals has seen substantial growth, driven by increasing health awareness and consumer interest in natural health products (Chopra et al. 2022). This surge underscores the need for rigorous toxicity evaluation to safeguard public health and maintain consumer trust (Effah et al. 2023).

The evaluation of toxicity in nutraceuticals is vital for several reasons. Primarily, it ensures consumer safety. Nutraceuticals, often available over the counter, might be perceived as inherently safe, but without proper testing, they could pose serious health risks (Morya and Danquah-Amoah 2017). For instance, excessive intake of certain vitamins can lead to toxicity (Gonçalves et al. 2018). Another crucial aspect is regulatory compliance. Various countries have different regulations governing nutraceuticals. Rigorous toxicity testing helps in adhering to these regulatory standards, ensuring products are safe for consumption (Ecobichon 1997). The toxicity evaluation helps in maintaining consumer confidence. A lack of trust in these products can significantly impact their market and acceptance. Hence, thorough testing is imperative for the sustained growth of the nutraceutical industry (Roberts et al. 2022).

19.1.1 Toxicity Evaluation Methods

Numerous techniques are used in toxicity evaluation to evaluate the safety and possible side effects of nutraceuticals. Various testing techniques are used at various phases of research and development, and each one offers insightful information about the dangers related to the intake of nutraceuticals. Toxicity evaluation in nutraceuticals employs a variety of methods, each with its own specific focus and application (Granato et al. 2020).

19.1.1.1 In Vitro Methods

When evaluating a substance's potential for harm to biological systems other than living beings, in vitro toxicity evaluation techniques usually employ cell-based or biochemical experiments (Barile 2019). These techniques are essential to toxicological research because they offer quick, affordable, and morally sound substitutes for conventional animal testing (Krishna and Krishna 2021). Numerous assays, such as those for cytotoxicity, genotoxicity, mutagenicity, carcinogenicity, and organ-specific toxicity, are used in in vitro investigations to assess different aspects of toxicity. In vitro methods provide important insights into the mechanisms underlying toxicity and aid in the early identification of potential hazards in the drug discovery and development process. These methods involve exposing cells, tissues, or isolated enzymes to test substances and measuring their effects on cellular viability, function, or molecular pathways (Zhang et al. 2016). Moreover, in vitro assays can be customized to target organs or certain toxicological endpoints, making it possible to screen for toxicity of a large number of substances more effectively and precisely (Table 19.1). All things considered, in vitro toxicity assessment techniques make a substantial contribution to our growing knowledge of chemical risks and the creation of safer and more efficient consumer goods, chemicals, and medications (Vo et al. 2019). The various methods recorded for the investigation using in vitro method are:

Table 19.1 In vitro toxicity results of common nutraceuticals.

Nutraceutical	Test method	Observed effect	References
Resveratrol	MTT assay	Reduced cell viability	Bernhard et al. (2003)
Curcumin	Brine shrimp test	Low toxicity	Harishkumar et al. (2019)
Omega-3 fatty acids	LDH release assay	Membrane damage in cells	Sundaram et al. (2020)
Quercetin	Comet assay	DNA damage in cells	Johnson and Loo (2000)
Lycopene	Cytotoxicity assay	Antioxidant activity	Alhoshani et al. (2022)
Spirulina	Apoptosis assay	No significant toxicity	Chu et al. (2010)

19.1.1.1.1 Cell Viability Assays Assays for determining the metabolic activity or viability of cells after exposure to nutraceuticals include MTT (3-(4,5-dimethylthiazol-2-yl)-2,5-diphenyltetrazolium bromide) and MTS (3-(4,5-dimethylthiazol-2-yl)-5-(3-carboxymethoxyphenyl)-2-(4-sulfophenyl)-2H-tetrazolium). These tests offer quantifiable information about the cytotoxicity of nutraceuticals and how they affect apoptosis, differentiation, and proliferation of cells (Dillard and German 2000; Rigacci and Stefani 2016).

19.1.1.1.2 Genotoxicity Assays Genotoxicity tests, such as the Ames or Comet assays, assess if nutraceuticals have the ability to cause DNA damage or mutations in cells. These tests evaluate the possibility of chromosomal abnormalities, DNA strand breaks, and gene mutations caused by nutraceuticals, as well as their genotoxic consequences (Sponchiado et al. 2016).

19.1.1.1.3 Oxidative Stress Assays Reactive oxygen species (ROS) or the antioxidant capacity of cells after being exposed to nutraceuticals are measured using oxidative stress assays. These tests evaluate how well the nutraceuticals can alter oxidative stress pathways, shield cells from oxidative damage, or strengthen antioxidant defense systems (De Leon and Borges 2020).

19.1.1.1.4 Mitochondrial Function Assays Assays for measuring mitochondrial function, such as the JC-1 or Seahorse XF Analyzer, assess how nutraceuticals affect the synthesis of ATP, membrane potential, and mitochondrial respiration. These tests shed light on how the nutraceuticals affect mitochondrial activity and cellular energy metabolism, two processes essential to cell survival and homeostasis (de Brito Monteiro et al. 2020).

19.1.1.1.5 Enzyme Inhibition Assays The capacity of nutraceuticals to inhibit particular enzymes involved in metabolic processes or cellular functions is evaluated by enzyme inhibition assays. These tests find possible interactions that may occur between metabolic enzymes, transporters, or receptors and nutraceuticals. These interactions could impact signaling pathways, medication metabolism, or nutrition absorption (Kuddus 2019; Nasri et al. 2014).

19.1.1.1.6 Organotypic Cell Culture Models Organotypic cell culture models, which imitate the form and functionality of certain organs or tissues in vitro, include 3D cell cultures and organ-on-a-chip systems. When assessing the impact of nutraceuticals on organ-specific toxicity, such as hepatotoxicity, neurotoxicity, or cardiotoxicity, these models offer a more physiologically appropriate platform (Hayden and Harbell 2021).

All things considered, in vitro techniques for assessing the toxicity of nutraceuticals provide important instruments for identifying possible side effects, clarifying the fundamental mechanisms of toxicity, and directing the creation of safer and more effective dietary supplements or functional foods. These tests support *in vivo* research and inform regulatory decisions about the use of nutraceuticals in human health and nutrition by offering crucial preliminary data on the safety profile of these products.

19.1.1.2 In Vivo Methods

In order to determine a chemical's potential toxicity, in vivo toxicity evaluation techniques examine the effects of the substance on living things, usually animals. These techniques are essential for assessing the safety of a wide range of substances, including chemicals, medicines, and nutraceuticals, since they offer important information about possible negative effects on biological systems. A variety of experimental designs are used in in vivo investigations, such as acute, subacute, subchronic, chronic, reproductive, developmental, and carcinogenicity studies, all of which are intended to assess distinct facets of toxicity over variable exposure times (Anadón et al. 2021). In vivo methods help to ensure the protection of human health and the environment by providing comprehensive data on the potential health risks associated with substance exposure through careful observation of physiological parameters, biochemical markers, histopathological changes, and other pertinent endpoints. These studies are crucial for providing information to regulators about the safety and usage of chemicals in consumer goods, pharmaceuticals, and environmental applications (Gosslau 2016). Presently, common methods are adopted to assess the toxicity levels of nutraceuticals.

19.1.1.2.1 Acute Toxicity Studies When assessing the immediate negative effects of nutraceuticals after a single high-dose exposure, acute toxicity studies are essential. In these investigations, experimental animals, like rats, are frequently given increasing dosages of the nutraceutical and closely observed for signs of toxicity over a brief period of time, often 24–72 hours. The parameters that are evaluated can be things like organ abnormalities, clinical symptoms, changes in body weight, mortality, and biochemical indications of toxicity. Finding the maximum tolerated dose (MTD) or median lethal dose (LD50), which offers crucial information for determining safe dosages for people, is the main goal of acute toxicity research. Furthermore, information on acute toxicity aids in the risk evaluation, hazard classification, and labeling specifications for nutraceutical goods. Overall, acute toxicity studies serve as a critical initial step in the comprehensive safety assessment of nutraceuticals, helping to ensure consumer safety and regulatory compliance. In acute toxicity tests, test animals, usually rats, are given a single large dosage of the nutraceutical and are then observed for a brief amount of time, usually up to 14 days. These investigations seek to ascertain the drug's LD50 (lethal dose, 50%) as well as any immediate negative consequences, including mortality, alterations in behavior, or anomalies in the body's systems.

19.1.1.2.2 Subacute Toxicity Studies The purpose of subacute toxicity studies is to evaluate the possible side effects of nutraceuticals after repeated exposure over a brief amount of time—usually between 14 and 28 days in experimental animals. These studies seek to assess the cumulative effects of long-term exposure to nutraceuticals while simulating realistic human consumption patterns. Animals are given different doses of the nutraceutical every day for the duration of the trial, and thorough evaluations are carried out, including body weight measurements, hematological and biochemical analysis, clinical observations, and tissue histological inspection. Subacute toxicity studies offer important information about the reversibility of negative effects linked to prolonged exposure to nutraceuticals, as well as possible systemic effects and target organ toxicity.

19.1.1.2.3 Subchronic Toxicity Studies Subchronic toxicity studies are important research projects that evaluate the possible negative effects of nutraceuticals after repeated exposure over a long period of time, usually 28–90 days, in test animals. The purpose of these studies is to shed light on the longer-term safety profile of nutraceuticals, thereby more accurately simulating the patterns of chronic exposure seen in humans (Hayes 2007). Animals are given daily doses of the nutraceutical under investigation for the duration of the study, and a full range of assessments, such as clinical observations, body weight measurements, hematological and biochemical analyses, urinalysis, and a thorough histopathological examination of tissues, are carried out. Subchronic toxicity studies play a crucial role in determining target organ toxicity, dose-dependent effects, and potential cumulative negative effects linked to extended exposure to nutraceuticals. The information produced by this research is essential for defining safe exposure thresholds, choosing suitable dosage schedules, evaluating the possible advantages and disadvantages of nutraceuticals, and assisting regulators in making choices about the safety of products and labeling specifications. To sum up, subchronic toxicity studies are an essential part of the thorough safety evaluation of nutraceuticals that guarantees consumer safety and advances public health.

19.1.1.2.4 Chronic Toxicity Studies Research on chronic toxicity is a crucial study carried out to assess the possible negative effects of nutraceuticals after sustained exposure over an extended length of time; these studies usually involve experimental animals and persist for at least a year. The purpose of these studies is to evaluate nutraceuticals' safety profile in environments that mimic long-term human use patterns. Animals are given daily doses of the nutraceutical for the duration of the study, and a wide range of assessments are carried out, such as routine tracking of clinical signs, body weight, food consumption, hematological and biochemical parameters, urinalysis, and in-depth tissue histopathology examination. Studies on chronic toxicity seek to determine any latent or delayed effects, cumulative toxicity, and carcinogenic potential linked to long-term exposure to nutraceuticals (Hayes 2007). Establishing safety thresholds, figuring out appropriate daily intakes, evaluating possible hazards and benefits, and guiding regulatory decisions about the approval, labeling, and marketing of nutraceutical goods all depend heavily on the data produced by this research. To sum up, studies on chronic toxicity are essential for guaranteeing the long-term safety and effectiveness of nutraceuticals, which protects the public's health and fosters customer trust.

19.1.1.2.5 Reproductive and Developmental Toxicity Studies Specialized studies known as reproductive and developmental toxicity studies are carried out to evaluate the possible negative effects of nutraceuticals on fertility, offspring development, and reproductive function. These studies are essential for assessing the safety of nutraceuticals, particularly in light of their possible effects on the developing fetus and human reproductive health. In most reproductive toxicity studies, the nutraceutical is administered to male and female animals in different doses prior to, during, and after mating, gestation, and lactation. Mating behavior, fertility, length of gestation, litter size, pup viability, and postnatal growth and development are examples of parameters that may be evaluated. To find any structural or functional abnormalities, histological examination of the reproductive organs and assessment of the hormones involved in reproduction may also be carried out (Boverhof et al. 2011). The goal of developmental toxicity research is to evaluate how exposure to nutraceuticals affects many aspects of prenatal and postnatal development, including organogenesis, neurobehavioral development, and embryonic and fetal growth. In such research, the nutraceutical is frequently administered to pregnant animals throughout crucial stages of organogenesis, and the offspring are then examined for behavioral abnormalities, growth retardation, and congenital defects. The results of research on reproductive and developmental toxicity offer vital insights for evaluating the safety of nutraceuticals during gestation and lactation, guiding risk mitigation tactics, and guaranteeing the preservation of mother and child well-being. All things considered, these studies play a crucial role in the thorough safety evaluation of nutraceuticals, supporting regulatory supervision, and evidence-based decision-making (Boverhof et al. 2011).

19.1.1.2.6 Genotoxicity Studies Studies on the potential for nutraceuticals to harm living cells' genetic material (DNA) are known as genotoxicity studies. Because genotoxicity can result in mutations, chromosomal abnormalities, and ultimately the development of cancer, these investigations are essential for assessing the safety of nutraceuticals. A variety of in vitro and in vivo tests are commonly used in genotoxicity testing to assess the genotoxic potential of the nutraceutical under study and identify different forms of DNA damage. In vitro assays may include tests for gene mutations (e.g. Ames test), chromosomal abnormalities, and DNA strand breaks (e.g. comet assay). In vivo experiments entail administering the nutraceutical to test subjects and evaluating whether it has any genotoxic effects on target organs such as the liver, bone marrow, or germ cells. These investigations also look at the possibility that reproductive cells can transmit genotoxicity to progeny. Genotoxicity study findings are essential for determining possible risks, evaluating the safety of nutraceuticals, and developing risk-reduction plans. Genotoxicity data are frequently needed by regulatory bodies to evaluate the safety of nutraceuticals before approving them for ingestion by humans. In general, genotoxicity studies are essential for guaranteeing the integrity and safety of nutraceutical goods, safeguarding the general public's health, and lowering the possibility of negative side effects.

19.1.1.2.7 Immunotoxicity Studies Studies on immunotoxicity are carried out to assess the possible negative effects of nutraceuticals on the immune system. These investigations are critical to determining the safety of nutraceuticals because immunotoxicity raises the risk of developing autoimmune illnesses, infections, and other immune-related conditions. A range of in vitro and in vivo tests are commonly used in immunotoxicity testing to evaluate the effects of nutraceutical exposure on various immune system components, such as immune cell populations, cytokine production, antibody responses, and immunological function. Using primary immune cells or immune cell lines, in vitro assays can measure phagocytic activity, cytokine production, and cell proliferation. In vivo investigations frequently entail administering the nutraceutical to test subjects and assessing immunological markers in a range of tissues and organs, including the lymph nodes, thymus, and spleen. The capacity of the immune system to mount a suitable defense against infections or antigens may also be evaluated by these investigations. Immunotoxicity study results are vital for determining potential immunosuppressive or immunostimulatory effects, evaluating the safety of nutraceuticals, and developing risk management plans. Immunotoxicity data may be required by regulatory bodies as part of the safety evaluation of nutraceuticals prior to their approval for human use. All things considered, immunotoxicity studies are essential to guaranteeing the integrity and safety of nutraceutical goods, safeguarding the public's health, and lowering the possibility of immunological side effects.

19.1.1.2.8 Special Toxicity Studies Special toxicity studies are an important subgroup of research conducted in the larger field of nutraceutical product safety assessment. Beyond the traditional acute, subchronic, chronic, genotoxic, reproductive, and immunotoxicity assessments, these specialist studies examine particular facets of toxicity with the goal of investigating potential negative effects on specific physiological processes, organs, or populations of interest. Such specialized studies include neurotoxicity studies, which analyze how nutraceuticals affect the central and peripheral nervous systems; cardiotoxicity studies, which look at how they affect cardiovascular health; hepatotoxicity assessments, which analyze potential adverse effects on the liver; and renal toxicity investigations, which look at how they affect kidney function. Moreover, studies on pulmonary toxicity look into possible effects on lung health; studies on ocular toxicity evaluate effects on vision and eye health; and studies on developmental neurotoxicity look at effects on the growing brain system. Furthermore, research on endocrine disruption looks into possible disruptions to the regulation of hormones. These investigations use a wide range of experimental models, endpoints, and techniques that are customized to the particular toxicity endpoints under consideration. In the end, results from specific toxicity studies offer invaluable information on possible risks connected to the use of nutraceuticals, making a significant contribution to evidence-based risk assessment, regulatory decision-making, and the general safety profile of nutraceutical products.

19.1.2 Clinical Trials

This is a critical phase in assessing the safety, effectiveness, and possible health benefits of nutraceuticals in humans in clinical trials. The purpose of these studies is to thoroughly evaluate the impact of nutraceuticals on a range of health outcomes, such as the prevention of disease, the relief of symptoms, and general well-being. In order to reduce bias and guarantee accurate results, clinical trials involving nutraceuticals usually adhere to a strict protocol and randomly assign participants to various treatment groups (e.g. nutraceutical intervention group, placebo/control group). Clinical trials can be designed as crossover studies, observational studies, cohort studies, RCTs, or observational studies, depending on the goals of the research (Table 19.2). Participants in the trial are constantly observed, and pertinent health factors, including clinical symptoms, illness development, biomarkers, and adverse events, are methodically evaluated throughout the trial duration (Banerjee 2014). Clinical trials for nutraceuticals frequently include a wide spectrum of participants, including people in good health, those who are at risk of contracting specific diseases, and patients who already have medical issues. The results of these studies offer insightful information that supports clinical practice, dietary recommendations, and public health policies regarding the safety, effectiveness, ideal dosage, and possible mechanisms of action of nutraceuticals. Additionally, clinical trials are essential for fostering consumer confidence in the use of nutraceuticals as part of a holistic approach to health and well-being, influencing healthcare decision-making, and creating scientific agreement (Borghi and Cicero 2017).

19.1.2.1 Phase 0 Clinical Trials

In phase 0 trials, sometimes referred to as exploratory or microdose trials, a limited number of healthy volunteers—typically fewer than 10—are given very low doses of the nutraceutical in order to evaluate its pharmacokinetics, metabolism, and initial safety profile. These studies offer early information on how the drug behaves in the human body and assist in guiding further clinical research (Vogelbaum et al. 2020).

Table 19.2 In vivo toxicity results in animal models.

Nutraceutical	Animal model	Observed effect	References
Ginseng	Rat model	No significant toxicity	Park et al. (2018)
St. John's Wort	Mouse model	Liver enzyme elevation	Bray et al. (2002)
Garlic extract	Rabbit model	Mild gastrointestinal upset	Papu et al. (2014)
Green tea extract	Rat model	Liver toxicity at high doses	Emoto et al. (2014)
Echinacea	Mouse model	Immune modulation	Spence (2002)
Milk thistle	Rabbit model	Protective liver effects	Refaie et al. (2019)

19.1.2.2 Phase I Clinical Trials

Phase I trials use a limited number of healthy volunteers (typically 20–100) to assess the safety, tolerability, and pharmacokinetics of the nutraceutical. The purpose of these trials is to ascertain the nutraceutical's MTD, dose-limiting toxicities, and initial safety profile when given at increasing dosages. In addition to closely monitoring participants for side effects, pharmacokinetic parameters including distribution, metabolism, excretion, and absorption are measured (Oaknin et al. 2020).

19.1.2.3 Phase II Clinical Trials

Phase II trials are meant to assess the nutraceutical's safety and initial efficacy in a larger sample size of participants, usually people with a particular medical condition or risk factors associated with the nutraceutical's intended usage. In order to evaluate the nutraceutical's safety profile, ideal dosage schedule, and potential therapeutic effects, these trials include giving it out at various doses or formulations. Adverse events, biomarker changes, and clinical outcomes pertinent to the target condition are tracked for participants (Xin et al. 2022).

19.1.2.4 Phase III Clinical Trial

Large-scale, randomized, controlled studies known as phase III trials are carried out to verify the safety, effectiveness, and ideal dosing schedule of the nutraceutical in a variety of patients with the intended medical condition. These studies assess the relative safety and effectiveness of the nutraceutical by comparing it to placebo or standard therapies. Phase III trials offer solid proof of the dangers and benefits of the nutraceuticals, which helps regulators make judgments about marketing and approval (Palacios et al. 2020).

19.1.2.5 Post-marketing Surveillance

Post-marketing surveillance, also referred to as Phase IV trials, is keeping an eye on the nutraceutical's safety after it is put on the market and is frequently used by customers. The purpose of these trials is to identify and evaluate long-term or uncommon side effects that would not have been seen in pre-market clinical trials (Alomar et al. 2020). Post-marketing surveillance entails gathering real-world information from patients, healthcare professionals, and regulatory bodies in order to constantly assess the safety profile of the nutraceutical and provide guidance for risk management plans.

19.1.3 Factors Influencing Toxicity of Nutraceuticals

Numerous factors, including dosage, exposure levels, interactions with other chemicals, and individual variability, might affect how poisonous nutraceuticals are. Comprehending these variables is essential for evaluating the safety of nutraceutical goods and reducing any possible negative consequences (Gupta et al. 2021) (Figure 19.1). An explanation of the variables influencing the toxicity of nutraceuticals is provided below.

19.1.3.1 Dosage and Exposure Levels

Nutraceuticals' dosage and exposure levels are important factors in guaranteeing their effectiveness, safety, and best possible health effects. The term "nutraceutical dosage" describes the number of bioactive compounds or active components in a supplement or functional food product. This amount is usually stated in serving sizes or standardized units like milligrams or micrograms. The bioavailability of the active ingredients, their pharmacokinetic characteristics, and the desired health consequences are some of the elements that must be taken into account when determining the optimum dosage of nutraceuticals (Filippini et al. 2021) (Table 19.3). The specific health issue being targeted, as well as individual factors like age, sex, weight, health status, and genetic

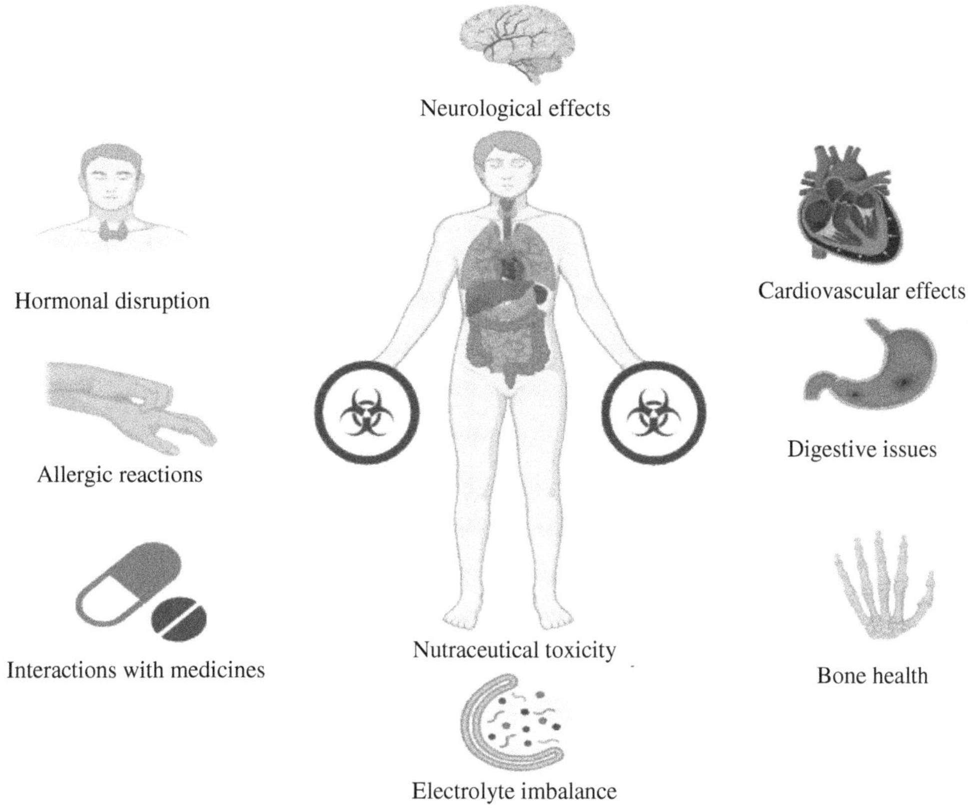

Figure 19.1 Harmful effects of nutraceutical toxicity.

predispositions, can influence the recommended dosage of nutraceuticals (Gonzalez-Suarez et al. 2016; Hernández and Gil 2021). Furthermore, a key idea in comprehending the effects of nutraceuticals is "dose-response" interactions; optimum health benefits are frequently noted within a specific dosage range, and going over prescribed levels may result in decreasing returns or possibly harmful effects (Taylor et al. 2020). The quantity of these substances that people are exposed to through diet, supplementation, or food product fortification is referred to as their exposure levels for nutraceuticals. In order to evaluate total nutritional intake, pinpoint possible causes of excessive or insufficient consumption, and guarantee adherence to safety and dietary criteria, exposure levels must be tracked. Furthermore, exposure assessment is essential for determining the possible health concerns linked to the use of nutraceuticals, especially for vulnerable groups including children, pregnant women, and the elderly.

19.1.3.2 Interactions with Other Compounds

Nutraceuticals' interactions with other substances are a complicated and important factor in both their safety and effectiveness. When taken with prescription drugs, dietary supplements, or different meals, nutraceuticals can interact with these substances in ways that change how the body absorbs, distributes, processes, and excretes them (Castejón-Vega et al. 2020). These interactions may increase or decrease the effects of either compound, which may have positive or negative therapeutic consequences. When nutraceuticals alter the distribution, metabolism, excretion, or absorption of co-administered drugs, this is known as a pharmacokinetic interaction (Anadón et al. 2021). Certain nutraceuticals have the potential to block or promote enzymes that are

Table 19.3 In vitro and in vivo efficacy studies of bio-based nutraceuticals for oral delivery.

Nutraceutical	Experimental models	Dosage	Assays	Key findings	References
Curcumin	Wistar adult male rats	50 mg/kg	Pharmacokinetics analysis	Increase in oral bioavailability	Khalil et al. (2013)
	Wistar male rats	50 mg/kg body weight	Pharmacokinetics analysis	Increase in oral bioavailability	Kakkar et al. (2010)
	Male Sprague—Dawley rats	250 mg/kg body weight	Pharmacokinetics analysis	Increase in oral bioavailability	Shaikh et al. (2009)
	In vitro: Caco-2 cells; in vivo: female CD-1 mice	In vitro: 20 μg/ml; in vivo: 240 mg/kg	In vitro: permeability test; in vivo: pharmacokinetics analysis	In vitro: increase in permeation rate across Caco-2 cells; in vivo: increase in oral bioavailability	Yu and Huang (2012)
	In vitro: HT29 cells; in vivo: Swiss Albino mice	In vitro: 20 μmol/l; in vivo: 25 mg/kg body weight	In vitro: cellular uptake; in vivo: pharmacokinetics analysis	In vitro: increase in cellular uptake; in vivo: increase in bioavailability	Anitha et al. (2014)
Epigallocatechin gallate (ECGC)	Caco-2 cells	0.0625, 0.125, and 0.25 mg/ml	Cellular uptake and permeability test	Increase in EGCG intestinal permeability and absorption through Caco-2 cells	Hu et al. (2012)
Fucoxanthin (brown seaweed)	Male KK-AY mice	0.20%	Anti-obesity effects	Decrease in body weight decrease in total white adipose tissue weight	Okada et al. (2011)
Quercetin	Wistar rats	25 mg/kg	Pharmacokinetics analysis	Increase in oral bioavailability	Penalva et al. (2017)
	Beagle dogs	45 mg	Pharmacokinetics analysis	Increase in plasma quercetin concentrations; increase in oral bioavailability	Li et al. (2011)
Resveratrol	CT26 colon cancer cells	20 and 40 μmol/l	*n.f.*	Decrease in CT26 colon cancer cell number; increase in apoptotic cell death	Jung et al. (2015)
	Male balb/c mice	25 mg/kg	Pharmacokinetics analysis	Increase in bioavailability	Ramalingam and Ko (2016)
Vitamin B12	Caco-2	1.5 μg/ml	Permeability test	Increase in vitamin B12 permeability	Zhang et al. (2015)
Vitamin D3	In vitro: Caco-2; in vivo: KM mice	In vitro: 400 μg/ml; in vivo: 2 mg/kg	In vitro: cellular uptake; in vivo: pharmacokinetics analysis	In vitro: increase in cellular uptake; in vivo: increase in oral bioavailability	Sun et al. (2012)

n.f., not found.

involved in drug metabolism, thereby modifying the medication's potency and metabolism. Direct interactions between other substances and the biological objectives of nutraceuticals, which modify their physiological effects, are known as pharmacodynamic interactions (Žuntar et al. 2020). Nutraceuticals possessing antiplatelet characteristics, for instance, have the potential to enhance the effects of blood-thinning drugs, hence raising the risk of hemorrhaging. Nutraceuticals may also interact with other nutrients to affect the overall nutritional status when they fight for absorption or utilization. Healthcare providers and customers must be aware of these interactions in order to maximize therapeutic results, reduce side effects, and guarantee the safe and efficient use of nutraceuticals in combination with other substances. In order to minimize dangers and optimize the advantages of nutraceutical use in promoting health and well-being, it emphasizes the significance of thorough medication evaluations, dietary assessments, and consultations with healthcare providers.

19.1.3.3 Genetic Variability and Individual Differences

Individual variances and genetic diversity are major factors in influencing how different people react to nutraceuticals. The biological effects, metabolism, and absorption of nutraceutical chemicals can all be influenced by an individual's genetic composition. Nutraceutical metabolism can be influenced by genetic polymorphisms in cytochrome P450 enzymes, which are involved in drug metabolism. Due to these genetic variants, people may be categorized as having a rapid, slow, or extended metabolism. This classification may have an impact on bioavailability as well as possible adverse effects or therapeutic benefits (Zahir 2022). Individual differences can also be attributed to variables other than heredity, such as age, sex, body weight, nutritional condition, and general health. For example, older people can require different nutrients and have different capacities for absorbing them than younger people. Hormonal profile variations associated with sex can also affect how some nutraceuticals affect a person's reaction. Furthermore, an individual's nutritional status—including shortages or excesses of particular nutrients—can affect the safety and efficacy of supplementing with nutraceuticals. Individual differences can also be attributed to variables other than heredity, such as age, sex, body weight, nutritional condition, and general health. For example, older people can require different nutrients and have different capacities for absorbing them than younger people. Hormonal profile variations associated with sex can also affect how some nutraceuticals affect a person's reaction. Furthermore, an individual's nutritional status—including shortages or excesses of particular nutrients—can affect the safety and efficacy of supplementing with nutraceuticals (Jia et al. 2020). Comprehending these genetic and individual variances is crucial to customizing nutraceutical suggestions and maximizing their advantages while lowering possible hazards. Nutraceutical applications for disease prevention and health promotion are increasingly being guided by personalized nutrition techniques that consider genetic factors and individual traits. A personalized and precision medicine approach to nutraceutical interventions is necessary to ensure that recommendations are in line with each individual's unique biological makeup and health situation, as highlighted by the recognition of genetic and individual variability (Morand et al. 2020).

19.1.3.4 Formulation and Manufacturing Processes

Nutraceuticals' overall quality, safety, and efficacy are significantly influenced by their formulation and manufacturing processes. Nutraceutical products come in a variety of formulations, such as liquids, powders, tablets, capsules, and functional foods (Puri et al. 2022). To guarantee the consistency, stability, and bioavailability of the active components in each formulation, particular production procedures are needed. Choosing the right excipients, such as coatings, binders, and fillers, can optimize medication distribution and improve absorption. These are some formulation concerns.

Furthermore, the physical, chemical, and release profiles of nutraceuticals can be affected by the selection of manufacturing techniques, such as blending, granulation, encapsulation, or emulsification. Throughout the production process, quality control measures—such as testing for potency, purity, and stability—are crucial to ensuring that nutraceutical goods fulfill regulatory standards and regulations (Puri et al. 2022). In addition, it is crucial to follow quality assurance procedures and good manufacturing practices (GMP) in order to reduce the possibility of contamination, adulteration, or variation in product composition. In general, providing high-quality nutraceutical products that are safe, trustworthy, and effective for consumers looking to improve their health and well-being requires careful attention to the formulation and production processes (Miao et al. 2023).

19.1.4 Ethical Considerations

Since the toxicity evaluation of nutraceuticals involves the use of living organisms or biological materials for research, ethical considerations are crucial. It is critical to protect the rights and welfare of study participants who are humans as well as study animals when they are involved in toxicity studies (Goel 2022). Strict ethical requirements, such as getting participants' informed consent, reducing risks and discomfort, and protecting participant privacy and confidentiality, must be followed during human clinical trials. The three R's (replacement, reduction, and refinement) should also guide animal studies, with the goals of minimizing animal usage, improving experimental procedures to lessen suffering, and, if feasible, substituting alternative methods for animal use (Sadgrove 2022). Furthermore, the public's trust and scientific integrity depend on openness in the reporting of study methods, findings, and potential conflicts of interest (Brody 1998). The requirement for ethical and responsible research methods that put the welfare of both people and the animals used in the evaluation process first is emphasized by ethical issues in the toxicity evaluation of nutraceuticals.

19.2 Emerging Trends and Future Prospects

The field of toxicity testing in nutraceuticals faces several challenges. One major issue is the complexity and variability of natural products, making standardization, and reproducibility of tests difficult. Also, translating results from animal models to humans remains a significant challenge, given the physiological differences between species. There is a growing need for developing more sophisticated in vitro models that can better mimic human physiology. Advancements in computational methods and alternative testing strategies, such as organ-on-a-chip and artificial intelligence, are promising areas that could enhance the accuracy and efficiency of toxicity evaluations. The future of nutraceuticals depends on continuous research to improve testing methods, ensuring these products are both effective and safe for public consumption. Advances in biotechnology, such as high-throughput screening and computational modeling, offer new ways to evaluate toxicity. These methods can provide more comprehensive data and reduce reliance on animal testing. There's a growing discussion on the need for harmonizing regulatory standards for nutraceuticals worldwide. This harmonization could facilitate better safety protocols and ease international trade. The environmental aspect of nutraceutical production, such as the impact of sourcing ingredients and waste generation, is gaining attention. The ecological toxicity of these products is an emerging field of study. After nutraceuticals are released into the market, ongoing surveillance is crucial to monitor any adverse effects in the larger population. This data is vital for reassessing safety and updating regulatory standards.

19.3 Conclusion

The evaluation of toxicity in nutraceuticals is a cornerstone in ensuring their safety and efficacy. This chapter highlighted the necessity of rigorous testing methods, from in vitro and in vivo studies to human clinical trials. The challenges in this field call for innovative approaches to enhance testing methodologies. The thorough assessment of nutraceuticals' toxicity is not just a regulatory requirement but a crucial step in safeguarding public health and sustaining consumer trust in these increasingly popular products. Nutraceuticals are frequently thought of as natural and harmless goods; toxicity screening is an essential step in determining any potential negative effects. To forecast possible toxicity and determine the proper dosage recommendations, it is necessary to comprehend metabolism and bioavailability. Studies conducted in vivo and in vitro offer important insights into the safety profiles of nutraceuticals, emphasizing the variety of effects seen at various doses and compounds. The market's expansion highlights the necessity of thorough toxicity screening to protect customer confidence and public health. The advent of cutting-edge technology, such as alternative testing methods and computer modeling, offers bright prospects for raising the precision and effectiveness of toxicity evaluations. Interactions with medications, long- and chronic-term toxicity, individual variability, developing technology, and global regulatory viewpoints should all be taken into account in future studies. Post-market surveillance and the environmental effect are crucial for a comprehensive knowledge of the safety of nutraceuticals. Nutraceuticals' future rests on further research to improve testing procedures and guarantee their efficacy and safety for general public use.

References

AlAli, M., Alqubaisy, M., Aljaafari, M.N. et al. (2021). Nutraceuticals: transformation of conventional foods into health promoters/disease preventers and safety considerations. *Molecules* 26 (9): 2540.

Alhoshani, N.M., Al-Zharani, M., Almutairi, B. et al. (2022). Antioxidant and anti-inflammatory activities of lycopene against 5-fluorouracil-induced cytotoxicity in Caco2 cells. *Saudi Pharmaceutical Journal* 30 (11): 1665–1671.

Alomar, M., Tawfiq, A.M., Hassan, N., and Palaian, S. (2020). Post marketing surveillance of suspected adverse drug reactions through spontaneous reporting: current status, challenges and the future. *Therapeutic Advances in Drug Safety* 11: 2042098620938595.

Anadón, A., Ares, I., Martínez-Larrañaga, M.R., and Martínez, M.A. (2021). Interactions between nutraceuticals/nutrients and nutrients and therapeutic drugs. In: *Nutraceuticals* (ed. K.G.R. Desmet, D. Ghosh, D. Bagchi, and S.J. Dey), 1175–1197. Academic Press.

Anitha, A., Deepa, N., Chennazhi, K.P. et al. (2014). Combinatorial anticancer effects of curcumin and 5-fluorouracil loaded thiolated chitosan nanoparticles towards colon cancer treatment. *Biochimica et Biophysica Acta (BBA)-General Subjects* 1840 (9): 2730–2743.

Aronson, J.K. (2017). Defining 'nutraceuticals': neither nutritious nor pharmaceutical. *British Journal of Clinical Pharmacology* 83 (1): 8–19.

Astashkina, A., Mann, B., and Grainger, D.W. (2012). A critical evaluation of in vitro cell culture models for high-throughput drug screening and toxicity. *Pharmacology & Therapeutics* 134 (1): 82–106.

Banerjee, A. (2014). Gastrointestinal toxicity biomarkers. In: *Biomarkers in Toxicology* (ed. R.C. Gupta), 269–277. Academic Press.

Barile, F.A. (2019). *Introduction to In Vitro Cytotoxicology: Mechanisms and Methods*. CRC Press.

Bernhard, D., Schwaiger, W., Crazzolara, R. et al. (2003). Enhanced MTT-reducing activity under growth inhibition by resveratrol in CEM-C7H2 lymphocytic leukemia cells. *Cancer Letters* 195 (2): 193–199.

Bhatia, L., Bachheti, R.K., Garlapati, V.K., and Chandel, A.K. (2020). Third-generation biorefineries: a sustainable platform for food, clean energy, and nutraceuticals production. *Biomass Conversion and Biorefinery* 12: 4215–4230.

Bhavani, M., Morya, S., Saxena, D., and Awuchi, C.G. (2023). Bioactive, antioxidant, industrial, and nutraceutical applications of banana peel. *International Journal of Food Properties* 26 (1): 1277–1289.

Borghi, C. and Cicero, A.F. (2017). Nutraceuticals with a clinically detectable blood pressure-lowering effect: a review of available randomized clinical trials and their meta-analyses. *British Journal of Clinical Pharmacology* 83 (1): 163–171.

Boverhof, D.R., Chamberlain, M.P., Elcombe, C.R. et al. (2011). Transgenic animal models in toxicology: historical perspectives and future outlook. *Toxicological Sciences* 121 (2): 207–233.

Bray, B.J., Perry, N.B., Menkes, D.B., and Rosengren, R.J. (2002). St. John's wort extract induces CYP3A and CYP2E1 in the Swiss Webster mouse. *Toxicological Sciences* 66 (1): 27–33.

de Brito Monteiro, L., Davanzo, G.G., de Aguiar, C.F., and Moraes-Vieira, P.M. (2020). Using flow cytometry for mitochondrial assays. *MethodsX* 7: 100938.

Brody, B.A. (1998). *The Ethics of Biomedical Research: An International Perspective*. USA: Oxford University Press.

Castejón-Vega, B., Giampieri, F., and Alvarez-Suarez, J.M. (2020). Nutraceutical compounds targeting inflammasomes in human diseases. *International Journal of Molecular Sciences* 21 (14): 4829.

Chopra, A.S., Lordan, R., Horbańczuk, O.K. et al. (2022). The current use and evolving landscape of nutraceuticals. *Pharmacological Research* 175: 106001.

Chu, W.L., Lim, Y.W., Radhakrishnan, A.K., and Lim, P.E. (2010). Protective effect of aqueous extract from Spirulina platensis against cell death induced by free radicals. *BMC Complementary and Alternative Medicine* 10 (1): 1–8.

Coppens, P., Da Silva, M.F., and Pettman, S. (2006). European regulations on nutraceuticals, dietary supplements and functional foods: a framework based on safety. *Toxicology* 221 (1): 59–74.

De Leon, J.A.D. and Borges, C.R. (2020). Evaluation of oxidative stress in biological samples using the thiobarbituric acid reactive substances assay. *JoVE (Journal of Visualized Experiments)* 159: e61122.

Dillard, C.J. and German, J.B. (2000). Phytochemicals: nutraceuticals and human health. *Journal of the Science of Food and Agriculture* 80 (12): 1744–1756.

Dixit, S., Shukla, A., Singh, V., and Upadhyay, S.K. (2021). Bioprospecting of natural compounds for industrial and medical applications: current scenario and bottleneck. In: *Bioprospecting of Plant Biodiversity for Industrial Molecules*. Wiley; Portico. https://doi.org/10.1002/9781119718017.ch3.

Ecobichon, D.J. (1997). *The Basis of Toxicity Testing*. CRC Press.

Effah, F., Adragna, J., Luglio, D. et al. (2023). Toxicological assessment of E-cigarette flavored E-liquids aerosols using Calu-3 cells: a 3D lung model approach. *Toxicology* 500: 153683.

Emoto, Y., Yoshizawa, K., Kinoshita, Y. et al. (2014). Green tea extract-induced acute hepatotoxicity in rats. *Journal of Toxicologic Pathology* 27 (3-4): 163–174.

Filippini, T., Malavolti, M., Whelton, P.K. et al. (2021). Blood pressure effects of sodium reduction: dose–response meta-analysis of experimental studies. *Circulation* 143 (16): 1542–1567.

Gautam, A., Morya, S., Neumann, A., and Menaa, F. (2022). A review on fruits bioactive potential: an insight into phytochemical traits and their extraction methods. *Journal of Food Bioactives* 19: 1–15.

Goel, M. (2022). *Nutraceuticals in Human Health. The Energy and Resources Institute (TERI)* 1–42.

Gonçalves, R.F., Martins, J.T., Duarte, C.M. et al. (2018). Advances in nutraceutical delivery systems: from formulation design for bioavailability enhancement to efficacy and safety evaluation. *Trends in Food Science & Technology* 78: 270–291.

Goncharov, N., Sobolev, V., Terpilowski, M. et al. (2019). Safety and toxicity evaluation of nutraceuticals in animal models. *Nutraceuticals in Veterinary Medicine* 1: 675–684.

Gonzalez-Suarez, I., Martin, F., Hoeng, J., and Peitsch, M.C. (2016). Mechanistic network models in safety and toxicity evaluation of Nutraceuticals. In: *Nutraceuticals* (ed. R.C. Gupta), 287–304. Academic Press.

Gosslau, A. (2016). Assessment of food toxicology. *Food Science and Human Wellness* 5 (3): 103–115.

Granato, D., Barba, F.J., Bursać Kovačević, D. et al. (2020). Functional foods: product development, technological trends, efficacy testing, and safety. *Annual Review of Food Science and Technology* 11: 93–118.

Grumezescu, A. (ed.) (2016). *Nutraceuticals*. Academic Press.

Gupta, R.C., Lall, R., and Srivastava, A. (ed.) (2021). *Nutraceuticals: Efficacy, Safety and Toxicity*. Academic Press.

Harishkumar, R., Reddy, L.P.K., Karadkar, S.H. et al. (2019). Toxicity and selective biochemical assessment of quercetin, gallic acid, and curcumin in zebrafish. *Biological and Pharmaceutical Bulletin* 42 (12): 1969–1976.

Hayden, P.J. and Harbell, J.W. (2021). Special review series on 3D organotypic culture models: Introduction and historical perspective. *In Vitro Cellular & Developmental Biology-Animal* 57 (2): 95–103.

Hayes, D.P. (2007). Nutritional hormesis. *European Journal of Clinical Nutrition* 61 (2): 147–159.

Hernández, A.F. and Gil, F. (2021). Nutraceuticals and adverse outcome pathways. In: *Nutraceuticals* (ed. K.G.R. Desmet, D. Ghosh, D. Bagchi, and S.J. Dey), 1159–1174. Academic Press.

Hu, B., Ting, Y., Yang, X. et al. (2012). Nanochemoprevention by encapsulation of (−)-epigallocatechin-3-gallate with bioactive peptides/chitosan nanoparticles for enhancement of its bioavailability. *Chemical Communications* 48 (18): 2421–2423.

Jia, Y., Khayatnezhad, M., and Mehri, S. (2020). Population differentiation and gene flow in Rrodium cicutarium: a potential medicinal plant. *Genetika* 52 (3): 1127–1144.

Johnson, M.K. and Loo, G. (2000). Effects of epigallocatechin gallate and quercetin on oxidative damage to cellular DNA. *Mutation Research/DNA Repair* 459 (3): 211–218.

Jung, K.H., Lee, J.H., Park, J.W. et al. (2015). Resveratrol-loaded polymeric nanoparticles suppress glucose metabolism and tumor growth in vitro and in vivo. *International Journal of Pharmaceutics* 478 (1): 251–257.

Kakkar, V., Singh, S., Singla, D. et al. (2010). Pharmacokinetic applicability of a validated liquid chromatography tandem mass spectroscopy method for orally administered curcumin loaded solid lipid nanoparticles to rats. *Journal of Chromatography B* 878 (32): 3427–3431.

Keservani, R.K., Sharma, A.K., and Kesharwani, R. K. (Eds.). (2020). *Nutraceuticals and Dietary Supplements: Applications in Health Improvement and Disease Management*. CRC Press.

Khalil, N.M., do Nascimento, T.C.F., Casa, D.M. et al. (2013). Pharmacokinetics of curcumin-loaded PLGA and PLGA–PEG blend nanoparticles after oral administration in rats. *Colloids Surfaces B Biointerfaces* 101: 353–360.

Kokova, V. (2023). Methods for determining subacute, sub-chronic, and chronic toxicity of chemical compounds. *Knowledge-International Journal* 59 (4): 257–261.

Krishna, G. and Krishna, M. (2021). Alternative in vitro models for safety and toxicity evaluation of nutraceuticals. In: *Nutraceuticals* (ed. K.G.R. Desmet, D. Ghosh, D. Bagchi, and S.J. Dey), 431–462. Academic Press.

Kuddus, M. (2019). Enzymes in food biotechnology. In: *Production, Applications and Future Prospects*, 883. Elsevier.

Li, W., Yi, S., Wang, Z. et al. (2011). Self-nanoemulsifying drug delivery system of persimmon leaf extract: optimization and bioavailability studies. *International Journal of Pharmaceutics* 420 (1): 161–171.

Lin, J., Sahakian, D.C., De Morais, S.M. et al. (2003). The role of absorption, distribution, metabolism, excretion and toxicity in drug discovery. *Current Topics in Medicinal Chemistry* 3 (10): 1125–1154.

McClements, D.J., Li, F., and Xiao, H. (2015). The nutraceutical bioavailability classification scheme: classifying nutraceuticals according to factors limiting their oral bioavailability. *Annual Review of Food Science and Technology* 6: 299–327.

McGonigle, P. and Ruggeri, B. (2014). Animal models of human disease: challenges in enabling translation. *Biochemical Pharmacology* 87 (1): 162–171.

Miao, M., Chen, L., and McClements, D. (ed.) (2023). *Bioactive Delivery Systems for Lipophilic Nutraceuticals: Formulation, Fabrication, and Application*. Royal Society of Chemistry.

Morand, C., De Roos, B., Garcia-Conesa, M.T. et al. (2020). Why interindividual variation in response to consumption of plant food bioactives matters for future personalised nutrition. *Proceedings of the Nutrition Society* 79 (2): 225–235.

Morya, S., Awuchi, C.G., and Menaa, F. (2022a). 14 Advanced functional approaches of nanotechnology in food and nutrition. In: *Environmental Management Technologies: Challenges and Opportunities* (ed. P. Chowdhary, V. Kumar, S. Kumar, and V. Hare), 257–272. CRC Press.

Morya, S. and Danquah-Amoah, A. (2017). Impact of whey derived bio-active components on resolving human health issues: as nutraceuticals. In: *Research Trends in Food Technology and Nutrition*, vol. 6, 176. AkiNik Publications.

Morya, S., Singh, N., and Awuchi, C.G. (2022b). *Health Hazards of Food Allergens and Related Safety Measures*. NY: CRC Press. Taylor & Francis.

Nasri, H., Baradaran, A., Shirzad, H., and Rafieian-Kopaei, M. (2014). New concepts in nutraceuticals as alternative for pharmaceuticals. *International Journal of Preventive Medicine* 5 (12): 1487.

Oaknin, A., Tinker, A.V., Gilbert, L. et al. (2020). Clinical activity and safety of the anti–programmed death 1 monoclonal antibody dostarlimab for patients with recurrent or advanced mismatch repair–deficient endometrial cancer: a nonrandomized phase 1 clinical trial. *JAMA Oncology* 6 (11): 1766–1772.

Okada, T., Mizuno, Y., Sibayama, S. et al. (2011). Antiobesity effects of Undaria lipid capsules prepared with scallop phospholipids. *Journal of Food Science* 76 (1): H2–H6.

Pagano, E., Romano, B., Izzo, A.A., and Borrelli, F. (2018). The clinical efficacy of curcumin-containing nutraceuticals: an overview of systematic reviews. *Pharmacological Research* 134: 79–91.

Palacios, R., Patiño, E.G., de Oliveira Piorelli, R. et al. (2020). Double-blind, randomized, placebo-controlled phase III clinical trial to evaluate the efficacy and safety of treating healthcare professionals with the adsorbed COVID-19 (inactivated) vaccine manufactured by Sinovac–PROFISCOV: a structured summary of a study protocol for a randomised controlled trial. *Trials* 21: 1–3.

Papu, S., Jaivir, S., Sweta, S., and Singh, B.R. (2014). Medicinal values of garlic (*Allium sativum* L.) in human life: an overview. *Greener Journal of Agricultural Sciences* 4 (6): 265–280.

Park, S.J., Noh, J., Jeong, E.J. et al. (2018). Subchronic oral toxicity study of Korean red ginseng extract in Sprague-Dawley rats with a 4-week recovery period. *Regulatory Toxicology and Pharmacology* 92: 83–93.

Penalva, R., Gonzalez-Navarro, C.J., Gamazo, C. et al. (2017). Zein nanoparticles for oral delivery of quercetin: Pharmacokinetic studies and preventive anti-inflammatory effects in a mouse model of endotoxemia. *Nanomedicine: Nanotechnology, Biology and Medicine* 13 (1): 103–110.

Puri, V., Nagpal, M., Singh, I. et al. (2022). A comprehensive review on nutraceuticals: therapy support and formulation challenges. *Nutrients* 14 (21): 4637.

Raies, A.B. and Bajic, V.B. (2016). In silico toxicology: computational methods for the prediction of chemical toxicity. *Wiley Interdisciplinary Reviews: Computational Molecular Science* 6 (2): 147–172.

Rajasekaran, A., Sivagnanam, G., and Xavier, R. (2008). Nutraceuticals as therapeutic agents: a review. *Research Journal of Pharmacy and Technology* 1 (4): 328–340.

Ramalingam, P. and Ko, Y.T. (2016). Improved oral delivery of resveratrol from N-trimethyl chitosan-g-palmitic acid surface-modified solid lipid nanoparticles. *Colloids and Surfaces B: Biointerfaces* 139: 52–61.

Refaie, A.M., Ghazal, M.N., Abo El-Azayem, E.H. et al. (2019). Impact of dietary supplementation of milk thistle (Silybum marianum) seed extract on doe rabbits performance. *Egyptian Journal of Nutrition and Feeds* 22 (2): 375–382.

Rigacci, S. and Stefani, M. (2016). Nutraceutical properties of olive oil polyphenols. An itinerary from cultured cells through animal models to humans. *International Journal of Molecular Sciences* 17 (6): 843.

Roberts, S.M., James, R.C., and Williams, P.L. (ed.) (2022). *Principles of Toxicology: Environmental and Industrial Applications*. John Wiley & Sons.

Sadgrove, N.J. (2022). Honest nutraceuticals, cosmetics, therapies, and foods (NCTFs): standardization and safety of natural products. *Critical Reviews in Food Science and Nutrition* 62 (16): 4326–4341.

Santini, A., Cammarata, S.M., Capone, G. et al. (2018). Nutraceuticals: opening the debate for a regulatory framework. *British Journal of Clinical Pharmacology* 84 (4): 659–672.

Schilter, B., Andersson, C., Anton, R. et al. (2003). Guidance for the safety assessment of botanicals and botanical preparations for use in food and food supplements. *Food and Chemical Toxicology* 41 (12): 1625–1649.

Shaikh, J., Ankola, D.D., Beniwal, V. et al. (2009). Nanoparticle encapsulation improves oral bioavailability of curcumin by at least 9-fold when compared to curcumin administered with piperine as absorption enhancer. *European Journal of Pharmaceutical Sciences* 37 (3-4): 223–230.

Shukla, S.J., Huang, R., Austin, C.P., and Xia, M. (2010). The future of toxicity testing: a focus on in vitro methods using a quantitative high-throughput screening platform. *Drug Discovery Today* 15 (23-24): 997–1007.

Singh, S.P. and Upadhyay, S.K. (Éds.) (2021). *Bioprospecting of Microorganism-Based Industrial Molecules*. Wiley. https://doi.org/10.1002/9781119717317.

Singh, S.P. and Upadhyay, S.K. (Éds.) (2023). *Microbial Bioreactors for Industrial Molecules*. Wiley. https://doi.org/10.1002/9781119874096.

Spence, K.M. (2002). In vivo evaluation of immunomodulatory properties of crude extracts of Echinacea species and fractions isolated from *Echinacea purpurea*. Doctoral dissertation. University of Southern Queensland.

Sponchiado, G., Adam, M.L., Silva, C.D. et al. (2016). Quantitative genotoxicity assays for analysis of medicinal plants: a systematic review. *Journal of Ethnopharmacology* 178: 289–296.

Sun, F., Ju, C., Chen, J. et al. (2012). Nanoparticles based on hydrophobic alginate derivative as nutraceutical delivery vehicle: vitamin D3 loading. *Artificial Cells, Blood Substitutes, and Biotechnology* 40 (1–2): 113–119.

Sundaram, T.S., Giromini, C., Rebucci, R., and Baldi, A. (2020). Omega-3 polyunsaturated fatty acids counteract inflammatory and oxidative damage of non-transformed porcine enterocytes. *Animals* 10 (6): 956.

Taylor, A.A., Tsuji, J.S., Garry, M.R. et al. (2020). Critical review of exposure and effects: implications for setting regulatory health criteria for ingested copper. *Environmental Management* 65: 131–159.

Tripathi, C., Girme, A., Champaneri, S. et al. (2020). Nutraceutical regulations: an opportunity in ASEAN countries. *Nutrition* 74: 110729.

Upadhyay, S.K. and Singh, S.P. (Éds.) (2021). *Bioprospecting of Plant Biodiversity for Industrial Molecules*. Wiley. https://doi.org/10.1002/9781119718017.

Upadhyay, S.K. and Singh, S.P. (2023). *Plants as Bioreactors for Industrial Molecules.* Wiley. https://doi.org/10.1002/9781119875116.

Vilas-Boas, A.A., Pintado, M., and Oliveira, A.L. (2021). Natural bioactive compounds from food waste: toxicity and safety concerns. *Foods* 10 (7): 1564.

Vo, A.H., Van Vleet, T.R., Gupta, R.R. et al. (2019). An overview of machine learning and big data for drug toxicity evaluation. *Chemical Research in Toxicology* 33 (1): 20–37.

Vogelbaum, M.A., Krivosheya, D., Borghei-Razavi, H. et al. (2020). Phase 0 and window of opportunity clinical trial design in neuro-oncology: a RANO review. *Neuro-oncology* 22 (11): 1568–1579.

Xin, Q., Wu, Q., Chen, X. et al. (2022). Six-month follow-up of a booster dose of CoronaVac in two single-centre phase 2 clinical trials. *Nature Communications* 13 (1): 3100.

Yu, H. and Huang, Q. (2012). Improving the oral bioavailability of curcumin using novel organogel-based nanoemulsions. *Journal of Agricultural and Food Chemistry* 60 (21): 5373–5379.

Zahir, F. R. (2022). Nutraceuticals in personalized and precision medicine–a preliminary scoping review indicates efficacy for disease treatment, general wellbeing, and defense against Covid-19.

Zhang, C., Ball, J., Panzica-Kelly, J., and Augustine-Rauch, K. (2016). In vitro developmental toxicology screens: a report on the progress of the methodology and future applications. *Chemical Research in Toxicology* 29 (4): 534–544.

Zhang, J., Field, C.J., Vine, D., and Chen LingYun, C.L. (2015). Intestinal uptake and transport of vitamin B12-loaded soy protein nanoparticles. *Pharmaceutical Research* 32 (4): 1288–1303.

Žuntar, I., Petric, Z., Bursać Kovačević, D., and Putnik, P. (2020). Safety of probiotics: functional fruit beverages and nutraceuticals. *Foods* 9 (7): 947.

Index

Herbal Nutraceuticals: Products and Processes, First Edition. Edited by Santosh Kumar Upadhyay and Sudhir Pratap Singh.

d